The Grand Strategy of the Byzantine Empire

The Grand Strategy of
the Byzantine Empire

EDWARD N. LUTTWAK

THE BELKNAP PRESS OF

HARVARD UNIVERSITY PRESS

Cambridge, Massachusetts, and London, England   2009

Library of Congress Cataloging-in-Publication Data

Luttwak, Edward.
  The grand strategy of the Byzantine Empire / Edward N. Luttwak.
    p. cm.
  Includes bibliographical references and index.
  ISBN 978-0-674-03519-5 (cloth : alk. paper)   1. Byzantine Empire—Military policy.
2. Strategy—History—To 1500.   3. Military art and science—Byzantine Empire—
History.   4. Imperialism—History—To 1500.   5. Byzantine Empire—History, Military.
6. Byzantine Empire—Foreign relations.   I. Title.
  U163.L86   2009
  355′.033549500902—dc22   2009011799

Contents

Maps

Preface

Once largely neglected, as if the entire Roman empire had really ended in 476, the eastern half that we call Byzantine by modern habit now attracts so much attention that it is even the subject of popular histories. While many are interested in the culture of Byzantium, it is the epic struggle to defend the empire for century after century against an unending sequence of enemies that seems to resonate especially in our own times. This book is devoted to one dimension of Byzantine history: the application of method and ingenuity in the use of both persuasion and force—that is to say, strategy in all its aspects, from higher statecraft down to military tactics.

When I first started to study Byzantine strategy in earnest, I had just completed a book on the strategy of the Roman empire up to the third century that continues to attract both inordinate praise and strenuous criticism. My original intention was simply to write a second volume to cover the subsequent centuries. What ensued instead was the discovery of an altogether richer body of strategy than the earlier Romans had ever possessed, which called for a vastly greater effort of research and composition. In the end, this lasted for more than two decades, albeit with many interruptions—some due to my not entirely unrelated work in applying military strategy in the field. There was one compensation for this prolonged delay: several essential Byzantine texts once available only as scarcely accessible manuscripts, or in antiquated editions replete with errors, have now been published in reliable form. Also, a considerable number of important new works of direct relevance to Byzantine strategy have been published since I started on my quest long ago.

For in recent years Byzantine studies have indeed flourished as never before. A great wave of first-class scholarship has illuminated many a dark corner of Byzantine and world history—and it has also inspired a climate of high-spirited generosity among the practitioners. Although I am more student than scholar in this field, I have experienced this generosity in the fullest measure.

Soon after I started reading for this book, circa 1982, George Dennis, whose translation of the *Strategikon* is the most widely read of Byzantine military texts, gave me an advance typescript of his work that would be published as *Three Byzantine Military Treatises*. Twenty-six years later, he sent me a typescript of part of his eagerly awaited edition of Leo's *Taktika,* which I urgently needed to complete this book; generosity is mere habit for George T. Dennis of the Society of Jesus. Walter E. Kaegi Jr., whose works illuminate the field, also gave me valuable advice early on.

Others whom I had never even met, but simply importuned without prior introduction, nevertheless responded as if bound by old friendship and collegial obligations. Peter B. Golden, the eminent Turcologist amply cited in these pages, answered many questions, offered valuable suggestions, and lent me two otherwise unobtainable books. John Wortley entrusted me with the unique copy of his own annotated typescript of Scylitzes. Peter Brennan and Salvatore Cosentino offered important advice, while Eric McGeer and Paul Stephenson and Denis F. Sullivan, whose work is here conscripted at length, read drafts of this book, uncovering errors and offering important advice. John F. Haldon, whose writings constitute a library of Byzantine studies in themselves, responded to a stranger's imposition with a detailed critique of an early draft.

Because what follows is intended for non-specialists as well, I asked two such, Anthony Harley and Kent Karlock, to comment on the lengthy text; I am grateful for their hard work, considered opinions, and corrections. A third reader was Hans Rausing, not a specialist but a profound and multilingual student of history, and to him I owe valuable observations. Stephen P. Glick applied both his encyclopedic knowledge of military historiography and his meticulous attention to the text, leaving his mark on this book. Nicolò Miscioscia was my able assistant for a season. Christine Col and Joseph E. Luttwak researched and graphically prepared all the maps, no easy task amidst endless revisions. Michael

Aronson, senior editor for social sciences at Harvard University Press, was the active proponent of my earlier book on Roman grand strategy a long time ago. It was with unending patience over two decades that he asked for this book as well, and his experienced enthusiasm is manifest in the physical quality of the publication, an effort in which he was ably assisted by Donna Bouvier and Hilary S. Jacqmin of the Press. It was most fortunate that they commissioned Wendy Nelson to serve as manuscript editor. With infinite care and talented discernment she uncovered many a stealthy error, and gently indicated infelicities in need of remedy. Finally, it is a pleasure to thank Alice-Mary Talbot, also here cited, Director of the Dumbarton Oaks Research Library and Collection, and the always helpful Deb Brown Stewart, Byzantine studies librarian at Dumbarton Oaks. I might have dithered forever instead of finally composing the text had I not met Peter James MacDonald Hall, who demanded the book and removed the excuse of all other work.

The Grand Strategy of the Byzantine Empire

PART ONE

The Invention of Byzantine Strategy

When the administration of the Roman empire was divided in the year 395 between the two sons of Theodosius I, with the western portion going to Honorius and the eastern to his brother Arkadios, few could have foretold the drastically different fates of the two halves. Defended by Germanic field commanders, then dominated by Germanic warlords, increasingly penetrated by mostly Germanic migrants with or without imperial consent, then fragmented by outright invasions, the western half of the empire progressively lost tax revenues, territorial control, and its Roman political identity in a process so gradual that the removal of the last imperial figurehead, Romulus Augustus, on September 4, 476, was mere formality. There were local accommodations with the invaders in places, even some episodes of cultural integration, but the newly fashionable vision of an almost peaceful immigration and a gradual transformation into a benign late antiquity is contradicted by the detailed evidence of violence, destruction, and the catastrophic loss of material amenities and educational attainments that would not be recovered for a thousand years, if then.[1]

Very different was the fate of the eastern half of the Roman empire commanded from Constantinople. That is the empire we call Byzantine by modern habit though it was never anything but Roman to its rulers and their subjects, the *romaioi,* who could hardly identify with provincial Byzantion, the ancient Greek city that Constantine had converted into his imperial capital and New Rome in the year 330. Having subdued its own Germanic warlords and outmaneuvered Attila's Huns in the supreme crisis of the fifth century that extinguished its western

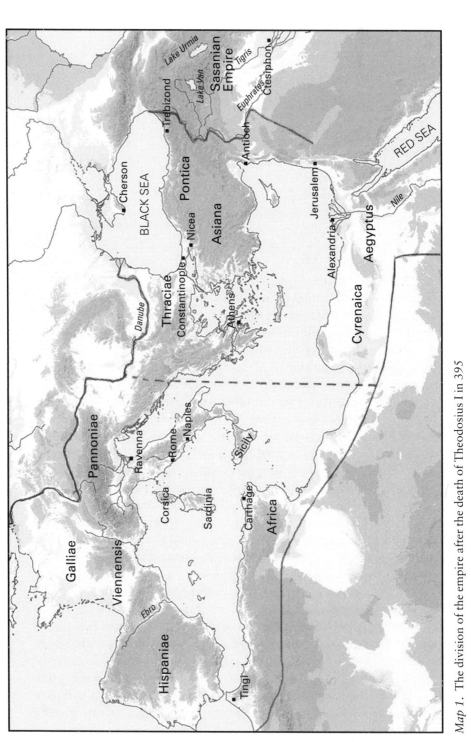

Map 1. The division of the empire after the death of Theodosius I in 395

counterpart, the Byzantine empire acquired the strategic method with which it resisted successive waves of invaders for more than eight hundred years by the shortest reckoning.

Again and again the eastern empire was attacked by new and old enemies advancing from the immensity of the Eurasian steppe, from the Iranian plateau homeland of empires, from the Mediterranean coasts and Mesopotamia, which came under Islamic rule in the seventh century, and finally from the reinvigorated western lands as well. Yet the empire did not collapse in defeat until the conquest of Constantinople in the name of the Fourth Crusade in 1204, to then revive once more in much-diminished form until the final Ottoman victory of 1453.

Sheer military strength was enough to provide ample security for the Roman empire when it was still undivided and prosperous, encompassing all the lands around the entire Mediterranean and reaching deep beyond them. Moderate taxation and voluntary recruitment were sufficient to keep fleets and some three hundred thousand troops in constant training in frontier forts and legionary garrisons, from which detachments *(vexillationes)* could be gathered in field armies to suppress rare internal rebellions or repel foreign invaders.[2] But until the third century, the Romans rarely had to fight to obtain the benefits of their military strength.

In every frontier province there were flourishing cities and imperial granaries to tempt the empire's neighbors, but they usually preferred a hungry peace to the certainty of harsh Roman reprisals or even outright annihilation. Commanding superior combat strength, the Romans at their imperial peak could freely choose between pure deterrence with retaliation if needed, which required only field armies, and an active defense of the frontiers that required garrisons everywhere, and both were tried in succession during the first two centuries of our era. Even later, when old and new enemies beyond the Rhine and Danube coalesced into mighty warrior confederations, while in the east formidable Sasanian Persia replaced its weaker predecessor Arsacid Parthia, Roman armies were still strong enough to contain them effectively with a new strategy of defense-in-depth.[3]

The Byzantines never had such an abundance of strength. In 395 the empire's administrative division—it was not yet a political division, for both brothers jointly ruled both parts—followed the boundaries between east and west first decreed by Diocletian (284–305), which bisected the entire Mediterranean basin into two almost equal halves. It was a neat division, but it left the eastern Roman empire with three separate regions on three different continents. In Europe the eastern boundary, marked off by the provinces of Moesia I and Praevalitania, now in Serbia and Albania, also encompassed the territories of modern Macedonia, Bulgaria, the Black Sea coast of Romania, Greece, Cyprus, and European Turkey—the ancient Thrace—with Constantinople itself. In Asia, imperial territory consisted of the vast peninsula of Anatolia, now Asiatic Turkey, as well as Syria, Jordan, Israel, and a slice of northern Iraq in the provinces of Mesopotamia and Osrhoene. In North Africa, the empire had the provinces of Egypt, reaching far up the Nile in Thebais, and the eastern half of modern Libya, composed of the provinces of Libya superior and Libya inferior, the earlier Cyrenaica.

This was a rich inheritance of productive and taxpaying lands for the first ruler of the eastern empire, Arkadios (395–408). Grain-exporting Egypt and the fertile plains of coastal Anatolia were especially valuable, and only the Balkans had recently been seriously damaged by the raids and invasions of Goths, Gepids, and Huns.

But from a strategic point of view, the eastern empire was at a great disadvantage as compared to its western counterpart.[4]

On its long eastern frontier, running some five hundred miles from the Caucasus to the Euphrates, it still had to face the persistently aggressive Sasanian empire of Iran, which had long been the most dangerous enemy of the united empire—but it could no longer summon reinforcements from the armies of the west. It has recently been argued that the Romans had an Iran complex dating back to the humiliating defeat at Carrhae of 53 BCE, while in reality the Sasanians were not especially expansionist.[5] Perhaps so, but their rulers styled themselves "King of Kings of *Eran* and non-*Eran*" (Šahan Šah Eran ud Aneran) and the Iran part alone encompassed Persia, Parthia, Khuzistan, Mesan, Assyria, Adiabene, Arabia, Azerbaijan, Armenia, Georgia, Caucasian Albania, Balaskan, Pareshwar, Media, Gurgan, Merv, Herant, Abarsahr, Kerman, Sistan, Turan, Makran, Kusansahr, Kashgar, Sogdiana and the mountains of Tashkent, and Oman on the other side of the sea—thereby including some actual Byzantine territory, important Byzantine dependencies in the Caucasus, Armenian client states, and central Asian lands that the Byzantines certainly never ruled but in which they had critical strategic interests, notably a succession of valiant allies.[6]

The situation in the northeast was almost as bad; the Byzantines had to defend the Danube frontier against successive invaders from the great Eurasian steppe—Huns, Avars, Onogur-Bulghars, Magyars, Pechenegs, and finally Cumans—all of them mounted archers inherently more dan-

gerous than the Germanic enemies of the western empire on the Rhine
frontiers. Even otherwise formidable Goths fled in terror from the Hun
advance—and that was before Attila had united the Hun clans and
added many foreign subjects, Alans, Gepids, Heruli, Rugi, Sciri, and
Suevi, to his strength.

Nor did the eastern empire have the safe hinterlands of the western
half: coastal North Africa, which was then fertile and exported much
grain, the entire Iberian Peninsula shielded by the Pyrenees, the southern Gallic provinces safely distant from the dangerous Rhine, and Italy
itself shielded by the natural barrier of the Alps. The geography of the
eastern empire was very different: except for Egypt and eastern Libya,
most of its territories were too near a threatened frontier to have much
strategic depth. Even Anatolia, which certainly shielded Constantinople
from overland invasion from the east, was mostly settled and productive along its Mediterranean and Black Sea coastal strips, both exposed
to attacks from the sea.

With more powerful enemies and a less favorable geography, the eastern empire was certainly the more vulnerable of the two.

Yet it was the western empire that faded away during the fifth century. In essence, the eastern, or Byzantine, empire so greatly outlasted
its western counterpart because its rulers were able to adapt strategically to diminished circumstances by devising new ways of coping with
old and new enemies. The army and navy, and the supremely important
tax-collection bureaucracy that sustained them both along with the emperor and all his officials, changed greatly over the centuries, but there is
a definite continuity in overall strategic conduct: as compared to the
united Romans of the past, the Byzantine empire relied less on military
strength and more on all forms of persuasion—to recruit allies, dissuade
enemies, and induce potential enemies to attack one another. Moreover,
when they did fight, the Byzantines were less inclined to destroy enemies
than to contain them, both to conserve their strength and because they
knew that today's enemy could be tomorrow's ally.

It was so at the beginning in the fifth century, when the devastating
strength of Attila's Huns was deflected with a minimum of force and a
maximum of persuasion—they attacked westward instead—and it remained so even eight hundred years later: in 1282, when the powerful
Charles d'Anjou was preparing to invade from Italy intent on conquering Constantinople, he was suddenly immobilized by the loss of Sicily to
explosive revolt, the result of a successful conspiracy between emperor
Michael VIII Palaiologos (1259–1282), King Peter III of distant Aragon,

and the master plotter Giovanni da Procida. Michael wrote in his memoirs: "If we should say that it is God who gave the Sicilians the freedom they now enjoy, but trusting in us to bring it about, we would be saying nothing but the strict truth."[7]

The epic survival of the Roman empire of the east was thus made possible by unique strategical success. This had to be more than just the winning of battles—no sequence of fortunate victories could have lasted eight centuries in a row. Indeed the empire suffered many defeats, some seemingly catastrophic. More than once the greater part of imperial territory was overrun by invaders, and Constantinople itself was besieged several times from its foundation in 330 to its ruinous seizure by the Catholic Fourth Crusade in 1204, after which it was not an empire that was restored but only the Greek kingdom that finally expired in 1453.

The strategical success of the Byzantine empire was of a different order than any number of tactical victories or defeats: it was a sustained ability, century after century, to generate disproportionate power from whatever military strength could be mustered, by combining it with all the arts of persuasion, guided by superior information. The current terms would be *diplomacy* and *intelligence,* if one could disregard their largely bureaucratic character in modern conditions—all use of those words in what follows is to be understood in inverted commas. Having neither a foreign ministry nor intelligence organizations as such, the Byzantine empire did not have professional, full-time diplomats or intelligence officers, only varied officials who sometimes performed those functions in between or along with other duties. To persuade foreign rulers and nations to fight against the enemies of the empire—most difficult precisely in times of weakness when such persuasion was most needed—was only the most elementary application of Byzantine diplomacy, though easily the most important.

As for intelligence, the emperor and his officials could not even keep systematic files, as far as we can tell, and espionage with all its eternal limitations was almost their only means of collecting intelligence. But however ill-informed they may have been by modern standards, the Byzantines still knew much more than most other contemporary rulers. For one thing, even though they did not have accurate maps—and it has been argued that the Romans could not even think in cartographic terms—their road building proves that they were perfectly well informed about routes and linear road distances.[8] That was quite sufficient to manipulate less informed foreigners, especially newly arrived steppe chieftains from the east.[9] The near contemporary Menander Protektor preserves the bitter complaint of a Turkic chief in 577:

> As for you Romans, why do you take my envoys through the Caucasus to Byzantium, alleging that there is no other route for them to travel? You do this so that I might be deterred from attacking the Roman Empire by the difficult terrain [high mountains hard for horses]. But I know very well where the river Danapris [Dniepr] flows, and the Istros [Danube] and the Hebrus [Maritsa, Meric].

That was a direct threat, because the three rivers mark the route to Constantinople along the steppe corridor that runs north of the Black Sea.[10]

Sometimes the empire's military strength was abundant enough to allow it to mount major offensives that conquered vast tracts of territory; then diplomacy was mostly employed to extract concessions from other powers intimidated by Byzantine victories—or at least to keep them from interfering. Sometimes the Byzantine army and navy were so weak—or their enemies so strong—that the very survival of the empire was made possible only by foreign allies successfully recruited long before, or just in time: more than once, bands of warriors from nations nearby or remote suddenly arrived to tip the balance and save the day.

In between these extremes, there commonly was a more balanced synergy, in which diplomacy guided by superior information was empowered by capable military forces, while military strength was in turn magnified by well-informed diplomatic action. All of that, and some good fortune too, were needed to preserve the Roman empire of the east, because it was inherently less secure than the Roman empire of the west that it would so greatly outlast.

Persuasion usually came first, but military strength was always the indispensable instrument of Byzantine statecraft, without which nothing else could be of much use—certainly not bribes to avert attacks, which would merely whet appetites if proffered in weakness. The upkeep of sufficient military strength was therefore the permanent, many-sided challenge that the Byzantine state had to overcome each and every day, year after year, century after century. Two essential Roman practices that the Byzantines were long able to preserve—as the western empire could not—made this possible, if only by a very small margin at times.

The first was a system of tax collection that was uniquely effective for the times and that none of the empire's enemies could begin to match. After a total budget was calculated—itself an invention of huge conse-

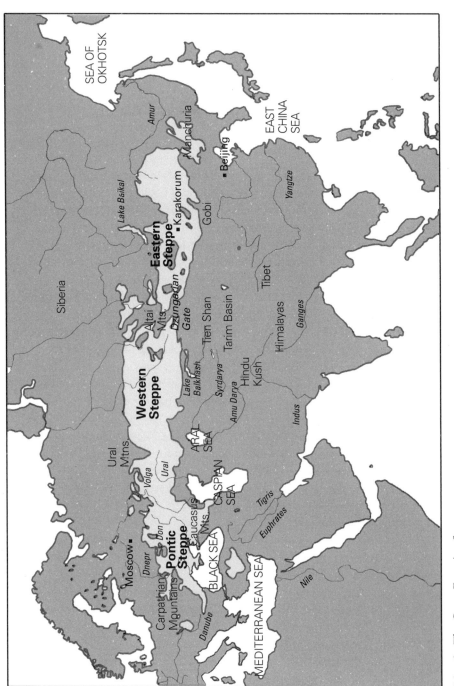

Map 2. The Great Eurasian Steppe

quence—the total amount of revenue to be provided by the principal tax, the land tax *(annona)*, was apportioned downward, first province by province, then city district by city district within each province, and finally down to individual plots of land in proportion to the estimated value of their output.[11] During the seventh century the top-down apportionment of an overall imperial budget seems to have ended, but the collection of the land tax assessed field by field continued in a bottom-up flow of revenue.[12]

There were many problems. Most obviously, the salaries of the evaluators, collectors, bookkeepers, auditors, inspectors, and supervisors were themselves a huge expense—those officials accounted for the greatest part of the imperial bureaucracy. In addition, officials accepted bribes, extorted illegal payments, and diverted revenues to their own pockets, judging by the many laws enacted by many emperors against those practices. There were also laws to safeguard the interests of small-holders, a class especially favored by emperors because they or their sons were deemed the most likely recruits, which tell us that wealthy landlords used their influence to divert tax collection from their broad acres to the plots of small-holders or even tenants.

Yet for all its faults, the fiscal machine that the Byzantines inherited had a decisive virtue: it worked year after year more or less automatically to supply vast amounts of revenue, mostly in gold. This income flow paid for the expenses of the emperor's court and of the entire civil bureaucracy but mostly served to sustain the armies and fleets. The resulting circulation of gold itself stimulated the development of the Byzantine economy: as salaried officials, soldiers, and sailors spent their money, they created a liquid market for farmers, craftsmen, and professionals of all kinds, who thus earned gold to pay for their taxes as well as their own market needs.[13]

From the strategic point of view, the most important consequence of regular taxation was regular military service. While most of their enemies had to rely on tribal levies, volunteer warriors, freebooters, or impressed peasants, with scavenging in the field to provide their supplies, the Byzantines could keep salaried imperial soldiers and sailors on duty all the year round, although they also had part-time reservists subject to recall.

That in turn allowed the vigorous revival of the second essential Roman practice that had decayed by the fifth century: systematic military training, both the individual instruction of new recruits and the regular exercise of unit and formation tactics. That may seem no more than what any army must do as a matter of course—how else would full-time soldiers pass their time? But most of those who fought the Byzantines were not full-time soldiers, they were levies summoned to the fight with no formal training, some with formidable, if narrow, traditional fighting skills, others with none. Besides, training as a continuous activity requires not only full-time forces, but also a serious degree of professionalism. Even today, most of the 150 or more extant armies both large and small barely train their recruits, who mostly receive only a couple of weeks of instruction in dress and ceremony, barrack-square drills, and the firing of personal weapons. After that, the recruits are assigned to units that now and then engage in mostly ritualistic exercises, and that hardly ever are combined in formations to carry out maneuvers—if realistic, they would only expose everyone's lack of training, so parade-ground theatricals are much preferred (I once witnessed a one-kilometer progression by a battalion of 42 tanks that kept in exact formation to the inch; weeks of training had been wasted on the tactically worthless show).

Over the centuries, the Byzantine army and navy had their cycles of institutional decay and recovery, but Byzantine survival through constant wars, often fought against superior numbers, could not have been possible without fairly high standards of training. It is characteristic of the Byzantine empire that when it was most immediately threatened in the year 626 by the converging forces of Sasanian Persia and the Avars, then both at the peak of their strength, and the remedy of emperor Herakleios (610–641) was the boldest of counteroffensives, everything started with vigorous training:

> [Herakleios] collected his armies and added new contingents to them. He began to train them and instruct them in military needs. He divided the army into two and bade them draw up battle lines and attack each other without loss of blood; he taught them the battle cry, battle songs and shouts, and how to be on the alert so that, even if they found themselves in a real war, they should not be frightened, but should courageously move against the enemy as if it were a game.[14]

Like their modern counterparts, and unlike traditional warriors, Byzantine soldiers were normally trained to fight in different ways, according to specific tactics adapted to the terrain and the enemy at hand. In that simple disposition lay one of the secrets of Byzantine survival. While standards of proficiency obviously varied greatly, Byzantine soldiers went into battle with *learned* combat skills, which could be adapted by further training for particular circumstances. That made Byzantine soldiers, units, and armies much more versatile than their enemy counterparts, who only had the traditional fighting skills of their nation or tribe, learned from elders by imitation and difficult to change. In describing the battle of the river Nedao of 454, in which the Huns were defeated by their Germanic subjects in revolt, the Gothic historian Jordanes describes how each nation fought: "One might see the Goths fighting with lances *(contis)*, the Gepids raging with the sword, the Rugi breaking off the spears in their own wounds, the Suevi fighting on foot, the Huns with bows, the Alani drawing up a battle-line of heavy-armed [cavalry], and the Heruli of light-armed warriors."[15]

Goths could certainly fight with swords as well, and the Gepids with lances, just as the classic Roman auxiliary trio of Balearic slingers, Cretan archers, and Numidian spearmen could fight with other weapons too. But while their enemies went into combat with a characteristic weapon or two, whether thrusting spear, sword, throwing javelin, dart, sling, lance, or composite reflex bow, by the sixth century Byzantine troops were trained to fight with all of them. Man for man this made them superior to most of the enemies they faced in battle and, along with unit exercises, endowed Byzantine armies with superior tactical and operational versatility.

To this, the Byzantines added the higher level of grand strategy, their own invention and not an inheritance from the past as were the fiscal system and the Roman tradition of training. There were no planning staffs, no formal decision processes, and no elaborate statements of "national strategy," which would have been alien to the mentality of the times. But there was an entire culture of strategic statecraft that emerged by the seventh century, and continued to evolve thereafter. It comprised a rich body of military expertise, well illustrated in surviving handbooks and field manuals that can still be read with interest; a sound tradition of intelligence, which inevitably is sparsely documented, though revealing traces do remain; and finally the most characteristic aspect of Byzantine strategic culture: the varied ways of inducing foreign rulers to serve imperial purposes, whether by keeping the peace or waging war against the enemies of the empire.

The Byzantines had to survive by strategy or not at all. We have already seen that the eastern empire was less favored in its geography and in its enemies than the western empire, and lacked the superior resources that the united empire had been able to deploy against its strongest enemies. Nor could obdurate resistance have sufficed. Sheer tenacity against all odds accounts for many a surprising outcome in war. It does happen that military forces seemingly superior by far are held, worn down, and finally repelled by defenders sustained by intangible and invisible strengths—whether regimental cohesion, exceptional leadership, intense religious faith, a stirring political ideology, or simply an amplitude of confidence in themselves. The Byzantine record includes many an episode of fierce resistance against vastly superior forces, none more splendid than the last fight of May 29, 1453, when the last emperor Constantine XI Palaeologus fought to the death against the armies of the Ottoman conqueror Mehemet II with five thousand loyal subjects in arms.

The loyalty that emperors could evoke from their troops was employed with better success in countless fights until the last, but obdurate resistance, no matter how sturdy, cannot explain the Byzantines' survival either—they often faced enemies much too strong to be long resisted by defensive combat alone. It was by creative responses to new threats—by strategy, that is—that the empire survived century after century. More than once, successive defeats reduced it to little more than a beleaguered city-state. More than once the great walls of Constantinople came under attack from the sea or by land, or both at once. But time after time, allies were successfully recruited to attack the attackers, allowing the imperial forces to regain their balance, gather strength, and go over to the offensive. And when the invaders were driven back, as often as not imperial control was restored over larger territories than before. The enemies of the empire could defeat its armies and fleets in battle, but they could not defeat its grand strategy. That is what made the empire so resilient for so long—its greatest strength was intangible and immune to direct attack.

Byzantine strategy was not invented all at once. Its initial elements emerged as a series of improvised responses to the unmanageable threat of Attila's Huns, the greater-than-expected threat in modern parlance. Ever since the imperial frontiers were first breached on a large scale under emperor Decius (249–251)—in one incident among many, in the year 250 a band of Franks crossed the Rhine and reached all the way to Spain—all manner of remedies had been tried. Some were ephemeral and some were lasting, some remedies were narrow and some were on a grand scale, notably the empire-wide fortifications and military enlargement of Diocletian, and the standing field army of Constantine.[16] For a century and a half, these incremental and exclusively military measures were not unsuccessful in protecting core imperial territories from incursions and territorial invasions, though at great cost to the taxpayers and to frontier populations left unprotected. But the incremental approach reached the end of its road with the arrival of Attila's Huns. For the specific tactical and operational reasons outlined in Chapter 1, military measures by themselves could no longer offer any hope of success.

That is when major strategic innovation occurs, not when it is first possible and perhaps much needed, but when all concerned finally accept that existing practices are bound to fail and that lesser remedies cannot suffice. This finally happened in Constantinople under Theodosios II (408–450),[17] when it became clear that no amount of military force within the realm of practicality could stop Attila's incursions, because they combined attributes previously believed to be mutually exclusive: they were both very fast and also very large. It was therefore useless to intercept them with small forces, no matter how mobile; and they penetrated deep in unpredictable directions, so that it was very hard to intercept them at all—and if the encounter did take place, the Huns could usually outfight their enemies anyway. The outcome of the military impasse was the emergence of a distinctly different strategic approach that was much less reliant on *active* military strength—it did require strong walls—thereby circumventing the military superiority of Attila and his similar successors.

What ensued over the next century, however, was not the straightforward consolidation of the new strategy, but rather a reversal of course and a return to a primarily military approach. With an army greatly strengthened by major tactical innovations learned from the Huns, with good leadership and good fortune, the empire reverted to an offensive military strategy of conquest under Justinian (527–565). Successful war in North Africa and Italy might have continued in spite of accumulating threats on other fronts, had the bubonic plague not arrived to wreck the entire Byzantine state and its army and navy. Recent evidence from the polar ice proves that it was the most lethal pandemic in history till then, and it is certain that the more densely inhabited empire, with its many crowded cities, suffered more than its enemies.

By the time Justinian died, the role of force had declined again, and the process continued under his successors, but it was only under Herakleios at the start of the seventh century that the distinctive grand strategy of the Byzantine empire was fully formed—just in time to overcome, if only just, the greatest crisis in its existence.

The invention of Byzantine strategy was therefore a long process, which started when Attila and his Huns, with numerous Germanic subjects, Alans, and assorted camp followers to swell their numbers, threatened to destroy the Roman empire of the east, having already undermined what was left of the western empire.

Who were the Huns? It has often been suggested that the Huns *(Hunni, Chunni, Hounoi, Ounoi),* unknown in the west until about 376 when they attacked the Goths, were arrivals from East Asia, the powerful Xiōngnú (or Hsiung-nu) nomad warriors who greatly troubled Han-dynasty China. They are described in some detail in a military report (in which the Roman empire is Da Quin, "great China," in recognition of its comparable civilization) incorporated in book 88 of the monumental dynastic history of the later Han empire, the *Hòuhànshū* compiled by the celebrated historian Fàn Yè.[18] There is some material evidence that suggests a connection—finds of iron cooking cauldrons of a specific design that can be attributed to both, which would have been used to cook their favorite horse-meat stews, among other things—but there is also chronological evidence that separates them, because the Xiōngnú are last heard of in what is now Mongolia or further east in historic Manchuria, some three centuries before the appearance of the Huns west of the Volga—an excessively long time for even the most leisurely of migrations.[19] As for the similarity in the sound of their names, it means nothing. With a monosyllabic language like Chinese, plausible identities and etymologies that meet the requirement of the sound alone "can be constructed from anything and for anything"; one example suffices: the English word *typhoon* is probably not from *da feng* ("big" and "wind") as confidently believed by people who speak both languages, but more likely from Arabic *tufan,* "storm" by way of Portuguese.[20]

The powerful Huns who suddenly became known to the Romans around 376 may have had no large origins at all, nor a specific ethnicity. They could have been, and probably were, formed just like many a better-documented warrior "nation," by a process of ethnogenesis around a fortunate Tungusic, Mongol, or Turkic clan, tribe, or war band. That is, success attracts camp followers to share in the plunder; the resulting numbers add strength that subjects weaker groupings and enslaves individuals, perhaps in large numbers. All additions of whatever sort enlarge the nation, within which individuals may retain separate subjective identities for as long as they desire, but which tends to become increasingly homogeneous with time, at a rate that depends for each assimilating group on the strength of its prior identity, and no doubt on the degree of its prior cultural, somatic, and linguistic similarity to the emerging common type. Just as collective success makes the nation, failure unmakes it, with disaffiliating groups either reverting to prior identities or embracing a new one, normally that of the more successful nation that arrives on the scene. In our own days, families of varied origins who lived in the Soviet Union acquired a Russian identity when that was the dominant nationality of a seemingly eternal empire, only to revert to prior ethnic identities when the Soviet Union declined even before it actually disintegrated, while some embraced entirely new identities after emigrating to Germany, Israel, or the United States.

Hugely controversial when applied to the Goths or more broadly "Germanic" populations, with everything from nineteenth-century Germanism, twentieth-century Nazi mythology, and twenty-first-century sociology thrown into the debate, the concept of ethnogenesis was originally introduced to describe much simpler processes in the Eurasian steppes.[21] They have no high mountains and remote valleys to shelter the weak, allowing them to preserve their identities, while the shared patterns of pastoralism flatten many differences anyway, so that immediate accommodation to stronger arrivals was followed by assimilation as a matter of course.

It was too soon for that process to have formed a common nation when Attila became the supreme ruler of diverse "Huns," Alans, Goths, Gepids, and assorted others, and his death undid the power of the Huns anyway. But there had already been much cultural integration by his time—the very name Attila is not Hunnish. After summoning proto-Chuvash and Old K'art'velian (less exotically, old Georgian), to scant effect, after dismissing perhaps too hastily the etymologies of Hungarian-nationalist historians: Attila = Atilla = Atil = Turkic "big river" = Volga, the unsurprising conclusion of the most eminent Hunologist is that Attila is Germanic or, if one prefers, Gothic: "little father."[22] There had been some assimilation no doubt, even perhaps of the "youth of Syria" who were captured in a 399 raid through the Caucasus, according to the poet Claudian in his masterpiece of invective against Eutropius—the eunuch consul whom Claudian unfairly blamed for the Hun irruption, in which cities were set on fire and youngsters were dragged off to slavery.[23]

Less biased sources confirm the raid itself and the enslavements—although one adds the very interesting information that other local youths volunteered to join the Huns to fight in their ranks.[24] That should not be a subject of wonder. The Huns were uncouth and pagans too, they had just pillaged, killed, and maimed their fellow citizens, perhaps friends or relatives. But for young men, or veteran soldiers for that matter, to join the Hun columns in a land just devastated by them was to go immediately from the category of the defeated and plundered to the category of the victorious, rich in the plunder loaded on their packhorses and wagons or tied behind them, including women.

That was and still is the essential mechanism of ethnogenesis. Success creates nations out of diverse groups, and then expands them by attracting volunteers. Soon enough, such expanding groups cease to be ethnically homogeneous but still preserve their original label, thus becoming pseudo-ethnic entities in greater or lesser degree. Thus after the Huns rose and fell and dispersed into other nations, it was the turn of the Avars to go from prestigious clan to a mighty power in the Balkans with many men, further swollen in numbers by their more numerous Slav subjects.[25] After expanding with success, eventually there was a first defeat in 626 before the walls of Constantinople that caused Slavic defections; other defeats further diminished the Avars over time, decisively so in 791 at the hands of Charlemagne himself. After that the Avars became smaller still—small enough to be attacked by the lesser power of the Bulghars, and soon they disintegrated entirely to be absorbed by other nations. By then, their former abode in what had been Roman Pannonia was occupied, as it still is, by the moderately successful Magyars, originally a tribe which became a nation by assimilating similar tribes that adopted its ethnic name, and who mostly still live in Magyarország, the country of the Magyars that only foreigners call Hungary.

Given the nature of ethnogenesis, what is formed by its processes of fusion, assimilation, subjection, and capture should be called, not a nation at all, for that does imply a degree of ethnic homogeneity, but rather a "state," for it is an essentially political entity after all. The only impediment is that some populations, such as the important Pechenegs, remained loosely affiliated tribes, clans, and war bands; they had an identity but not overall chiefs or common institutions, so "nation" they must be after all. Such indeed were also the Huns, a large nation by the time Attila came to rule over them as sole king, endowing them with the essential institutions of a state, and making them far more powerful than before.

CHAPTER I

Attila and the Crisis of Empire

The extraordinary prominence of Attila's Huns in the annals of antiquity is most remarkable for they had so much competition. Their time of unity and power until Attila's death in 453 came just after the great Germanic invasions that would eventually extinguish the Roman empire in the west. Comprehensively fortified and once well-garrisoned, the Rhine and Danube frontiers had protected the European provinces of the Roman empire for almost four centuries. Thousands of watchtowers, connected by palisades or even stone walls where there was no river barrier, with patrols to link them and hundreds of garrisoned forts in support, formed a continuous barrier in northern England and right across Europe, from the North Sea estuary of the Rhine in modern Holland to the delta of the Danube on the Black Sea coast of modern Romania.[1] No thin linear defense could stop powerful invasions, but everyday security from raids and robbery was well provided by the imperial frontier system, the *limes*.

The decay, abandonment, and final collapse of the Rhine and upper Danube frontiers was the great catastrophe of the age for the citizens of the empire, who were left exposed to pillage and ruin, if not worse fates. That prolonged tragedy is reflected in virtually every contemporary text that still survives, not just histories and chronicles but also poems, letters, saintly biographies, and writings on quite other subjects by way of incidental comment. The invaders described or just deplored in those writings included the Germanic Alamanni, Burgundians, Riparian Franks, Salian Franks, Gepids, the powerful Greuthungi and Thervingi

Goths, Heruli, Quadi, Rosomoni, Rugi, Sciri, Suevi, Taifali, and the original Vandals as well as Alan horsemen of Iranic origin and the probably Slavic Antae.

Yet Attila's Huns were deemed a more terrible threat than any of them, and were more vividly remembered in the aftermath, as indeed they still are today—more so than Alaric's Goths who sacked Rome in 410, or the proverbial Vandals who inflicted a greater disaster by cutting off North Africa's grain supply to Italy.

For contemporary ecclesiastical writers the Huns were the great scourge of God and Attila himself the Antichrist, unless he was depicted as the most terrible of human barbarians in miracle stories—one featuring the historical pope Leo I:

> For the sake of the Roman name he undertook an embassy and traveled to the king of the Huns, Attila by name, and he delivered the whole of Italy from the peril of the enemy.[2]

Because they were identified with the Massagetae of Herodotus, the more ancient horrid people of the steppe, the Huns inevitably became the protagonists of the apocalyptic war of Gog (the Goths) and Magog in Ezekiel. Another ecclesiastical voice, Ambrosius, the later saint and the first of the still highly politicized bishops of Milano, omitted God and Magog but came to the same end-point:

> The Huns threw themselves on the Alans, the Alans on the Goths, the Goths on the Taifali and Sarmati. Expelled from their homeland, the Goths have expelled us from Illyricum, and it is not yet ended . . . we are at the end of the world.[3]

It is also suggestive that the wars of the Huns and the Goths, the massacre of the Burgundians of King Gundahar in 437, and Attila himself were still vividly remembered centuries later even very far from any lands they ever crossed. In the Old English poem *Widsith* the hero declaims: "I visited Wulfhere and Wyrmhere; there battle often raged in the Vistula woods, when the Gothic army with their sharp swords had to defend their ancestral seat against Attila's host."[4] Even in distant Iceland, Attila was remembered in the Old Norse poem "Lay of Hloth and Angantýr," in which Attila appears as Humli, king of the Huns and grandfather of Hloth. It is part of the *Hervarar Saga* in which there is also a battle of Goths and Huns precipitated by Attila's marriage to Gudrun. In the *Volsunga Saga*, Attila is killed by Gudrun, who had been forced to marry him, a story derived from the older *Atlakvið a*, "The

Lay of Atli," or from the longer version in *Atlamál hin groenlenzku,* the "Greenland Ballad of Atli"—thus we know that his fame reached that most remote of places, Ultima Thule.

More famously, Attila is the Etzel of the *Nibelungenlied,* the medieval German epic that Wagner turned into music and spectacle: the murdered Siegfried's vengeful wife Kriemhild marries Etzel, king of the Huns, and bloody mayhem ensues. In the earlier Latin epic poem *Waltharius* by Ekkehard of St. Gall, Alphere, king of Aquitaine, had a son named Waltharius, who is given as a hostage to Attila, king of the Huns, when he invades Gaul. In Attila's service, Waltharius wins great renown as a warrior, before fleeing with much gold from his court.[5]

In deference to the then-fashionable theory that individuals are insignificant as compared to historical processes, and also in obedience to the Marxist theory of stages, an important modern historian portrayed Attila as a bungler of minor importance; and while the greatest authority on the Huns disagreed, he nevertheless then went on to compare him to the ephemeral Gothic warlord Theoderic Strabo ("the Squinter"), who in 473 extorted two thousand pounds of gold from the eastern emperor Leo.[6]

Contemporary and later public opinion as retained in the sagas disagreed. While Attila himself is not depicted as particularly heroic—the heroes are Germanic—the stories show that Attila's Huns were believed to be exceptionally powerful, more powerful than any other kingdom or nation.

That was also the opinion of far more analytical sources, starting with the professional military officer and sober factual historian Ammianus Marcellinus, who assessed the strategic importance of the Huns even before Attila: "The seed and origin of all the ruin and various disasters that the wrath of Mars aroused . . . [the catastrophic Roman defeat at Adrianople on August 9, 378] we have found to be this. The people of the Huns . . . [who] exceed every degree of savagery."[7]

Thus it was by way of interposed fugitives-in-arms that the Huns first burst on the Roman scene. In 376 vast numbers had arrived at the well-guarded Danube frontier—men, women, and children—some Iranic Alans but mostly Germanic Gepids and much more numerous Tervingi and Greuthungi Goths, all begging to be admitted within the safety of imperial territory. There were many formidable warriors among them, not only Germans with spears and swords but also Alan horsemen with armor and lances. Yet all had been terrified into panicked flight by the Huns who had advanced upon them from the steppe further east. The Romans then knew nothing about the Huns, but they had known the Goths and Gepids since the mid-third century, first as dangerous raiders by land and by sea, then as almost peaceful neighbors who mostly came to the frontier to trade and to offer their mercenary service to the imperial army. After agreeing to let them enter on condition that they serve the empire, Roman officials failed to deliver promised grain supplies, eventually provoking a revolt that emperor Valens came to suppress with the field army of the east. He was defeated and killed, two thirds of the army of the east were destroyed, and Romans then learned that the Goths, Gepids, and Alans, who had been strong enough to defeat them, had themselves fled like frightened sheep before the Huns.

Writing more than a century later, by way of incidental and thus especially revealing comment, the greatest historian of the age, Prokopios of Caesarea, offers the following abbreviated history:

> The Roman Emperors of former times, by way of preventing the crossing of the Danube by the barbarians who live on the other side, occupied the entire bank of this river with strongholds, and not the right bank of the stream alone, for in some parts of it they built towns and fortresses on its other bank. However, they did not so build these strongholds that they were impossible to attack, if anyone should come against them but they only provided that the bank of the river was not destitute of men, since the barbarians had no knowledge of storming walls. In fact the majority of these strongholds consisted only of a single tower, and they were called appropriately "lone towers" *(monoturia)*, and very few men were stationed in them. At that time this alone was quite sufficient to frighten off the barbarian clans, so that they would not undertake to attack the Romans. But at a later time Attila invaded with a great army, and with no difficulty razed the fortresses; then, with no one standing against him, he plundered the greater part of the Roman Empire.[8]

In his works, except for the scandalous *Anecdota*, Prokopios unfailingly explains his controversial contentions, but he did not consider it necessary to justify his judgment that Attila's Huns were a qualitatively different and greater threat. Evidently that was the common opinion of his times, when the Huns, long since stripped of their subjects and camp followers, had scattered, some to return to the steppe where they were absorbed by the more successful Turkic groupings of Avars, Ogurs, Onogurs, and Bulghars.[9]

There was an excellent reason for the unique reputation of the Huns. With their hardy Mongolian ponies, they introduced an entirely new and highly effective style of warfare into the Roman world, which was destined to be adopted and adapted to form the basis of the emerging Byzantine army, which thereby came to differ fundamentally from its classic Roman predecessor.[10] This new style of war was first described with commendable precision by Ammianus Marcellinus, whose reliability is enhanced by considerable professional military expertise as both a combat soldier and a staff officer. From his essay on the Huns of the late fourth century, we can first of all extract a valid description of their tactics:

> You would not hesitate to call them the most terrible of all warriors, because they fight from a distance with missiles [arrows] having sharp bone, instead of their usual [metal] points, joined to the shafts with wonderful skill; then they gallop over the intervening spaces and fight hand to hand with swords, regardless of their own lives; and while the enemy are guarding against wounds from the sharp [sword] points, they throw strips of cloth plaited into nooses over their opponents and so entangle them that they fetter their limbs and take from them the power of riding or walking.[11]

These are the common tactics of all accomplished steppe warriors, which would become abundantly familiar to the Byzantines as the Huns were followed down the centuries by the Avars, the first Turks, the Onogur-Bulghars, Magyars, Pechenegs, Cumans, Mongols, and finally the Mongol-Turkic subjects of Timur, our Tamerlane. And these are tactics that the Byzantines would eventually learn to imitate very successfully (lariats aside), and even improve.

First there are the rapid volleys of arrows discharged from the exceptionally powerful bows discussed below—weapons that could kill even from "a distance," as less powerful bows could not. As for the bone arrowheads, they would not be less lethal than metal ones if sturdy enough, and the text indicates that Hun arrows were exceptionally well made, that their sharp points would not separate on impact.

If the enemy did not attack, it would suffer mounting losses to the arrows. If it did attack, it could not come to grips with the mounted Huns, who did not have to stand their ground—if they did, it was because they were confident of victory, and to attack them was probably imprudent.

If the enemy withdrew to avoid further casualties, this would allow the Huns to ride them down, killing with both their arrows and their swords. (It is unspecified if they were straight, or curved sabers—the words in the text are first *ferro,* "iron" the most generic of terms, and then *mucro,* sword edge or sword-point).

Next, if there is no retreat, once enemy ranks are depleted enough, the charge and melee follow, the Huns wielding swords in one hand, throwing entangling lariats or lassos with the other. Unlike the bow of the Huns, the lasso was not a new weapon—it was widely used by steppe peoples, the Alans, and even by the Goths if not other Germanic warriors,[12] but only a few were likely to use it as well as steppe warriors and herdsmen, who must control their horses without walls or fences, using only lariats (the *urga* of the Mongols, Turkic *arqan*), a rope loop at the end of a pole, and hobbles.

But the greatest skill of the Huns, all our sources concur, was with their primary weapon, the composite reflex bow: "Shapely bows and arrows are their delight, sure and terrible are their hands; firm is their confidence that their missiles will bring death, and their frenzy is trained to do wrongful deeds with blows that never go wrong." Thus wrote Gaius Sollius Modestus Apollinaris Sidonius, who was twenty years old when Attila invaded the northern parts of his native land of Gaul.[13] He was no military expert—elsewhere he praises the Gallic notable Marcus Flavius Eparchius Avitus, one of the ephemeral last emperors of the west (455–456), as the "equal of the Huns in javelin-throwing [*jaculis*]," which was not one of their skills.[14] But there is no doubt that Hun archery was an innovation in warfare, partly because it was combined with exceptional horse mobility at all levels—tactical, operational, and strategic—and partly because of their new weapon.

The Composite Reflex Bow

Early versions of the "the Scythian bow" had been known since ancient times, but the Hun bow would not have attracted so much attention had it not been the distinctly more powerful weapon that was to be used in war until the sixteenth century across the entire span of Asia, from the Ottoman empire to Japan.[15] There are many variations, and no attested example, fragment, or credible depiction of a Hun bow has survived—though a historian of the period has confidently asserted that the Hun bow was asymmetric, even proffering its exact dimensions.[16] That it was longer above the handle than below is certainly possible, and it is true that such asymmetry allows a longer and therefore potentially more powerful bow that will still clear the horse's neck when held vertically upright directly ahead of the rider. It should be noted, however, that in the only mounted archery we can observe today, in Japan's Yabusame competitions at the Meiji shrine, in Kamakura, and other ceremonial venues, where the methods of the Ogasawara and Takeda schools have been perpetuated since the twelfth century and all bows are asymmetric, very few riders hold their bows upright when using them, for the excellent reason that it is easier to hold the bow at an angle. Asymmetry is just a matter of preference, and we have no evidence that the Huns so preferred; incidentally, the amply illustrated Mongol bows were perfectly symmetrical.

Even if we had fully attested illustrations of Hun bows, they would not tell us much, because the appearance of these weapons is so misleading: when viewing unstrung examples in museum cases today, all we see are long, slim fusiforms seemingly made of painted wood but that actually consist mostly of thin layers of dried horse sinew and bone plates. The composite bow accumulates energy on both counts when the string is withdrawn, and any functional example is so powerful that it reflexes and reverses itself when unstrung.[17] More fully, there are five elements: a wooden core, which in itself would be a simple "self bow"; a belly—the side toward the archer—made of keratin, the outer and more elastic layer of horn, usually bovine; the multilayer sinew backing that provides much of the tension, added layer by layer as each one dries; the "ears," straight extensions attached at the end of each curved limb to increase energy accumulation; and the handle, either built up in the center or made as a separate piece with the two limbs then inserted or spliced on. Animal glues made from the collagen skimmed off boiled hide or sinew hold together the horn belly, the wooden core, and the sinew back.

Bovine horn plates can compress by 4 percent before yielding, as opposed to 1 percent or so for the best woods; the preferred horn from European or Indian cattle, or better, Asiatic water buffalo, was split and then steamed or boiled to make it pliable and more easily cut and shaped. The dried sinew layers of the highly stretched back of the bow have roughly four times the tensile limit of wood. Taken from animal tendons, either the hind legs or the back-strap, the threads of sinew must be applied in a matrix of hide or sinew glue, as in the making of modern fiberglass.[18]

This is obviously a far more elaborate process than the manufacture of self bows, which only requires the selection of a straight and elastic wooden stave; or of the reflex bow, obtained by cutting a curved stave of wood that is reversed when strung; or the compound bow, made by binding together more than one stave of wood—the celebrated English and Welsh longbow, though made from a single stave of yew, was effectively a compound bow because the stave was cut from the radius of the tree, so that the elastic and tensile sapwood became the back and the heartwood, which resists compression, formed the belly; or even the sinew-backed wooden self bow of the American Indians.

Because their bows were so hard to manufacture, even when Germanic populations of Goths and Gepids lived and fought together with their Hun overlords for decades, they still did not adopt the bow as their weapon, presumably because they lacked expert bowyers, who were probably not abundant even among the Huns themselves. In 1929, the greatest scholar of the Huns, Otto J. Maenchen-Helfen, visited the Barlyq-Alash-Aksu region of Tuva, the old Tannu-Tuva of stamp collectors, now the Tyva Republic of the Russian Federation. There he encountered old men who told him that in the 1870s and 1880s there were only two men who could still make composite reflex bows. Connoisseurs will detect more than one political undertone in his comment: "The idea that each . . . archer could make his own bow could have been conceived only by cabinet scholars who never held a composite bow in their hands."[19]

Deceptive in appearance, the composite reflex bow hides its power. Tension and compression forces are minimal within the wooden core, allowing more of the energy stored in the bow by withdrawing the string to accelerate the arrow, rather than the mass of the limbs themselves. Both the wooden core and the matching horn plates are grooved to double the gluing surface; the glue joints are subject to shear rather than tension when the bow is drawn, increasing its relative strength. Finally, the ears act as static recurves, loading all the energy into the middle third of the limb as the bow is drawn. Also, as the bowstring is withdrawn, the effective length of the string increases, making it easier to withdraw the string further.

If properly cured, glue extracted from hide or sinew is stronger than all but the most advanced of contemporary adhesives, but it is hygroscopic—it absorbs moisture from the air, even if improved with tannin extracted from tree bark, an ancient and effective practice of Asiatic bowyers. For this reason alone, the mounted archers of the Eurasian steppe could not prosper in wetter northern climates, limiting the geographic reach of their conquests. In the *Hervarar Saga,* the wise king Gizur, Gizurr, or Gissur (whose title was incorporated in the *"Sveriges, Götes och Vendes Konung"* of Swedish kings until 1973), taunts the Huns on the eve of the final battle of the Huns and the Goths: "We fear neither the Huns nor their horn-bows."[20] That is an echo of the historical and devastating defeat of the Huns by their former Germanic subjects in the battle of Nedao in what was Pannonia and is now Serbia in 454. It is certainly true that mounted archers weak in the use of other weapons were apt to suffer catastrophic reverses if unable to avoid battle when it rained.

Very hard to manufacture, the composite reflex bow is also very hard to use with any accuracy, because its power makes it correspondingly resistant. Unlike swords, spears, or even self bows, it is therefore useless in the hands of novices, who could not even string it—for the tension of the sinew backing must first be reversed. It was evidently a composite reflex bow that the far-traveling Odysseus had left in his rustic palace of Ithaca when he sailed to Troy—the bow that none of Penelope's suitors could even string, the bow with which Odysseus began their execution:

> In his great cunning he bade his wife set before the wooers his bow and the grey iron [target] to be a contest for us ill-fated men and the beginning of death. And no man of us was able to stretch the string of the mighty bow; nay, we fell far short of that strength. But when the great bow came into the hands of Odysseus, then we all cried out aloud not to give him the bow, how much soever he might speak; but Telemachus alone urged him on and bade him take it. Then he took the bow in his hand, the much-enduring, goodly Odysseus, and with ease did he string it and send an arrow through the iron. Then he went and stood on the threshold and poured out the swift arrows.[21]

The Ithaca provincials had tried to string the bow with brute strength, by forcing it to curve enough to receive the string—easy to do if one has at least three hands, two to pull back the limbs into position, one to tie or loop the string on each ear—but impossible with only two. Odysseus knew how to string reflex bows such as his own: he strung it "with ease" by first pulling back each limb into position with a "bastard string" looped on a wooden stick tied to the stave, only then slipping on the true string on the reversed bow, to finally remove the stick and bastard string to commence his execution of the suitors.

Once strung, the composite reflex bow is still too resistant to be employed with any accuracy without much practice, preferably starting in childhood, with yet more practice needed to use the weapon usefully on horseback and on the move.

That is the main reason why even the very first personal firearms— thick-barreled harquebusiers, or still heavier muskets that had to be supported on tripods, much slower to reload down the muzzle with powder, packing, ball, and more packing, and also less accurate—nevertheless supplanted both the Welsh longbow and the superlative Ottoman composite bow as soon as they were available in numbers. (Another reason was that loud firearms could frighten enemies and terrorize untrained horses.)[22] Kings and warlords with gold in hand could quickly convert it into military strength by raising regiments of musketeers—a week's training was quite enough to master the weapon. By contrast, the supply of capable archers was inelastic—their training had to have begun years before. Moreover, some infantrymen and rather more cavalrymen simply could not master the bow, which requires some talent as well as intense training. For missiles, they had to rely on stone-throwing slings, which bowmen also carried as a reserve weapon to be used if the supply of arrows were exhausted, or if conditions were so wet that bows would be ruined.

There is no question, however, that mounted archery could be taught and learned, given plenty of time and much effort: the Byzantine mounted lancers and archers who replaced the heavy infantry as the core of the army during the sixth century were not children of the steppe, merely well-trained.

The composite reflex bow that is so hard to manufacture and so hard to use redeems itself with its performance in trained and talented hands. The maximum range records set by Ottoman archers, notably the celebrated 482 yards achieved in 1795 by Mahmoud Effendi, secretary to the Ottoman ambassador in London before several members of the Royal Toxophilite Society, are irrelevant, because those were flight arrows with no penetrating power or accuracy.[23]

There is also evidence of Mongol archery at its best in the Uighur-Mongol (uigarjin) inscription on a famous granite stele circa 1224/1225 found by the polymath G. S. Spassky, initially read by local lamas and reported in 1818 in the *Sibirsky Vestnik,* and now in the Hermitage museum in St. Petersburg: "When, after the conquest of the Sartaul [Muslims] people, Genghis Khan assembled the noyans [chiefs] of all the Mongol ulus in the place called Bukha-Sujihai, Yesungke [his nephew] shot an arrow 335 *sazhens.*" That is roughly 400 meters, but *sazhens* or *alds* in modern Mongolian is the unspecified length of a man's open arms, and there is also a patriotic estimate of 536 meters.[24] That arrow could not have had any penetrating power either. It is indicative that in the archery event of the contemporary Eriin Gurvan Naadam festivity in Mongolia, men discharge their arrows from 75 meters away while women discharge their arrows from 60 meters away. That, however, understates the useful range of composite reflex bows, because the competition strongly emphasizes the rate of fire: men have to discharge forty arrows and women twenty, high numbers with resistant bows, as these must be.

What is certain is that ranges of military value were still phenomenal as compared to ordinary self bows: an *effective* (potentially killing) range of up to 150 meters, especially relevant when bowmen could volley their arrows into dense formations of unarmored men or horses; an *accurate* range of up to 75 meters, especially relevant in ambushes and sieges, when bowmen in the role of snipers had opportunities to aim carefully at single targets; a *piercing* range of up to 60 meters, against most forms of scale (sewn on), mail (interlocked rings), or lamellar (linked plates) armor.[25]

The composite bow of the Huns was as powerful as the Welsh longbows that slaughtered French armored cavalrymen at Agincourt in 1415, but unlike those six-foot weapons it was handy enough to be used on horseback. It was the penetrating power of their arrows that utterly surprised the Romans when they first encountered Huns with their composite reflex bows. The resulting shock can be sensed in the contemporary account of Ammianus Marcellinus. When they first appeared, the Hun bows overthrew previous certitudes, as men confidently relying on shields and body armor were pierced by arrows from ranges till then thought impossible. The Huns could launch their arrows with the minimum of accuracy needed to hit someone in a dense mass of soldiers even while riding fast, even at a full gallop and laterally or even backward. They could therefore calmly approach their enemies to discharge their arrows at the piercing range of a hundred yards or so, or much closer to defeat armor, while already turning back to ride out of reach again, only to repeat the attack time after time.

Any infantry armed with no better missile weapons than javelins, slings, or plain wooden bows was badly outranged, and left helpless if caught out in the open with no protection against Hun arrows. Roman light cavalry was better off only insofar as it could flee the scene, while "heavy" cavalry trained for the charge could easily disperse but not actually defeat the mounted archers of the steppe, who had no reason to stand their ground if charged. And for that, too, charging cavalry needed good armor protection to remain alive after their momentum was spent, because the arrows launched by well-made composite bows could penetrate scale and mail armor at fifty yards if not more.

The Huns thus had a net tactical superiority in open-field operations in dry weather, the most frequent scene of important battles. But they were at a disadvantage in very wet weather, in rugged terrain unfriendly to horses, in dense woods that neutralized their missiles, and also in siege operations for which they lacked the technology until later days under Attila, and the logistic staying power—especially when they relied on a mass of Germanic subjects less self-reliant than themselves. Tactically, therefore, the military strength of the Huns would have been largely confined to battles in the steppe, had it not been for their abilities at the higher levels of strategy.

The Operational Level

Tactical strength is the basic building block of military power, but battles are decided at the higher, operational level of strategy in which all forces on both sides interact, and in which tactical achievements alone may not mean much. For example, in frontal combat, if the defenders of a particular sector are more tenacious than those on either side of them, they will only contrive their own isolation, eventual encirclement, and capture if they persist in holding on when their comrades on either flank withdraw. Conversely, a unit that fought hard and suffered casualties to advance more than the units on either side of it may be ordered to withdraw and abandon the territory it had just won, if it is viewed as a vulnerable salient hard to defend and easily cut off by the enemy.

These examples are drawn from linear ground combat in the manner of the First World War because they are the simplest to visualize, but the operational level of strategy is present in all forms of warfare and can be far more subtle in negating, confirming, or magnifying tactical achievements and strengths.[26]

That was so with the tactical superiority of the Huns—it was magnified at the operational level by agile maneuvers made possible by superior mobility, which exceeded the usual mobility of competent cavalry. "They are almost glued to their horses," wrote Ammianus Marcellinus,[27] who is amplified by Sidonius:

> Scarce has the infant learnt to stand without his mother's aid when a horse takes him on his back. You would think the limbs of man and beast were born together, so firmly does the rider always stick to the horse, just as if he were fastened in his place; any other folk is carried on horse back, this folk lives [on horseback].[28]

This time Sidonius is not led astray by poetical needs—he is describing quite accurately the routinely superlative riding skills of the horsemen of the steppe, like the Mongol and Tuvan riders one may still admire today, who are the product of a horse-centered culture in general, and of the specific practice related by Sidonius—children start riding as soon as they can walk, well before they can lift themselves onto a pony.

In the contemporary horse races of the Mongol Eriyn Gurvan Naadam festival, up to a thousand horses can compete, and their jockeys are all under thirteen, with a prevalence of younger children: the minimum age is five. And this to race two-year-old horses over 16 kilometers, and seven-year-olds over 30 kilometers, very long distances indeed, especially given that there is no prepared course but only the open grassland, which is not especially flat nor lacking in rodent holes. As for contemporary evidence of mounted archery as opposed to just horsemanship, the Yabusame archers mentioned above gallop down a 255-meter-long track at high speed, controlling their horses with their knees alone, while using both hands to draw their arrows back beyond their ears before release.

Over short distances, the steppe riders could be outraced with ease by Western jockeys on thoroughbreds, but their seat is infinitely more secure, allowing them to do much more from horseback than simply ride. I have witnessed very accurate shooting with incongruous AK-47 assault rifles ahead, sideways, and rearward by Mongol horsemen racing at a full gallop, just as their predecessors once did with the bow, simply turning to aim as if they were in a swivel seat, without the slightest unease of imbalance.

Most important for combat, the unity of man and horse allows them to habitually ride in a melee, which they do when chasing down and catching untamed horses with their *uurga* pole nooses. Riders and horses are so confident in each other that there is no dread of the lethal pileups that terrify Western jockeys.

These same riding skills can be confidently attributed to the Huns, whose "extraordinary rapidity of movement" was first noted by Ammianus Marcellinus, who also pointed out the operational-level implication—it allowed exceptionally agile maneuver:

> They enter the battle drawn up in wedge [or anvil] formations *(cuneatim)* . . . . And as they are lightly equipped for swift motion, and unexpected in action, they purposely divide suddenly into scattered bands and attack, rushing about in disorder here and there, dealing terrific slaughter.[29]

Thus their plan of action could not be "read" from their battle line; the later field manual *Strategikon* (book XI, 2) warned that when fighting steppe peoples it was essential to send scouts to probe all around their formations because there was no saying how deep they might be, concealing their true numbers.

The operational method described by Ammianus Marcellinus amounted to a fluid sequence of unexpected actions, as groups of warriors moved in and out of range, sometimes to launch arrows from a safe distance that could nonetheless pierce mail and other light armor, sometimes charging in for close combat once enemy formations were broken up. In ancient battles, the defeated could usually flee safely because they could outrun any infantry in arms by dropping their shields. The Spartan mother might tell her son, "Come back with your shield, or on it," but Archilochus had more practical advice:

> Some lucky Thracian has my noble shield:
> I had to run; I dropped it in a wood.
> But I got clear away, thank God! So hang
> The Shield! I'll get another, just as good.[30]

Victorious cavalry could pursue and cut down fleeing enemies, but not if they were just as well-mounted—besides, simulated flight to lure enemies into prepared ambushes was so common a cavalry tactic that no prudent commander would allow headlong pursuits. As we shall see in Part III, Byzantine military manuals advised extreme caution when pursuing fleeing enemy cavalry, especially in broken terrain. But Hun riders were lightly equipped, as Ammianus Marcellinus noted, having neither metal armor nor heavy lances, so they could outrun fleeing cavalry as well as defeated enemies on foot; they also had less to fear from ambushes because of their tactical agility. Unless there was dense forest, or sharply contoured high ground, or a well-walled city within closest reach, death or capture awaited those whom the Huns defeated. That too is a reason why Ammianus Marcellinus wrote: "You would not hesitate to call them the most terrible *(acerrimos)* of all warriors."

The Level of Theater Strategy

Results at the operational level are also provisional, because the winning or losing of battles can be nullified, confirmed, or magnified by the wider struggle in the entire geographic context. For example, battles won within a confined geographic setting are far more likely to be definitive than the same battles won at the edge of an extensive theater of war, within which the defeated have room to retreat in depth, and fall back on their own core territories to regroup, recruit, resupply, recuperate, and eventually counterattack. That is the chief reason why in modern times the *Wehrmacht* was distinctly more successful in invading small Belgium than immense Russia, and why the deepest of all offensives of Sasanian Persia against the Byzantine empire, which reached all the way to the shore opposite Constantinople in 626, was ultimately defeated, and the Sasanian empire with it; had the Sasanians been content with the narrower lands of Byzantine Syria, they might have won their war.

Geographic distance, as enhanced by terrain obstacles and a lack of usable resources (starting with water), or to the contrary, as alleviated by roads and bridges as well as usable resources along the way, becomes the "strategic depth" that protects the invaded—insofar as it is not overcome by mobility—of humans, animals, carts, or wagons.[31]

Very high speeds were attainable in ideal conditions. With its relays of fresh horses, in favorable weather over easy terrain with good roads, the Byzantine official mail could deliver messages at speeds that could reach 240 Roman miles, or 226 statute miles, or 360 kilometers, within 24 hours.[32] That was almost ten times faster than the rate of advance of an expeditionary army or even of cavalry formations, because they too would not long remain effective without their supplies of food, tentage, tools, extra arrows, and spare clothing, carried on pack animals at best, but more likely carts or even slower ox-drawn wagons.

It has been estimated that pack mules and horses can have sustainable average speeds of up to 3.5 miles, or 5.2 kilometers, per hour, if in very disciplined convoys over easy terrain. But their load capacity has been estimated at an average of only 152 pounds, or 69 kilos, as opposed to the 400 pounds that a single ox can pull, or the short ton, 2,000 pounds or 907 kilos, that can be loaded into a four-ox wagon.[33]

Ten such wagons could therefore replace 130 pack animals—an important consideration, because pack horses and even mules are difficult to manage in large numbers, and their need for secure pasture or forage and water can easily become a severe constraint on the conduct of a campaign. Oxen also need food and water but do not wander off and need not be hobbled or watched. Hence ox-drawn wagons were normally indispensable for large forces moving with their supplies for serious campaigns. Oxen, however, are distinctly slower, as I know from personal experience, with a maximum speed of two and a half miles or 4 kilometers per hour in favorable conditions, and they cannot exceed a total of 20 miles or 32 kilometers per day, because they need 8 hours of grazing and another 8 hours of cud-chewing and resting.[34] Those too are only theoretical numbers, because oxen, mules, and especially horses do not last long if worked to their maximum limits, so that Byzantine expeditionary forces, more substantial than light-cavalry scouting units with their own spare horses, were unlikely to exceed 15 miles or 24 kilometers per day except across flat terrain with decent roads.[35]

To set that in context, we may note that when entirely unopposed, the supply convoys of the German army of the Second World War were expected to move beyond the last railhead at those same speeds of up to 15 miles or 32 kilometers per day, with 12 miles or 19 kilometers being more likely. They also mostly relied on animal transport, notwithstanding the dashing motorization shown in propaganda newsreels, but their wagons had rubber wheels and they were pulled by two horses, not oxen, thus achieving maximum speeds of as much as 20 miles or 32 kilometers per 24 hours over good roads in flat country, in good weather, with well-trained and healthy horses—but only for one day, followed by a day of rest.[36]

The Huns had a much greater mobility advantage over their more settled enemies than the Wehrmacht's ground forces would have had. Although they too had horse-drawn wagons for their families and possessions, and without the advancement of rubber wheels (Ammianus Marcellinus in a florid moment, XXXI, 2.10), their fighting forces even on the largest scale, like those of other horse-centered steppe cultures, moved at the rate of the horse, not of carts or wagons—as much as 50 miles or 80 kilometers per day in favorable conditions—and more routinely twice as fast as the upper estimate of Byzantine theater-scale mobility. In other words, the speed of Hun expeditionary forces as a whole approximated the speed of Byzantine light-cavalry patrols at their best.

Not even hardy Mongol horses can keep up such speeds when carrying a fighting man and his weapons, equipment, and rations, but neither did they have to. If the Huns were like their steppe successors, as all evidence indicates and none contradicts, they too rode in a "vast herd" of horses rather than as conventional cavalrymen with their individual horse and a single remount at most.[37] By switching horses at frequent intervals long before they were tired, by distributing loads very lightly among several horses, if not a dozen or more, by keeping the next two intended remounts entirely unencumbered, large numbers of Huns could move across favorable terrain at rates of 30, 40, or even 50 miles per day for quite a few days in a row.

The resulting advantage at the level of theater strategy was very great. The Huns could reach a distant location, launch their attacks to achieve their aim of force-destruction or plunder, and withdraw out of reach of whatever reaction they provoked thereby. That is the perfectly normal form of any raid, including the raids that would become routine operations for the Byzantines, and the subject of a specific field manual.[38] Indeed, raids must be just as old as warfare. But in every case, to have the relative action–reaction speed advantage that a successful raid absolutely requires, the raiding force must be small, or light, or have access to superior vehicles that the enemy lacks, or else achieve complete strategic surprise, as in the case of the mass boat raid of Kievan Rus' against Constantinople in 860, when little was known of this very new state, and nothing was known of its Viking tactics. Such rare exceptions aside, raids will happen and will succeed but cannot do much damage because the forces involved must be small, relative to the full strength of either side—commando teams as opposed to entire brigades or divisions in modern parlance.

That, however, was not true of the Huns or of the other mounted archers of the steppe. Because of the enormous 2-to-1 speed advantage of all-cavalry forces with multiple horses per man, they could raid on the scale of entire armies achieving corresponding results—not only quantitatively but also qualitatively, to the point that the raid could become something else entirely, not an incursion but an invasion.

Quantity could become quality because the speed advantage was so great that it could overcome lower-level tactical and operational deficiencies. For example, a force of mounted archers is almost useless in thick woods, hence an enemy could do well by defending a frontage especially selected to include as much woodland as possible. But that takes time, and by moving fast the Huns could arrive in force before the enemy was deployed into the woodland frontage, when it was still moving toward it through more open terrain, unprotected from their arrows.

The same was true of the other major tactical disadvantage of the Huns—their lack of siege technology—until later days under Attila, when there were Roman defectors in his camp who taught the Huns how to construct beam-elevated mobile fighting posts, large swinging-beam rams, protected scaling ladders, and "all manner of other engines," and their relative inferiority in besieging cities even after that, if only because they had no supply trains to feed their numerous camp followers.[39]

If a walled city was properly prepared to resist investment, with food and water stocked for a long siege, and the walls and towers adequately manned all along their circuit, there is not much that mounted archers could do. And if they had Roman defectors with them who knew how to undermine walls or assemble siege engines, the officers sent to reinforce the garrison for the occasion would know how to countermine, and how to attack siege engines.

But again, that takes time, certainly weeks if not months—time that could easily be denied by the swift arrival of the Huns before preparations were properly completed. That is how in 441–447 the Huns conquered each one of the major fortress cities that formed the central axis of Roman power in the Balkans, running from the hinterlands of Constantinople in Thrace all the way up to Sirmium (Sremska Mitrovica, in Vojvodina, Serbia), a straight-line distance of six hundred kilometers, by way of Serdica (Sofia, Bulgaria), Naissus (Nish, Serbia), Viminacium (Kostolac), Margus (near Dubravica), and Singidunum (Belgrade). Of these, Naissus and Serdica were taken last, along with Ratiaria on the Danube (Arcar, Bulgaria), as a prelude for raids in Thrace on the approaches to Constantinople.[40]

Another perspective on the role of sheer speed in giving a major strategic advantage to the Huns comes from a letter of 399 of the contemporary observer, Eusebius Hieronymus, now better known as St. Jerome, who never set eyes on a Hun in arms from his hermitage in Bethlehem, but who tells us that he had informants, and who certainly knew how to write seductively, as befits a man who made a career out of persuading wealthy Roman ladies to finance his altruistic, indeed saintly, projects:

> While I was seeking a dwelling suitable for so great a lady [Fabiola visiting from Rome, very rich, divorced, remarried but in penance for this sin] . . . suddenly messengers flew this way and that and the whole eastern world *(oriens totus)* trembled. We were told that swarms of Huns had poured forth from the distant Maeotide [Sea of Azov], midway between the icy river Tanais [the Don, but that would be excessively accurate] and the savage tribes of the Massagetae, where the gates of Alexander [the "Caspian Gates"?] keep back the barbarians behind the rocky Caucasus.

The geography is dubious, but the ensuing strategic observations are insightful:

> Flying hither and tither on their swift steeds, said our informants, these invaders were filling the whole world with bloodshed and panic. . . . Every-

where their approach was unexpected, they outstripped rumor by their speed [thereby preserving strategic surprise even after launching their attack]. . . . The general report was that they were making for Jerusalem. . . . The walls of Antioch [Antakya, Turkey], neglected in the careless days of peace, were hastily repaired. . . . Tyre [Sur, Lebanon], desirous of cutting herself from land, sought again her ancient island [the citadel on a narrow tongue of land]. We too were compelled to prepare ships . . . as a precaution against the enemy's arrival; [we had] to fear the barbarians more than shipwreck . . . for we had to think not so much of our own lives as of the chastity of our virgins.[41]

They were indeed a great preoccupation of the saintly Jerome. The incursion was real enough—the Huns came through the Caucasus in 399 and raided through Armenia, Mesopotamia, and Syria and into Anatolia as far as Galatia, before retreating with their booty, captives, and volunteers.[42] And the point is that even if there had been powerful Roman field forces in the region—an impossibility after the great loss of mobile forces at Adrianople in 378 and too many troubles since—it would still have been virtually impossible to intercept the Huns. They maneuvered in different directions much too fast—"flying hither and tither on their swift steeds" as Jerome wrote.

One can visualize the sequence—in perfectly ideal circumstances: the Huns are detected early, moving in a given direction; imperial messengers are sent to alert Roman command posts at record speed, outpacing the Huns. But because the raiders are very many and not few, to avoid a debacle sizable forces must be assembled to intercept their expected line of advance. During each day required for that, the Huns could move thirty miles or more if columns loaded with loot were sent back separately, while Roman forces could do more than twenty.

In this specific case there was a much better solution, of course. A Syriac source states that the Huns also raided down the Euphrates and the Tigris, evidently not realizing that they were approaching the well-garrisoned "royal city of the Persians" (Ctesiphon, some 35 kilometers south of Baghdad): "The Persians chased them and killed a band. They took away all their plunder and liberated eighteen thousand prisoners."[43] The solution was not to do the same—the Persians only outpaced that particular band because it was overloaded with plunder and captives, and because the Huns had made it easy for them by venturing too near their capital and military headquarters.

The solution, rather, was for the Romans and the Persians to resolve or simply set aside their differences to jointly garrison and close off the only two passages through the Caucasus Mountains that had tolerable pastures and elevations for a mass of horses: the Darial Pass, between what is now Russia and Georgia, and the "Caspian Gates" at Derbent, now in Daghestan, Russia, a narrow coastal strip between the mountains and the Caspian Sea. Having suffered jointly, the two empires did just that in the "fifty-year" peace treaty of 562.

These then were the tactical, operational, and theater-strategic advantages of the Huns, the first mounted archers of the steppe to reach the west, who were destined to have many successors: the Avars and their bitter enemies of the first Turkic steppe empire or qaganate (khanate); the Bulghars and Khazars who separated to form their own qaganates; the Magyars, Pechenegs, Cumans, and finally the Mongols. But the Huns had the inestimable advantage of surprise in the fullest sense even beyond strategic—of cultural surprise, so to speak, because they were the first of their kind to reach the west.

Processes and Personality: Attila

For all their strengths, the Huns became a threat to the survival of the eastern empire only when Attila ruled them, circa 433–453, declining into bands of migrants, freebooters, and mercenaries after his death. By uniting under his leadership diverse Hun clans and all others who were with them willingly or otherwise, he added the element of mass to their superior abilities as individual warriors; and he added a focused strategic direction to their tactical, operational, and theater-strategic advantages. It is true that even under Attila the Huns remained raiders rather conquerors, but on such a grand scale that they could endanger even an empire.

Attila's rise to power is efficiently described by Jordanes, and/or his chief source Cassiodorus:

> For this Attila was the son of Mundzucus [Mundiuch] whose brothers were Octar and Ruas, who are said to have ruled before Attila, although not over the very same domain. After their death, he succeeded to the Hunnic kingdom, together with his brother Bleda. In order . . . [to be] . . . equal to the expedition he was preparing, he sought to increase his [dynastic] strength by murder. Thus he proceeded from the destruction of his own kindred [and potential rivals] to the menace of all the others. . . . Now when his brother Bleda, who ruled over a great part of the Huns, had been slain [in 445] by his treachery, Attila united all the people under his own rule. Gathering also a host of the other [ethnicities] which he then held un-

der his sway, he sought to subdue the foremost nations of the world, the Romans and the Visigoths.[44]

Attila united the clans of the Huns under his undisputed command by a combination of dynastic legitimacy, or at least a respected lineage, insofar as the Huns were not especially wedded to dynastic principle; the equitable sharing out of the revenues of loot and tribute; and that careful construct which is called charismatic leadership. In the eyewitness account of Priskos of Panium, who was attached to a Byzantine delegation sent to negotiate with him in 449, we can recognize Attila's use of specific techniques to enhance his authority that were by then ancient, but effective withal; indeed, the very same techniques were used not long ago by other "great men of history." A dinner party has just started:

> When all were seated in order, a wine waiter came up to Attila and offered him a wooden cup of wine. He took the cup and greeted the first in the order [there is a status order, hence status competition—which can only be adjudicated by Attila]. The one who was so honored with the greeting stood up, and it was the custom that he did not sit down until he had either tasted the wine or drunk it all and had returned the wooden cup to the waiter.

This was drinking under observation—as in Stalin's drinking bouts, through which members of his court were kept off balance and infantilized by the alternation of honors and humiliations.

> Attila's servant entered first bearing a plate full of meat, and after him those who were serving us placed bread and cooked foods on the tables. While for the other barbarians [Hun lords] and for us there were lavishly prepared dishes served on silver platters, for Attila there was only meat on a wooden plate. He showed himself temperate in other ways also. For golden and silver goblets were handed to the men at the feast, whereas his cup was of wood. His clothing was plain and differed not at all from that of the rest [ordinary Huns], except that it was clean. Neither the sword that hung from his side nor the fastenings of his barbarian boots nor his horse's bridle [were] adorned, like those of the other [Hun lords] with gold or precious stones or anything else of value.[45]

We are reminded of Adolf Hitler eating his soup and veggies in his plain brown uniform while all around him generals and field marshals glittering with the medals he had given them feasted on meats and champagne. It is not all modesty, however—in front of the people, the leader must be enhanced by ceremony:

> As Attila was entering, young girls came to meet him and went before him in rows under narrow cloths of white linen, which were held up by the hands of women on either side. Those cloths were stretched out to such a length that under each one seven or more girls walked. There were many such rows of women under the cloths, and they sang [Hunnish] songs.[46]

This is very different from the rolling drums, giant banners, and flaming torches of Nuremberg rallies—the Huns had their own language of signs to proclaim power, derived from shamanistic ceremony rather than military parades or Wagnerian opera; in our own times, the nominally communist Kim Il Sung of North Korea, who was in reality the great shaman of a cult of his own person, was greeted on public occasions by panoplies of virginal young girls fervently singing his praises.

But the leader is also a man of the people, or at least of some people; their wives can demonstrate proximity to power, while remaining at their proper elevation, and thus constitute a bridge between ordinary folk and the great man:

> When Attila came near to the compound of Onegesius [his chief underling, Hunigasius, Hunigis?] . . . [his] wife came out to meet him with a crowd of servants, some carrying food. And others wine. . . . In order to please the wife of a close friend, he ate while sitting on his horse, the barbarians who were accompanying him having raised aloft the platter which was of silver.[47]

Whatever its sources—legitimacy, the distribution of plunder, charismatic techniques, terror—Attila's authority over the Huns allowed him to unite them under his orders, and with them in turn, to impose obedience on the Alans, and on the Gepids, Heruli, Greuthungi or Ostrogoths, Rugi, Sciri, and Suebi—all of them Germanic warrior nations formidable to the Romans, but Attila's obedient subjects. Their agriculture helped to feed the Huns, who remained horse-nomads averse to farming, and their warriors had to follow Attila in his campaigns, adding the weight of their great numbers to the peculiar fighting skills of the Huns.

Finally, Attila contributed his own considerable statecraft to Hun military strength. He relied on violence, of course, but carefully controlled violence: instead of unleashing his armies from the first, he would usually start by using force in small doses—in sharp but localized attacks, not to gain territory or even weaken his enemy, but to set the stage for coercion and extortion. Attila did fight one large and costly campaign two years before his death in 451, but it was very much the exception—

he could usually obtain what he wanted with the mere threat of violence, without actually having to expend his forces in large-scale combat.

In sharpest contradiction to his image as a savage warrior, whether in Icelandic sagas or the contemporary imagination, Attila was a great believer in negotiations. He often demanded the dispatch of envoys to his encampment, and often sent envoys to Constantinople and to Ravenna, seat of what remained of the western empire. A modern historian described him as a diplomatic "bungler" and catalogued his errors.[48] Perhaps so, but for a nomad king to concurrently negotiate coercively with both Roman empires, while an invasion of the Sasanian empire of Persia through the distant Caucasian mountains was mooted at his court, was at least to bungle on a huge scale—nothing like it had been seen before, nothing like it would seen again, until the Mongols dominated all the Russias while actually ruling China. In a very interesting passage, Priskos, the man of letters attached to the eastern delegation, listens attentively to the opinions of the experienced envoy who headed the western empire's delegation; at the time both sets of negotiations were not faring well:

> When we expressed amazement at the [extortionate demands of Attila], Romulus, an [envoy] of long experience, replied that his very great good fortune and the power which it had given him had made [Attila] so arrogant that he would not entertain just proposals unless he thought that they were to his advantage. No previous ruler of Scythia [the steppe lands] or of any other land had ever achieved so much in so short a time. He ruled the islands of the Ocean [the Baltic Sea] and, in addition to the whole of Scythia, forced the Romans [of both empires] to pay tribute. He was aiming at more than his present achievements and, in order to increase his empire further, he wanted to attack the Persians.[49]

Deliberately mingling and confusing force and negotiations, Attila normally proposed peace talks as soon as he invaded. That too was a way of dividing his enemies, for in each case the war party in Constantinople or Ravenna was denied the clarity of an all-out war with no alternative.

It was also part of his method to justify his demands with legal, or at least legalistic, arguments. While Priskos was with him in 449, Attila was claiming from the western Romans a set of gold cups pawned by a fugitive as his own rightful booty, and from the eastern Romans the return of a number of escaped prisoners. It hardly mattered if Attila's arguments had any legal merit. Even a thin veneer of plausibility was quite enough because he was not trying to persuade a court of law, but instead wanted to divide the counsels of his opponents. When facing Attila, the peace party always had a legalistic argument, however weak, to accept his demands. On the other hand, he did respect the rules of the diplomatic game. Most notably, he held himself bound by the unwritten law that already then assured the immunity of envoys, even under extreme provocation.

In all these ways, Attila transformed the tactical, operational, and theater-level advantages of his mounted archers and Germanic warriors into a combination of mass *and* fast strategic mobility that was extraordinary for the times, and added statecraft too.

From his well-built headquarter village at an unknown location somewhere across the middle Danube in Hungary, or better in the Banat now in Romania (my birthplace, but it does fit the evidence[50]), he could freely choose to send his forces in a southeast direction to attack Thrace and Constantinople, some eight hundred straight-line kilometers away and twice that overland, more or less. Or he could send them in a westward direction to attack Gaul, where Roman lives continued sometimes grandly as the empire was fading ever more, some fourteen hundred kilometers in a straight-line direction and perhaps two thousand overland. Or else he could send his forces in a southwest direction into Italy, which still had riches to loot, via the northeast passage to Aquileia (near modern Trieste) that altogether avoids the Alpine barrier unfriendly to horses. Or, finally, having greater strength than the Huns had in 399, he could replicate their much-longer-range but highly profitable offensive by sending forces eastward, across the Dnepr and Don through the Caucasus to Armenia and Cappadocia, then turning through Cilicia all the way back to Constantinople. That is certainly a very long way round, three thousand kilometers overland at least, but such an expedition could have been an excellent prelude to a direct attack on Constantinople by luring away its defenders. Even grander all-cavalry expeditions were launched by the Mongols, who had no advantage in mobility over the Huns.

Only the last of these possibilities was left untried by Attila. In the years 441–447, Attila did send his forces across the Danube to seize the ill-prepared fortress cities from Sirmium down to Serdica, as mentioned above, then continuing down into Thrace to Arkadioupolis (Lüleburgaz), within a hundred kilometers of Constantinople, and peeling off southwest to Kallipolis (Gallipoli, Gelibolu) on its famous peninsula. While his chronology is unreliable, Theophanes Confessor, the most substantial of Byzantine chroniclers, reported the expedition and its results:

> Attila . . . overran Thrace. Theodosius [II, 408–450] . . . sent out Aspar [Flavius Ardabur Aspar, an Alan and the empire's highest-ranking officer as *magister militum,* Master of Soldiers,] with his force together with Areobindos and Argagisklos[51] against Attila, who had already subdued Ratiaria, Naissos, Philippoupolis [Plovdiv, Bulgaria] Arkadioupolis Costantia [Constanța, Romania] and very many other towns, and had collected vast amounts of booty and many prisoners. After the generals had been thoroughly defeated [by the Huns] . . . Attila advanced to both seas, to that of the Pontos [the Black Sea] and to that which flows by Kallipolis [the Sea of Marmara] and Sestos [Eceabat], enslaving every city fort except for Adrianople [Edirne] and Herakleia [Marmara Ereğli]. . . . So Theodosius was compelled to send an embassy to Attila and to provide 6,000 pounds of gold to secure his retreat, and also to agree to pay an annual tribute of 1,000 pounds of gold for him to remain at peace.[52]

That transaction of the year 447 was the point of departure of the negotiations in which Priskos participated in 449.[53]

The impact of these events was very great. Even the sparse *Chronicle* of Marcellinus Comes, who was writing in the sixth century, recalls the invasion:

> A mighty war, greater than the previous one, was brought upon us by king Attila. It devastated almost the whole of Europe [the province Europa] and cities and forts were invaded and pillaged. King Attila advanced menacingly as far as Thermopylae. Arnigisclus [Arnegisklos Magister militum per Thracias, Aspar's regional subordinate] fought bravely in Dacia Ripensis alongside the Utum [Vit] river and was killed by king Attila, when most of the enemy [a Hun war band weighted down with plunder] had been destroyed.[54]

This raid inflicted much strategic and political damage, for heavily taxed citizens had been left unprotected. Only then—too late—was Attila paid off, ensuring that more taxes would have to be collected from the ravaged lands. The policy implication was obvious: either avoid paying Attila by destroying him in the true Roman style, with a huge and successful expedition, costly as that would be, or else pay him off *before* he invades.

The westward offensive came in 451.[55] Attila's forces swept through what is now Germany and France, crossing the Rhine in April, possibly intent on attacking the Visigothic kingdom of Toulouse. That history is intertwined with the famous story of Attila's claim to half the western empire because Justa Grata Honoria, sister of the western emperor Valentinian III (425–455), had supposedly sent him her ring in order to be rescued from a forced marriage to a worthy bore, after a scandalous affair with her steward Eugenius, who was duly executed for his effrontery. This story has everything—sexual scandal and treacherous intrigue in melancholy Ravenna in the final twilight of empire. Even eminent historians have not resisted its admittedly irresistible appeal—in part because Honoria's better-documented mother, Galla Placida, really was a formidable figure—but alas, it must all be dismissed as Byzantine court gossip.[56]

The other story was consequential: Attila was advancing on Gaul with "an army said to have numbered five hundred thousand men"; our source, Jordanes, or his source, Priskos, was careful to insert "said" *(ferebatur)*. But the true number must have been exceptionally large, even though it was not really an army *(exercitus)* but rather a great number of Hun, Alan, and Germanic warrior bands. They were all under Attila's strategic direction—that is how they reached Gaul—but not under his operational control except for Attila's own battle force mentioned by Jordanes. That was due not to insubordination but to operational necessity, because with very large numbers, separate columns had to peel off and range widely to find enough food and forage. Jordanes writes: "He was a man born into the world to shake the nations, the scourge of all lands, who . . . terrified all mankind by the dreadful rumors noised abroad concerning him."[57]

The aim was to terrorize, preferably in order to dissuade resistance—both to conserve forces and also because Attila must have preferred to receive gold delivered neatly packed by envoys pleading for his retreat, instead of having to extract gold in bits and pieces from lucky looters among his own followers. Or if not, failing dissuasion, the aim was to terrorize in order to demoralize, so as to induce men to seek safety in flight rather than to stand firmly in his path. It seems that Attila did succeed in terrorizing Gaul, or at least the poet Sidonius:

> Suddenly the barbarian world, rent by a mighty upheaval, poured the whole north into Gaul. . . . After the warlike Rugian come the fierce Gepid, with the Gelonian close by; the Burgundian urges on the Scirian; forward rush the Hun, the Bellonotian, the Neurian, the Bastarnian, the Thuringian, the Bructeran, and the Frank, he whose land is washed by the sedgy waters of the Nicer [Neckar, to be excessively accurate]. Straightway

falls the Hercynian [Black] forest, hewn to make boats, and overlays the
Rhine with a network of its timber; and now Attila with his fearsome
squadrons has spread himself in raids upon the plains of the Belgian.[58]

Because of his panic, or more likely because of his poetical needs,
Sidonius has included among Attila's men the long-defunct Bastarnae,
Bructerii, Geloni, and Neurii, and also the "Bellonoti," who never ex-
isted at all.

At that point, a powerful army should have arrived from Italy to fight
Attila, but there were no more Roman armies. Instead there was only
the *magister militum,* master of soldiers of the western empire, Flavius
Aetius, who crossed the Alps into Gaul "leading a thin, meager force of
auxiliaries without real soldiers" *(sine milite).*

Thus we encounter Aetius, another greatly romanticized figure ("The
Last of the Romans"), who arrives in the early summer of 451 with his
tiny band of low-grade troops, to defeat the most numerous and powerful of enemies. He was a veritable expert on the Huns: as a youth he had
been a hostage at the Hun court before Attila, and later he had procured
and successfully commanded Hun mercenaries, and therefore knew their
tactics and ruses.[59] Aetius was evidently hoping to recruit allies among
the very invaders of Gaul to prevent the new invasion, and he was successful—but according to our poet only because his great hero Avitus
succeeded in recruiting the most powerful of them all, Theodoric I,
bastard son of Alaric and king of the Vesi, later called Visigoths; he
joined the fight against "the lord of the earth who wishes to enslave the
whole world."

Attila could have evaded interception easily enough—at least his core
Hun forces were faster than his foes—but evidently accepted battle in
the Campus Mauriacus somewhere in the Loire Valley not far from
Troyes and northeast of Orleans, which had resisted his attack.
Jordanes reports that, in what is now usually called the battle of
Chalons in place of the earlier Catalaunian Fields, Aetius and Theodoric
jointly commanded forces of "Franks, Sarmatians (Alans), Armoricans
[Bretons], Liticians [?], Burgundians, Saxons, Riparian [Franks],
Olibriones [former Roman soldiers praised as the best of auxiliaries],
. . . and some other Celtic or German nations" as well the many Goths
of Theodoric and the few Romans of Aetius.[60] Attila also had his own
Goths to fight for him, the Ostrogoths, as well as "countless" Gepids
and "innumerable people of diverse nations," including Burgundians—

who thus fought on both sides, reasonably enough because they were probably a nation only in the eyes of others, while themselves valuing only clan and tribal identities.

In the great battle that ensued, Theodoric was killed, Aetius fought hard, many casualties were suffered ("the fields were piled high with bodies"), and Attila withdrew his forces—or more likely just his own Hun battle force—into a camp barricaded with wagons, "like a lion pierced by hunting spears, who paces to and fro before the mouth of the den and dare not spring, but ceases not to terrify the neighborhood (vicina terrere) by his roaring. Even so this warlike king at bay terrified his conquerors."[61] But there was no all-out Gothic assault, no last stand. Instead Attila was left quite free to retreat at leisure across central Europe to return to his capital, with none in pursuit.

Jordanes explains the mystery quite simply: Thorismund, Theodoric's eldest son and his successor as head of the Visigothic kingdom of Toulouse, was eager to attack but first consulted Aetius because he was "older and wiser"—as indeed he was, at least for himself and the empire, if not for Thorismund:

> Aetius feared that if the Huns were totally destroyed by the Goths, the Roman Empire would be overwhelmed, and urgently advised him to return to his own dominions to take up the rule which his father had left. Otherwise his brothers might seize their father's possessions and obtain the power over the Visigoths. . . . Thorismund accepted the advice without perceiving its double meaning.[62]

We may therefore see in Aetius a proto-Byzantine, so to speak, unless the situation is judged too simple to require any particular talent for statecraft: if the power of Huns were "wiped out now, the Western Empire would be hard put to it to defend itself against the kingdom of Toulouse." Simple enough, yet the very same historian immediately proceeds to charge Aetius with both duplicity and naiveté, a rare combination indeed, because Attila was far from grateful in the aftermath and came back to attack again.[63] Perhaps the statecraft involved was not so simple after all. It was not a matter of winning Attila's sentimental gratitude but of the inherent advantage of a balanced balance of power: it was much better for the enfeebled remnant of the Roman empire to have two powers in existence that would not combine against it, because each one could destroy it easily enough, than to have just one power. With two, it might be possible to persuade one to fight the other in the interests of the empire—as had just happened; with one, there was no avoiding subjection or destruction.

Jordanes depicts Attila after the battle as a wounded lion, modern historians have also categorized what happened as a crushing defeat.[64] What happened next, however, suggests an altogether different explanation of the same evidence: as usual, Attila had planned a raid—on a very large scale, but still an incursion rather than an invasion. Having encountered too much opposition to make the raid profitable, he called it off to return home, after suffering far from irreparable battle losses. In Jordanes we read that 180,000 had died on both sides.[65] Neither he nor his source could have known the true number, but whatever that was, the losses on Attila's side are likely to have been much more among his Germanic allies fighting on foot than among his own Huns fighting on horseback—long-range arrow fighters who could avoid loss by evasive maneuver, as infantry in tight formations cannot.

That is the only possible explanation for what ensued: in September of that same year, 451, having just returned from Gaul, Attila sent a raiding force of Huns across the Danube. There was a new emperor in Constantinople, Marcian (450–457), who was refusing to pay the annual tribute and action was called for, but if Attila had been thoroughly defeated in Gaul with heavy losses, he could hardly have mounted an attack on a new front, with no interval to recuperate, no time to induct the next age cohort into his forces. Nor was this a minor or short-range raid: the one thing we know of how its magnitude was perceived is that Marcian summoned an Ecumenical Council at Nicea (İznik), a pleasant lakeside town inland from the Propontis (Sea of Marmara) but hurriedly moved it to Chalcedon (Kadıköy), directly across the water from Constantinople.[66] (It was at this council that the dispute over the nature of Christ became an irreparable breach between Chalcedonian human-and-divine, and non-Chalcedonian Monophysite churches, whose persecution had deeply divided the empire by the time Islam arrived in the seventh century.)

The difference between İznik and Kadıköy is that from the latter all the assembled bishops could have been brought immediately into the safety of the walled capital, even with rowboats—memories of how the Huns had arrived unannounced all the way to Kallipolis (Gelibolu) must have been fresh.

Even more telling, in the very next year, 452, Attila mounted his third offensive in a third direction, this time advancing southwest to cross into Italy where Trieste now rises at the head of the Adriatic, where the Julian Alps are reduced to hills that decline into the sea, forming no barrier to horses. From there westward toward the depth of Italy, the first target was Aquileia, a very major city, with a mint and an imperial palace. Ausonius had placed it ninth in his ranking of the cities of the empire *(Ordo urbium nobilium)*, praising its "most celebrated port." This rich prize was well defended by formidable walls that had resisted strong attacks before. Ammianus Marcellinus, an expert of sieges, described it as a "well-situated and prosperous city, surrounded by strong walls"; he notes that his hero emperor Julian (361–363) "recalled reading and hearing that this city had indeed oftentimes been besieged, but yet had never been razed nor had surrendered."[67] Nevertheless Julian's forces fighting Constantius II (340–361) in civil war did besiege the formidable defenses with all manner of advanced techniques but without success. A century before that Maximinus Thrax (235–238), on his march on Rome, had made an all-out effort to take the city with his capable and ingenious Pannonian troops:

> The soldiers . . . remained out of range of the arrows and took up stations around the entire circuit of the wall by cohorts and legions, each unit investing the section it was ordered to hold. . . . The soldiers kept the city under continuous siege. . . . They brought up every type of siege machinery and attacked the wall with all the power they could muster, leaving untried nothing of the art of siege warfare. . . . They launched numerous assaults virtually every day, and the entire army held the city encircled as if in a net, but the Aquileians fought back determinedly, showing real enthusiasm for war.[68]

The city did not fall, and at length the disenchanted troops killed Maximinus instead of the valiant Aquileians.

Siegecraft was hardly the specialty of Attila's Huns, yet they were up to the task: "Constructing battering rams and bringing to bear all manner of engines of war, they quickly forced their way into the city, laid it waste, divided the spoil and so cruelly devastated it as scarcely to leave a trace to be seen."[69] This was not purposeless destruction—it was designed to dissuade resistance. After hearing what happened at Aquileia, known for the strength of its fortifications, the authorities in all the cities in Attila's path, all the way to Mediolanum (Milan) and Ticinum (Pavia), thought it best to open their gates without resistance.

The vast plain bisected by the river Po that forms the core of northern Italy has had its rare famines, but has always been one of the richest regions of the planet in movable property, when not especially and newly ravaged—and it had not been invaded since Alaric crossed it in 408 on his way to Rome. Attila's gains must have been immense, as city after city bought its immunity or was thoroughly looted. Then, as recounted, Pope Leo arrived from Rome to negotiate with Attila, along with the former prefect Trygetius and the very wealthy ex-consul Gennadius Avienus.[70] One may presume that the trio did not forget to bring gold with them, and not a small amount either, if only because there must have been a great many captives to ransom.

These are not the deeds of the spent force depicted by modern historians, even the best of them: "Attila's campaign was worse than a failure . . . the loot may have been considerable but it was bought at too high a price, too many Hunnic horsemen lay dead in the towns and fields of Italy. A year later Attila's kingdom collapsed."[71]

It is true that Attila died in his bed in the following year, 453, supposedly after a drunken feast to celebrate his marriage to a new, young, and beautiful wife—a lubricious tale that may be true—why else be a conqueror? And it is true that his sons quarreled, fatally ruining his empire. But the decline-and-fall narrative is pure determinism, and factually dubious: the loot was just "considerable"? Too many Hunnic horsemen lay dead? There is no such evidence. But there is contrary evidence of vigor: newly returned from Italy, Attila demanded the resumption of the agreed annual tribute from Constantinople:

> Attila . . . sent envoys to Marcian, the Emperor of the East, threatening to devastate his provinces because what the previous Emperor, Theodosius [II] had promised had not been sent and in order that he might appear still more cruel to his enemies.

That is again Jordanes, repeating a lost fragment of Priskos, but we do have the continuation in the original version:

> When Attila demanded the tribute agreed by Theodosios [II] and threatened war, the Romans replied that they were sending envoys to him, and they sent Apollonius . . . who held the rank of general (strategida, strategos). He crossed the Danube but was not given admittance to the barbarian. For Attila was angry that the tribute, which he said was agreed with him by better and more kingly men, had not been brought, and that he would not receive [the envoy] since he scorned the one who sent him. . . . Then he left having accomplished nothing.[72]

This is definite, if negative, evidence that it was Attila's intention to wage a major war against the empire: he did not send a threat with

Apollonius as he did in the past when playing the extortionist—he sent nothing, he did not even receive him. Having successfully breached the formidable walls of Aquileia, it is not inconceivable that Attila planned a siege of Constantinople, and not just plunder raids. After all, he still had the undiminished tactical and operational superiority of his mounted archers, he still had a monopoly of pure-cavalry fast offensives in great depth, and judging from the furious fighting that would soon break out between Attila's sons, and between their Hun contingents and the Goths, Gepids Rugi Suevi Alani and Heruli in revolt, he still had numerous subjects of all origins who were effective warriors.

Had there been a war with the eastern empire, there is every reason to believe that Attila would have won again as he did in 447, in the specific sense that he could have persisted in inflicting damage until bought off. Instead he died, but by then his greater-than-expected threat had evoked a series of improvised reactions that soon combined into something much broader, and very long lasting as it turned out.

CHAPTER 2

The Emergence of the New Strategy

In facing Attila, the mild and scholarly Theodosios II (408–450, son of the first eastern emperor Arkadios), his assertive sister, his vigorous wife, and the experienced civil officials of his court were doubly constrained.[1] They had no military forces that were tactically effective against the Huns, and they had more urgent priorities on other fronts.

As always, the strongest foreign power by far was Sasanian Persia, with which relations had been exceptionally peaceful until 420 in the time of the shah Yazdgird; there was a sharp deterioration under his successor, Bahram V (420–438), though not especially because of him.[2] There was a revival of the ancient quarrel over the Armenian lands, and a new quarrel over religion. While often described as "buffer states" by modern historians, the evidence indicates that the autonomous existence of Armenian states between the two empires was more conflict-inducing than conflict-buffering, as both contended for authority over the fractious *nakharars*, the petty rulers of narrow valleys who made up political Armenia.[3]

The religious quarrel was quite new, and provoked by a sharp rise in orthodox militancy by the Christian empire and Zoroastrian militancy by the Sasanians, whether coincidentally or reciprocally is not clear, though there is much evidence of the increased persecution of pagans and Jews, the inquisition of non-Greek churchmen suspected of christological deviations, and the condoning of violent attacks on non-Christians and their places of worship. (In 415, Christian enthusiasts outraged by the paganism of Hypatia the philosopher dragged her from her carriage to the Caesareum church, stripped her of course, killed her, tore her body in pieces, and removed them elsewhere for burning in pious deference to the sanctity of the place.)

Theophanes the Confessor, himself an enthusiast withal, deplored excessive zeal:

> Abdaas, bishop of the capital city of Persia [Ctesiphon], driven by his zeal for God, but not applying this zeal where it was appropriate, set fire to the Temple of Fire [the Zoroastrian temple in the political capital of that faith]. When the emperor [the Shah] learned of this, he decreed that the churches in Persia be destroyed and punished Abdaas with various torments. The persecution lasted for five years.[4]

Fighting duly broke out in Armenia and Mesopotamia around Nisibis (now Nusaybin in southeast Turkey), the strongly fortified city that was the classic focus of warfare between Romans and Persians. It continued with no dramatic results until 422, when the *magister officiorum* (master of offices, the highest administrative post), Helion, arrived to negotiate peace; there had been Hun attacks across the Danube, and Bahram V also may have been under pressure on his central Asian border. The prior status quo was restored unchanged. There was more trouble in Armenia after that—its condition was chronic—but no war until 441.

The arrival of a new Sasanian ruler was usually marked by military initiatives—no doubt they were useful to affirm his authority—and Yazdgird II, who succeeded Bahram in 438, three years later duly launched his attack at Nisibis in the usual way, until the *magister militum per Orientem* (highest military commander east of Constantinople), Anatolius, arrived in the usual way to negotiate a peace treaty. Once again the prior status quo was restored. There was no more fighting while Theodosios lived, in part because Yazdgird himself lived till 457, but troops had to remain available to defend the Persian front anyway, because no peace could long outlast their absence. Unlike Attila's incursions, a Sasanian invasion could lead to the permanent loss of imperial territory, hence that frontier retained its priority.

The second front was in *Africa*, North African territory corresponding to modern Tunisia and coastal Algeria that did not even belong to the eastern empire, whose boundary stopped in Libya.

In October 439, Vandals and Alans who had arrived via Spain under their formidable warlord Gaiseric seized Carthage, capital of *Africa*, a major source of grain for Rome and central Italy.[5] Valentinian III's western empire was directly damaged, but Carthage was a major port with much shipbuilding, a fleet was under construction, and the eastern empire was also threatened. Constantinople was much farther away and well defended, but with a fleet favored by the prevailing westerly winds Geiseric could cut off the grain supply of Egypt as well, by attacking Alexandria.

It was a most difficult time.

The stark chronicle of Marcellinus Comes is eloquent in its brevity; under the ninth indiction, "consulship of Cyrus alone," corresponding to September 440/August 441, we read: "The Persians, Saracens [= Bedouin of Mesopotamia], Tzanni [ancestors of the Mingrelians of Georgia], Isaurians [mountaineers of southeast Anatolia], and Huns left their own territories and plundered the land of the Romans."[6]

Nevertheless action was deemed imperative. In 440 Gaiseric's new fleet had attacked Sicily, the second source of grain for Italy after *Africa*, and both empires agreed to send fleets against him in 441. According to Theophanes, the eastern expedition was mounted on the largest scale:

> Theodosios [II] . . . sent out eleven hundred cargo ships with a Roman army commanded by the generals Areobindos, Ansilas, Inobindus, Arintheos and Germanus [i.e., a large force, of the order of thirty thousand or even fifty thousand men, between sailors and troops]. Gizerich was struck with fear when this force moored in Sicily [on its way to Carthage, some three hundred kilometers away] and he sent an embassy to Theodosios to discuss a treaty.[7]

Next year's entry, for the year 5942 since the creation, explains why the grand fleet never reached Carthage and instead returned quickly to Constantinople: "While the fleet was waiting in Sicily as we have mentioned, . . . for the arrival of Gizerich's ambassadors and the emperor's commands, Attila, in the meantime, overran Greece."[8] But the expedition was not wasted: it seems that Gaiseric was thoroughly intimidated—at any rate, he never attacked Alexandria or any other eastern possession, and did not attack at all until 455, when his expedition sacked Rome, apparently inflicting more damage than Alaric in 410. In the *Liber Pontificalis*, the potted hagiography of Leo I includes: "After the Vandal disaster he replaced all the consecrated silver services throughout all the *tituli* [parish churches], by melting down six [silver] water jars . . . which the emperor Constantine had presented, each weighing 100 pounds . . . he renewed St. Peter's basilica."[9]

Intimidation had worked with Gaiseric as far as the eastern empire was concerned—even his conquest and sacking of Rome was actually precipitated by a court intrigue. But intimidation failed with Attila—he had nothing to fear even from an all-out attack, the land equivalent of the eleven-hundred-ship expedition that had stopped in Sicily.

In later times, the Byzantines would have an excellent diplomatic remedy against enemies from the steppe: again and again they persuaded different steppe powers to fight each other instead of attacking the empire. But Attila's empire was too large for that—the Byzantines could not reach behind it to find new allies. The most profound historian of the Huns wrote that it was a "thankless task" to determine the geographic extent of Attila's power, to do so would "clash with long-cherished myths."[10] He then rejected more expansive estimates—including Mommsen's—to settle for a rather modest empire stretching from central Europe to the shores of the Black Sea. As it happens, we have negative evidence that disproves the eminent historian: there is no sign hat any independent steppe power existed west of the Volga, that is, within reach of Byzantium. So either Attila ruled from the Danube to Volga or he might as well have done so, because there was no other power in the vast space that the Byzantine could seduce into attacking the Huns.

In the distant future of the eleventh century, the Turkic nomad Cumans (actually Qipchaqs, or Polovtsy in Russian) were persuaded to attack their no longer useful predecessors as Byzantine allies, the Turkic nomad Pechenegs. Since the ninth century, in exchange for regular payments, the Pechenegs had been of great help against the Turkic qaganate of the Khazars on the Volga, itself a former ally of great importance, against Kievan Rus' farther west on the Dniepr, which remained more enemy than friend even after conversion to Christianity, and also against the Magyars who were moving in between. Before the Magyars became a nuisance to the empire and were driven off northward by Pecheneg pressure into what became Hungary or Magyarország, they too had made themselves useful by attacking the Bulghars, who in turn had greatly helped the empire in the seventh century by attacking the formidable Avars, before becoming a major threat themselves.

In between these major steppe powers, there were lesser nations, tribes, and war bands that also alternated between fighting against the empire and fighting for the empire. All were subject to the dynamics of pastoralism in the steppe: because of the relentless natural increase of unmolested herds, there were perpetual struggles over pasture that made it easy for Byzantium to find allies; and nomads who had plenty of meat, milk, leather, and horn, but nothing else, had a perpetual need of gold to purchase grain and everything else.[11]

The entire steppe corridor west of the Volga, which runs below the forests and over the Black Sea all the way to the Danube, thus became a permanent arena of Byzantine diplomacy, which was routinely successful in converting the very multiplicity of potential enemies into its own remedy. But in the time of Attila that did not happen, either because of an improbable absence of other peoples on the steppe, or because his power really did extend much farther east than the Danube. Either way, from a diplomatic point of view Attila's empire might as well have spanned the entire distance to Vladivostok, because when Byzantium most urgently needed allies further east who might be persuaded to move westward to attack the Huns from the rear, there were none to be found, neither large nor little.

That left no choice but to resort to an inferior, though still useful, form of diplomacy: instead of using gold to induce others to attack the Huns, it had to be used to buy them off. Keeping both infantry and cavalry at home, Theodosios II instead sent envoys to negotiate with Attila to induce him to stay out of imperial territory in the future. It was more effective than sending new forces to be defeated like the old, and cheaper than the lost tax revenues of ravaged provinces. There had been earlier annual payments to Attila of several hundred pounds of gold, and the tribute was increased to 2,000 pounds of gold a year, but it was not paid until 447, when a comprehensive agreement was reached that required the lump-sum payment of 6,000 pounds of gold, and future annual payments of 2,100 pounds of gold per year. Enormous sums? Six thousand pounds of gold at today's prices would be worth $75,072,000, but of course ancient gold was relatively more valuable. Priskos of Panium certainly thought that the payment was disastrously large:

> To these payments of tribute and the other monies which had to be sent to the Huns they forced all taxpayers to contribute, even those who for a period of time had been relieved of the heaviest category of land tax through a judicial decision [legal exemption] or though imperial liberality. Even members of the Senate contributed a fixed amount of gold according to their rank. To many their high station brought a change of lifestyle. For they paid only with difficulty what they had each been assigned . . . so that formerly wealthy men were selling on the market their wives' jewelry and their furniture. This was the calamity that befell the Romans after the war, and the outcome was that many killed themselves by starvation or the noose.[12]

A modern historian who disliked the wealthy dismissed this passage as mere rhetorical excess and/or evidence of class solidarity with high-bracket taxpayers. He also offered some valid comparisons: the 2,000 pounds of gold per year that Leo I (457–474) paid to the Goth Theodoric "Strabo" in 473, and the one-time payment of 2,000 pounds of gold and 10,000 pounds of silver, and the 10,000 solidi per year (139 pounds) that Zeno (474–491) agreed to pay him; in another comparison, Leo's failed expedition against the Vandals of *Africa* of 468 cost no less than 100,000 pounds of gold.[13]

To this one may add more evidence from a surviving fragment of the historian Malchus: "Whereas the governor of Egypt is usually appointed for a payment of fifty pounds of gold, as if the country had become richer than before, he [Zeno?] appointed him for almost five hundred pounds."[14] That was not a colossal and outrageous salary—the text lends itself to misinterpretation—but rather the opposite, a capital payment made to the treasury in exchange for an annual salary (unspecified), the exact equivalent of a modern annuity.[15] Attila's 6,000 pounds could thus have been covered by the annuity investments of six officials of the highest rank, not a small amount therefore but not enormous either. Evidently Priskos was outraged by the payment of tribute, or perhaps it was another rhetorical pose, because payoffs to barbarians had been standard operating procedure for the Romans even at the height of their power.

In the event, the remedy was successful: Attila did not attack the eastern empire, instead attacking westward. By 451 he was in Gaul. The year before, Theodosios had been succeeded by the talented Marcian (450–457), who refused to pay the annual tribute, as we saw, but by then Attila was committed in the west and no ill consequence ensued.

Had Theodosios II pleased Priskos and the traditionalists by making peace with Persia, leaving Gaiseric's Vandals alone, accommodating Isaurians, Tzanni, and any other troublesome tribesmen, to assemble all the forces of the eastern empire to confront Attila with maximum strength, it is almost certain that the imperial army would have been destroyed, and the empire with it, for there would have been nothing left to stop Persians, Vandals, inland and frontier tribesmen, as well as the Huns and their subjects from seizing imperial territory.

That is the conclusion one would reach theoretically, considering the tactical, operational, and theater-strategic advantages of Attila's forces, and the mass of their subject warriors.

And that is also the conclusion one would reach empirically, on the basis of the only relevant evidence there is—which is quite sufficient however: given that the Salic Franks, Alans, Bretons, Liticians [?],

Burgundians, Saxons, riparian Franks and former Roman auxiliaries, very numerous Vesi Goths and the few Romans of Aetius, succeeded only in repelling Attila's army in the battle of *Campus Mauriacus,* and not in destroying it or even damaging it enough to prevent the subsequent invasion of Italy, it is reasonable to conclude that the eastern army would have been defeated.

Instead of gambling with the survival of the empire, the extraordinary threat of Attila's Huns was contained without large-scale warfare until it passed away with no lasting injury. A new strategic approach was thereby affirmed, which marked another transition from Rome to Byzantium: diplomacy first, force second, for the costs of the former were only be temporary, while the risks of the latter could be all too final.[16]

Under this strategy, varied means of persuasion were employed, but gold was consistently the most important. Combined with effective military forces to set limits to extortion, in the ensuing centuries many dangerous enemies were successfully paid off—that is, the cost of tribute was less than the double cost of resisting incursions and invasions, both in military expenditures and in the damage inflicted on civilian lives and property.

Economically, the payment of tribute was not deflationary. The circulation of gold, from taxpayers to the imperial treasury, from the treasury back to the taxpaying economy by way of imperial salaries and payments, was only briefly diverted when tribute was paid. The Huns and all their successors inevitably used their tribute gold to buy necessities and baubles from the empire—special arrangements were negotiated for border markets—hence the gold exported to the Huns returned to circulate within the empire rather quickly, except for the minute fraction retained for jewelry. To be sure, tribute converted products that could have been consumed locally into unrequited exports, reducing the standard of living within the empire. But the payment of tribute did not depress production, in fact it probably stimulated economic activity by increasing the velocity of the circulation of gold.

From a strategic point of view, the payment of tribute was an effective way of exploiting the empire's greatest comparative advantage: its financial liquidity.

Egypt was more fertile and parts of Mesopotamia also, Persia was better placed for long-range trade, having access to both the Central Asian routes to China and the Persian Gulf route to India and the spice islands: others too had advanced crafts, but the wealth of nations is one thing, the wealth of states quite another. It depends on their extractive capacity, their ability to collect revenue, in which the empire had the superior system, as we have seen. Even after the catastrophe of 1204, the diminished state restored in Constantinople by Michael VIII Palaiologos (1259–1282), a Greek kingdom that was imperial only in name, still had more gold in its treasury than any kingdom of Europe, simply because it routinely collected taxes, as they could not.

The Tactical Revolution

Another response to Attila's greater-than-expected threat was entirely different, but it too marked a transition from Rome to Byzantium.

In a huge simplification, it has often been written that after the devastating battle of Adrianople of 378, the cavalry displaced the infantry as the primary arm of the Roman army. Actually it was the solid and stolid heavy infantry of the classic legions that was displaced, not cheaper foot soldiers in general—and that process was well under way more than a century before Adrianople. In the time of Gallienus (260–268), the emperor's massed cavalry force became the most effective form of military strength at a time of acute crisis, equally useful to swiftly repel foreign incursions or to suppress internal revolt before it could spread; his *dux equitum*, commander of the cavalry, Aurelianus, unsurprisingly became emperor in 270. It was also before Adrianople, under Constantine, who died in 337, that standing mobile forces for the empire as a whole, *comitatenses,* were added to the provincial frontier forces.[17]

As opposed to these large and complicated changes, whose authorship and timing are still the subject of research, the tactical revolution was perfectly straightforward: having found no effective way of defeating the Huns with their existing forces of infantry and cavalry, the Byzantines decided to copy the Hun mounted archers, adding some armor to make them more versatile. That was no easy accomplishment. In the absence of a steppe culture of hunting and warfare, in which instruction in riding and archery begins in early childhood, it required veritable training programs both intensive and prolonged to convert recruits into skilled horsemen and skilled archers, and especially into skilled mounted archers.

One year of training was not considered sufficient to turn out combat-ready troopers; one may note incidentally that contemporary American and British troops can be sent into combat within six months of recruitment,. But of course the composite reflex bow is a much harder weapon to use than contemporary rifles, especially on a moving horse.

Indeed, provisions were made for soldiers who could not muster the needed skills; some cavalrymen were armed with the sling, while some archers served as infantrymen.

There is no clear evidence on how and when the transformation occurred, but by the time Justinian came to power in 527, the most effective forces of the Byzantine army were certainly its units of mounted archers. Even if they lacked the fullest riding skills and endurance of the steppe riders, they had compensating advantages of their own: body armor that made them more resilient, a lance strapped to their back that they could pull out to mount a charge, and very thorough combat training. For this we have the eyewitness account of Prokopios of Caesarea, who is defending the new missile cavalry from ill-informed criticism, and from snobs nostalgic for the *hoplites* of classical Greece who fought it out hand-to-hand and despised those who only fought by launching arrows from afar:

> There are those . . . who call the soldiers of the present day "bowmen" [despised in Homer], while to [soldiers] of the most ancient times they wish to attribute such lofty terms as "hand-to-hand fighters," "shield-men," and other [prestigious] names of that sort; and they think that the valor of those times has by no means survived to the present, . . . [but] . . . the [ridiculed] Homeric bowmen . . . were neither carried by horse nor protected by spear and shield. In fact there was no protection at all for their bodies. . . . Least of all could they participate in a decisive struggle in the open. . . . But the bowmen of the present time go into battle wearing corselets [chest and upper-back armor] and fitted out with greaves which extend up to the knee. From the right side hang their arrows, from the other the sword. And there are some who have a [lance] also.[18]

These troopers could therefore fight in close combat as well, and not only with their missiles from afar, as Homeric bowmen did with their simple wooden self bows, acquiring a cowardly tint thereby. That was true of the Huns also—they too could fight with sword and spear, and their archery could be dismounted well, contrary to their caricature as centaurs who did everything, truly everything, on horseback and who could barely walk, let alone fight on foot.

There was more than tactics to the tactical revolution—it had clear strategic implications. The old heavy infantry of the legions—foot soldiers trained and equipped to firmly hold their ground, to dislodge others from their ground, and to relentlessly kill enemy soldiers in close combat face-to-face, was most suitable for "attrition" warfare aimed at destroying the enemy, at some proportionate cost in casualties. The tacit assumption was that once that enemy was destroyed, there would be peace.

The Byzantines knew better. They knew that peace was a temporary interruption of war, that as soon as one enemy is defeated, another would take up his place in attacking the empire. Hence the loss of scarce and valuable soldiers to inflict attrition was irreversible, while the strategic gains could only be temporary. Even the destruction of the enemy was not a definitive gain, because in the unending war, yesterday's enemy could become the best ally. Because they fully recognized this, as their military manuals clearly prove, the Byzantines rejected the old Roman striving to maximize attrition that was embodied in the measured pace, elaborate armor, heavy throwing spears, and short stabbing swords of the classic legions—veritable meat grinders. Whenever possible, the Byzantines tried to avoid the frontal attacks and rigid stands that inflict and cost high casualties, relying on maneuver instead to fight the enemy by raiding on the offensive and by ambushing on the defensive, by containment, by outflanking, and by enveloping, different ways of winning by disruption rather than destruction. They therefore favored the more mobile and more flexible cavalry over the infantry, because the cavalry was better suited for all forms of maneuver, at least in open country, and could usually retreat safely under pressure instead of being trapped into last stands.

The tactical revolution was therefore a major military innovation that transcended the tactical level—it amounted to a new style of war that left unchanged only siege warfare, and also such rare combat as took place in rugged mountains and forests. That remained the domain of the light infantry, as is still the case today for the most part.

Offensively, the new style of war was employed most successfully in Justinian's wars of conquest in North Africa against the Vandals from 533, and then in Italy against the Ostrogothic kingdom, and also in the renewed warfare started by Sasanian Persia in 540—though Byzantine mounted archers had certainly taken part in the previous war started by them in 502, under the fortunate and talented Anastasios I (491–518).

Defensively on the other hand, the great test came when the Avars arrived, the first steppe power of great consequence since Attila's Huns. Formed in the usual way of ethnogenesis around a Mongoloid core from Inner Asia, probably the Jou-jan or Juan-juan or Ruan-ruan of the Chinese sources, they accumulated Turkic and other subjects as they moved westward.[19] They were mounted archers just like Attila's Huns had been, but much better equipped with armor and thrusting lances as well, and more accomplished in other forms of warfare, including siegecraft. We have specific information about Avar equipment in the most important Byzantine military manual, known as the *Strategikon* of [emperor] Maurikios, which is discussed in detail in Chapter 11. It contains the first reference to stirrups—a very major innovation—and describes various items of Byzantine equipment as of the "Avar type." Perhaps it was from their Chinese antecedents that the Avars acquired the designs that the Byzantines eagerly copied. But mounted archery they already had from the Huns, neutralizing its Avar equivalent or near enough, and that made all the difference: the Avars could be confronted in open-field combat, which had not been the case with the Huns unless badly outnumbered or in very wet weather.

By 557 the Avars had reached the Volga boundary between the steppe north of the Caspian in what is now Kazakhstan, and the Pontic steppe north of the Black Sea. In 558 or possibly 560, they sent an embassy to Constantinople with assistance from the Caucasian Alans, who introduced them to the Byzantine commander in nearby Lazica, in contemporary southern Georgia. When Justinian was informed, he summoned the Avar delegation to Constantinople. Menander Protektor relates:

> One Kandikh by name was chosen to be the first envoy from the Avars, and when he came to the palace he told the Emperor of the arrival of the greatest and most powerful of tribes. The Avars were invincible and could easily crush and destroy all who stood in their path. The Emperor should make an alliance with them and enjoy their efficient protection. But they would only be well-disposed to the Roman state in exchange for the most valuable gifts, yearly payments and very fertile land to inhabit.[20]

Menander goes on to write that Justinian was then old and feeble, but that he would have "crushed and utterly destroyed them," if not by war then by wisdom, had he not died shortly thereafter, concluding, "Since he could not defeat them, he followed the other course." That he did, but only in part, because while there were certainly gifts, including gold, no "fertile land" changed hands and the Avars continued to wander westward.

Though an acute observer and less enamored of heroic poses than Priskos, Menander was certainly wrong in theorizing that Justinian would have "utterly destroyed them" if he could have.

At that point, in the steppe corridor west of the Avars there were Turkic Utrigurs and Kutrigurs who periodically threatened Byzantine possessions in Crimea and along the Black Sea coast, there were dangerous Slavic Antae ahead of them, and a greater mass of Slavs *(Sklavenoi)* pressing against the Danube frontier and infiltrating as far as central Greece.[21] The Avars did became a great threat a generation later, but in 558 or 560 it would have been against all the rules of Byzantine state-craft to risk a great army and accept the certainty of many casualties to "utterly destroy" a potential enemy that was more immediately a potential ally. The Avars did in fact proceed to attack, rout, and subject Utrigurs, Kutrigurs, Antae, and many Slavs. When the Avars did finally attack the empire circa 580, the new style of war proved itself. Under emperor Maurikios (582–602) there were reverses at first—for reasons of operational command rather than tactics—but Byzantine forces strong in mounted archers successfully attacked the Avars around 590.

The useful if often chronologically challenged near-contemporary historian Theophylact Simocatta (= snub-nosed cat) has preserved an account of the Roman offensive that started on the bank of the Danube opposite Viminacium (now Kostolac, Voivodina, Serbia) and continued toward the river Tisza on the Banat border:

> [Priscus, commander of the army in the Balkans] . . . strictly marshaled three forces for the Romans. Next he firmly committed the wings to split apart and thus admit the Avars, so that the barbarians would be cooped up in the middle as the forces surrounded [them], and would fall into unexpected disasters. [3.3] Then in such a manner the barbarians were out-generalled and nine thousand of the opposite enemy forces were slain. . . . [3.4] On the tenth day the general heard that the barbarians had again arrived for an engagement; when the day grew light, he equipped the Romans, drew them up in good order, and moved into battle. [3.5] And so Priscus mobilized his forces in three divisions again whereas the Barbarian . . . [formed] a single division. And so Priscus occupied the advantageous land in the locality and having the might of the wind as an assistant, he clashed with the Avars from a height and with his two wings outfought the enemy. [3.6] Since a swamp was spread below that locality, he drove the barbarians towards the waters. For this reason the barbarians were beaten back amidst the shallows, had the ill fortune to confront the swamp, and drowned most horribly.[22]

Of an earlier fight, Theophylact writes, "The Romans laid aside their bows and combated the barbarians at close quarters with their spears" (2.11)—evidently that was an exception due to peculiar conditions of terrain (wooded?) or weather (wet), and that is why it was noticed. As for the norm, in 3.5 above the reference to the "might of the wind as an assistant" proves that the Byzantines were relying on their bows, because of course arrows are impeded by contrary winds, deflected by crosswinds, and enhanced by a straight downwind.

The Avars were indeed formidable. Overcoming their defeats of the 590s, in 626 they reached and besieged Constantinople with great numbers of Slav subjects in coincidence, or planned conjunction, with the deepest of all Sasanian offensives, which reached the shore opposite Constantinople. After they failed in 626, the Avars lost control of many of their Slav subjects, who were efficiently subverted by Byzantine agents—including the tribes that became the Croats and the Serbs with consequences that endure till this day. Nevertheless the Avars remained a threat until they moved north into what is now Hungary, where they were finally defeated in a decisive way by Charlemagne in 791, soon to disperse and assimilate after Bulghar attacks as well. Overall, the Avars were a severe but manageable threat for the Byzantines, who were not outclassed as they had been by the Huns, because they too had mastered the difficult art of mounted archery. That revolution may have been tactical, but it had strategic implications.

There was a major problem, however, whose consequences could also be strategic: mounted archery is not only a very demanding skill but also very perishable both individually and, more important, institutionally. Unless it is learned in childhood, it is emphatically not one of those skills that degrade hardly at all over time, such as bicycling. Weapon trainers are familiar with the sharp difference between the retention of shooting skills with pistols and with rifles; pistol skills degrade so quickly that without serious monthly practice the average pistol shooter becomes a danger to his colleagues, whereas a well-trained rifleman can retain his skills with just an annual refresher. Mounted archery is like pistol shooting, only more so; there is contemporary evidence for that in the regular sequence of refresher training absolutely required even by the most senior riders in the Yabusame mounted archery events.

Therefore, whenever the overall material and moral state of the Byzantine army prevented regular and intensive training, the mounted archers would lose their edge much faster than other soldiers, and if new recruits were not patiently inducted into the art, or too few of them were, the army as a whole could soon lose this capability. It has been plausibly suggested that this factor alone had an important influence in the inability of the Byzantine forces to contain the Turkic Seljuks at the end of the eleventh century, incidentally adding a technical explanation for the downfall of an army that had been victorious in all directions as late as 1025.[23]

Intelligence and Covert Action

In the Byzantine field manuals examined in Part III, the commanders were invariably urged to do their utmost to gather intelligence by all means available, given the special importance of information in the Byzantine style of war. Commanders were to use not just one means of gathering information, but all three: (1) light-cavalry and foot patrols to probe the enemy with hit-and-run surprise attacks, to test his morale and skill, and to provoke the enemy to send out more of his forces, exposing them to observation and assessment—in other words, *reconnaissance* in modern terms, performed by smaller, faster, but well-armed combat units operating ahead of the main forces; (2) the stealthy exploration of terrain and enemy forces farther out, deeper in enemy-controlled territory, by small teams of soldiers on foot or on horseback, who were to avoid any form of combat that would interfere with their primary duty of seeing and reporting back—in other words, *scouting* in modern terms, a clandestine mission performed by undisguised soldiers with light weapons and no armor who keep out of sight by hiding in the terrain; and (3) intelligence gathering much deeper into enemy-controlled territory, if possible in his encampments and fortresses, and even the seat of government, by "covert" agents, not hidden in the terrain but protected by false identities as merchants, innocent civilians, or even enemy soldiers or officials—in other words, *espionage* in modern terms. In addition to infiltrated agents, there were also "secret friends," "agents-in-place" in modern parlance, enemy citizens and presumably officials or military chiefs recruited to provide inside information.[24]

The field manuals made these distinctions and explained their necessity. Reconnaissance patrols by light cavalry units—*prokoursatores,* "those who run forward" was the most usual term—were too large for proper scouting, to observe the enemy as it is, because they were apt to induce either responsive reinforcement or prudent withdrawals. Scouts who start fighting the enemy are unlikely to fare well, with their light armament and lack of numbers, and fail in their duty to observe and report back. Nor can clandestine scouts become useful spies by just walking out of the woods or down the mountain to enter the nearest town— they are trained as soldiers, not as covert agents, whose very different selection, training, and management are also mentioned in the field manuals.

Intelligence gathering in a broader sense—the striving to understand the mentality of foreign nations and their leaders and not just their immediate intentions, to assess their military strength in the round including its sustenance and not just what forces are in the field and where—was very much a Byzantine concern. Our knowledge of Attila and his Huns largely derives from the detailed account of Priskos of Panium, who was invited to join a Byzantine delegation, evidently for the purpose of reporting back on the culture of the Huns—a practice already ancient when Tacitus wrote his *Germania*. But aside from such explorations of peoples and land rendered as literature, whatever their purpose might have been, there was also systematic espionage. By nature espionage must be poorly documented—and the accusation of espionage is poorly correlated with its actuality, as I could personally testify. But there is a famous complaint by Prokopios that illustrates how Byzantine espionage had operated in his day, until it was supposedly ruined by Justinian's parsimony:

> And the matter of spies is as follows. Many men from ancient times were maintained by the State, men who would go into the enemy's country and get into the Palace of the Persians, either on the pretext of selling something or by some other device, and after making a thorough investigation of everything, they would return to the land of the Romans, where they were able to report all the secrets of the enemy to the magistrates. And they, furnished with this advance information, would be on their guard and nothing unforeseen would befall them.

Justinian is then accused of having destroyed the system "by refusing to spend anything at all." Many mistakes ensued, according to Prokopios, including the loss of Lazica (the southern part of modern Georgia), "the Romans having utterly failed to discover where in the world the Persian king and his army were."[25]

Covert operations are a natural extension of espionage, and they had a very natural place in the Byzantine style of war, as a particularly economical way of reducing or even avoiding combat and attrition. Normally their aim was to weaken the enemy by subversion, that is, an induced transfer of loyalties. Field commanders were urged to get in touch with and send gifts and promises to the chiefs of foreign allies or auxiliaries in the enemy camp, or even to his own officers if they had some autonomy, as with the garrison chiefs of frontier fortresses. Beyond the battlefield, there were persistent efforts to recruit and reward lesser dynasts, officials, and subordinate tribal chiefs to serve the empire in preference to their sovereigns, for whatever reason, from personal resentment, jealousy, or greed to enthusiasm for the Christianity of the true orthodox church.

The tasks given to subverted enemy chiefs might be to dissuade counsels of war against the empire, or to promote the merits of fighting for the empire, or simply to argue the virtues of friendship with the empire, all of which might be viewed as wise statecraft anyway. Subversion was more difficult when the conflict of loyalties could not be masked, when it was unambiguously disloyal.

In one most dramatic case examined in Chapter 15, Shahrbaraz, the very successful commander of the Sasanian army that had penetrated all the way to the shore opposite Constantinople in 626, who was officially and openly contacted at the time for a failed battlefield negotiation, evidently remained in touch thereafter by covert means, and later—in military circumstances that had become very adverse for the Persians—overthrew the shah of the time to make peace with the empire.

This was subversion on a strategic scale, achieved by a combination of official and private negotiations conducted with skill and tact for which there is some evidence, and above all by victories on the battlefield that had changed the military balance—it was still very useful, however, in saving further fighting. It is extremely unlikely that Shahrbaraz could have been subverted by pure bribery: a successful army commander who had recently conquered the richest trading cities of the empire was unlikely to be wanting for gold, which the Byzantine emperor of the time, Herakleios (610–641), sorely lacked in any case—he had to seize and melt down Church plate and vessels to pay his soldiers. Even in dealing with much less exalted personages, it was the Byzantine method, as we shall see, to wrap bribery in flattery and to present it as a spontaneous gift motivated by imperial benevolence, all of which made it much easier for the subverted to accept payment and act accordingly.

In a famous case, however, there was little opportunity to flatter or disguise, and what was wanted from the subverted was not his favorable counsel in the enemy's court, which might have been felt to be no treason at all, but the assassination of his ruler Attila, of whom he was an habitual intimate. The episode was recorded by Priskos of Panium, normally a reliable eyewitness but in this case strongly biased against the protagonist: Chrysaphius, whose rank is variously recorded as *cubicularius* (chamberlain of the bedchamber) or *spatharius* (ceremonial sword bearer), but who was all-powerful regardless of rank as the particular favorite of Theodosios II (408–450). Although universally reviled in our sources as a eunuch (surnamed Tzoumas) of low birth and an irreligious extortionist (according to the patriarch and future saint Flavian[26]), Chrysaphius is in good standing among strategists. He has as good a claim as any to the invention of Byzantium's new strategy, whereby the direct use of military force to destroy enemies was no longer the first instrument of statecraft, but the last. Priskos, who mostly calls him "the eunuch," blamed him for the unmanly policy of paying off Attila, ignoring its low cost and high effectiveness.

The pursuit of the broad and long-term policies that any grand strategy requires does not exclude specific actions to exploit special opportunities. Confronted by the world-historical phenomenon of Attila, whose individual abilities greatly magnified the power of the Huns, and correctly calculating, as events would soon show, that without Attila the Huns would be greatly weakened, Chrysaphius set out to bribe Attila's trusted official Edeco, or Edekon, who was in Constantinople as an envoy, to kill his master. Priskos relates his careful procedure:

> The eunuch asked if he had unrestricted access to Attila. . . . When Edeco replied that . . . together with others selected from among the leading men [he] was entrusted with guarding Attila (he explained that on fixed days each of them in turn guarded Attila under arms), the eunuch said that if he would receive oaths, he would speak greatly to his advantage; there was, however, need of leisure for this, and they would have it if Edeco came to dinner . . . without . . . his fellow ambassadors.

Edeco came to dinner at the eunuch's residence. With Vigilas (an official translator under the orders of Chrysaphius) interpreting, they clasped hands and exchanged oaths, and Edeco swore that he would not reveal what would be said to him, even if he did not work toward its achievement.

> Then the eunuch said that if Edeco should . . . slay Attila and return to the Romans he would enjoy a life of happiness and very great wealth. Edeco promised to do this and said that for its accomplishment he required money—not much, only fifty pounds of gold to be given to the force acting under his orders, to ensure that they cooperated fully with him in the attack.

There was a problem, however. Attila had instituted standing security precautions:

> [Edeco explained] that since he had been away, he, like the others, would be closely questioned by Attila as to who amongst the Romans had given him gifts and how much money he had received, and because of his [fellow envoys] he could not hide the fifty pounds of gold.[27]

It was agreed at Edeco's request that Vigilas would travel to Attila's court with him, ostensibly to collect Attila's reply in the current negotiations, but actually to receive instructions as to how the gold was to be sent.

It was a good try and sound statecraft—that history is made by impersonal processes and that individual rulers are irrelevant is discredited neo-Marxist dogma. But the necessary secrecy of covert operations makes them even more subject to error than other state actions. Chrysaphius, the supposed master of deceit whose cunning outmaneuvered many enemies at court, was outmaneuvered by the supposedly simple barbarian who obviously meant to reveal the plot to Attila all along, and only asked for the gold to exhibit it as evidence.[28]

When Attila received the mission headed by the official envoy, the unknowing Maximinus, he did not disclose his knowledge of the plot, though he dropped an enigmatic hint. Maximinus "gave him the letters from the Emperor and said that the Emperor prayed that he and his followers were safe and well. [Attila] replied that the Romans would have what they wished for him." Attila then set out to intimidate Vigilas, by suddenly launching a furious attack over the side issue of the return of Hun fugitives who had defected to the Romans, directing his remarks at him rather than Maximinus:

> When Vigilas replied that there was not one [Hun] fugitive . . . amongst the Romans, . . . Attila became even more angry and abused [Vigilas] violently shouting that he would have impaled him and left him as food for the birds if he had not thought that it infringed the rights of ambassadors to punish him in this way.[29]

Attila's court was not unequipped to rule an empire, for secretaries then read the names of the Hun fugitives "which were written on papyrus."

Attila's next move was to tell Vigilas to depart immediately, ostensibly to bring the list to Constantinople but actually to give him an opportunity to fetch the gold for the plot. Edeco duly arrived at the delegation's tent to take Vigilas aside, confirm his willingness to proceed, and tell him to bring the gold to reward his men.

To further set the stage for the exposure to come, a message arrived from Attila declaring that no member of the delegation was to ransom any Roman prisoner, or buy a slave or anything else except food. That removed the need for any large amount of gold for the delegation.

While awaiting the return of Vigilas, Maximinus and Priskos joined Attila in a long trip to the north; it was then or later, very possibly as part of a calculated softening up, that they witnessed the capture of a

Hun sent from Roman territory to spy—Attila ordered him to be impaled; next, two slaves who had killed their masters were gibbeted alive; then a formerly "gentle and friendly" Hun chief, Berichus, came to take back a horse that he had previously given to Maximinus, and was uncivil.

When Vigilas returned with a party that included his son, he was arrested and the gold was duly found. Attila cut short his denials, ordering that his son be killed with a sword unless Vigilas confessed. That he did. "He burst into tears and lamentations and called upon justice to use the sword on him, not upon an innocent youth. Without hesitation he described [the plot] . . . all the time begging that he be put to death and his son be sent away."[30] For Attila, the exposure of the plot offered further opportunities for extortion, starting with another fifty pounds of gold to ransom the son of Vigilas. Soon a Hun delegation arrived in Constantinople, with the original bag in which Chrysaphius had placed the fifty pounds of gold sent to Edeco, and large new demands, starting with the head of Chrysaphius.

The unfortunate outcome of the attempt to deal with the problem of Attila at his source (though in the end Vigilas was liberated) did not dissuade future attempts at covert operations, though mostly for purposes of subversion rather than murder. In 535, when the army of Justinian (527–565) had totally destroyed the power of the Vandals of North Africa and was sent to destroy the Ostrogothic kingdom of Italy, the landing of Byzantine troops from reconquered Sicily to the mainland was preceded by secret negotiations with Theodahad, nephew and unworthy successor of Theodoric the Great, who was to surrender his kingdom in exchange for fine estates elsewhere; earlier the secret negotiations had been with Theodoric's daughter and former regent Amalasuntha, who was to remove herself and the Gothic treasury to Constantinople. And seven centuries later, it was the entire island of Sicily that was subverted from papal authority and Angevin power.

Fortress Constantinople

It was remarked above that the strategic geography of the eastern empire was generally less favorable than that of the western empire. But there was a huge exception, for the geographic setting of its capital city of Constantinople was exceptionally favorable, and the city itself, built on a promontory jutting out into the Bosporus with the sea on three sides, was exceptionally defensible.

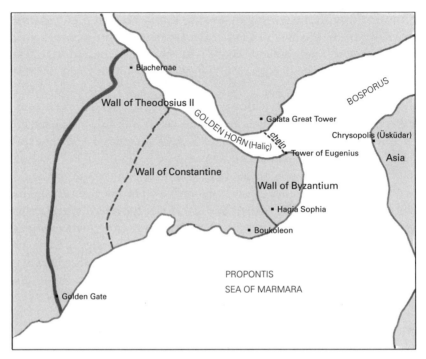

Map 3. The defenses of Constantinople

True, no natural barrier protected its landward side, whose defense therefore required fortified walls, which were duly built on a grand scale. Nor was there a river or an abundance of nearby springs to supply water as in Rome, so that aqueducts had to reach very far over difficult terrain, and their capacity was often inadequate nonetheless.[31] During sieges, once aqueducts were cut, there was only the finite supply of cisterns; in the worst of times, as in 717 when an exceptionally prolonged siege by the jihadi army of Maslama bin Abdul-Malik was correctly anticipated, the city was partially evacuated. But obviously this defect was less than fatal, because Constantinople prospered and grew, even though it never had the one million cubic meters a day supplied by Rome's magnificent aqueducts—the Christian suspicion of bathing certainly helped to reduce demand. As for cisterns, they could hold a great deal of water—just three in place by the sixth century held one million cubic meters.[32] Almost a hundred public and private open and covered cisterns are known, including the spectacular underground cistern of Justinian that has become a major tourist attraction as the Yerebatan Sarayi. No cisterns could suffice for exceptionally prolonged sieges, but they also required exceptional logistics to feed the besiegers, and indeed Maslama's men eventually starved.

For the rest, the logistic position was excellent. By 430, when the city's population had reached some 250,000, the authorities did not find it difficult to supply the 80,000 daily free food rations originally offered by Constantine to populate his city.[33] With its ports at the entrance of the Bosporus, Constantinople had ready access to seaborne supplies from the Aegean Sea and the entire Mediterranean on one side, and from the Black Sea on the other. At the meeting point of Europe and Asia, Constantinople had the hinterland of Thrace on one side, and on the other the fertile shores of the Propontis (the Sea of Marmara) at the western edge of Anatolia. The Bosporus is only seven hundred meters across at its narrowest, between Kandilli and Asiyan some eight miles up the Thracian shore, but until the contemporary bridges were built, the cliffs on the Anatolian side made it much easier to transit through the low-lying shore at the western entrance to the Bosporus, where the adjacent cities of Chrysopolis (Üsküdar) and Chalcedon (Kadıköy) faced Constantinople directly across the water only a mile away. The narrow strait also provided another bounty—large seasonal catches of tuna and other pelagic predators that followed the mackerel's migrations into and from the Black Sea; along with sedentary species, which included enough sturgeon to make caviar a common food, they made fish abundant and very cheap as compared to Rome.[34]

To starve out the city in a siege, an enemy therefore had to control Thrace to cut off the supplies of livestock and produce that reached Constantinople overland, and this did happen several times at the hands of the Avars and, more persistently, the Bulghars before the Ottoman occupation at the end. But to cut off the city from its European hinterland was of no avail in itself, unless the Anatolian shore was also controlled, because foodstuffs and livestock could and did also arrive from there by vessels as small as one-man rowboats. The Asian shore of the Sea of Marmara did briefly fall under Persian control in 626 when the Avars and their Slav subjects held the Thracian side, and many parts of it were held by successive Arab expeditions in 674–678 and during the greatest Arab offensive of 717, but until the final Ottoman conquest it was only threatened by occasional incursions from the Anatolian side and not solidly occupied. Even when both Thrace and the Asian shore were in enemy hands, as in 626, Constantinople could still be supplied by ship at least from the Aegean side, if the Bosporus traffic was cut off.

All this meant that Constantinople could not be starved into surrender in the usual way of most successful sieges in antiquity, but it was only toward the end that it was reduced to a city-state, for which it was enough to survive sieges. And it was only after the restoration of 1261 that Constantinople was nothing more than the capital of a Greek kingdom, whose twin hinterlands could be held with land power alone.

Otherwise, for the greater part of its existence until the military collapse of the later eleventh century, Constantinople was the capital of an empire with possessions scattered in the Aegean and the Mediterranean, and with important outposts at the far end of the Black Sea. It was therefore its maritime dimension that was most important strategically, and in that respect its natural endowment was more than favorable, it was uniquely advantageous for reasons both obvious and not.

In the obvious category there is the Golden Horn (Haliç), a narrow inlet some four miles long sheltered from the wind by the sharp hills of the northern shore and also by the lower hills of Constantinople itself. This was the finest natural harbor known to antiquity because it was calm in all weather, and because the entrance has no shallows and could be navigated without a pilot. Only 240 meters across at the narrowest point, from the time of Leo III (717–741) the rather broader entrance of the Golden Horn could be closed to enemy vessels by an iron chain floated on barrels that was secured to the Tower of Eugenius on the city side, and to the Megalos Pyrgos (great tower) on the far side, now Galata, which contained the elevation gears.

Along the shore of the Golden Horn there were the landing stages, quays, slipways, shipyards, and beaching slides used by the Byzantine navy, merchant vessels, and local ferries. But the Propontis (Sea of Marmara) shore was also sheltered from winds from the north, and several harbors were built along that shore that gave more direct access to the heart of the city, including the landing stage of the palace of Bucoleon, on the edge of the acropolis at the end of the promontory that housed both the imperial palace and the Hagia Sophia (Great Church of the Holy Wisdom), which still stands there.

The less obvious maritime attribute of the city was the "Devil's current." The Black Sea receives proportionately much more water from its great rivers than the Mediterranean receives from its scant rivers; there is therefore a surface current through the Bosporus that varies greatly in strength, with maximum speeds of up to four meters per second, or eight knots, and four knots quite common. That alone made it very difficult or simply impossible for any ancient fleet to land directly onto the shore of the acropolis jutting out into the Bosporus current.

In addition, because of its large inflows of river water, the Black Sea is less saline than the Mediterranean, and because of osmotic pressure there is a subsurface current up the Bosporus. The interplay of the two currents, along with the winds often channeled from the north by the Bosporus, results in a great deal of turbulence that in turn gives a great advantage to sailors with local expertise. This was a factor in the defeat of foreign fleets that came to attack Constantinople.

With such an excellent base, insofar as the Byzantine navy was effective—and it had its cycles of decay and revival—it could defend Constantinople not only by keeping away enemy fleets and preventing landings, but also by bringing reinforcements from anywhere in the empire, both in its own warships and transports and in enlisted merchant vessels. That did happen—it was by sea that Herakleios arrived from Carthage with his followers in 610 to claim the imperial office—and was another reason why the city was never conquered in more than eight hundred years of wars, until dissension destroyed its resistance in 1204.

Constantinople was also the greatest base complex of the Byzantine army, with its standing forces of horse and foot guards, and workshops for the manufacture of armor, weapons, uniforms, and footwear, and the imperial breeding stables. Accordingly the Byzantine navy was much more often employed to convey imperial forces from Constantinople to active fronts around the empire than to bring in troops to reinforce the city garrison.

The sources for the history of the Byzantine navy are very poor.[35] Not much is known of its recruitment of sailors and marines, its ship designs over the centuries, the management of its fleets, its tactics, and its weapons, including "Greek fire," which was useful indeed but undeserving of its mythic reputation, as we shall see in Chapter 13. But even our fragmentary knowledge of individual naval actions and expeditions is sufficient to determine all that needs to be determined here—that when Constantinople was in good working order as a political capital and naval base, as it mostly was, it was secure from naval attack even on the largest scale—as in 717 when the Arab offensive mobilized all the vessels in all the seaports of the eastern Mediterranean to fill the Propontis with warships and transports.

What gave the city such security was the juxtaposition of the perennial calm of the Golden Horn that kept the Byzantine navy's warships and transports quite safe in all weathers, with the often turbulent conditions immediately outside, which ensured the good sea-keeping qualities of its vessels and the skill of its sailors, while making life difficult for enemy crews, especially if they came from warmer lands with calmer waters.

The triangular promontory of Constantinople has no natural barriers to defend its base on the broadest, landward side; its modest elevations on the northern edge and the acropolis do not exceed fifty meters or so. Therefore all the successive cities, starting with the Greek polis of Byzantion founded in the seventh century BCE, had to have a wall along the base. Roman Byzantium had a sturdy wall, which was demolished by Septimius Severus (193–211) after his civil war in 196 CE but was soon replaced.

When Constantine established his Nova Roma, he started to build a wall reinforced with towers much farther out, running from the Plateia Gate on the Golden Horn to what became the Gate of St. Aemilianus on the Propontis side, and bulging outward. That perimeter ambitiously enclosed an area some five times larger than the Severan wall, yet it was not ambitious enough, because by the start of the long reign of Theodosios (408–450) the city had expanded beyond Constantine's wall into the exurb known as the Exokionion, which was exposed to marauders.

In 408 an earthquake damaged parts of the Constantinian wall, and construction began on what would eventually become the Theodosian Wall (Theodosianon Teichos) some 1.5 kilometers farther out and extending some 5.5 kilometers (3.5 miles) from the Propontis coast to the suburb of Blachernae near the Golden Horn, now the wards of Ayvansaray and Balat. From the beginning, it was not just a wall but a complete defensive system, consisting of three walls, two of them reinforced with ninety-six towers each, a roadway, and a moat—a formidable combination of synergistic barriers.

Constructed of alternating layers of stone and brick, a technique that also has aesthetic merit but may have been intended to increase resistance to earthquakes, the principal wall, or Mega Teichos (great wall), is 5 meters across at the base and 12 meters high. Its ninety-six towers are spaced out at intervals of 55 meters, half the lethal range of composite reflex bows. Varying in height between 18 and 20 meters, each tower had a battlemented platform of alternating crenels and merlons, so that artillery and thrusting rods to push away ladders could be operated through the crenels, while the merlons in between provided protection for wall guards. In each tower, an upper chamber with arrow slits and embrasures that was entered directly from the wall's walkway formed an enclosed fighting compartment, which was more resilient than the top platform open to bombardment by catapults and the harassment of plunging arrows. There was also a lower chamber at street level that was used for storage.

Some 15 to 20 meters ahead of the Mega Teichos there was an outer wall (Exo Teichos or Proteichisma), which was 2 meters across at the base and 8.5 meters high including the battlemented walkway. It too had ninety-six towers that were sited in correspondence to the halfway point between the towers of the Mega Teichos. These smaller towers also had a fighting compartment with a battlemented terrace on top.

The moat (souda), which started roughly 15 meters ahead of the outer wall, was 20 meters wide and 10 meters deep. The 15-meter strip between the perimeter and the moat was terraced, and a paved road ran along its entire length, so that inspection patrols could move rapidly and safely by day and by night, guarded over from the outer wall. Next to the roadway, at the edge of the moat, there was a crenellated fighting wall 1.5 meters high, of use to shelter inspection patrols if they came under missile attack and for sniping against enemy scouts and skirmishers, rather than in major siege operations when the outer and great walls would be manned in strength. In addition to small posterns easily walled off during sieges and five military gates, there were five gates for public transit that opened to bridges that crossed the moat, including the substantially preserved "golden gate" (chryse pyle), originally a triumphal arch on the Via Egnatia that was incorporated into the Theodosian Wall, forming the ceremonial entrance to the capital; on either side were added towers that would be decorated with bronze elephants and winged Victories.

The suburb of Blachernae was not originally enclosed by the Theodosian Wall, which stopped some 400 meters from the Golden Horn. A single wall bulging out in a semicircle was added in 626–627 to enclose Blachernae during the reign of Herakleios at the time of the Avar-Persian siege, and an outer wall was added in 814 under Leo V when the Bulghars were the direct threat; the elaborate Byzantine wall segment now to be seen there was built in the twelfth century under Manuel I Komnenos.

As with Istanbul now, Constantinople was subject to frequent earthquakes, and some came at particularly inconvenient times. On November 6, 447, just when the Huns were approaching the city, an earthquake destroyed large parts of the wall. Extant inscriptions in Greek and Latin record that Kyros, the prefect of the city, was able to restore the wall within sixty days with the help of one of the circus factions of chariot-racing fans, the reds. The Gate of Rhegium (Pyle Regiou), also known as Pyle Rousiou, "gate of the reds," presumably because they helped repair it in 447, is now known as the new Mevlevihane Gate (yeni Mevlevihane kapisi). Kyros was hailed as a new Constantine; that may explain the entry in Marcellinus Comes, under the fifteenth indiction, the consulships of Ardabur and Calepius: "3. in the same year [as Attila's offensive] the walls of the imperial city, which had recently been destroyed in an earthquake, were rebuilt inside three months with Constantine the praetorian prefect in charge of the work."[36]

What is described as the Theodosian Wall was therefore much more than that—it was a complete defensive system.

The moat was the first formidable barrier. The standard technique to cross moats was to lower prefabricated wooden bridges across the gap, but that was not easy to do across a width of 20 meters; the other technique was to keep dropping bundled and weighted branches or fascines into the moat until infantrymen could run across the improvised ford, but a great many fascines would be needed, given the depth of 10 meters, and even then there would be no firm surface for battering rams, mobile assault towers, swing ladders, and other siege engines that must reach the wall, or just about, to be effective. In this case, the distance between the outer edge of the moat and the wall was 35 meters, or 38 yards—too far for siege engines other than stone-throwing or arrow-launching artillery.

In addition, the depth of the moat made it difficult to use the technique that was generally found to be most effective in antiquity: tunneling to reach the wall, mining through its foundations while propping it up, doping the wooden props with inflammable resin, and then setting them on fire to collapse the wall. Given the 10-meter depth and an allowance for softened mud at the bottom, that made for a very deep tunnel—and one easily flooded.

Because the towers on both walls were 55 meters apart, with the outer-wall towers placed at the midpoint between main-wall towers, and given the 15- to 20-meter gap between the two walls and the 10-meter height difference, it can be calculated that any assaulting infantry that could cross the moat would come within lethal arrow range from at least four towers and two 50-meter wall segments. That would accommodate a total in excess of 300 archers; if skilled, not already exhausted, and well supplied with arrows, that number could suffice to hold up an army of thousands, but of course there would also be artil-

lery on both sides. The Byzantines regularly employed artillery in their
field forces, as we shall see, and certainly more so in the defense of fortresses, and especially the most important of all.[37]

The great wall, the space in between the walls, and depending on circumstances, the roadway in front of the outer wall, could all serve to
shift the defensive effort from one segment of the perimeter to another
more rapidly than the attackers could move on the outside, with no prepared roadways and walkways.

The enemies that prompted the construction of the Theodosian Wall
preferred horses to boats, but there had been Gothic sea raids even in
the third century. The Avars in 626 came with large numbers of Slav
boatmen, and from 674 the attacking Muslim Arabs arrived by sea
from the ports of the Levant.

Currents and winds greatly helped to defend the city, as noted, but
from the time of the first Constantine there were seawalls as well. Under
Theodosios II, the land wall system was supplemented by seawalls that
lined the shores on both the Propontis, or Sea of Marmara, side and
along the Golden Horn. The latter was 5,600 meters long from the
Blachernae land wall down to the cape of St. Demetrius, while the
Propontis wall was 8,460 meters long, not counting the inner walls of
the several harbors.

The first seawalls were just that, walls and not high. Under the impulse of the Arab attacks of 674–677 and then again in 717 the seawalls
on both sides of the promontory were repaired, reinforced, and elevated
in places, but it was under Michael II (820–829) that large-scale reconstruction to a greater height was started, in reaction to the Arab maritime conquest of Crete in 824. Then and later, towers were also added,
all to no avail in 1204 when the Venetians landed on the Propontis side.

Taken as a whole, the moats, walls, and towers of the Theodosian
land wall amounted to a very effective "force multiplier" in modern
parlance; useless in themselves, they could greatly magnify the defensive
capacity of an adequate, well-trained, and well-armed garrison. But for
the most part—there was a startling exception in 860—Constantinople
came under attack only in times of acute crisis in the empire as a whole,
for enemies could scarcely reach it otherwise. In such times, notably after a major battlefield defeat, imperial forces were not likely to be present in good condition and in large numbers to form a strong garrison
for the city. A dedicated *tagma* (battalion, more or less) "of the walls"
(ton teikhon) was established under Constantine V (741–755), but it
would have needed many more men than an ordinary *tagma* of a thou-

sand to fifteen hundred men to guard such an extensive system of fortifications.

It was precisely in such times of crisis and disorganization, when enemies were most likely to reach the Theodosian Wall, that its "force multiplier" effect was most valuable, for it could offset even severe deficiencies in the number of the defenders; in 559, 601, 602, and 610, citizen corporations including the Blue and Green racing fans were mobilized to man the walls—and in 559 even the senators, or at least their retinues, were summoned.[38] In the supreme test of 626, a garrison of perhaps twelve thousand, including fully trained soldiers sent by Herakleios, was sufficient to defend Constantinople when it came under attack by the well-equipped and formidable Avars accompanied by vast numbers of Slavs, while a Persian army was camped on the opposite shore, denying any reinforcements from either continent.

Any properly built fortification can be a force multiplier, but the Theodosian Wall system was by far the most effective fortification in the world for a thousand years from its inception, and its magnitude was such that it had strategic significance in itself.

The seawalls were also an impressive achievement of construction, and certainly useful against raiders. But the maritime security of the city necessarily required naval power and the ability to control the seas, not Mediterranean-wide but certainly the approaches to Constantinople and the mouth of the Bosporos. It was internal dissension that caused the downfall of 1204, but it was the lack of a functioning navy that made the city immediately vulnerable.

It was noted at the start that the eastern empire was disadvantaged as compared to its western counterpart because of its lack of strategic depth. That is why Constantinople had to function concurrently as the magnificent capital of a great empire and as a fortress that had to look to its own protection, much like Paris in the modern epoch of German unity, simply because it was too near to the Rhine, exposing it to siege in 1870, a close call in 1914, and conquest in 1940. In between the two world wars, the French attempted to increase the effective strategic depth in front of Paris by constructing the most elaborate linear fortification known to history, the Maginot Line, which was undefeated in 1940 but circumvented through Belgium.

The Byzantines tried to do the same from the fifth century by adding another fortified perimeter between Constantinople and the northern threat: the Long Wall (Makron Teikhos), also known as the Wall of Anastasios, which extended for 45 kilometers from the Sea of Marmara, starting at a point six kilometers beyond Selymbria (Silivri), to the Black Sea coast at what is now Evcik İskelesi. The wall is named for the fortunate and economical Anastasios I (491–518), but perhaps he only completed, repaired, and enhanced a prior wall that may have dated back to Leo I.[39] Where it is best preserved it is a construction 3.3 meters across at the base, some 5 meters high, and complemented by a moat, towers, fortified gates, forts, a rectangular camp perimeter, and an inner roadway that allowed smooth riding from sea to sea, secure from ambushes when the wall was manned. It made Constantinople and its hinterland "almost an island instead of a peninsula, and for those who wish provides a very safe transit from the so-called Pontus [Black Sea] to the Propontis [Sea of Marmara], while checking the barbarians . . . who have poured forth over Europe," in the words of the admiring Evagrius Scholasticus.[40]

It was the great virtue of the Long Wall that it formed a defensive perimeter 65 kilometers beyond the Theodosian Wall, giving depth to the defense of Constantinople. If properly manned by sentries and patrols, the Long Wall could stop bandits, small group of marauders, and localized attacks. On a larger scale, it offered a secure base for field armies sent to intercept enemies at a dignified distance from the capital, instead of allowing them to come right up to its walls.

The great defect of the Long Wall was that the geography was unfavorable: to provide 65 kilometers of depth, it extended for 45 kilometer in length and required a commensurate garrison of at least ten thousand to provide an adequate number of sentries, patrols, and reaction units. In the preface to one of Justinian's laws it is noted that two rather senior officials were in charge of the wall, which implies a substantial establishment.[41] That is certainly the reason why the Long Wall was abandoned in the early seventh century if not before—it needed too many troops. So it was with little strategic depth that Constantinople was successfully defended by the garrisons of the Theodosian Wall for almost eight hundred years.

Justinian's Reversal Reversed: Victory and Plague

Flavius Peter Sabbatius Iustinianus, Justinian I, Justinian the Great, Saint "Justinian the Emperor" of the Orthodox Church, was born a peasant's child in what is now Macedonia, yet came easily to the throne, having long served as assistant, understudy, co-emperor, and increasingly the effective ruler for his uncle Justin I (518–527).

When he was formally enthroned in 527, seventy-seven years had passed since the end of the reign of Theodosius I, and its strategic innovations had been absorbed, consolidated, and institutionalized to good effect. The empire was much stronger than it had been in 450, but still needed the Long Wall and the Theodosian Wall to protect Constantinople, not against large-scale invasions but rather against plunder raids from across the Danube and the robberies of Balkan marauders.

As had been true since its inception in the third century, the Sasanian empire of Persia remained a permanent strategic threat, undiminished by mutual respect, frequent negotiations, and formal treaties, including the "endless peace" of 532. Persistent vigilance and a readiness to deploy reinforcements quickly were always necessary, if often insufficient, to contain Sasanian power in the Caucasus, across contested Armenia, and down to southern Syria.

On the other hand, there no longer was any rival power north of Constantinople below or beyond the Danube, while across the Adriatic, the Ostrogothic kingdom of Italy only desired good relations with the empire, and part of its elite even wanted reunion under the empire. The Vandals and Alans who had conquered *Africa* in the last century were still there, but no longer threatened naval expeditions against Egypt. As for the dangers of the great Eurasian steppe, the nearest warlike nomads were the Turkic Kutrigurs in what is now the Ukraine, at worst a nuisance rather than an irresistible force as Attila's Huns had been.[42]

More powerful steppe enemies were on their way, but by the time of Justinian the warriors of the steppe had irreversibly lost their tactical superiority. The imperial army had undergone its own tactical revolution, having thoroughly mastered the difficult techniques of mounted archery with powerful composite reflex bows, while still retaining close-combat skills with sword and thrusting lance. Even if their archery could not quite match the best that the Hun mercenaries with them could do, Byzantine troopers could no longer be outfought. The steppe warriors had also lost much of their operational superiority, because the cavalry had become the primary force of the imperial army, it had adopted agile tactics, and what individual riders may have lacked in virtuoso horsemanship could be compensated by the greater resilience of their disciplined and cohesive units.

This also meant, of course, that the imperial army now had tactical and operational superiority over the Vandals and Alans of *Africa* and the Ostrogoths of Italy. The Alans were primarily horsemen, Vandals and Goths were formidable fighters at close quarters, fully capable of organizing major expeditions, and not unskilled in sieges, but all now found themselves lacking in missile capability and battlefield mobility. Prokopios of Caesarea, who was there, reports how Belisarios, Justinian's celebrated commander, explained the difference that made:

> Practically all the Romans and their allies, the Huns [Onogur mercenaries], are good mounted bowmen, but not a man among the Goths has had practice in this branch, for their horsemen are accustomed to use only spears and swords, while their archers enter battle on foot and under cover of the heavy-armed men [to ward off cavalry charges]. So the horsemen, unless the engagement is at close quarters, have no means of defending themselves against opponents who use the bow, and therefore can easily be reached by the arrows and destroyed; and as for the foot-soldiers, they can never be strong enough to make sallies against men on horseback.[43]

This was only tactics not strategy, but without this advantage it may be doubted that Justinian would have embarked on his plan of reconquest, first of North Africa in 533–534 and then of Italy from 535.

Modern historians almost unanimously assert that he was excessively ambitious, that his conquests overextended the empire, true enough in retrospect, though perhaps only because of unforeseeable catastrophe. But not even his harshest critics consider Justinian a fool, or irrational or incapable of sober calculation. In the celebrated mosaic of San Vitale in Ravenna, Justinian looks at us earnestly but we see the calculating ambition more than the moral fervor.[44]

And there was an inescapable fact: the impossibility of sending really large armies by sea. In the largest expedition that could be mounted, Belisarios set out from Constantinople in the summer of 533 with some ten thousand infantry and eight thousand cavalry carried in five hundred transport ships manned by thirty thousand crewmen, and escorted by ninety-two war galleys.[45] It was certainly a most impressive armada, but eighteen thousand soldiers were not that many to take on the Vandals and Alans, let alone the Ostrogoths, whose fighting manpower was sustained by the resources of the whole of Italy.

But it could be done, if only just, with the tactical and operational advantages of maneuver with forces of mounted bowmen. In the prior phrases of the passage quoted above, Belisarios himself clearly stated as much: "[because of their mounted archers] the multitudes of the enemy could inflict no injury upon the Romans by reason of the smallness of their numbers."[46] That also required a successful theater strategy, of course, and generalship in general. Justinian was famously well served by talented field commanders, especially the eunuch Narses, who was the better tactician perhaps, and the more celebrated, infinitely resourceful Belisarios of the many stratagems and ingenuities—he is still remembered today by unlettered Romans for his improvised floating mills, powered by the current of the Tiber, that ground corn into flour during the siege of 537–538.

Successful stratagems are the classic force multipliers, and it was with Belisarios that they first became a Byzantine specialty, to so remain for centuries to come, along with his systematic avoidance of attrition and maximum exploitation of maneuver.

In the detailed account of both the Vandal and the Gothic wars left by his secretary Prokopios, certainly an admirer but not uncritical, we read how Belisarios was always willing to undertake longer marches on more perilous routes to avoid the expected direction and reach instead the enemy's flank or, better, into his rear, and we read how he was willing to hazard the most risky stratagems to avoid direct assaults. To win with few against many, he replaced the mass he lacked with high-payoff, high-risk maneuvers and bold surprise actions, *coups de main* that all would approve of in the successful aftermath but that were gambles indeed.

One example will suffice. In 536 Belisarios succeeded in seizing Naples after a siege of twenty days, not by assaulting its stout walls but with a bold stroke that could have ended very badly. A soldier moved only by curiosity had descended into the underground aqueduct leading to Naples, whose water flow had of course been cut at the start of the siege. He continued to explore until he reached a segment too narrow for a man—which promised to continue right through the walls into the city. When word reached Belisarios, he promptly offered a large reward to induce the man and his companions to scrape away at the rock very quietly, until the passage was wide enough:

> Selecting at nightfall about four hundred men . . . he commanded them all to put on their [chain mail] corselets, take in hand their shields and swords, and remain quiet until he should himself give the signal. Selecting two commanders, he ordered them to lead the four hundred men into the city, taking lights with them. And he sent with them two men skilled in the use of the trumpet, so that as soon as they should get inside the circuit-wall, they might be able both to throw the city into confusion and to notify their own men what they were doing. . . . [He] also sent to the camp [of his main forces], commanding the men to remain awake and to keep their arms in their hands. At the same time he kept near him a large force—men he considered most courageous.

It was a meet precaution, because "of the men who were on their way to the city [through the long, narrow, dark tunnel] above half became terrified at the danger and turned back." Belisarios succeeded in shaming them back into the tunnel, and to cover the proceedings he sent Bessas, an officer of Gothic birth, to have a shouting match with the Goths manning the nearest guard tower. The four hundred had to keep going inside the narrow aqueduct until they finally reached a roofless [segment] from which they could climb out. Then "they proceeded toward the wall; and they slew the garrison of two of the towers before the men in them had an inkling of trouble."[47]

After that, Belisarios was able to send his own picked men to climb the unguarded wall with ladders, to conquer Naples without a costly assault. Had the four hundred been detected, they could all have been lost—and they were no minor expendable force, in an expeditionary army so small altogether that in the aftermath Belisarios could spare only three hundred men to garrison Naples, then as now the largest city south of Rome.[48]

Stratagems aside, it was mostly its archery as well as good tactics that enabled the Byzantine army to regularly defeat enemies in larger numbers. According to an authoritative reconstruction of two major battles of the Italian campaign, at Tadinae or Busta Gallorum on the Via Flaminia in what is now Umbria in 552, and at the river Casilinus, now Volturno, near Naples in 554, the Byzantine forces commanded by Narses included assorted foreign contingents of Lombards, Heruls, and even Persians, but in both cases it was the bowmen of the imperial army that made the critical difference in the critical phase of the fight with their volleys of powerfully lethal arrows.[49]

In sum, the army's tactical and operational superiority was the *sufficient condition* for the two campaigns of North Africa and Italy; the *necessary condition* was the negotiated peace with the Sasanian Persians, as Justinian himself explained by way of incidental comment in the text of a new law on the administration of Cappadocia:

> We have undertaken such great labors, incurred so much expense, and fought such great wars, in consequence of which God has not only granted Us the enjoyment of peace with the Persians and the subjugation of the Vandals, the Alani, and the Moors, as well as enabled Us to recover all Africa and Sicily, but has also inspired Us with the hope of again uniting to Our dominions the other countries which the Romans lost by their negligence, after they had extended the boundaries of their Empire to the shores of both oceans, which countries We shall now, with Divine aid, hasten to restore to a better condition.[50]

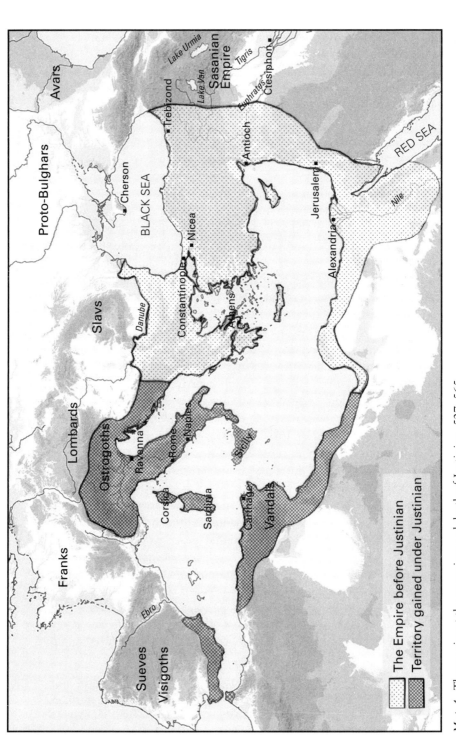

Map 4. The empire at the accession and death of Justinian, 527–565

Avars

Proto-Bulghars

Slavs

Franks

Lombards

Sueves

Visigoths

Ostrogoths

Ravenna

Rome

Naples

Corsica

Sardinia

Sicily

Carthage

Vandals

Ebro

Danube

Cherson

BLACK SEA

Constantinople

Nicea

Athens

Trebizond

Lake Urmia

Lake Van

Sasanian Empire

Tigris

Euphrates

Ctesiphon

Antioch

Jerusalem

Alexandria

Nile

RED SEA

The Empire before Justinian

Territory gained under Justinian

Italy was hardly restored to a better condition *(in melius convertere)* by being liberated from the Ostrogoths in fighting that lasted until 552 through many destructive vicissitudes; and from 568 the Lombard invasion started a new round of destructive fighting, if only after Justinian's death in 565, and long after the unforeseeable catastrophe that invalidated all his strategic plans.

Whatever the future held, Justinian achieved his ambitions almost in full, for his forces conquered North Africa from Tunis through coastal Algeria to what is now the northern tip of Morocco, thus reaching the Atlantic, and across the straits a coastal slice of the Iberian Peninsula in what is now southeast Spain, and all the islands, the Balearics, Corsica, Sardinia, and Sicily, and all of Italy. Except for a tract of the Iberian coast and the southern coast of Gaul where no rival naval power existed, the entire Mediterranean was once again a Mare Nostrum for all practical purposes, with none to contest the Byzantine navy. Nor was this the achievement of a military adventurer, but rather the military dimension of broader political ambitions.

The Justinian who became emperor in his mature forties was notoriously indefatigable, demonstrably very intelligent, unchallenged by rivals, unfettered by conventions—he felt free to marry a woman with the social status of an ex-prostitute—and possessed of two more attributes that empowered him greatly: a full treasury, and a particular talent in finding the especially talented to serve him. All of this could have made Justinian an even more successful version of Anastasios, who ruled for twenty-seven years, built a great deal, including the Long Wall and the fortress city of Dara, lost no wars, reduced taxes, yet supposedly left 320,000 pounds of gold in the treasury for his successor Justin.[51]

But Justinian had much larger aims. Even before he started his military conquests, Justinian set out to codify all the extant *constitutiones*, imperial pronouncements with the force of emperor-made laws, from the time of Hadrian. Theodosios II had also issued a codification, but it was incomplete, while Justinian's code, already published in 529, which implies that it was started as soon as he gained the throne, collated all the *constitutiones* in the Theodosian code with those in two unofficial collections adding more recent laws, including his own, to produce the *Codex Iustinianus,* in twelve books. The lawyer Tribonian, another of Justinian's highly talented appointees, was in charge, and he was the chief author of the *Pandectae, Pandektes,* or *Digesta,* the jurisprudential treatise that followed the *Codex,* which contains in fifty books the legal opinions on all manner of cases of thirty-nine legal experts, notably Ulpian and Paulus. Once issued with official authority, the *Digest* became in effect an additional code of jurist-made law, not dissimilar from the body of English Common Law—except that Romans were involved, hence the thing is organized. Tribonian and his colleagues next produced a much shorter work, the *Institutiones* in four books, a manual of legal training. By 534 the *Codex Iustinianus* was issued in a new edition with corrections and additions, including Justinian's laws in the interim, but after that his new laws, *novellae,* were collected in a separate compilation that included 168 laws, mostly in Greek, by the time Justinian died in 565.

The sum total has been known as the *Corpus Juris Civilis* since the sixteenth century. Long before then, by the end of the eleventh century, it was rediscovered in Italy to form the foundation of legal studies at Bologna and of the first real university with them, and of the Western jurisprudence that now extends worldwide. The continued use of untranslated Latin in American courts (*sine die, nole prosecutere, ad litem, res judicata,* etc.) symbolizes a much deeper persistence: these terms all come from the *Digesta* of the *Corpus Juris Civilis.* The Japanese commercial code contains no Latin but a great deal of Ulpian via Tribonian and the European codes from which it was derived. If lawyers had saints, Justinian would certainly be their patron saint, as he and his wife Theodora are saints of the Orthodox Church celebrated each year on October 14.

Equally vast and equally successful was Justinian's ambition in the realm of public works. Prokopios wrote an entire book, *Peri Ktismaton* (On Buildings), to describe the churches, fortresses, and all else that Justinian built or enhanced—sometimes attributing to him the edifices of other emperors, but we do know that under Justinian dozens of fortresses and other fortifications were built, or substantially rebuilt, in many parts of the empire, and that thirty-nine churches were built or rebuilt in Constantinople alone, including the vast Hagia Sophia, whose immense floating dome still amazes visitors, and whose design is reproduced with varying degrees of felicity in thousands of churches all over the world. From the detailed description in Prokopios of how the Hagia Sophia was built, we learn that the men chosen by Justinian in person to build a radically innovative main church for him, Anthemios of Tralles and Isidore of Miletus, used mathematical engineering to calculate the dynamics of the delicately counterweighted dome.[52] Once again the talented Justinian had found exceptional talents to realize his inordinate ambitions, and the evidence remains intact in Istanbul to prove that he was fully successful, just as he was in his inordinately ambitious juris-prudential project, whose influence is much wider now than it was at his death in 565.

So why were Justinian's military ambitions different? That they were not grossly unrealistic we know from the simple fact that the mari-time expedition of 533 sent to conquer *Africa* was neither shipwrecked nor defeated on arrival, so that what is now Tunisia and coastal Algeria were duly conquered. The conquest of Italy from the Ostrogoths that started in 535 was a much more demanding undertaking, but it too was successfully completed in May 540, when Belisarios entered the Ostrogothic capital and last refuge of Ravenna to accept the surrender of King Witiges, or Vitigis, and his wife Mathesuentha.

As noted above, modern historians explain that Justinian's military ambitions were different because they exceeded the capacity of the em-pire to sustain them. One year after Belisarios ceremoniously concluded his Italian war, in May 540, because no powerful garrison was stationed in Italy to control them, the Goths were able to start fighting again, and with increasing success once Totila became their king. One estab-lished explanation is that Justinian did not replace Belisarios and his army because he was "afraid of the threat that a mighty general could pose."[53] Even Rome was lost in 546, in a war that continued until 552. And because Sasanian Persia had repudiated the "endless peace" treaty to also start fighting in 540, continuing with interruptions until 562, the empire had to sustain two long and large-scale wars on two very widely separated fronts, so that in 559 there were hardly any troops in Con-stantinople to fight off an incursion of Kutrigurs and Slavs. That was certainly evidence of overextension, and presaged an inability to defend the Danubian frontier and the Balkan Peninsula with it, and therefore Greece also, from Avar invasions and Slav occupations.

The charge of overextension therefore implies a charge of strategic in-competence, or more simply a lack of ordinary common sense: having himself inherited a war with the perpetually aggressive Sasanians when he came to the throne, Justinian had to know that the Persian front had to be well guarded in peace as in war. What military strength was left would be needed for the "northern front" of the empire, from Dalmatia to the Danube, which was not under attack in 533 but which was bound to be attacked again sooner or later, as the turbulence of peoples continued beyond the imperial frontiers. That northern front was in-deed the primary defense perimeter of the empire, which protected the valuable sub-Danubian lands all the way to the Adriatic, and shielded

Greece as well as Thrace and therefore Constantinople itself. The northern front also contained prime recruiting grounds for the imperial army, including the village near the fort of Bederiana where Justinian himself was born and lived his first years when he was still Flavius Peter Sabbatius.

To launch expeditions far away, even to conquer the rich grain fields of *Africa* and the hallowed first Rome, while neglecting the defense of the very hinterland of the imperial capital, was therefore a strategic error so obvious that it betokens a foolish mind—not the mind of the Justinian we know.

It is true, of course, that history is the record of the crimes and follies of mankind, and many a foolish war of conquest has been launched since 533, so Justinian would not be alone if he did forget the overriding need to protect his own birthplace and capital.

But there is an altogether different explanation, formed by evidence in part as old as the event, and in part very new—so new that it is not yet incorporated in the broader research on Justinian and his wars, let alone more general histories.[54] Entirely new historical evidence of large significance is very rare, and it is invariably the product of fortunate digging. That is true in this case also, even if the evidence itself is neither epigraphic nor numismatic, or conventionally archaeological, for it is found in the DNA of skeletons and in ice cores.

First the old evidence: in book 2, chapter 22, of Prokopios's *History of the Wars*, we read:

> During these times [from 541] there was a pestilence, by which the whole human race came near to being annihilated. Now in the case of all other scourges sent from Heaven some explanation of a cause might be given by daring men. . . . But for this calamity it is quite impossible either to express in words or to conceive in thought any explanation. . . . For it did not come in a part of the world nor upon certain men, nor did it confine itself to any season of the year, so that from such circumstances it might be possible to find subtle explanations of a cause, but it embraced the entire world. . . .
>
> It started from the Aegyptians who dwell in Pelusium.
>
> Then it divided and moved. . . . And in the second year it reached Byzantium in the middle of the spring, where it happened that I was staying at the time. . . . With the majority it came about that they were seized by the disease without becoming aware of what was coming. . . . They had a sudden fever. . . . And the body showed no change from its previous color, nor was it hot as might be expected when attacked by a fever, nor did any inflammation set in. . . . It was natural, therefore, that not one of those who

had contracted the disease expected to die from it. But on the same day in some cases, in others on the following day, and in the rest not many days later, a bubonic swelling developed . . . not only in [the groin] . . . but also inside the armpit, and in some cases also beside the ears. . . . There ensued for some a deep coma, with others a violent delirium. . . . Death came in some cases immediately, in others after many days, and with some the body broke out with black pustules about as large as a lentil and these did not survive even one day but all succumbed immediately. With many also a vomiting of blood ensued . . . and straightaway brought death.[55]

In chapter 23, we come to the demographic consequences:

Now the disease in Byzantium ran a course of four months, and its greatest virulence lasted about three. And at first the deaths were a little more than the normal, then the mortality rose still higher, and afterwards [the number of] dead reached five thousand each day, and again it even came to ten thousand and still more than that.[56]

Three months, or ninety days, of the greatest virulence, at 5,000 a day comes to 450,000; if we take the 10,000 estimate, we reach 900,000, and Prokopios mentions a still higher daily mortality, yielding seemingly impossible numbers.

When writing as a historian and not as a polemicist, Prokopios is generally deemed a trustworthy source by his modern colleagues, but on the subject of the pandemic there were two different reasons to suspect him greatly. First, in an age without statistics there were no mortality figures to peruse and incorporate in a text, while impressionistic assessments of the effects of epidemics are notoriously misleading—anyone who read prose accounts of AIDS in the United States, when that disease first attracted general attention, would never guess that it had scant demographic effects.

The second reason has always been cited but acquired greater resonance with the arrival of structuralist approaches to the study of texts. Like any sane person, Prokopios immensely admired Thucydides, and tried to emulate his language, by then a millennium removed from the common Greek of his day. And it so happens that Thucydides wrote of the plague of his own days most poignantly in what is now edited as his book 2, in ways that Prokopios clearly strove to emulate in the text now edited as his own book 2, including its origination: "The disease began, it is said, in Ethiopia beyond Egypt, and then descended into Egypt. . . . Then it suddenly fell upon the inhabitants of the Peiraeus." Then comes the carefully qualified, very detailed description of the symptoms (". . . men were seized first with intense heat of the head . . .")[57] which

Prokopios was clearly intent on emulating. Hence his testimony is discounted as a literary exercise.[58]

Of course, it was universally accepted that there was a pandemic and a very severe one, not only because Prokopios was trusted that far, but also because all other extant contemporary and retrospective texts mention it, some in some detail.[59]

One such writer is Evagrius Scholasticus of Antioch, a well-educated lawyer who had started his elementary education in 540, a year before the pandemic began, and who lost his wife, daughter, grandson, and other relatives in a later recurrence that he himself survived. Written circa 593, his *Ecclesiastical History* offers a description of the pandemic "in its 52nd year." He starts with the origins: "It was said, and still is now, to have began from Ethiopia." Then he describes the symptoms: ". . . in some it began with the head, making eyes bloodshot and face swollen. . . ."[60]

The well-educated Evagrius also explicitly refers to Thucydides, but uncontaminated sources also depict an unprecedented catastrophe, notably the *Chronicle of the Pseudo-Dionysius of Tel-Mahre,* which was written in Syriac, or late eastern Aramaic, in northern Mesopotamia in the eighth century, but which preserves a lost contemporary source, a book on the pandemic written by the prelate and historian John of Ephesus. Under the Seleucid year 855 (= 543/544 CE) the text reads, "There was a great and mighty plague in the whole world in the days of the emperor Justinian," and then proceeds with a plangent jeremiad of laments:

> over corpses which split open and rotted in the streets with nobody to bury [them];
> over houses large and small, beautiful and desirable which suddenly became tombs . . . ;
> . . .
> over ships in the midst of the sea whose sailors were suddenly attacked . . . and became tombs . . . and they continued adrift in the waves;
> . . .
> over bridal chambers where the brides were adorned, but all of a sudden there were just lifeless and fearsome corpses . . .
> . . .
> Over highways which became deserted.[61]

The chronicler then lists the affected provinces of the empire: all the Egyptian provinces and Palestine as far as the Red Sea, Cilicia, Mysia, Syria, Iconium (Konya, central Anatolia), Bithynia, Asia (western

Anatolia), Galatia, Cappadocia. The text records what "we saw" on a journey from Syria all the way to Thrace beyond Constantinople:

[villages] . . . void of their inhabitants;

staging-posts on the roads [checkpoints, and so relay posts of the imperial courier service] full of darkness and solitude filling with fright everyone who happened to enter and leave them;

cattle abandoned and roaming scattered over the mountains with nobody to gather them;

flocks of sheep, goats, oxen and pigs which had become like wild animals . . . ;

fields in all the countries through which we passed from Syria to Thrace, abundant in grain which was becoming white and stood erect, but there was none to reap them.[62]

This is no mere literary emulation but rather the description of a demographic catastrophe of the order of five thousand or ten thousand deaths a day, just as Prokopios had written, which if true would have killed half the population of the empire in a very short time. And if that was indeed true, it would also have been an institutional catastrophe: when half the soldiers of cohesive army units become casualties, those units do not lose half their combat capability but all of it, or almost. All components of the imperial military system, tax-collection offices, central administrative commands, weapon workshops, supply depots, fortress-construction teams, warships and fleets, and army units everywhere, would have been in the same predicament, with their surviving personnel much more likely to have scattered to flee the pandemic or tend to their sick families, or shocked into immobility, or weakened by the disease, or simply demoralized, so that a 50 percent mortality would lead to a more than 50 percent incapacitation.

That therefore was the old narrative evidence, which, if true as to the dimensions of the demographic collapse, would immediately explain why Justinian's military capabilities declined so drastically from 541, irremediably ruining his ambitious plans.

But the old evidence cannot be conclusive, because it is devoid of credible numbers, and it has been rejected on that basis. As an example among many, an especially productive modern historian much relied upon in these pages has written as follows:

Scholarly orthodoxy, influenced by graphic and emotive eye-witness accounts by the contemporaries Procopius and John of Ephesus, accepts that the plague effected a catastrophic and irreversible loss of life within the Roman empire, perhaps as much as one third of the population overall and even more in Constantinople and other very large, and unhealthy cities; at best it must be seen as one cause among many.[63]

In other words, Prokopios exaggerates, and exaggeration means "perhaps as much as one third of the population." In the account of Justinian in the latest edition of the most authoritative survey of late antiquity, the principal evidence is presented—including fiscal legislation necessitated by the death of many taxpayers—but the implication is that it was just another disaster ("there were other disasters, notably earthquakes, one of which destroyed the famous law school at Berytus") whose consequences were incremental: "Justinian's difficulties were increased by a severe outbreak of bubonic plague."[64]

The new evidence, which comes in two parts, definitely proves that Prokopios was correct: it was not just another outbreak of disease, not just another disaster soon assuaged, it was a pandemic of historically unprecedented lethality.

First, a study published in 2005 contains the first definitive evidence obtained through DNA analysis that the disease of Justinian's pandemic was caused by an exceptionally virulent and exceptionally lethal biovar of *Yersinia pestis*, the bubonic plague.[65] That disease is entirely different from the plague narrated by Thucydides, indeed from any known plague till then. When *Yersinia pestis* reappeared as the agent of the Black Death from circa 1334 in China and from 1347 in Europe, some residual acquired immunity would have persisted, but it was an entirely new pathogen for the populations of the empire in 541, and therefore they had no acquired immunity as opposed to much less prevalent natural resistance.

This made the pathogen exceptionally virulent; that is, its ability to cause the disease was very high; for practical purposes, a bite from a flea carrying *Yersinia pestis* in 541 would ensure infection, which is certainly not the case with established pathogens, because many people have acquired immunities against them. Infection rates of 90 percent or more were therefore possible for people in contact with fleas, which was practically everyone in antiquity. Justinian had the disease, as did our witness Evagrius, among other survivors. For virulence is one thing, lethality another. In fact, for obvious reasons, very virulent diseases are not very lethal: common influenza biovars kill very few of their very many victims.

That would not have been true of the biovar of *Yersinia pestis* in 541 because it was entirely new for the affected population. By way of comparison, in the avian influenza outbreaks of 2003–2006, caused by the new A/H5N1 pathogen, a cumulative total of 263 humans were infected in Indonesia, Vietnam, and ten other countries—in other words, the virulence of this pathogen is extremely low, considering the many millions in contact with infected poultry in those countries and elsewhere. But the lethality of the disease was very high indeed: 158 of the infected humans died—that is, 60 percent, sixty times higher than the average lethality of cholera, a fearsome disease withal and still the greatest killer. And that, of course, is what caused worldwide alarm, in spite of the insignificant virulence of a disease that can be caught only by eating raw parts of infected animals or exchanging fluids with them, or very bad luck.

In the 541 pandemic a lethality of 50 percent or more was just as likely as with A/H5N1 recently, because the biovar in both cases was entirely new and the population therefore lacked acquired immunities. Hence what seemed unlikely, if not impossible, to serious historians—who reasonably considered mortality rates of one-third of the population exaggerated because of the much lower lethality of other known pandemics—was eminently probable, and at even higher rates than 30 percent.

A second stream of new evidence indicates that what could have happened, did happen. Climatology is now infected by partisan polemics, but ice core studies that show rising carbon dioxide levels in the atmosphere over the last ten thousand years are undisputed. An "anthropogenic" explanation recently proposed by an eminent climatologist with much persuasive evidence is that agricultural deforestation, which replaces natural greenery with bare planted fields, and increasing livestock herds, especially of methane-producing cattle, have at least measurably contributed to rising levels of carbon dioxide over the last several thousand years. In that context, carbon dioxide levels in the ice show two abrupt and drastic declines, one of which correlates with circa 541, providing external evidence of an unprecedented demographic collapse that caused the widespread reversion of cleared fields to natural greenery and the predation of abandoned cattle—imperial territories still contained populations of wolves, bears, lions, and cheetahs, and also Caspian tigers in eastern Anatolia.[66] The climatological evidence is more decisive than the archaeological evidence available so far, but the latter is perfectly consistent; a recent overview concludes:

> The expansion of settlement that had characterized much of rural and urban Syria in the fifth and early sixth centuries came to an abrupt end after the middle of the sixth century. There is evidence that housing starts al-

most ceased, although renovations and additions to houses did continue in rural areas.[67]

The latter, incidentally, are readily explained by the formation of new families out of surviving fragments of pre-plague families. Taken together, the new biological evidence and the climatological theory compel a reassessment of Justinian and his policies. He could have been just as successful in his military ambitions as he was in his jurisprudential and architectural endeavors. It was *Yersinia pestis* that wrecked the empire, drastically diminishing its military strength as compared to enemies that were less affected because they were less infected because they were less urbanized, or because they were less organized to begin with, hence less vulnerable to institutional breakdown.

Quite suddenly, with frontiers denuded of their defenders—the disappearance of coinage from Byzantine military sites on the frontiers of Syria and Arabia has long been attested, if misunderstood[68]—strongholds abandoned, once prosperous provinces desolate, its own administrative machinery greatly enfeebled, the empire found itself in a drastically altered world, in which the nomads of the steppe and the desert were greatly favored as compared to empires, and in which the less urbanized Persian empire was relatively favored also.

Still, what Justinian did would not have been done by his successors. It was his policy to totally destroy the power of the Vandal conquerors of *Africa* and he succeeded. Therefore when the native tribes started raiding from the desert and the hills of the Aurès, there was no subdued Vandal militia to resist them, let alone a dominated Vandal client-state, so the overburdened imperial army had to fight them instead. Likewise there were promising opportunities for a quietly negotiated acquisition of Italy instead of an invasion and all-out war to destroy the Ostrogothic power. As noted, the landing of Byzantine troops from reconquered Sicily to the mainland of Italy in 535 was preceded by secret negotiations with king Theodahad; there was talk of him remaining as the client-ruler of a dependent state, or the award of landed estates yielding 86,400 solidi a year—the income of 43,200 poor men. Perhaps 100,000 solidi could have done it, or a compromise on Theodahad's kingship. Justinian's successors would have done it, he did not—before the pandemic.

After it, there was no choice but to revert to the embryonic Theodosian strategy whose "diplomatic" dimension was based on a simple arithmetic of war and peace. When at peace, the empire's economy was exceptionally productive by the standards of the time, generating tax revenues that could allow the payment of large subsidies to its aggressive neighbors to keep them quiet—gold that would very soon return to the empire anyway to pay for all the goods those neighbors craved and did not know how to produce.

After the Turkic Kutrigurs of the Pontic steppe under a leader called Zabergan mounted raids in 558 that penetrated Greece and approached Constantinople, indulging in the usual outrages that allowed Agathias Scholasticus to indulge himself and his readers ("well-born women of chaste life were most cruelly carried off to undergo the worst of all misfortunes, and minister to the unbridled lust of the barbarians . . . etc. etc."),[69] Justinian called back Belisarios from retirement (he was 53) to repel them with ceremonial palace guards, three hundred veterans, and a mob of volunteers, but then took more decisive action:

> Justinian at this time was applying pressure to Sandilkh, the leader of the Utigurs [another Turkic tribe]. He made continual attempts to rouse him somehow to war against Zabergan, sending a stream of embassies and trying various means to provoke him. . . . Justinian added in his own messages to Sandilkh that if he destroyed the Kutrigurs the Emperor would transfer to him all the yearly tribute—monies were paid by the Roman Empire to Zabergan. Therefore, Sandilkh, who wished to be on friendly terms with the Romans, replied that utterly to destroy one's fellow tribesmen was unholy and altogether improper, "For they not only speak our language, dwell in tents like us, dress like us and live like us, but they are our kin [Ogur Turks], even if they follow other leaders. Nevertheless we will deprive the Kutrigurs of their horses and take possession of them ourselves, so that without their mounts they will be unable to pillage the Romans." This Justinian asked him to do.[70]

The alternative of waging war could be very successful tactically and operationally, but even in total victory the only definite result would be the cost of it, while the benefit would only be temporary, as the demise of one enemy merely makes room for another. Theophylact Simocatta, who was born a generation after Justinian and who lived to see the destruction of Sasanian Persia and its replacement by the armies of Islam, inserted the argument in a speech he attributed to a Persian envoy to Maurikios (582–602). The envoy is arguing that Rome would not benefit if Persian power is utterly destroyed:

> It is impossible for a single monarchy to embrace the innumerable cares of the organization of the universe . . . for it is never possible for the earth to resemble the unity of the divine and primary rule. . . . Therefore even

though the Persians were to be deprived of power, their power would immediately transfer to other men. . . . Sufficient proof is the insane unreasonable ambition of a Macedonian stripling: Alexander . . . he attempted to subjugate the temporal universe to a single unitary power. But, sooner than this, affairs proceeded once more divided up into a leadership of multiple tyranny, so to speak. . . . Accordingly, what prosperity would events devolve upon the Romans if the Persians are deprived of their power and transmit mastery to another nation?[71]

It is hard to imagine that the empire could have overcome the ensuing century of acute internal crises and devastating invasions without its new strategy. It generated disproportionate power by magnifying the strength obtainable from greatly diminished forces, and by combining that military strength with the means and techniques of persuasion—the essence of diplomacy, to which we now turn.

PART TWO

Byzantine Diplomacy: The Myth and the Methods

Like most myths, the myth of Byzantine diplomacy—infinitely cunning, habitually treacherous, sometimes lost in meandering intrigue—is a tale spun around a kernel of truth.[1] In the first place, diplomacy as now understood did not yet exist. Least important is that the word itself had yet to be coined by the Benedictine monk Dom Jean Mabillon in his *De Re Diplomatica* of 1681 to describe the examination of documents in order to ascertain their origin, meaning, and authenticity.[2] By way of the examination of international treaties, Mabillon's *diplomatia*—our diplomacy—acquired its present meaning, which embraces all forms of communication between states, notably the presence of resident ambassadors in foreign capitals, which in turn requires some sort of foreign-affairs secretariat to read and react to their dispatches.

That was yet another invention of the Italians of the Renaissance, whose states and statelets were routinely exchanging ambassadors by the middle of the fifteenth century; the first documented resident envoy served Luigi Gonzaga, ruler of Mantova, at the court of the Holy Roman Emperor Luis of Bavaria from 1341.[3] Italian circumstances both favored and required the novel practice of resident envoys. That a Florentine gentleman could easily mingle with fellow Italians at the papal court in Rome, that the Venetian envoy in Milan needed only one reliable messenger riding back and forth to have his dispatches answered within the week, made resident ambassadors a practical proposition. That chronically insecure Italian states, whose deadly enemies might be only a day's walk away, needed timely information on every shift in attitude, made resident ambassadors especially useful.

The circumstances of the Byzantine empire were entirely different. Until its very last years of precarious survival as an Ottoman dependency, it had no friends or foes within easy reach. On the contrary, it often had to deal with distant powers with which it shared no common language or customs, including steppe peoples still on the move. Even if there was a capital of some sort where a representative of the emperor could have resided, he could not hope to blend in with the local elite to monitor moods and decisions from the inside, nor could he hope to report his findings in a timely fashion. Instead of residing, Byzantine envoys had to travel, sometimes far indeed.

CHAPTER 3

Envoys

An extreme example of long-distance diplomacy, which fortuitously turned out to be of the greatest possible significance half a century later, was the three-year mission of Zemarchos, envoy of emperor Justin II (565–578), to the grand potentate Yabghu qagan ishtemi, who appears as the khaganos Sizaboul or Silziboulos in our Greek source. "Khaganos" was the Greek version of a chief of chiefs (themselves khans or qans), and Ishtemi was the ruler of the western division of the Türk qaganate, a very new but already vast steppe empire, routinely but misleadingly described as the Kok or "blue" Turk, which would mean eastern empire under the Turkic color code for the four directions (white is west, hence White Russia, etc).[1] Having expanded at a phenomenal rate since 552, when these early Turks revolted against their Jou-jan (or Ruan-ruan) masters in Mongolia, by the time Zemarchos reached them they had advanced right across Central Asia, engulfing dozens of nomadic tribes and the settled populations of river valleys and oasis cities.

As with both earlier and later mounted archers of the Eurasian steppe with their hardy horses and powerful composite bows, tactical effectiveness was elevated into strategic power to the extent that charismatic and skillful chiefs could unify clans, tribes, and nations to fight together instead of fighting each other. Evidently, the founder, T'u-wu in the Chinese sources, who led the 552 revolt and his son Bumin (T'u-men) the Il [regional] qagan had the required leadership talent in abundance, because father and son rapidly elevated the former serfs of the Jou-Jan or

Ruan-Ruan into the ruling class of a steppe empire of subjected Sabirs, Utigurs, Kutrigurs, Ogurs, and Onogurs, some of whom would later emerge as independent powers inimical or friendly to Byzantium, or more often both. The Yabghu qagan Ishtemi was Bumin's brother.

That was the political variable. The unchanging military parameters were that the mounted archers of the steppe did well against less-agile enemies in battles large and small, but in addition their first irruptions also terrorized civilians. That was useful, because terrorized civilians would plead, press for, or force the negotiated surrender of cities on terms to avoid massacre or at least unlimited despoliation. That is how nomads could conquer well-fortified cities without even besieging them in earnest, and that is how the Türk qaganate conquered the Central Asian cities of the silk route.

Of all the steppe empires that emerged from central and northeast Asia, this first Turkic state was to be the largest but for the Mongol six centuries later, and this episode inaugurated the alliance that would save the Byzantine empire in the next century under emperor Herakleios. The westward expansion of the Türk qaganate had inevitably collided with the northern outposts of the Sasanian Persian empire across the Amu Darya (Oxus River). After their earlier cooperation had broken down, the Yabghu qagan Ishtemi, or "Sizabul," sent envoys to Constantinople "carrying greetings, a valuable gift of raw silk, and a letter." The envoys offered direct sales of silk that would bypass Sasanian territory, Sasanian duties, and Sasanian middlemen, "enumerated the tribes subject to the Turks and asked the Emperor for peace and an offensive and defensive alliance"—implicitly aimed against the Sasanians.[2]

Silk was no longer of great interest—local production had already started within the empire. An alliance, on the other hand, was of the greatest possible interest. Until its destruction in the seventh century, the containment of Sasanian Persia, the only equal empire and more than equal in aggression, was always the highest Byzantine strategic priority and it was to take up his most welcome offer that Zemarchos was sent to "Sizabul" in August 569.[3]

It was a very long journey whose route we do not know except that it crossed the land of the Sodgians, centered on Samarkand in modern Uzbekistan. To reach it, Zemarchos with his guards and servants had to sail to the far side of the Black Sea, cross what is now southern Russia, southern Kazakhstan, and western Uzbekistan, some two thousand miles in a straight line and overland far more than that. It was still a long way from there, another thousand miles or so in a straight line, to the seat of Sizabul, in a valley of the Ek-tagh (translated as "golden mountain" in the Greek text, but probably Aq Tag or "white mountain") in the Altai region of southern Siberia where the borders of modern China, Kazakhstan, Mongolia, and the Russian Federation converge, or perhaps further south in the Tekes river valley of Dzungaria in modern China's Xinjiang Province.[4]

Our only source, Menander, reports an incident along the way to Aq Tag that may have alarmed Zemarchos but that is reassuring for us because it proves beyond any doubt that this account of momentous events in the most remote of exotic lands is indeed authentic:

> Certain others of their tribe appeared, who, they said, were exorcisers of ill-omened things, and they came up to Zemarchus and his companions. They took all of the baggage that they were carrying and placed on the ground. Then they set fire to branches . . . chanted some barbarous words . . . making noise with bells and drums, waved above the baggage the . . . boughs as they were crackling with the flames, and, falling into a frenzy and acting like madmen, supposed that they were driving away evil spirits.[5]

This is definitive evidence that Zemarchos had indeed traveled very far, because the exact characteristics of shamanistic ceremonies still practiced today in Mongolia can be recognized in Menander's account; and the text could not have been copied from Herodotus or any other literary antecedent we know of. Evidently Zemarchos included the episode in his written official report because such ethnography was already normal operating procedure for Byzantine envoys.

Zemarchos found Sizabul in a tent sitting on a golden throne, and after handing over the customary gifts proffered the emperor's military alliance and asked for Sizabul's in return. The Turkic style of war could be applied on the largest scale with organized supplies and siege engines, but its most basic building block was the band of mounted archers that could set off at a moment's notice to mount a raid large or small. That is what Sizabul did to affirm forthwith the new alliance in the most tangible way possible:

> [He] decided that Zemarchus and twenty followers and attendants should accompany him as he was marching against the Persians. . . . When they were on the march and encamped in a place called Talas [after the river of the same name Taraz in the Zhambyl region of southern Kazakhstan] an envoy from the Persians came to meet Sizabul. He invited the Roman and the Persian [envoys] to dine with him. When they arrived, Sizabul treated

the Romans with greater esteem. . . . Moreover, he made many . . . vehement accusations . . . against the Persians.[6]

That led to a bitter altercation—no doubt the intended result—and Zemarchos was sent on his way home while Sizabul supposedly prepared to attack the Persians.

Again the route cannot be determined—when Menander writes of an "enormous, wide lake" it could be either the Aral or the Caspian sea—but it was certainly a very dangerous, as well as very long, journey. In the steppe there were potentially hostile peoples once the Byzantine party journeyed beyond the limits of Sizabul's influence, and traveling as they did just above the northern boundaries of Sasanian power, Zemarchos and his men were exposed to Persian intruders sent specifically to capture them—indeed, Menander reports that Zemarchos was at one point warned of a Persian ambush ahead. He therefore sent ten of his porters carrying silk on the expected route to make it seem that he was coming next, then instead took another route bypassing the place where he believed the Persians were waiting to ambush him.

It was not until he reached the shores of the Black Sea that Zemarchos returned to the world of organized travel that Menander could describe with precision: "He took ship to the river Phasis (the modern Rioni, in Georgia), and took another ship to Trapezus (Trabzon in Turkey). He took the (horse relay) public post to Byzantium, came before the Emperor and told him everything."[7] Evidently Zemarchos redacted a detailed report that Menander was able to examine to write his own account.

The expedition of Zemarchos to the Altai or Yulduz mountains— either way some five thousand miles there and back in a straight line and perhaps twice that by land and sea travel—was an extreme case, as noted, but long-distance relations with foreign powers were more or less the norm for Byzantium, making resident embassies impractical.

It is not surprising, therefore, that no corps of professional diplomats was ever established by the empire, nor was there its necessary counterpart of a specialized foreign-affairs office. Court officials, soldiers— Zemarchos was a very senior one as *magister militum per orientem*— scholars, bureaucrats, and prelates were all sent at different times to negotiate with foreign rulers. There was a distinction of rank, however: envoys to Sasanian Persia invariably held the highest administrative rank, *Illustris,* while the Maximinus who negotiated with Attila held the lower rank of *spectabilis.*[8]

Even though diplomacy was not a exclusive profession, a sixth-century manual shows that there could still be some selection and training. There are no surprises when it comes to selection:

> The envoys we send out should be men who have the reputation of being religious, who have never been denounced for any crime or publicly condemned. They should be naturally intelligent and public spirited enough to be willing to risk their own lives . . . and they should undertake their mission eagerly and not under compulsion.

A specific attitude is recommended:

> Envoys should appear gracious, truly noble, and generous to the extent of their powers. They should speak with respect of both their own country and that of the enemy and never speak disparagingly of it.

That was a needed warning: in the sixth century almost any place that a Byzantine envoy could visit would be backward in the extreme as compared to Constantinople, or else a ruined remnant of earlier glories.

The most interesting suggestion comes at the end: "An envoy is generally tested before being sent on a mission. A list of topics is presented to him, and he is asked how he would deal with each one of them under various assumed circumstances"—scenario-based role playing, in modern parlance.[9]

Byzantine envoys had to risk their lives each time they went on a mission—that was inevitable, given the hazards of almost any navigation even in the most familiar parts of the Mediterranean, and the perils of overland travel to any power with which the empire did not share a border. The eternal logic that dictates that alliances are best made with the unfriendly neighbors of unfriendly neighbors, meant that any intervening territory between Constantinople and its allies, or potential allies to be recruited, was likely to be hostile to Byzantine envoys. If, on the other hand, the intervening territories were ungoverned—a rare condition at present that requires the term "failed state" but far more common in antiquity—then the envoy's escort would have to contend with fierce tribes, predatory nomads, and roaming bands of freebooters.

Roman western Europe was overrun at the start of the fifth century; the traditional version from the chronicle of Prosper of Aquitaine has Vandals and Alans crossing the frozen Rhine on New Year's Eve, December 31, 406 (*Wandali et Halani Gallias traiecto Rheno ingressi II k. Jan*).[10] But there were also Sueves with them, and many more Goths and Franks were already inside the crumbled frontiers. Because several powers and potentates had replaced one empire, there was much more traffic of envoys back and forth than ever before. Given the acute insecurity of the times, it was a heroic diplomacy, sometimes celebrated as such. The hagiography of Germanus of Auxerre by Constantius shows how the future saint dissuaded Goar, "most fierce king" of the especially exotic Alani mounted warriors of Iranic origin:

> Already the tribe had advanced, and iron-clad horsemen filled the whole route, yet our priest . . . reached the king himself . . . and stood before the armed general amongst the throngs of his followers. Through an interpreter, he first poured forth a prayer of supplication, next he rebuked the one who rejected him; finally, thrusting forth his hand, he seized the reins of the bridle, and thus brought to a halt the entire army.[11]

Saints have special powers, but others who had only gold to trade, or the residual strength of remaining Roman garrisons, or other allies in hand who would fight, could also tame the newcomers into new balances of power, with as little as possible by way of transferred lands or revenues.

A celebrated example is Marcus Maecilius Flavius Eparchius Avitus, wealthy landlord, Gallic aristocrat, even emperor for a year in 455. His successful epistolary negotiation with the Goths ensconced in Toulouse was celebrated by his poetical son-in-law Sidonius Apollinaris: "The reading of his page tamed a savage king . . . will future nations and peoples believe this? That a Roman letter revoked what a barbarian had conquered."[12] Avitus would die on the road while trying to flee back to his Gallic villa from Piacenza, because once-safe journeys on well-maintained, well-guarded highways had become perilous adventures.

Byzantine envoys, also, would find safety only when reaching the territory of organized powers, even if extremely hostile. For by then the principle of the absolute immunity of envoys already had ancient authority and was almost universally respected, even by the otherwise famously ferocious. Attila the Hun, whose strategy, as we saw, required frequent exchanges of envoys with all the powers within his very long reach, was thoroughly conversant with the norms regulating the dispatch and reception of envoys, which he respected even under the extreme provocation of Chrysaphius's attempt to procure his assassination.

The principle of absolute immunity was already so well established that Menander Protektor found it remarkable that even the fiercest barbarians would fail to respect it. He recounts that the probably Slavic

Antae, then still in the Pontic steppe running north of the Black Sea, and the unfriendly neighbors of the Kutrigurs, were "ravaged and plundered" by the Avars; they sent Mezamer, evidently their leading war chief as an envoy to the Avars to ransom captives and presumably to seek an accord. It seems that Mezamer was not "gracious" as envoys should be:

> Mezamer was a loudmouth braggart and when he came to the Avars he spoke arrogantly and very rashly. Therefore, [a] Kutrigur who was a friend of the Avars and had very hostile designs against the Antae, when he heard Mezamer speaking more arrogantly than was proper for an envoy, said to the Khagan [qagan, the supreme ruler], "This man is the most powerful of all amongst the Antae and is able to resist any of his enemies whomsoever. Kill him, and then you will be able to overrun the enemy's land without fear." Persuaded by this the Avars killed Mezamer, setting at nought the immunity of ambassadors and taking no account of the law.[13]

That would have been the *jus gentium,* the "law of the nations," not Roman-made law but the customary law of nations, which extended across the Byzantine-Sasanian sphere and its surroundings, but was not recognized by the Türk qaganate or the Avars. They followed different rules under the law of hospitality, still practiced today by the few remaining Bedouin nomads and mostly by the Pathans or Pashtuns of Pakistan and Afghanistan, under their overcelebrated *Pashtunwali* code (honor goes only so far in regulating human affairs, and disregards the humanity of nonfighters, including all women). Under the law of hospitality, the obligation of offering hospitality to those who seek it—with an informal limit of time—includes the obligation of protecting the guest from hazards, to the point of fighting and dying for him if necessary.

That is why when the Avars first sent a delegation to Constantinople in 558 or 560, they did so under the sponsorship of the Alans:

> They came to the Alans and begged Saronius, the leader of the Alans, that he bring them to the attention of the Romans. Saronius informed Germanus' son Justin, who at that time was general of the forces in Lazica, about the Avars, Justin told Justinian, and the Emperor ordered the general to send the embassy of the [nation] to Byzantium.[14]

In their minds, the Alans had accepted the absolute duty of protecting them, and they assumed that the Byzantines would not kill or harm them in deference to their friendly relations with the Avars—and that is why the envoy Kandikh, finding himself in the imperial palace, unimaginable splendor for a tent dweller, nevertheless felt free to boast and threaten. Equally, it is obvious that the Avars simply did know then that their envoys did not need Alan or any other protection, being amply safeguarded by the *jus gentium* of absolute diplomatic immunity. And two generations later, after having ample opportunity to know the law of diplomatic immunity, they still did not respect it. In June 623, when the qagan of the Avars was to meet with emperor Herakleios in 623 in Thrace to conclude a peace agreement in a festive setting, the qagan tried to capture him instead while sending his men on a plunder raid:

> The Chagan of the Avars approached the Long Wall with an innumerable throng, since, as it was supposedly rumored, peace was about to be made between the Romans and Avars, and chariot races were to be held at Herakleia [no relation] . . . at about hour 4 of this lord's day [5th of June] the Chagan of the Avars signalled with his whip, and all who were with him charged and entered the Long Wall . . . his men . . . plundered all whom they found outside [Constantinople] from the west as far as the Golden Gate [of the Theodosian Wall].[15]

That is a distance of sixty-five kilometers, a raid in depth but also a cover for the attempt to capture Herakleios: "The barbarian, transgressing the agreements and oaths, suddenly attacked the emperor in a treacherous manner . . . the emperor took to flight and returned to the city."[16]

The qagan of the Turks, Menander's "Sizabul," also relied on an intermediary and the derived law of hospitality when sending its first envoys to Constantinople:

> Maniakh [the leader of the Sogdians of the oasis-cities of Central Asia] said that he himself was very willing to go along with envoys from the Turks, and in his way the Romans and Turks would become friends. Sizabul consented to this proposal and sent Maniakh and some others as envoys to the Roman emperor.[17]

Although by then the Sogdian silk-route cities of Central Asia had come under his control as his qaganate expanded westward, Sizabul sent his envoys under Maniakh's protection, on the presumption that the Byzantines would not want to offend their Sogdian host, given the traditionally good relations between the two sides. They were united in their resistance to Sasanian aggression, not to speak of the very ancient acquaintance of Greeks and Sogdians—eight centuries before, in 324 BCE, Alexander the Great's companion Seleucus, who fought in India and founded a lasting dynasty, married the Sogdian Apama.

One law excluded the other. Just as a Pathan or Pashtun chief today who will protect anyone who asks for his hospitality would have no compunction in violating diplomatic immunity, when the Byzantines later on angered the Türk qagan by concurrently negotiating with their enemies the Avars, the Byzantine envoy was harassed and his life was threatened. In this sense, Avars and the first Turks of the qaganate remained exotic creatures; the Avars did not last long enough to change, but the Turks certainly evolved when it came to the rules of diplomacy. Or at least some of them did, notably the Seljuk Sultans who first conquered and then lost much of Anatolia at the cusp of the eleventh and twelfth centuries. They were the most dangerous enemies of Byzantium at the time, but in their dealings with envoys and emperors they habitually added refined courtesy to the scrupulous observance of the rules. The aggressive Sasanians had also respected the rules, but the Seljuks who had started in the savage steppe learned civility in the fullest sense—and the Byzantines reciprocated, inciting the fury of the Crusaders bent on uncompromising holy war.

On May 21, 1097, Kilij Arslan, the Seljuk sultan of Iconium (Konya), was defeated by the men of the First Crusade outside his new capital of Nicea (İznik) when he was withdrawing with his remaining forces. Left to its own devices, the city's Seljuk garrison prudently surrendered to the Byzantine emperor Alexios Komnenos (1081–1118), who infiltrated his men into the city to raise his flag over it, seizing Kilij Arslan's court, treasure, favorite wife, and children.

The Crusaders, who had fought for seven weeks and three days, suffering many casualties, were outraged at losing the opportunity to sack the city—its Christian population notwithstanding; but the eyewitness who compiled the *Gesta Francorum et aliorum Hierosolymytanorum* records their greater outrage at the treatment that Alexios Komnenos reserved for Kilij Arslan's captured wife ("Sultana") and children: because of the emperor's "evil intent" *(iniqua cogitatione)* they were protected from the Franks, royally cared for, and returned without a ransom (II, 8).

In contrast to the dangerous but civil Seljuk enemy, the most consistently useful allies of the empire during the tenth century were the wild Pechenegs of the Pontic steppe north of the Black Sea, the latest gathering of Turkic mounted archers and horse herders to arrive in that region.[18] The obscurity of their origins is such that the best source extant is a Tibetan translation of an eighth-century Uygur account, which places them in Central Asia before they too moved westward, thus coming within the sphere of Byzantine diplomacy. The Pechenegs vigorously fought the enemies of the empire for due reward, and in the next century bands of them would willingly serve in Byzantine armies. But they were evidently savages and dangerous to deal with, judging by the procedures recommended in the tenth-century manual of statecraft attributed to the emperor Constantine VII Porphyrogennetos (913–959) now known as the *De Administrando Imperio:*[19]

> When an imperial agent [delivering tribute gifts] is dispatched from here with ships of war, [to the Pontic steppe] . . . when he has found [the Pechenegs], the imperial agent sends a message to them by his man, himself remaining on board the ships of war, carrying along with him and guarding in the ships of war the imperial goods. And they come down to him, and when they come down, the imperial agent gives them hostages of his men, and himself takes other hostages of these Pechenegs, and holds them in the ships at sea, and then he makes agreement with them; and when the Pechenegs have taken their oaths to the imperial agent according to their "zakana" [customary law], he presents them with the imperial gifts . . . and returns [aboard ship].[20]

This account irresistibly evokes the precautionary rituals that attend the exchange of money for illegal drugs. By contrast, a radically different set of procedures, treaty making with the Sasanians, presumes the prior existence of protocol practices so well established that they were not themselves the object of any reported negotiation, and that, incidentally, persist till this day. Thus the bargaining for the "fifty-year" peace treaty in 561–562 was extremely intense, with every point at issue the subject of arduous negotiations. But when a comprehensive treaty was finally agreed on, both sides knew what to do:

> The fifty-year treaty was written out in Persian and Greek, and the Greek copy was translated into Persian and the Persian into Greek. . . . When the agreements had been written on both sides, they were placed side-by-side to ensure that the language corresponded.[21]

The first eleven clauses were substantive: closing off the Caspian Gates at Derbent to barbarian invaders; nonaggression by allies; trade only through specified customs posts; use of the public post by envoys and their right to trade; forcing barbarian merchants to remain on highways and pay customs duties; rejection of defectors; damage settlements for private offenses; no new fortifications except for Dara (near Oğuz of modern Turkey, the fortress city of Anastasios); no attacks on each other's subject nations; numerical limits on the Dara garrison; and double indemnities for damage inflicted by frontier cities on each other, with a cumulative time limit of one and half years for payments. The twelfth clause invokes God's grace for those who abide by the treaty and God's enmity for those who do not. That was the one-and-only God, for the Ahura Mazda of the Zoroastrians was also an only God, albeit not omnipotent. Then there is one more clause, whose language nondiplomats might dismiss as redundant and picayune:

> The treaty is for fifty years, and the terms of the peace shall be in force for fifty years, the year being reckoned according to the old fashion as ending with the three-hundred-and sixty-fifth day.

That is, with no leap year—this being the kind of detail whose significance diplomats understand, easy to overlook but possibly consequential.

At that point letters were to be sent by both rulers, Justinian and Khusrau I Anushirvan, to ratify everything that the envoys had agreed to. Prior letters of authorization for the talks from each ruler were then exhibited. But protocol required more: the texts of the two documents in Greek and Persian were now "polished" to insert language of equivalent force in each language (not "try" but "strive," not "same" but "equivalent," and so on). Then:

> They made facsimiles of both. The originals were rolled up and secured by seals both of wax and of the other substance used by the Persians, and were impressed by the signets of the envoys and of twelve interpreters, six Roman and six Persian. Then the two sides exchanged the treaty documents.[22]

The Byzantines received the Persian text and vice versa. Then an unsealed Persian translation of the Greek original was given to the Persians and vice versa. Only then was the procedure complete.

In spite of all such diplomatic professionalism, there were no professional diplomats, and the varied officers and officials who were enlisted to serve as envoys could report to any one of a number of high officials, or to the emperor himself. No official was exclusively in charge—there was no minister for foreign affairs. There had been no such thing in the bureaucratic hierarchy of the undivided Roman empire, and none was ever added. The interpreter and amateurish secret operative Vigilas, who served Chrysaphius so poorly in the failed assassination plot against Attila, was one of the many subordinates of the official at the top of the contemporary bureaucratic hierarchy, the *magis-*

ter officiorum, the master of offices. He was in charge of messengers and interpreters and *agentes in rebus,* "agents for things"—often misunderstood or misrepresented as secret agents, but actually junior officials destined for promotion, whose elite status was affirmed by laws limiting their numbers: 1,174 in the year 430 according to a law of Theodosios II, and 1,248 under Leo I (457–474).[23] Either number could have staffed a proper foreign ministry, with geographical bureaus, country desks, and functional divisions for trade and so forth. As it was, with no foreign ministry, let alone an intelligence organization, the *agentes in rebus* (or *magistriani,* after the official they served) went on to serve in all the different departments controlled by the master of offices, whose duties were exceedingly varied.

In the early fifth-century list of civil and military officials, military units and senior military officers of the eastern and of the western empire known as the Notitia Dignitatum—a bureaucratic compilation no doubt somewhat removed from realities on the ground but nonetheless illustrative—we find the diverse units, offices, and staffs under the control of the "illustrious Master of the Offices" for the eastern half of the empire:[24]

> 7 formations *(scholae)* of palace guards, which he supplied, paid, and supervised but did not command in war: First shield-bearers *(scutariorum prima),* Second shield-bearers *(scutariorum secunda),* Senior tribal guards *(gentilium seniorum),* Shield-bearing archers *(scutariorum sagittariorum),* Shield-bearing Cuirassiers *(scutariorum clibanariorum),* junior light-arms *(armaturarum iuniorum),* and Junior tribal guards *(gentilium iuniorum);*
> the quartermasters and torch bearers (a dark palace could be dangerous);
> 4 departments *(scrinia)* for records, correspondence, petitions, arrangements;
> the staff of the palace audiences;
> 15 arsenals *(fabricae infrascriptae)* for the production of shields, cuirasses, spears and other weapons.[25]

The master of offices could scarcely administer all of this by himself, what with torchbearers, letter writers, ushers and clerks, a substantial establishment of palace guards, and finally the western world's very first true factories, an entire "military-industrial complex" of fifteen arsenals that produced all the weapons and armor issued to the army. He was duly staffed by a corps of *agentes in rebus* with more of the same seconded to him, by an aide, and a number of assistants, two for the factories, three for the department of barbarian affairs, three for Oriens, one for Asiana, one for Pontica, one for Thraciae and Illyricum, one inspector of the public post, an inspector for all the provinces, and finally the one thing that belongs to a foreign ministry: interpreters for various peoples *(interpretes diversarum gentium)*.

We can see, therefore, that the master of offices could not possibly have been a proper foreign minister, for sheer lack of time. Admittedly the same was not true of the Byzantine *logothetes tou dromou,* who only inherited a few of its functions, when the position of master of offices was both stripped of its vast executive powers by the end of the reign of Leo III (717–741) and elevated to head of the Senate. That was a ceremonial remnant of the Roman Senate of old, and its head was the emperor's representative in his absence—a dangerous role if ever exercised in earnest.[26] It is reasonable to deduce that Leo III, who had come to power by forcing the abdication of Theodosios III (715–717), but then had to contend with dangerous rebellions, viewed the master of offices as too powerful and best kicked upstairs to enjoy empty honors.

Logothetes tou dromou literally means "accountant of the course," that being a literal translation and foreshortening of *cursus publicus,* the Roman and Byzantine system of imperial mail and transportation. It had its ups and downs over the centuries, but in good times it provided both a freight service *(platys dromos)* of wagons pulled by oxen and a fast service *(oxys dromos)* of horses and mules for imperial officials and their bags. As noted in Chapter 1, oxen universally sleep eight hours, chew the cud eight hours, and will only pull at two or two and a half miles an hour on level ground, so at best the *platys dromos* might move one metric tonne a hundred miles in five days, if there were a span of at least ten oxen, with as many as eighteen needed on rising roads. By contrast the mounted travelers of the *oxys dromos* could move very much faster, because there were fresh horses for them at relay and rest stations *(stathmoi).* Prokopios described how the system worked at its best:

> Horses to the number of forty stood ready at each station . . . and grooms in proportion . . . were detailed to all stations. traveling with frequent changes of the horses, which were of the most approved breeds . . . [riders] . . . covered, on occasion, a ten-days' journey in a single day.[27]

That would be 240 Roman miles or 226 statute miles or 360 kilometers, no ordinary speed for a riding traveler; even with relays of good and fresh horses, half of that was more likely. (By way of comparison, in

1860 Pony Express riders with twenty-pound mailbags covered 250 miles a day on the 1,966-mile, or 3,106-kilometer, route from St. Joseph on the Missouri to Sacramento, California, with 190 relay stations for fifty riders and five hundred horses.) Prokopios praised the system in his polemical *Anecdota* to harshly denounce Justinian for having abolished some routes, reduced the number of remount stations, and replaced horses with mules and too few of them at that, ruining the system. Actually the *dromou* functioned rather well except in times of most acute crisis, and it was one more difference between the empire and its neighbors.

For both Byzantine envoys and important foreigners invited as official guests, overland travel outside the empire was at best a very slow adventure, while inside the empire it was usually mere routine and distinctly faster.[28] The only extant detailed documentation of such a journey goes back to the fourth century (between 317 and 323): Theophanes, wealthy landowner and official from the important city of Hermopolis Megale (near today's el-Ashmunein) in upper Egypt, used the cursus publicus to travel to Antioch (today Turkey's Antakya). He started on April 6 in Nikiu (Pshati) and arrived on May 2, averaging forty kilometers a day, with a minimum of twenty-four kilometers per day when crossing the roadless Sinai desert, and a maximum of more than a hundred on the final stretch of good highway in Syria.[29]

Use of the *dromou* could make all the difference for overland travelers in a hurry. Only inordinate wealth could privately provide fresh remounts spaced out all along a journey, but much less money might corruptly buy a permit. Only the master of offices and later the *logothete tou dromou* could issue permits, and only to officials traveling on official business. Naturally there was bribery. John Lydos or Lydian, the sixth-century bureaucrat whose *On the Magistracies of the Roman Constitution* is a relentless sequence of administrative notes and yet quite droll in places, writes that the chief of the investigators *(frumentarii)* of the prefecture was always supposed to be present in the permit office

> to make a host of inquiries and find out the reason why many people are provided . . . with the so-called official authorization . . . [to] use the *cursus publicus*. These inquiries were made although the so-called *Magister (officiorum)* is also the first to sign the official authorization for the use of the *cursus*.[30]

In other words, not even the highest officials could be trusted with permits so easily marketable for large sums. These *frumentarii* incidentally have been misunderstood or misrepresented as forming an imperial security service (a Roman FBI, MI5, or DST), and thus being counterparts to the equally imaginary imperial intelligence service of the *agentes in rebus;* if so, the empire would have been poorly protected indeed, because the total number of *frumentarii* was absurdly small for such a task, not more than a few hundred.[31]

Supervision of the dromou was just one of the duties and powers of the logothete. He was in charge of the bureau that looked after visiting barbarians, which operated a guesthouse for foreign envoys *(apkrisarion),* and of the bureau of interpreters inherited from the master of offices' *interpretes diversarum gentium,* whose chief would be a *megas diermeneutes.* These posts all do suggest a ministry of foreign affairs, but then there were purely domestic duties, including the protection of the emperor, the supervision of security measures in certain provinces, and ceremonial duties. A bureau to conduct the logothete's business *(logothesion)* with a specified staff is attested from the ninth century and was renamed *sekreton* in the twelfth, with no particular reference to foreign affairs. The one thing that the logothete did not do was conduct negotiations with foreign powers, so that even if he were the emperor's advisor on foreign affairs—a subject that was bound to come up at court every day because every day some power somewhere was threatening some part of the empire—he was not an executive minister who could implement a foreign policy.[32]

So the Byzantines had no foreign minister, and no professional diplomats. Yet the distinguishing characteristic of Byzantine grand strategy from the beginning to the end was precisely its very great emphasis on the arts of persuasion in dealing with foreign powers. Persuasion is of course the essential purpose of all diplomacy, with or without the machinery of resident embassies and foreign ministries, and most of what the Byzantines did by way of persuasion had been done by others long before them, just as it is still being done by today's modern states.

To frighten off potential aggressors by threatening punishment—deterrence as we now call it—is a practice as old as humanity, with war cries and arms waving serving exactly the same function as displays of nuclear weapons did during the Cold War. So is the offer of gifts or outright tribute to buy off enemies whom it would be more costly to fight, even in the certainty of victory. That a supposedly decadent "Byzantium" had to pay off its enemies, relying for its security on cowardly gold in place of the fighting iron of the Romans in their best days, is just one of the false distinctions between the two. The evidence shows that the Romans of all periods were uninhibited by heroic pretensions: even at their strongest, from Augustus in the first century to Marcus Aurelius in the second, they preferred gold to iron whenever enemies were more cheaply bought off than fought.[33] The list of known payments is very long: the annual subsidies of the eastern government to the Huns set circa 422 at 350 pounds of gold, increased to 700 pounds in 437 and tripled in 447 to 2,100 pounds—much less gold paid to much more effect than the 4,000 pounds paid by the western government to the Visigoths of Alaric in 408 followed by 5,000 pounds of gold in 409, along with 30,000 pounds of silver, 4,000 silk tunics, 3,000 scarlet-dyed skins, and 3,000 pounds of pepper.[34] All these payments—which did not prevent next year's sack of Rome—came from the hugely diminished, hugely impoverished city of those years, illustrating just how much gold and silver had been accumulated over centuries of imperial depredations followed by even more profitable taxation. A modern historian has carefully differentiated between six categories of payments, the intentions behind them and the results. His conclusion was that the danger of unlimited blackmail because of weakness was averted, and that gold was a "flexible and cost-efficient instrument of foreign policy"—in conjunction with iron of course, for deterrence, compellence, and punishment.[35]

The Byzantines continuously relied on deterrence—any power confronting other powers must do so continuously, if only tacitly—and they routinely paid off their enemies.

But they did much more than that, using all possible tools of persuasion to recruit allies, fragment hostile alliances, subvert unfriendly rulers, and in the case of the Magyars, even divert entire migrating nations from their path. For the Romans of the republic and the undivided empire, as for most great powers until modern days, military force was the primary tool of statecraft, with persuasion a secondary complement. For the Byzantine empire it was mostly the other way around. Indeed, that shift of emphasis from force to diplomacy is one way of differentiating Rome from Byzantium, between the end of Late Roman history in the east, and the beginning of Byzantine history.[36]

The all too obvious reason for this fundamental change was the relative weakness of the Byzantine empire: its military strength was often insufficient to cope with its multiplicity of enemies. But there was also a positive reason to rely on diplomacy: the Byzantines commanded more effective tools of persuasion than their predecessors or rivals, including the Christian religion of the true orthodox faith.

CHAPTER 4

Religion and Statecraft

That almost all the Byzantines we know of were intensely devout Christians is beyond question, but so is the empire's persistent use of religion as a source of influence over foreign rulers and their nations. For the devout, there was no cynicism or contradiction in this, not even when opportunistic turncoats such as captured Turkic raiders or uncomprehending barbarians from the steppe were eagerly baptized. If it did not help them spiritually, conversion to the Byzantine religion could at least help the empire materially, and it alone was the defender of the True Orthodox Church, which was in turn the only gateway to the eternal life according to its own doctrine. To strengthen the empire was therefore to advance Christian salvation.

With its magnificent churches, stirring liturgies, melodious choirs, tightly argued doctrines, and clerics highly educated for the times, the Byzantine church attracted entire nations of converts, the Russians most importantly. Some fought the empire vigorously all the same, but others were predisposed to cooperation or even alliance by conversion, and even if they would concede nothing to the emperor as secular head of the Church, the authority of the patriarchs of Constantinople was less willingly denied, though they were imperial appointees. Even in the twilight years of the city-state that lingered till 1453, the Russians willingly accepted the guidance of notable patriarchs such as Philotheos (1364–1376).[1]

When Byzantine missionaries set out from the ninth century to convert the neighboring Bulghars, Balkan Slavs, Moravians, and the Scandinavian rulers of Kievan Rus' to greatest effect, they were saving souls from paganism—reason enough for all their efforts. But as a matter of inherent consequence, they were also recruiting potential allies. True, conversion to the Orthodox faith did not prevent strenuous warfare against the empire by the Christianized Bulgarians or Kievan Rus', but even after recognizing the Bulgarian church as autocephalous in 927, Byzantine diplomacy could and did exploit the authority of the patriarch of Constantinople over local churchmen to enlist help or at least dissuade hostility.

The Byzantines may also have benefited at times from religious inhibitions against attacks on their Christian empire. Even the inflamed Latins of the Fourth Crusade, who were about to attack, conquer, and loot Constantinople, were so inhibited—or at least their leaders feared that they were, because on April 11, 1204, on the eve of the final assault:

> It was announced to all the host that all the Venetians and every one else should go and hear the sermons on Sunday morning; and they did so. Then the bishops preached to the army . . . and they showed to the pilgrims that the war was a righteous one; for the Greeks were traitors and murderers, and also disloyal, since they had murdered their rightful lord, and were worse than Jews. Moreover, the bishops said that . . . they would absolve all who attacked the Greeks. Then the bishops commanded the pilgrims to confess their sins . . . and said that they ought not to hesitate to attack the Greeks, for the latter were enemies of God. They also commanded that all the evil women should be sought out and sent away from the army to a distant place.[2]

We cannot know what did not happen when there were no murderous prelates at hand to preach the sanctity of attacking fellow Christians, but what can be documented is the role of Constantinople as a devotional center in the diplomacy of Byzantium. It can also be shown that the city's religious credentials were deliberately enhanced as a matter of imperial policy.

When Constantine originally established his capital, it had no particular claim as a pilgrimage destination. It did contain the emperor—more than merely the secular head of the Church, for emperors could and did pronounce on doctrine—and also the ecumenical patriarch, the most senior of his clerical appointees, who followed only the bishop-patriarch of Rome in the order of precedence established at the ecumenical Council of Chalcedon in 451, before becoming the first among all Orthodox patriarchs, after the schism of five centuries later.

But as a new city Constantinople could not begin to compete with the

Christian prestige of Rome, with its many churches and seat of the successors of St. Peter, nor with Alexandria, Antioch, or Jerusalem. The patriarchates of Alexandria and Antioch came after Constantinople in the Chalcedonian order, but both had been episcopates long before Constantinople and had much older churches. The patriarchate of Jerusalem came last in precedence, but only there could pilgrims visit the sites of the birth, life, and death of Jesus, from the Church of the Nativity in nearby Bethlehem to the Holy Sepulcher near the Temple Mount. If only because their Jewish precursors had periodically journeyed from all parts of the empire and beyond to celebrate the major festivals in the temple at Jerusalem, pilgrimage was exceedingly important as an act of faith, and Constantinople could not aspire to religious significance—with its inevitable political dimension—unless it too could attract pilgrims.

That was the challenge met and overcome by emperors and patriarchs. With vast efforts and large expenses, Constantinople was turned into the Christian city par excellence, and a pilgrimage destination in the class of Rome or Jerusalem, and for long periods more visited than either.

First came the construction of churches, preeminently the new Hagia Sophia, the Church of the Holy Wisdom, preeminent mosque after the conquest of 1453, secularized in 1935 and since then Istanbul's most visited monument. The previous Hagia Sophia, already the second church on the site, was burned down in the Nika revolt of January 532. By order of Justinian the new edifice was very deliberately designed by Anthemios of Tralles and Isidore of Miletos as an instant wonder of the world, with an astonishingly vast and high dome, 31.87 meters across and 55.6 meters (182 feet) above the ground, supported by the new device of pendentives pierced by windows, so that it seems to float high above the visitor as if by magic, or miraculously. (In a further miracle of aesthetics, more than a century after the conquering Ottomans first plastered over the interior mosaics to turn the Hagia Sophia into a mosque free of forbidden images, the architect Koca Mi'mār Sinān Āġā [1489–1588] added the four tall, relatively thin cylindrical minarets that strike a perfect contrast with the massive rotundity of the original building. That clash of civilizations at least achieved a splendid architectural fusion.)

Some three hundred churches were eventually built in Constantinople, but from its first inauguration on December 27, 537, it was the Hagia Sophia, above all other attractions, that attracted pilgrims to

Constantinople. With its vast open interior under the lofty dome uninterrupted by supporting columns, its very structure mysteriously upheld by a mathematically calculated counterweighted tension that required no interior buttressing, with its entire ceiling overlaid in gold, multicolored marbles, and polychrome mosaics (an unknown art form for many visitors) and decorated with gorgeous silks as well as hung paintings, it was for many centuries and by a long measure the most impressive building in the world. To many believers, of course, it was more than that, it was a godly wonder, a fit domicile for the holy wisdom itself. After a detailed description of the innovative design and unprecedented method of construction, Prokopios of Caesarea recorded the reaction of the first visitors:

> Whenever anyone enters this church to pray, he understands at once that it is not by any human power or skill, but by the influence of God, that this work has been so finely turned. And so his mind is lifted up towards God and exalted, feeling that He cannot be far away, but must especially love to dwell in this place. . . . Of this spectacle no one has ever had a surfeit, but when present in the church men rejoice in what they see, and when they leave it they take proud delight in conversing about it.[3]

Newcomers who had known of it only from hearsay, upon arriving in Constantinople would first go to see the Hagia Sophia, no matter what was their business in the city. But many pilgrims traveled especially to worship there—and kept coming for centuries, adding to the prestige of the empire in the many lands to which they returned.

But even the most impressive architecture and opulent decorations were not as powerful in attracting pilgrims as famous saintly relics. In Orthodox as in Catholic Christianity, saints are the approachable intermediaries, each one of them apt to evoke particular local or societal loyalties—many of the faithful had and have their own special saint, to whom they turn for their most intimate worship, for whom they might give donations, and whose burial or bodily relics they will strive to visit, to show respect but also to benefit from the spiritual emanations they generate.

Relics could therefore attract devotees even from far away, enhancing the religious establishments that had them. Some began as shrines for a tomb or relic in the first place, while others acquired relics if they could afford the cost—there was a lively trade, and prices could be high because in addition to their spiritual value and the institutional prestige they conferred, relics earned income from pilgrim donations. Rulers who commanded the scene shared in all those benefits, and the Byzantine empire certainly did because its international standing was increasingly enhanced among Christians near and far by the growing accumulation of important relics in its capital. The relics are individually enumerated in a twelfth-century collection of reports about Constantinople redacted in Skalholt, in most remote Iceland.

As evidence of the attractive power of the Constantinople relics, in relating the history of the Danish kings the *Knýtlinga Saga* describes the long sojourn in Constantinople of Erik Ejegod (Ever-Good) en route to the Holy Land (he would die in Paphos, Cyprus, in July 1103); according to the *Gesta Danorum* of Saxo Grammaticus, when King Erik was preparing to leave Constantinople to resume his journey, the emperor asked him what he most wished to receive as a parting gift. Erik replied that he desired only holy relics. He was given the bones of St. Nicholas and a fragment of the True Cross, which he sent home to Roskilde and to a church in his native Slangerup in North Zeland.[4]

Not all relics were of equal value, for there is a hierarchy of saints, starting with the first disciples. Moreover, recognizable limbs always outranked fragmentary tissue. While the ultimate attraction was a splinter of the True Cross, a well-preserved arm or leg attributed to a first-league saint was very highly rated as well. Emperors as well as clerics spared no effort or expense to acquire these "helping hands for the empire,"[5] though hands were not enough: heads, arms, legs, hearts, noses, mere fragments of tissue, and indeed every part of the body other than the predictable exception, were hugely in demand. When the arm of St. John the Baptist, stolen in Antioch, arrived in Constantinople in 956 by imperial barge in the final leg of its journey, it was received by Patriarch Polyeuktos and the assembled Senate of high officials in their best robes, amidst candles, torches, and burning incense, before being taken to the palace rather than to any church, monastery, or shrine—the emperor Constantine VII Porphyrogennetos wanted its protection for himself. By the time the city fell to the Latin conquerors of the Fourth Crusade in 1204, there may have been more than thirty-six hundred relics of some 476 different saints in Constantinople, including the aforementioned arm, which is still to be seen encased in Venetian silverwork, but now unworshipped, in the Topkapi museum of modern Istanbul.[6]

Relics were most important, but its collections of especially revered religious images, or "icons," also enhanced the religious attraction of Constantinople. Except for the furiously controversial interval of iconoclasm during the eighth and ninth centuries, Orthodox ritual has always been characterized by the great devotional importance attributed to icons—depictions of Jesus, the Virgin Mary, the apostles and other saints, most often painted tablets but also portable or fixed mosaics. In this regard, the Hellenic proclivity for imagery evidently prevailed over abstract Jewish monotheism with its stern prohibition of images of gods, which still resonated in the writings of the early fathers of the Church.

As with relics, not all icons were equal. Most were just paintings or mosaics that might be appreciated for their decorative or educational value—the Byzantine-made, or Byzantine-inspired mosaics of Cefalu, Monreale, and the Capella Palatina in Sicily effectively summarize much of the bible—but themselves had no inherent sanctity. Some icons, however, were said to be miracle-working, holy emanations in themselves. Their possession conferred religious authority as relics did, and its increasing inventory of holy icons also contributed to the plausibility of Constantinople as a holy city.

The most revered of all painted Byzantine images was an icon of the Virgin Mary holding the child Jesus Christ and pointing to him as the source of salvation—the Hodegetria, "She Who Shows the Way"—supposedly painted by Saint Luke the Evangelist, the disciple of St. Paul, to whom two of the New Testament books are attributed by believers. According to what Nikephoros Callistos Xanthopoulos, in the early fourteenth century, claimed was a quote from a fragment from the sixth-century ecclesiastical historian Theodorus Lector, but most likely was a fabrication by Xanthopoulos himself, the St. Luke Hodegetria was sent from Jerusalem to Pulcheria, daughter of the emperor Arkadios (395–408). Held in the Monastery of the Panaghia Hodegetria in Constantinople, it was taken out, paraded, and even displayed on the walls of the city to ward off enemies in times of great danger, and although it survived the Latin sack of 1204, it disappeared after the Ottoman conquest of 1453.

Luke was a saint but still human, while the most sacred images of all were the *acheiropoieta*, "images not painted by hands," icons that came into existence miraculously and were miracle-working in themselves. A post-Byzantine example best conveys the intensity of faith that such images can evoke and also their political significance—a most incongruous pairing for some but not for the Byzantines. The Kazan Theotokos, "Our Lady of Kazan," whose underground hiding place was reportedly revealed to a little girl by the Virgin Mary herself on July 8, 1579, was all the more easily accepted as not man-made because Kazan was a recently conquered Tatar and Muslim city with no Christian antecedents whatever. Credited with repelling the Polish invasion of 1612, the Swedish invasion of 1704, and Napoleon's invasion of 1812, it could not defeat Japan in the disastrous 1904–1905 war, nor keep the Bolsheviks from seizing power, because the Kazan Theotokos was stolen for its bejeweled frame and reportedly destroyed on June 29, 1904, evoking perfectly accurate predictions of immense disasters to come.

In 1993, what was said to be the same icon surfaced again and was given to Pope John Paul II, who venerated the image for eleven years ("It has accompanied me with a maternal gaze in my daily service to the Church") while tenaciously trying to negotiate its return to Kazan by himself in person. That would necessitate his visit to the Russian Federation—a visit that the politically adept Polish prelate eagerly desired, and that the Moscow patriarchate and the Kremlin were determined to deny. In the end, Russian obstinacy triumphed, and the Vatican returned the icon unconditionally in August 2004. On its next feast day, July 21, 2005, according to the Western calendar, Patriarch Alexis II of Moscow and All Russia (a.k.a. ex-KGB agent "Drozdov") and Mintimer Shaymiev, the nominally Muslim president of Tatarstan, placed the Virgin Theotokos in the Annunciation Cathedral of the Kazan Kremlin.

The Byzantines would have understood and even sympathized with the cold-blooded political calculations of everyone involved in this affair—while at the same time believing most sincerely in the Theotokos.

The acheiropoieta were most important doctrinally because they reconciled the desire to possess powerful spiritual instruments with the prohibition of graven images in Exodus 20:4. By way of a compromise after deadly controversies, post-iconoclastic Orthodox doctrine condemned idol-image worship *(latreia)* while prescribing reverence *(dulia)*, such as might be given a king, though the Virgin Mary rated *hyperdulia*.

But the acheiropoieta not made by humans were different because they could effect miracles in themselves, including their own miraculous reproduction. By far the most important of these was the Mandylion—the face and neck of the living Jesus impressed on a towel originally sent to (the historical) King Abgar V of Edessa in Osrhoene by Jesus himself in lieu of a personal visit. Repeatedly conquered with the city, lost and refound in 944, the Mandylion was brought to Constantinople and solemnly installed in the great palace by emperor Romanos I Lekapenos (920–944), whom it did not help to keep his throne. There it rested as the city's premier image and relic until it finally disappeared in the sack of 1204, unlike the similar Veronica, "true image," of Jesus, which is briefly and reservedly displayed in St. Peter's of Rome on each Palm Sunday.

Holy relics and sacred images were only part of the overpowering experience that awaited pilgrim worshippers and all visitors who attended services in the great churches of Constantinople and especially the Hagia Sophia.

The first Russian historical narrative, *Povest Vremennykh Let* (Tale of Bygone Years), dubbed *Primary Chronicle* in English—an amazing mingling of fragmentary historical facts, outright fiction, devotional writing, and frolicsome ribaldry—records under the year 6495 from the creation, 987 by our calendar, the combined "multimedia" impact of magnificent architecture, gilded mosaics, candlelit icons, gorgeous priestly robes, aromatic incense, and that other great Byzantine accomplishment, liturgical choral music, still today as stirring as any music can be. A delegation had supposedly been sent by Vladimir I, the Varangian (= Scandinavian) ruler of Kievan Rus', to search for a suitable faith for himself and his people, fellow Scandinavians and native Slavs no longer satisfied by Perun the Slav thunder god or the imported Nordic deities. Upon its return, the delegation reported as follows, according to the *Primary Chronicle* under the year 6494 (= 986):

> When we journeyed among the [Muslims] we beheld how they worship in their temple, called a mosque, . . . the [Muslim] bows, sits down, looks hither and tither like a man possessed, and there is no happiness among them but only sorrow and a dreadful stench. Their religion is not good. Then we went among the [Catholics] and saw them perform many ceremonies in their temples; but we beheld no glory there. Then we went on to [Constantinople] and the Greeks led us to the edifices where they worship their God, and we knew not whether we were in heaven or on earth. For on earth there is no such splendor or such beauty, and we are at a loss how to describe it. We only know that God dwells there among men, and their service is fairer than the ceremonies of other nations. For we cannot forget that beauty.[7]

There was nothing accidental about this encounter with Orthodox religion. The year before, in 986, a Byzantine missionary, revealingly described as a "scholar" in the *Primary Chronicle*, had supposedly journeyed to Kiev to present himself and preach at Vladimir's court. Nor was he the first. Byzantine missionaries had been coming for some time—Vladimir's grandmother Olga had already been individually converted, and had herself been entertained with much pomp and circumstance in Constantinople. Vladimir's delegation was also received with the most elaborate ceremonials, in what was obviously a carefully scripted tour complete with the dramatic finale of an imperial audience:

> Then the emperors Basil [II, 976–1025] and Constantine [VIII, his brother and nominal co-emperor] invited the envoys, and said, "go hence to your native country" and thus dismissed them with valuable presents and great honor.[8]

Religious recruitment was not just a tool of diplomacy. The Byzantines were altogether too devoted to their faith not to view evangelism as their religious duty—even if it did not guarantee imperial influence over the converts. The Christianized, increasingly Slavic, Bulgarians were not a bit less troublesome than their pagan Turkic Bulghar predecessors had been, and as Christians themselves, their most successful tsars even challenged the primacy of the Byzantine emperor over the Christian world.[9]

In the case of Kievan Rus' at least, the benefits of conversion for the emperor himself were rapid and substantive. In the wake of the delegation's glowing report, or independently of it for all we know, Vladimir converted himself and his people in 988. By way of explanation, the *Primary Chronicle* unreliably recounts that he sacked the important Byzantine outpost of Cherson in the Crimea, threatening to do the same to Constantinople unless he was given Basil's sister Anna in marriage:

> After a year had passed in 6496, Vladimir marched with an armed force against Kherson, a Greek city. . . . Vladimir and his retinue entered the city, and he sent messages to the Emperors Basil and Constantine, saying, "Behold, I have captured your glorious city. I have also heard that you have an unwedded sister. Unless you give her to me to wife, I shall deal with your city as I have with Kherson." When the emperors heard this message, they were troubled and replied, "It is not meet for Christians to give in marriage to pagans. If you are baptized, you shall have her to wife."[10]

The more plausible version is different: Bardas Phokas, scion of the richest and most powerful family in the empire, disgraced high commander *(domesticus)* of the eastern armies, and a veteran soldier of heroic reputation and gigantic stature, rose in revolt against Basil II, then still young and not yet victorious, proclaiming himself emperor on August 15, 987. His own family and other aristocratic families having rallied to his cause, as did the eastern troops of Anatolia, early in 988 Bardas Phokas advanced on Constantinople. Two years earlier Basil II had been badly defeated by the Bulgarians, and the western troops he had commanded were still in a weakened condition. That left Basil II almost defenseless when Bardas Phokas invested Constantinople by sea and by land from Chrysopolis, today's Üsküdar, just across the Bosporus, and nearby Abydus (Çanakkale). All seemed lost for Basil II, but in the months that followed the outbreak of the revolt, he successfully negotiated with Vladimir I to obtain his military help:

> The Emperor fitted out some ships by night and embarked some [Rhos] in them, for he had been able to enlist allies among the [Rhos] and he had made their leader, Vladimir, his kinsman by marrying him to his sister, Anna. He crossed with the [Rhos], attacked the enemy without a second thought and easily subdued them.[11]

Thus in the spring of 988, six thousand Varangian *(Vaeringjar)* warriors came from Kievan Rus'; e destined to remain in the imperial service, they formed the initial contingent of the Varangian guard—the celebrated elite corps of imperial bodyguards that was to attract recruits directly from Scandinavia, even Iceland and, later, Saxons from England after the defeat of 1066, and Normans too.[12] Basil II personally led the Varangians against the rebels, first defeating their forces at Chrysopolis, and then again on April 13, 989, at Abydus, where Bardas Phokas himself died, apparently of a heart attack.

Vladimir was not yet baptized in early 988 when he sent the Varangians, and he may have attacked the Byzantine coastal possession of Kherson in Crimea just before converting. But in the moment of supreme crisis he did provide vital help to the emperor and the head of his church. Vladimir may have had his own purely secular reasons for helping Basil II. It has also been suggested that he had an inherited treaty obligation. True, under a treaty of 971 between Vladimir's father Svjatoslav and emperor John Tzimiskes (969–976), Svjatoslav promised to defend the empire against all adversaries. But the treaty had been signed under duress in the wake of utter defeat, and Svjatoslav himself was killed by Pechenegs before he could return to Kiev. It is not credible that the son would help Basil II merely to honor such a treaty. It is more likely that it was the process of conversion, and the resulting dialogue between the imperial court and Kiev, that provided the favorable context in which Basil II could request and obtain the troops that saved his throne.

More broadly, conversion expanded the Christian-Orthodox ambient within which the empire was at least assured of a central position. Instead of being alone in a world of hostile Muslims, inimical Monophysites, exotic pagans, and the western followers of dubious papal doctrines, by the end of the tenth century the Byzantines had engendered an Orthodox commonwealth of autocephalous churches, which were destined to increase in number.[13] That in turn widened the cultural sphere of the Byzantines, and even the market for their artifacts—in Russian museums one can still admire bright and colorful icons purchased in Constantinople.

CHAPTER 5

The Uses of Imperial Prestige

The metropolis of Constantinople with its spiritual and earthly attractions was itself a most powerful instrument of persuasion, at least before and after the miseries of the seventh and eighth centuries, when successive sieges, recurring pandemic plague, and the especially severe earthquake of 740 temporarily reduced it to a shrunken remnant. Even so, Constantinople remained the largest city within the sphere of European civilization, as it had been since the fifth-century decline of Rome's population.

It was also by far the most impressive city, with its spectacular maritime setting on a promontory projecting into a strait and its array of majestic palaces and churches. To enhance the effect, official visitors were carefully guided in their movements around the city, to expose them to its most impressive vistas and sometimes to glimpses of well-equipped soldiers on parade.

That the Byzantines were immensely proud of their capital is to be expected, but what mattered for their diplomacy was its impact on foreign visitors, and that was all the more overwhelming because so many of them came from a world of huts, tents, or yurts. We have the rarity of a report by the writer Jordanes of the reaction of the Gothic king Athanaric in the late fourth century—and that was before Justinian (527–565) added magical Hagia Sophia and much else that impressed later visitors:

> Theodosius . . . in the most gracious manner invited [King Athanaric] to visit him in Constantinople. Athanaric very gladly consented and as he en-

tered the royal city exclaimed in wonder "Lo, now I see what I have often heard of with unbelieving ears," meaning the great and famous city. Turning his eyes hither and thither, he marvelled as he beheld the situation of the city, the coming and going of the ships, the splendid walls, and the people of divers nations gathered like a flood of waters streaming from different regions into one basin. So too, when he saw the army on parade he said "Truly the Emperor is a god on earth, and whoso raises a hand against him is guilty of his own blood.

That was the intended effect, and the text of Jordanes—supposedly an abridgment of a lost collaborationist history by Cassiodorus, who served the Gothic king Theodoric—duly records that even after Athanaric's death his whole army continued in the Roman service, "forming as it were one body with the imperial soldiery."[1]

Its names alone show that the prestige of the city was immense and far-reaching. To the Slavs nearby in what is now Bulgaria and Macedonia, or farther away in Russia, Constantinople was Tsargrad, the "city of the emperor," the capital of the world, even the outpost of God on earth. In distant Scandinavia and most remote Iceland, it was Miklagard, Mikligardr, or Micklegarth, the "great city" immensely admired in the sagas.

The emperor himself was the focus of elaborate court rituals performed by officials in resplendent robes, to better overawe foreign envoys at court. If that was not enough, there was a period when hydraulic machinery elevated the imperial throne just as visitors approached, and activated lions that stamped their tails and roared convincingly enough to shock and awe the unprepared.[2] That was little more than childish foolery, but there was much preparation and careful stage management in the dealings of Byzantine emperors with the envoys of the many and varied powers, nations, and tribes they encountered over the centuries, including non-Christians and schismatics unmoved by their religious authority. Much of what they did was calculated to preserve and enhance the prestige of the imperial court even as it was being exploited to impress, overawe, recruit, even seduce. Unlike troops or gold, prestige is not consumed when it is used, and that was a very great virtue for the Byzantines, who were always looking for economical sources of power.

The court was thus an instrument of persuasion in itself, as well as many other things: it was the sole focus of political, legislative, and administrative power; the site of the treasury from which gold flowed out to the civil and military servants of the emperor, and also to foreign allies, clients, auxiliaries, and sometimes plain blackmailers; the palatial setting of an unending cycle of private and public ceremonies enlivened by the resplendent silk robes of high officials, each signifying a rank; the ideal destination of ambitious youths from all over the empire in search of official careers—some of them especially castrated to join the eunuchs of the palace. At times the court was also the venue of artistic, literary, and scholarly endeavors, but always it was the seat of the emperor himself, sacred to Orthodox Christians as God's secular vicar on earth, and the most important man on earth for many non-Christians too, both nearby and very far.

For visiting potentates and chieftains who had only known the rude pleasures and brutish manners of wooden halls, yurts, or rough-hewn forts, the Byzantine palaces and court with their stately audiences, processions, and ceremonies must have been unimaginably impressive, startling visions of unearthly elegance.. Detailed accounts of how foreign potentates were received are contained in an invaluable compilation of court ceremonial and much else attributed to the emperor Constantine VII Porphyrogennetos, which is usually known by its Latin title *De Cerimoniis Aulae Byzantinae*, but is here cited as the *Book of Ceremonies*.[3]

Of particular interest is the reception of Muslim envoys in 946; they knew not only huts and tents but also the monumental Umayyad mosque in Damascus, the exquisite (and purely Byzantine) Dome of the Rock, and the court in Baghdad, and were not so easily impressed. They came in the name of the Abbasid caliph, still the supposed ruler of all Islam, but by then the caliphate was entirely powerless, and the envoys who arrived in May and then in August 946, to discuss truces and prisoner exchanges, represented less cosmic but very real powers: frontier warlords and more substantial regional rulers. Among the former was the emir of Tarsos or Tarsus in Cilicia (near modern Turkey's Mersin) on the empire's southeast frontier, whose summons to jihad were sometimes heeded far and wide across the Muslim world; his fellow jihadi and competitor the emir of Amida (Diyarbakir in modern Turkey, Amed in Kurdish) facing the empire's east-central frontier; the altogether more powerful Buyid or Buwayhid *(Āl-i Būya)* ruler Ali, heterodox Shi'a military potentate from Iran who had just seized control of Baghdad, whose great strength was the sturdy infantry of his fellow Daylami highlanders;[4] and Ali Abu Al-Hasan ibn Hamdan, of the very heterodox Nusayri or Alawite sect better known by his soubriquet Sayf ad-Dawlah, "Sword of the Dynasty," meaning of the caliphate, but in fact founder of his own Hamdanite power in Syria, whose eventual defeat marked the ascent of Byzantine fortunes during the tenth century.

(He remains famous among Arabs, but mostly as the sometime patron of the supremely gifted, irreverent, and pugnacious poet Abou-t-Tayyib Ahmad ibn al-Husayn, universally known as "al-Mutanabbi," the would-be prophet, for he so proclaimed himself at one point in just one of his wild escapades).

From the *Book of Ceremonies* we learn just how elaborate were the preparations for receiving these Arab envoys.[5] Existing palace furniture and decorations, magnificent enough for other visitors, were deemed insufficient, and therefore wreaths, silver chandeliers, a golden plane tree set with pearls, embroideries, hangings, and other ornaments were borrowed from churches and monasteries, while the Hagia Sophia and the great Church of the Apostles contributed their well-robed choirs to the proceedings. That too was not considered decorative enough, so the eparch, or prefect of the city, borrowed additional ornaments from travelers' hostels, old-age homes, more churches, and the shops of the silversmiths; he was also given the more normal task of supervising the decoration of the processional route through the city and the Hippodrome.

When the moment came, there was a row of imperial standards on either side of the steps leading up to the palace; the chief oarsmen held two standards and the commander of the Hetareia palace guard held the emperor's own preeminent gold-embroidered silk standard. Inside the palace, Roman scepters, diptychs, and military ensigns were ranged on either side of the throne; and the borrowed silver organs of the green and blue circus factions were added to the golden imperial organ. Silk draperies transformed the arboretum into a strolling reception area, while precious robes, enamel, silverware, Persian carpets, laurel wreaths, and fresh flowers added to the display. The floors were strewn with laurel, ivy, myrtle, and rosemary with roses in the principal reception hall.

The degree of magnificence of the robes of court officials was strictly determined by their rank; but on this occasion less exalted officials were given more resplendent higher-ranking robes, and even humble palace servants down to the bath attendants, literally "soapers," the *saponistai,* were kitted out in fancy capes.[6]

The emperor Constantine VII Porphyrogennetos did not entrust such grave matters to his officials—he personally intervened to provide especially sumptuous robes for the Muslim envoys as well, whose collars were encrusted with "precious stones and huge pearls":

> It is against the rules for a non-eunuch . . . to wear a collar like that, either with pearls or with precious stones, but for display, and for this one occa-

sion only, they were directed to wear these ornaments by Constantine the Christ-loving lord.[7]

This particular episode can be interpreted in two diametrically opposed ways: is Constantine with his antiquarian passion lost in foolish ritualism? Or is it a calculated psychological move to dress up the Muslim envoys as well, so as to engulf them in the splendid celebrations, instead of leaving them out as shabby spectators? Both is the only right answer, especially considering what ensued after the first grand reception: many days passed without any actual negotiations. Instead there was a banquet enlivened by the two choirs, except for intervals of organ music as each course was served. When the envoys rose, they received gifts in gold and in kind, and there were tips for their retinues.

The envoys were next entertained at the Hippodrome with a special performance, the Feast of the Transfiguration on August 8 was celebrated with extra pomp, and there was another full-dress banquet on August 9, with a variety show. It was standard protocol at the time to include eighteen Muslim prisoners in the emperor's Easter Sunday and Christmas banquets, no doubt with symbolically proselytizing intentions—at different times, Muslim prisoners were variously executed, mutilated, tortured, or held in very decent conditions to be exchanged, with an apparent evolution toward better treatment—although in the year 995 the Mu'tazilite theologian 'Abd al-Jabbar bin Ahmad al-Hamadhani al-Asadabadi (d. 1025) bitterly complained:

> During the early years of Islam, when Islam was strong and they were weak, they used to take care of their war prisoners, so they could exchange them. . . . But [later, when stronger, they] disregarded the Muslims, insisting that the rule of Islam stopped to exist.[8]

That was wild exaggeration. The shift in the balance of power in favor of the Byzantines during the tenth century was a matter of degree, while the prisoner exchanges (fida') had started in Umayyad times from circa 805.[9] As for the custom of allowing some prisoners to dine in banquets, this is first attested in the *Kletorologion* of Philotheos, circa 899.[10] Forty were seated for the August 9 banquet with the two envoys of the emir of Tarsos—a prisoner exchange was being negotiated. Again there were after-dinner gifts: five hundred silver miliaresia of 2.25 grams for each of the two envoys; three thousand for their retinue, and a thousand for the forty prisoners and banquet guests, and a sum was also sent to the other prisoners not invited to the banquet. The total value of all these gifts was not great, but they did help to instill the idea that it was more enjoyable and more profitable to negotiate with the emperor than to fight him.[11] For the Muslim envoys of 946 themselves, it was obvious that only further negotiations could again give them access to the court with its gifts and its banquets. Moreover, Byzantine prestige was enhanced more broadly by the wide circulation of reports by the envoys, who were obviously greatly impressed.[12]

Once seen and experienced, the life of the court was not willingly given up without first securing an entitlement to experience it again. There were amenities, comforts, banquets, decorous entertainment, literary declamations at times, the ladies could wear their best at their own occasions, and there was always gossip, educated conversation, guarded talk of policies, and furtive talk of politics.[13]

Above all, there was the immanent presence of power, whose magnetic attraction is felt by all in some degree and scorned only by those with no access to it anyway. In contemporary Washington, D.C., even able people accept poorly paid positions in the executive office of the president for the sake of its immediate proximity to the seat of power, even if they are unlikely to see the president in the flesh from one year to the next. White House identity cards are often worn outside the office in apparent forgetfulness, casually dangling in full view. And in the quest for office, even expensive professionals eagerly donate their services to presidential candidates during the interminable electoral campaigns. In the court of Constantinople the attraction of power was much greater because it was a power unlimited by laws, regulations, audits, parliamentary interventions, or judicial review: the emperor could castrate, blind, behead, and provide succor; promote to any position and demote and exile; give the most valuable gifts and confiscate, endow a man with a rich estate or take away all his possessions. From an individual perspective, that was infinitely more power than any U.S. president can have.[14]

There was therefore a great striving to attain access to the court from every part of the empire and from foreign parts also, as chieftains and princes came asking for support against their enemies foreign or domestic, or they came for entertainment and for ceremonial gifts, while others came in quest of titles and offices with accompanying emoluments— a steady income from the most reliable source that then existed. In exchange, all these claimants offered all manner of things, military alliances or just the temporary loan of their forces, job lots of warriors for the emperor's guard, or just their own body and loyalty for military service. That is how the emperor Justin I, uncle and patron of Justinian, started his career—if Prokopios at his least trustworthy is to be trusted, for he was eager to denigrate Justinian, and humble origins had not yet become a desirable trait in public life:

> When Leon was holding the imperial power in Byzantium [circa 462] three young farmers, Illyrians by race, Zimarchus, Dityvistus, and Justinus from [Bederiana], men who at home had to struggle incessantly against conditions of poverty and all its attendant ills, in an effort to better their condition set out to join the army. And they came to Byzantium, walking on foot, and themselves carrying cloaks slung over their shoulders, and when they arrived they had in these cloaks nothing more than the [dry biscuit] which they had put in at home; and the Emperor enrolled them in the ranks of the soldiers and designated them for the palace guard [the newly established *Excubitores,* a select unit of 300]. For they were all men of very fine figure, three fine-looking men.[15]

Coming from the hamlet of Taurisium, near the fort of Bederiana, far from Constantinople, near today's Macedonian capital of Skopje, the three were presented as starveling rustics and barbarians to the intended readers of Prokopios—Zimarchus and Dityvistus were Thracian names—but they were certainly not foreign barbarians, because Justin's language was Latin, at any rate what passed for Latin in Bederiana.

Many foreigners also came to guard emperors from their domestic enemies, and to fight for the empire, and not only hungry young peasants like Justin—the gold to be had at the imperial court was certainly a powerful incentive to serve, even for well-fed chieftains. Before the discovery of the vast gold deposits of the Americas, Siberia, Transvaal, and Australia, gold was altogether more rare than it is today, and correspondingly more valuable in relation to other goods. Only the emperor in Constantinople could command a steady supply, derived from the circulation of fiscal gold, gathered into his treasuries as tax payments and then paid out in salaries that ultimately generated the money incomes that would in turn be taxed.

The very currency of the empire was a source of prestige. From its first issue by Constantine (306–337) until its debasement under Romanos Argyros (1028–1034), the *solidus* (whence our term *soldier*) the later Nomisma, was the preferred currency of traders much beyond imperial frontiers, because of its constancy: it was struck at 72 to the Roman pound for a weight of 4.544 metric grams of 955–980/1,000 gold. That too was a rarity:. The emperor's solidi were almost pure gold. To the author of the *Saga of Harald Hardrade,* collected and edited by Snorri Sturluson (1179–1242) for his chronicle of the kings of Norway now known as the *Heimskringla,* it was enough to see a significant amount of gold to know where it must have come from. Two kinglets are robustly vying for leadership by displaying their wealth in gold:

> Then Harald had a large ox-hide spread out, and turned the gold out of the caskets upon it. Then scales and weights were taken and the gold separated and divided by weight into equal parts; and all people wondered exceedingly that so much gold should have come together in one place in the northern countries. But it was understood that it was the Greek emperor's property and wealth; for, as all people say, there are whole houses there full of red gold. The kings were now very merry. Then there appeared an ingot among the rest as big as a man's hand. Harald took it in his hands and said, "Where is the gold, friend Magnus, that thou canst show against this piece?" All that king Magnus could produce was a single ring.[16]

This anecdote is evidence even if the episode never happened (why would Magnus enter the competition with only one ring to show?), because Harald, son of Sigurd, nicknamed Hardråde ("hard ruler"), was a fully historical figure who definitely found gold in Byzantium. Born in Norway in 1015, he died in battle at Stamford Bridge in what is now Greater London in 1066, in a failed attempt to conquer England just before his distant Norman kin tried it with better luck. In between, Harald had lived in Kiev as a warrior captain for its ruler prince Yaroslav, had served as an officer of the Varangian guard in Constantinople, and had successfully returned to claim the throne of Norway after a brief detention in France: held as a suspected marauder because of all the gold he was carrying, he was released when a letter arrived from Constantinople confirming that the gold was his severance bounty.

Foreigners frequently attacked the empire in the hope of seizing some of its gold or extracting it in tribute, and as often served the empire loyally to earn its gold. But there was also another attraction: the possibility of acquiring imperial titles refulgent with the immense prestige of the imperial court, some of which came with an annual salary and precious robes of office, with or without the obligation to perform civil or military duties.[17] The craving of foreign chieftains for titles and robes is discussed in *De Administrando Imperio.* Otherwise full of good and hardheaded advice on how to deal with foreign powers, the text is deliberately misleading in a rather silly way on this particular issue:[18]

> Should they ever require and demand, whether they be Chazars, or Turks, or again Russians, or any other nation of the northerners and Scythians, as frequently happens, that some of the imperial vesture or diadems or state

robes should be sent to them in return for some service or office performed by them, then thus you shall excuse yourself.[19]

What follows is a tedious peroration claiming that God himself sent the robes of state and diadems for exclusive use by the emperor on festival days, so that they cannot possibly be handed over. It is the "as frequently happens" that gives the game away: titles and the robes of office that went with them were routinely given to "northerners and Scythians" for services rendered, and of course the emperor's own robes were neither requested nor given.

Salaried titles without duties, sinecures that is, became annuities in modern terms when sold to raise capital sums, and they could be especially valuable gifts for useful foreigners. But even titles that came with no position or salary or robes of office were much in demand, for they signified imperial recognition and an implied promise of some continued access to the court with its banquets, ceremonies, and entertainments. *Patricius,* for example, a rank once reserved for the more ancient families of the first Rome, was also available to especially favored foreigners by the seventh century. But no single honorific title could possibly have sufficed to accommodate the great diversity of competitive claims for honors. The *Book of Ceremonies* lists a great many titles suitable for foreign potentates. Derived from all sorts of antecedents, some are easily decoded and others not:

Exousiaokrator, exousiarches, exousiastes [variations on "outside" ruler]; archon of archons, archegos, archegetes, archon, exarchon [from an ancient term for ruler or high official, loosely: "prince"]; pro(h)egemon, hegemonarches, hegemon, kathegemon [variants of overlord]; dynastes, prohegetor, hegetor, protos, ephoros [Spartan overseer]; hyperechon, diataktor, panhypertatos, hypertatos, koiranos, megalodoxos [great rule-giver]; rex [king]; prinkips [Roman *princeps* = first citizen, the title that Augustus favored by way of dissembling his vast powers, later "prince"]; doux [*dux,* regional commander, later duke]; synkletikos, ethnarches [tribal chief]; toparches [same]; satrapes [originally a Persian governor]; phylarchos [tribal chief]; patrarchos, strategos, stratarches, stratiarchos, stratelates [four variants of "general"]; taxiarchos, taxiarches [infantry formation commander]; megaloprepestatos [magnificent]; megaloprepes, pepothemenos, endoxotatos [most esteemed]; endoxos, periphanestatos, periphanes, peribleptos, peribleptotatos [variations on distinguished]; eugenestatos, eugenes [two versions of well-born]; ariprepestatos, ariprepes, aglaotatos, aglaos, eritimotatos, eritimos, gerousiotatos, gerousios, phaidimotatos, phaidimos, kyriotatos, kyrios [both "lord"]; entimotatos, entimos, pro(h)egoumenos, hegoumenos [currently abbot];

olbiotatos, olbios, boulephoros, arogos, epikouros, epirrophos, aman-tor.[20]

This great diversity was obviously useful, because it hopelessly con-fused the hierarchy of ranks. If a chieftain proudly bearing the mag-nificent title of *megaloprepestatos* encountered a most distinguished *megalodoxos,* both could feel that they had received the greater honor from the emperor, and both could therefore feel impelled to show the greater loyalty.

The imperial court could benefit from a confused disorder of titles but its elaborate ceremonials needed clarity and order. These could hardly be improvised, because for each ceremony many people had to be in the right place at the right time, and in the right order of precedence. A strict protocol was accordingly imposed on all things, including the ex-act wording of official greeting and welcoming statements. They could not just be made up on the spot without risking misunderstandings that might even be dangerous. Unless it was their purpose to cause offense, the many foreign envoys who came to the court needed help to prepare their formal statements, and learn their steps for the elaborately staged ceremonials; and such help was duly provided.

The *Book of Ceremonies* preserves the text of the salutations to the emperor that were expected from visiting envoys and potentates, com-plete with spaces for the appropriate names, and the text of the replies prescribed by protocol. The salutations, which must have required a fair amount of drilling beforehand to avoid errors, imply the use of the Greek language by all, evidently through interpreters when needed from papal envoys:

> The foremost of the Holy Apostles protect you: Peter the keyholder of heaven, and Paul teacher of the nations. Our spiritual father [name] the most holy and ecumenical patriarch, together with the holiest bishops, priests and deacons, and the whole clerical order of the holy Church of the Romans send you, Emperor, faithful prayers through our humble persons. The most honored princeps of the elder Rome with the leading men and the whole people subject to them convey to your imperial person the most faithful obeisance.[21]

The emperor is too exalted to reciprocate the greeting. The logothete does so for him—this being the *logothetes tou dromou,* in charge of dealings with foreign envoys, as we saw:

> How is the most holy bishop of Rome, the spiritual father of our holy Em-peror? How are all the bishops and priests and deacons and the other

clergy of the holy church of the Romans? How is the most honored [name] princeps of the elder Rome?

This last being a piece of antiquarianism in true Byzantine style, or perhaps a deliberately slighting reminder of Rome's reduced condition, for of course there had not been an emperor there to protect the pope for half a millennium.

Next came the greetings of the envoys of the ruler of the Bulgarians, for centuries the most important neighbors of Byzantium and often the most dangerous, especially after their conversion to Orthodox Christianity—for then Bulgarian rulers could even contest the imperial throne as competing defenders of the faith. In the *Book of Ceremonies,* compiled when the Bulgarian state was becoming more powerful, its envoys were instructed to use a greeting that was specifically meant to deflate the pretension of the Bulgarian ruler that he was the emperor's equal:

> How is the Emperor, crowned by God, the spiritual grandfather *(pneumatikos pappos)* of the Prince *(archon)* by God *(ek theou)* of Bulgaria? How is the empress *(augousta)* and mistress *(despoina)*? How are the emperors, the sons of the great and high Emperor, and his other children? How is the most holy and ecumenical patriarch? How are the two Masters *(magistroi)*? How is the whole senate? How are the four Logothetes? [The *logothete tou dromou,* in charge of the postal service and dealings with foreign envoys, the *logothete ton oikeiakon,* in charge of the civic economy and security of Constantinople, the *logothete tou genikou* in charge of taxation, and the *logothete tou stratiotikou,* chief paymaster.]

The logothete's reply to the Bulgarian envoys again suggests that its ruler's status is subordinate to that of the only true emperor in Constantinople: the ruler of Bulgaria—though claiming to be an emperor—becomes a "grandson" and the Byzantine emperor, his nominal grandfather:

> How is the spiritual grandson *(pneumatikos engonos)* of our holy Emperor, the ruler by God of Bulgaria? How is the Princess *(archontissa)* by God? How are the *Kanarti keinos* and the *Boulias tarkanos,* the sons of the ruler by God of Bulgaria, and his other children? How are the six great Boyars *(Boliades)*? How is the common folk?

Since 945, as noted above, the most important Muslim potentate for the Byzantines was the Ali ibn Hamdan or "Sayf ad-Dawlah." Muslim envoys could hardly be expected to invoke the favor of Jesus and his apostles for the emperor, but they too were drilled into a well-mannered greeting that made good use of the common foundation of Jewish monotheism in both religions:

> Peace and mercy, happiness and glory from God be with you, high and mighty Emperor of the Romans. Wealth and health and longevity from the Lord, peacemaking and good Emperor. May justice and great peace rise in your reign, most peaceful and generous emperor

The logothete's responsive greeting was elaborately polite:

> How is the most magnificent *(megaloprepestatos)* and most noble *(eugenestatos)* and distinguished *(peribleptos)* Emir of the Faithful? How is the Emir and the Council *(gerousia)* of Tarsos? . . . How are you? How were you received by the Patrician and General of Kappadokia? [the Byzantine authority in the territory that envoys from Syria would have had to cross] How were you treated by the imperial aide *(basilkos)* who was sent to look after you? Did anything unfortunate or distressing occur on your journey? Leave cheerfully and delighting in the fact that today you dine with our holy Emperor.

The reference to events "unfortunate or distressing" is perfectly understandable. To reach Constantinople from Sayf ad-Dawlah's capital of Aleppo overland, the envoys had to cross the frontier zone, the arena of raids and counter-raids, ambushes, surprise attacks, robbery, and livestock rustling by frontier forces, bands of jihadists, wild borderers, roaming bandits, and smugglers—except that these were highly interchangeable categories.

The text continues with greetings and responses from and to the envoys of the emirs of Egypt, Persia, and Khorasan—corresponding to parts of modern northeast Iran, northwest Afghanistan, Tajikistan, Turkmenistan, and Uzbekistan—among other rulers.

One can well appreciate the psychological purpose of these ceremonial exchanges. With almost all the powers involved, tension was almost constant and armed conflict very frequent. Then as now, Muslim rulers obedient to their religion had to view all non-Muslim states on the planet as part of the land of war, *dar al harb,* that Muslims were destined to conquer before the day of redemption. Hence no permanent peace *(salaam)* with a non-Muslim power could be, or indeed can be, religiously legitimate. The Muslim claim on Byzantine lands was accordingly unlimited. All that was allowed to the believers was the interruption of war for a truce *(hudna),* a temporary, pragmatic arrangement to gain time, for a week, year, or generation—until jihad could be resumed. But while a *hudna* lasted there were negotiations to be conducted, and both sides had an interest in conducting mutual relations with civility, which was duly achieved in spite of the ferocity of the fighting before and after.[22]

It had been no better before Islam on the empire's Mesopotamian frontier, where the Sasanians were constantly dangerous and periodically launched vast offensives, including the last from 603 onward that succeeded in wrecking both empires with fatal effect.

As for the empire's northern, Danubian, or Balkan front—imperial frontiers moved north or south with the balance of power—by then fully Christian Bulgaria was no better as a neighbor. When they were really powerful, its tsars were not content with partial territorial gains and tried to claim the Byzantine throne and all the empire for themselves. Other enemies that preceded or alternated with Bulgaria—the Huns, Avars, Kievan Rus', Magyars, Pechenegs, and Cumans—could be almost as dangerous, even if they had no pretensions on the imperial throne.

So when envoys arrived at court, war with their principals had just ended, was still under way, or could imminently begin. It was just as well to begin talks with an exchange of pleasantries before plunging into the negotiations at hand, with their inevitable recriminations and implied or outright threats. The language prescribed by court protocol was rigidly formal and hardly encouraged spontaneous exchanges, but it could at least prevent unintended slights and embarrassing gaffes.

CHAPTER 6

Dynastic Marriages

Even without a diplomatic service or foreign ministry, the Byzantines could and did exploit every tool of diplomacy, and this naturally included dynastic marriages intended to cement relations with powerful foreigners.[1] That had not been a Roman practice, for lack of valid counterparts, but for the Byzantines there was the precedent of the dynastic marriages between the rival Hellenistic autocracies established by the successors of Alexander the Great. Initially ruled by his direct subordinates and then by their descendants or near enough, these Greek-speaking kingdoms not infrequently made peace agreements by marriages, though more frequently they warred, with or without divorces.

Matters were rather more delicate for the emperor of the Romans. For himself, for a sister, or for his palace-born children, intermarriage with lesser mortals was inconsistent with the claimed position of the emperor as God's viceroy on earth and overlord presumptive of all Christians, who must exist on a higher plane than all other rulers. Besides, the notion of consigning the daughter or sister of an emperor to the bed of a barbarian, howsoever Christian, or to a nomad's tent, even if filled with golden treasure, or worse still a Muslim harem, was revolting, offending both Greek racial pride and Christian propriety.

Things were easier when emperors or their sons married the daughters of foreign potentates. Justinian II, dubbed the "slit-nosed" (*rhinotmetos)*, who ruled from 685 only to be dethroned, symbolically mutilated, and exiled to the remote outpost of Cherson in Crimea in 695, formed a dynastic alliance with the Khazars who ruled the adjacent steppe. He married the sister of the qagan, Busir Glavan (Ibousiros

Gliabanos to the Greeks), who took the name Theodora—though it was with the help of the Bulghar qan or khan Tervel that he eventually regained the throne in 705 to misrule until 711, when he was overthrown again.

A century later, Leo III (717–741), to seal his alliance with the steppe empire of the Khazars against the Muslim Arabs, whom they separately vanquished on their respective fronts, arranged the marriage of his son and successor Constantine V (741–775) to the qagan's daughter, who took the name Irene—her son and his successor, Leo IV (775–780), was nicknamed "The Khazar." Incidentally this Irene is remembered for two rather contrary accomplishments. The first was that, upon embracing Christianity, she acquired a reputation for intense piety. Under the year 6224 since the creation, that is, 731/732 CE, Theophanes Confessor records: "In this year the emperor Leo [III] betrothed his son Constantine to the daughter of the Chagan. . . . He made her a Christian and named her Irene. She learned Holy Scripture and lived piously, thus reproving the impiety [iconoclasm] of those men."[2]

Her second accomplishment was that she introduced to the Byzantine court her national dress, a well-decorated caftan—the horse-nomads' long coat that can be opened in front to mount the horse—which came to be called *tzitzakion* at the Byzantine court. Starting out as nomadic outerwear, it migrated to the very summit of middle-Byzantine court costume, for the *tzitzakion* was worn by the emperor himself and only on the most solemn occasions. This was explained much later by Constantine VII Porphyrogennetos (912–959), himself a keen antiquarian: "You must know that the *tzitzakion* is a Khazar costume that appeared in this God-protected imperial city since the empress of Khazaria."[3]

In spite of this precedent, the official version was that the imperial family would not marry into lesser ruling families, no matter how great their pretensions. No requests were anticipated from religiously inimical Muslim powers; the steppe powers were in no sense anti-Christian, but they too were to be refused. In *De Administrando Imperio* there is the crib of a suggested reply to fob off such requests

> [if] any nation of these infidels and dishonorable tribes of the north shall ever demand a marriage alliance with the emperor of the Romans, and either take his daughter to wife, or to give a daughter of their own to be the wife to the emperor or the emperor's son.

To this "monstrous and unseemly" demand, a typically arch reply is suggested:

[A] dread and authentic charge and ordinance of the great and holy Constantine is engraved upon the sacred table of the universal church of the Christians, Hagia Sophia, that never shall an emperor of the Romans ally himself in marriage with a nation of customs differing from and alien to those of the Roman order, especially with one that is infidel and unbaptized . . .[4]

Nothing could be more categorical—except that what directly follows is an exception:

. . . unless it be with the Franks alone; for they alone were excepted by that great man, the holy Constantine, because he himself drew his origin from those parts . . . [and] because of the traditional fame and nobility of those lands and races.

That was entirely spurious—Constantine never left instructions on marriage, and in any case he was born in Moesia Superior (now southern Serbia) whereas the Frankish confederacy emerged in the lower Rhine valley—but the fiction did justify dynastic alliances with the strongest power of west, the Francia of Charlemagne and his descendants, then the East Francia that became the *Regnum Teutonicum,* the Kingdom of Germany, in the tenth century with the Ottonian dynasty.

In 781 Irene, widow of Leo IV "The Khazar" (775–780) and regent for her only son, the ten-year-old Constantine VI, arranged his betrothal to Rotrud, the six-year-old daughter of Charlemagne, still "king of the Franks" and not yet crowned emperor, as he would be in 800, but already the ruler of much of western Europe. There was as yet no significant friction between the two empires, but with Charlemagne still expanding his reach and increasingly active in Italy, collisions were highly predictable, because the Byzantines still possessed the southern coastal enclaves of Naples, Reggio in Calabria, and Brindisi in Puglie, and also Venice as the residue of the extinct exarchate of Ravenna, and the port towns of the Dalmatian coast of the Adriatic—though Istria at its head already belonged to the Franks. A precautionary dynastic alliance with the most powerful western potentate since Roman times was certainly prudent.

Eschewing the barbarian sound of "Rotrud," the Byzantines named her Erythro and sent the eunuch Elissaios to educate her in the Greek language and court manners. But in 786, when she was still only eleven, the formidable and scheming Irene broke off the engagement for reasons unknown—as for Constantine VI, he would end his life deposed and blinded by will of his mother.

In the absence of a dynastic alliance, relations with Charlemagne did not prosper, although direct warfare was avoided till much later.

Charlemagne's acceptance of the title of *Imperator Augustus* at his crowning by Pope Leo III on Christmas Day, December 25, 800, was a direct challenge to Byzantine supremacy, regardless of his own intentions. His official biographer Einhard or Eginhard or Einhart, monk, Frankish historian, and Charlemagne's dedicated courtier, entirely blamed Pope Leo III for the deed:

> The [Roman populace] had inflicted many injuries upon the Pontiff Leo, tearing out his eyes and cutting out his tongue, so that he had been compelled to call upon the King for help. Charles accordingly went to Rome, to set in order the affairs of the Church . . . and passed the whole winter there. It was then that he received the titles of Emperor and Augustus [*Imperator Augustus*], to which he at first had such an aversion that he declared that he would not have set foot in the Church the day that they were conferred, although it was a great feast-day, if he could have foreseen the design of the Pope. He bore very patiently with the jealousy which the Roman emperors [of Constantinople] showed upon his assuming these titles, for they took this step very ill; and by dint of frequent embassies and letters, in which he addressed them as brothers, he made their haughtiness yield to his magnanimity, a quality in which he was unquestionably much their superior.[5]

It is true that the pope and the Roman Church had a more urgent need of a western emperor to protect them than Charlemagne had need of a title—by then his personal preeminence and his hegemony within continental western Europe were both unchallenged. The recent emperors of Byzantium had become heretical in Roman eyes because of their iconoclasm, but their even greater offense was that they were too far away to safeguard the popes from the savagery around them, not all of it barbarian—it was a Roman gang sent by disgruntled relatives of his noble predecessor Adrian I that attacked the commoner Leo III, driving him to escape to Charlemagne.

The Byzantine view of Charlemagne's coronation, as a calculated political act by both sides, is much more plausible:

> [After he was attacked, Pope Leo] sought refuge with Karoulos, king of the Franks, who took bitter vengeance on his enemies and restored him to his throne, Rome falling from that time onwards under the authority of the Franks Repaying his debt to Karoulos, Leo crowned him emperor of the Romans in the church of the holy apostle Peter after anointing him with oil

from head to foot and investing him with imperial robes and a crown on 25 December.[6]

Irene, effectively emperor from 797 to 802 as regent for her son, would not compromise the imperial primacy by recognizing Charlemagne as Imperator Augustus. What next ensued is both attested by the best source for the period and also hard to believe:

> In this year, on 25 December, . . . [800] Karoulos, king of the Franks, was crowned by Pope Leo. He intended to make a naval expedition against Sicily, but changed his mind and decided instead to marry Irene. To this end he sent envoys the following year.[7]

Moreover, the long-anticipated territorial conflict had started over Venice and its surroundings—Istria on the other side of the Adriatic had already been claimed by Charlemagne's father, Pippin III, in 789.

Irene's successor, Nikephoros I (802–811), reached a peace agreement in 803 but still refused to recognize Charlemagne's imperial title. Fighting later resumed and continued until under emperor Michael I Rangabe (811–813) a new peace agreement was reached in 812 whereby Venice and Istria were returned to the empire, and an imperial title was allowed to Charlemagne: not *Imperator Augustus* or *Imperator Romanorum* but at least the awkward and temporary-sounding *Imperator Romanorum gubernans imperium,* or Emperor of the Romans Governing an Empire; Charlemagne and his secretariat were content with *Imperator et Augustus* plain, and *rex* of the Franks and Lombards, leaving "Emperor of the Romans" to Michael I and Byzantium.[8]

That Frankish marriage never took place, but others did. Most notably, emperor John Tzimiskes (969–976) agreed to wed Theophano, proffered as his niece, to the son of Otto I, king of Germany and Italy, the future emperor Otto II. Negotiations had begun under his predecessor, Nikephoros II Phokas (963–969), who had spurned the proposal, provoking the acerbic tit for tat of Otto's irascible negotiator, Liutprand of Cremona, who also wrote a polemical account of the negotiations.[9] This was more than a dynastic marriage, it was a *strategic* marriage, an integral part of a war plan.

Under his predecessor Nikephoros II Phokas, the two empires had been colliding in Italy, but Tzimiskes wanted to resume the offensive at the opposite extremity of the empire, against the Muslim Arabs. The marriage of Theophano and Otto was celebrated in Rome on April 14, 972, apparently putting an end to confrontation in the west. In the same year, Tzimiskes launched his successful campaign to drive back the Muslim Arabs. Ioannis Scylitzes is brief: "The cities which . . . had been appropriated by the Emperor [Nikephoros] and made subject to the Romans had now kicked up their heels and thrown off Roman domination; so the Emperor set out against them and advanced as far as Damascus."[10]

There would be many more dynastic, strategic, and increasingly exotic marriages with powers old and new. Isaac I Komnenos (1057–1059) married Catherine of Bulgaria, a daughter of the long-dead tsar Ivan Vladislav; Michael VII (1071–1078) went much farther afield to marry Maria of Alania, daughter of King Bagrat IV of Georgia of the millennial Bagration clan—and she was also taken as legitimizing spouse by Michael's successor Nikephoros III Botaneiates (1078–1081), who overthrew her ex-husband (who was generously allowed to retire as a monk, thereupon starting a new career that culminated with his installation as metropolitan archbishop of Ephesus).

Ioannes II Komnenos (1118–1143) also went far afield, marrying Piroska—civilized into Irene—daughter of King Ladislaus I of Hungary, gaining nothing thereby but entanglement in Hungarian quarrels; Manuel I Komnenos (1143–1180) married Bertha of Sulzbach, sister-in-law of Conrad III of Germany, and after her death in 1159, Maria of Antioch, daughter of Raymond of Antioch, a French noble from Aquitaine.

All such distant connections were exceeded by Michael VIII Palaiologos (1259–1282), reconqueror of Constantinople from the Latins, and the Ulysses among emperors for his unending series of stratagems. In addition to seven legitimate children, including his successor Andronikos II (1282–1328), he had two known illegitimate daughters, both of whom he married into the geographically most expansive empire in history. By 1279 the successors of Temujin, the Cinggis Qan or Genghis Khan (Oceanic Ruler) of the Mongols had conquered eastward even the southern part of China as well as Korea, westward as far as Hungary, and southwest from Central Asia into Afghanistan, Iran, and Iraq. Everywhere agile Mongol horsemen outmaneuvered superior numbers to inflict devastating defeats, as in the battle of Wahlstatt ("battlefield") near Liegnitz in historic Germany (now Poland's Legnica) known to every German schoolboy; there on April 9, 1241, Henry II the Pious was killed along with most of his Polish, Moravian, and Bavarian forces and a few Knights Templar and Hospitaller by what was imagined to be the Mongol army, but was only a secondary column. Others were wiser: by 1243 the Seljuk Turks, who had been fighting Byzantium for almost two hundred years, became obedient Mongol vassals. Those who resisted were destroyed: an army under Temujin's grandson Hülegü destroyed both the Ismailis of Syria and the remnants of the Abbasid caliphate, sacking and ruining Baghdad in 1258.[11]

Consolidation rapidly ensued on both sides of Constantinople as the descendants of Temujin—Cinggis Qan organized enduring states that must be defined as "Cinggisid" rather than simply Mongol, because they recruited increasingly from local populations, remaining Mongol only in their higher leadership, and not for very long.

In the east, as subordinate *il-qan* of the ruler of all the Mongols, Hülegü established a state that stretched from what is now western Afghanistan to eastern Turkey by way of Iraq, encompassing all of Iran; this il-qanate also dominated the Seljuk rulers in Anatolia, who became its subjects to avoid destruction. On the other side of the Caspian and Black seas, the entire vast expanse of the steppe from what is now Moldavia all the way east to what is now Uzbekistan, and north to encompass much of Russia came under the domination of the western army or "horde" (from *orda*, Mongol for "camp," hence the chief's camp, and his army).[12] It is still remembered today by all Russians as the *Zolotaya Orda*, the Golden Horde—a later blanket term for the successive Mongol and Turkic powers that collected tribute from Russian towns and potentates as late as 1476, and whose last remnant was the Giray qanate of Crimea, which lingered until 1783. When first established, the Mongol state dominated the peoples of Central Asia, the Volga Bulghars, and the Qipchaq of the Pontic steppe north of the Black Sea known as Cumans to the Byzantines, as well as the Russians even north of Moscow.

Mongol raiders from both Cinggisid states reached imperial territory, but the same Michael VIII Palaiologos who would discomfit Charles d'Anjou by supporting Peter of Aragon at the other end of the Mediterranean, was fully up to the challenge. His illegitimate daughter Euphrosyne Palaiologina was successfully married to Nogai, son of Baul son of Jochi son of Cinggis Qan himself, indefatigable commander of the western army who never claimed formal leadership but dominated the western Orda all the same.

His other illegitimate daughter, Maria Despina Palaiologina, was betrothed to a greater man than Nogai, Hülegü the destroyer of Baghdad, but upon his death married instead his son and successor Abaqa or

Abakha, another great-grandson of Cinggis Qan and Hülegü's successor as ruler of the il-qanid state. Though widely separated, the two sisters were therefore married to husbands who were relatives.

Driven by the dynamics of intra-Mongol competition, both Cinggisid states were expansive, at least in directions where there was grass for horses (which spared both mountainous central Europe and Egypt), and so their forces collided in the Caucasus, where the two powers naturally met.[13]

It was not all-out war but only a jurisdictional dispute, in theory at least, because all territories under Cinggisid control across twelve thousand miles of Eurasia were supposed to be the collective possession of the clan of Temujin's descendants; but Nogai was leading his men as usual and lost an eye in fighting the forces of his bother-in-law Abaqa. The reaction of the two sisters is not recorded.

Michael VIII Palaiologos had certainly succeeded. Neither daughter was merely lost to the harems of busy warriors. Both delivered. At one point Nogai Qan provided four thousand horsemen to fight for Michael in Thessaly; more important, no power to the north could freely contemplate attacks on the emperor without fearing a visitation by Cinggisid outriders.

As for Abaqa Qan, he tried to convert his Muslim subjects to Buddhism, the pacifist religion that the warlike Mongols somehow found most congenial. Maria Despina Palaiologina was an influential figure, and neither the Seljuks nor other Turkish chieftains could attack her father with impunity in Anatolia. In Istanbul, in the Fener quarter facing the Golden Horn, stands the only Orthodox church that was not converted to a mosque after the conquest of 1453—Panaghia Muchliótissa, "All Saints of the Mongols," rebuilt by Maria Despina when she returned to Constantinople after Abaqa's death. Whatever else may be said of them, the Byzantines were not provincial.

CHAPTER 7

The Geography of Power

Given the linear mentality of Romans and Byzantines—they thought in terms of routes from place to place rather than spaces, and relied on itineraries rather than maps—this chapter should perhaps be entitled "the ethnography of power." Curiosity about foreign peoples was a Greek virtue that the Romans did not really share until they evolved into Byzantium. In spite of the modern academic fashion for seeing nothing but hostility and prejudice in them, Byzantine writings show that they were greatly interested in foreign cultures and customs, as entire nations are not even today.[1] True, new information about foreign peoples was filtered through an accumulation of prior myths—including Gog and Magog, the Amazons, and the noble savage forever being reinvented to castigate softness and worse. But Byzantine soldiers also collected much real information about enemy tactics and weapons, while Byzantine envoys reported assiduously on the great variety of peoples they encountered, with sufficient accuracy to be the principal source on many of them. Christianity certainly helped to combat prejudice—not only because of its universal embrace but also because it dissuaded its followers from bathing, and therefore removed the barrier of smell that greatly inhibited Roman intimacy with barbarians.

The *Book of Ceremonies* of Constantine Porphyrogennetos specifies how the recipients of official correspondence were to be addressed under contemporary protocol rules, and the value of the seal for each letter (thousands of Byzantine seals are still preserved, all that remains of as many lost documents). The long list of appellations illustrates the vast geographic scope and detailed reach of Byzantine diplomacy.[2] Not counting intermittent contacts with powers much farther away in Asia, the Byzantine diplomatic horizon reached a thousand miles east from Constantinople to the Caspian shore, more than a thousand miles westward across Europe, more than five hundred miles north to Kievan Rus', and as far south as Egypt.[3]

The order of precedence in the *Book of Ceremonies* in part reflected the hierarchy of real power, and in part was set by traditional protocol—hence the pope in Rome comes first:

> To the Pope of Rome *(eis ton papan Romes)*. A one-solidus gold bull. "In the name of the Father and of the Son and of the Holy Spirit, our one and sole God. [Name] and [Name], Emperors of the Romans, faithful to God, to [Name] most holy Pope of Rome and our spiritual father *(pneumatikon patera)*."

For the patriarchs of Alexandria, the patriarch of Antioch, and the patriarch of Jerusalem, the wording is the same but with the omission of "our spiritual father." On the other hand, their letters were sealed with three gold solidi.

The first of the secular rulers, in what clearly was an order of precedence, was the Abbasid caliph in Baghdad, supposedly ruler of all Islamic lands everywhere but by then reduced to a mere figurehead for competing regional powers—both emirates that conceded a nominal authority to the caliph, and sultanates and rival caliphates that did not. When the text was written, the Hamdanid emirate headquartered in Aleppo was by far the most important Muslim power for the Byzantines. By then the empire was more feared by the Muslims than in fear of them, so there was no danger that courtesy would be misunderstood as a sign of weakness:

> To the first counselor *(protosymboulon)* of the Emir of the Faithful [*Amermoumnes* = leader of the believers]. A four-solidus gold bull.
> To the most magnificent and most noble and distinguished [Name] First Counsellor and Guide of the Agarenes [= Arabs, from Hagar, Abraham's repudiated concubine] from [Name] and [Name] faithful Autocrats, Augusti and Great Emperors of the Romans [Name] and [Name] whose faith is in Christ the Lord, Autocrats, Augusti and Great Emperors of the Romans to the most magnificent, most noble and distinguished [Name] First Counsellor and Guide of the Agarenes.

After the Muslims come the Transcaucasian rulers. The peculiar topography of the Caucasus, whose deep valleys are separated by high mountains impassable in winter and scarcely trafficable even in summer, certainly favored cultural, linguistic, and political fragmentation to an extreme degree.[4] To this day the region remains the uneasy home of many diverse populations with sharply differentiated languages, religions, and somatic features. If each had its own state, there would be many more than the present Armenia, Azerbaijan, Georgia, and the seven Caucasian Republics of the Russian Federation—of which the largest, Dagestan, with two and a half million people, contains ten nationalities rated as major though thirty languages are commonly spoken. Indeed, there already are more unrecognized states, including Abhazia, Southern Ossetia, and Nagorno Karabak.

In Byzantine times as in our own, Caucasian powers fought each other over very little territory, with warriors motivated by overblown versions of identity politics as well as a bandit ethos that remains ineradicable. With a well-placed tower of roughly hewn stones and a band of warriors, any petty chief could become the ruler of a segment of his own valley, while chiefs who could dominated an entire valley were on their way to becoming princes. Some Caucasian rulers were significant potentates while others were precarious chieftains, but the Byzantines could not afford to ignore even minor rulers, because any one of them could open or close a passage through the mountains that might make all the difference in war. All were vulnerable to assault or siege, but it was much better to avoid time-consuming combat by offering soothing communications and gifts.

During the long Byzantine struggle with Sasanian Persia, which had ended three centuries before, the Caucasian rulers could also be foes, if only because they were as much Persian in material culture as Christian in religion. The empire collaborated more easily with them against the culturally alien Muslim Arabs. Most Caucasian rulers were conversant with Byzantine culture, many were personally acquainted with the attractions and rewards of the Constantinople court, and while some had their own autocephalous churches for which they claimed an earlier antiquity, they had no religious inhibitions in accepting the primacy of the empire in every other way.

Accordingly, most Caucasian potentates acted in the twin capacity of native ruler, usually with the Armenian title *isxan* (*archon* in protocol Greek, loosely equivalent to our *prince*), and concurrently as imperial officials, often with the high rank of *kouropalates* (mayor of the palace). That was the third highest of all ranks during the ninth and tenth centuries, preceded only by *caesar*, which was mostly reserved for the emperor's family, and *nobilissimus*.

When the *Book of Ceremonies* was compiled, one Caucasian ruler was grand indeed: the kouropalates David III, or Davit' to the Georgians who claim him, more often known in history as David of Tao or Taron or Tayk, of the Armenian-Georgian Bagratid family. Rulers of Armenia, of Iberia to the north (now in western Azerbaijan), and later of Georgia, the highest Bagratids held the rank of kouropalates. They ruled territories mostly in modern Georgia and Armenia from circa 966 to David's murder in the year 1000, when his territories were absorbed into the Byzantine empire; their descendants emerged again as local rulers, eventually becoming nobles of the Russian empire until the Bolshevik revolution.[5] Although the text was compiled just before he came to prominence, David of Tao would have fitted the next prescribed form of address better than any contemporary Caucasian ruler:

> To the Prince of Princes *(archon ton archonton)* of Great Armenia.
>   A three-solidus gold bull.
>   "Constantine and Romanos, whose faith is in Christ the Lord, Autocrats, Augusti and Great Emperors of the Romans to [Name] most renowned ruler of Great Armenia and our spiritual son."

There was also a specifically Armenian ruler of the Artzuni (or Ardzruni) family who ruled Vaspurakan, not in the Caucasus at all, but to the south of it, in modern Turkey all around Lake Van. The rulers of Vaspurakan were under some degree of Bagratid suzerainty most of the time but they were addressed as independent rulers. It was not only overlords who warranted their own specified forms of address but also the many petty rulers who divided contemporary Armenia between them, the *archontes* of Kokovit, Taron (which was destined to expand), Moex, Auzan, Syne, Vaitzor, Chatziene, and "the three Princes of the Servotioi, *who are called black boys (maura paidia)."*

North of Armenia was Iberia, the Greek and Roman name for the ancient Georgian kingdom of Kartli, whose inhabitants still call themselves Kartveli as opposed to the Mingrelians, Laz, and Svans of other parts of Georgia. Its ruler also had the high rank of kouropalates.

Historic Iberia was not larger than modern Belgium or Taiwan, but given Caucasian proclivities it was still too large to have a single ruler, so the four archontes of Veriasach, Karnatae, Kouel, and Atzara were also recognized. Indeed, the entire Caucasian region is not larger than Greece but it was divided by several other states in addition to historic Armenia and Kartli, or Georgia: Alania, which corresponds more or less to modern Ossetia within the Russian Federation; Abasgia, more or less modern Abkhazia, whose secession from Georgia is recognized only (disinterestedly no doubt) by the Russian Federation; and Albania within the modern Republic of Azerbaijan.

Even today the eastern side of the Caucasian region contrives to be more fragmented than the western portion, whether in the Dagestan Republic with its thirty languages or Azerbaijan with its enclaves. Hence the archontes of many more localities were also recognized, including Azia, the modern Derbent of Dagestan where a Sasanian fortress still stands, "where the Caspian Gates are located."[6] Petty ruler he may have been, but the archon of Azia controlled the strategic passage par excellence, the easy coastal route between the southern steppe and northwest Iran.

After the many petty Caucasian rulers, the list continues with ecclesiastics of Christian churches not in communion with the Orthodox church headed by the patriarch of Constantinople: the Katholikos of Armenia (still the title of the head of the Armenian Apostolic Church—not in communion with Rome—who resides in Echmiadzin in the Republic of Armenia); the Katholikos of Iberia (the predecessor the modern patriarch of the Georgian Apostolic Autocephalous Church); the Katholikos of Albania (the lapsed title of an extinct church).

Now comes an interesting repetition, with more elaborate salutations to the already saluted pope in Rome, and a richer seal of two solidi of gold instead of one. There is a ready explanation. The *Book of Ceremonies* was not a coherent authorial work but rather a compilation of extracts from archival documents, and all the more valuable for that. In this case, the compilers inadvertently included salutations from two different letters, probably redacted on different dates.

In western Europe, power was much less fragmented than in the Caucasus, if not much more stable. The post-Carolingian fragmentation that would continue for centuries with the formation of ever smaller states had not progressed very far by the tenth century:

> To the King [*rex*, a title much inferior to *Augustus-Basileus*] *of Sazonia* [= Saxonia = eastern Germany];
>   to the King of Vaioure "this is the land of those called Nemitzoi" [Slavic for "Germans"]
>   to the King of Gallia [in 987 Hugh Capet, Count of Paris and Duke, was crowned King of France—a much smaller dominion than modern France];
>   to the King of Germania [Otto I, of the Liudolfing family, who started his own Ottonian dynasty; he was crowned in 936 by the archbishop of Mainz, the primate of Germany, in Charlemagne's Aachen cathedral, in a clear sign of imperial aspirations[7]].

Protocol for all the aforementioned. "In the name of the Father and of the Son and of the Holy Spirit, our one and sole true God. Constantine and Romanos, Emperors of the Romans, faithful to God, to [Name] distinguished King, desired spiritual brother *(pepothemenos pneumatikos adelphos)*."

That was before the declared schism between the Churches of Rome and Constantinople, allowing a full spiritual brotherhood, if so desired. It was only on February 2, 962, after the compilation of the *Book of Ceremonies,* that the scandalous Pope John XII, much in need of protection, accorded the imperial title *(romanorum imperator augustus)* to Otto I in Rome. Ten days later, Otto gave John what he needed: a written security guarantee for the papal territories, the *Diploma Ottonianum.* The emperor promptly marched out of Rome to fight the enemies of the pope but was too successful: Pope John XII now feared for his independence and secretly sent envoys to the still-pagan Magyars and to Constantinople asking them to fight Otto. The secret leaked out and Otto returned to Rome on November 963 to convene a synod of bishops that deposed John.

The offense to the emperor in Constantinople of crowning another in Rome was undiminished by the pope's almost immediate second thoughts, but in 972 John I Tzimiskes (969–976) recognized the title of the most powerful ruler of the west. Otto's convincing argument was that from 966 he had been sending his allies to attack Byzantine possessions in southeast Italy, Langobardia Minor, today's Calabria and Puglie, which had been regained from the Muslim Arabs in 876. Local Byzantine forces under the strategos of each region handily defeated these attacks, but John I Tzimiskes was preparing a large offensive against the Hamdanids of Aleppo, and instead of diverting his forces to fight Otto I in strategically secondary Italy, he preferred to reach an agreement, including a dynastic alliance.

Next we encounter a salutation to a nonexistent ruler: "To the Prince [*prinkips*] of Rome." When the *Book of Ceremonies* was redacted there was no *princeps* or emperor in Rome and there had not been one for more than half a millennium; it is not clear if this was mere antiquarianism or a barbed reminder to the pope.

"To the Emir of *Africa*" (*Ifriqiya* to the Arabs, the Roman province of *Africa,* and the modern Republic of Tunisia); *Africa* was ruled by the Aghlabids until 909 and then by the Zirids—Berbers serving the Fatimid caliphate of Egypt. Christianity was proudly reaffirmed in greeting these Muslims:

Constantine and Romanos, whose faith is in Christ the Lord, Autocrats, Augusti and Great Emperors of the Romans to the most esteemed *(endoxatatos)* and most noble *(eugenestatos)* Exousiastes of the Muslims." A two-solidus gold bull.

"To the Emir of Egypt." That was the *Ikhshid,* or governor, under the suzerainty of the caliph in Baghdad until the Fatimid conquest of 972. As Ismaili "Sevener" Shi'a, the Fatimids claimed the title of caliph for themselves and would have rejected the subordinate title of emir.

After that, two Italian powers were recognized, one especially obscure—"the Prince *(archon)* of Sardania," that is, Sardinia, but at the time there were four independent rulers on the island; and the other destined for splendid wealth and glorious renown: "the *doux* [= *dux, doge*] of Venice."

Venice started on its upward path from village-state to city-empire as a Byzantine dependency. Its Byzantine *doux* or local governor had evolved almost imperceptibly into the *doge* of an independent maritime empire by the time the *Book of Ceremonies* was compiled. In 723–727, Venice was still ruled from the Ravenna headquarters of the Byzantine exarch ("outside ruler," viceroy). Next there is the first Venetian ruler, Ursus (until 738), with the old Roman title *dux* (originally combat leader, later regional commander, eventually duke), who is followed by Dominicus, Felix Cornicula, and Deusdedit, son of Ursus, and three others, including Deusdedit, again until 756, all of whom have the Byzantine military title of *magister militum.* In between, the exarchate of Ravenna was overrun and extinguished in 751. No more Byzantine titles are recorded in coins or inscriptions after 756, even though Byzantine power persisted until at least 814. Only after that did the doges of Venice emerge as heads of a fully independent oligarchical republic with increasingly important overseas outposts and ventures: in 1204 the Venetians, under Doge Enrico Dandolo (1192–1205) had the lion's share in the sack of Constantinople by the assembled forces of the Fourth Crusade.

When the *Book of Ceremonies* was redacted, and for another nine hundred years after that, Italy remained a geographic expression without unitary government. There were instead local potentates, both backwoods lords, indistinguishable from bandit chiefs, and the more urbane governments of the major cities. One was Amalfi, then a considerable maritime republic, which would remain independent until the Norman conquest of 1073. Hence the princes of Capua, Salerno, Amalfi, and Gaeta are recognized along with the *dux* of Naples.

Next comes a more exotic and much larger power, the Khazar qaganate:[8]

> To the [*Chaganos*] of Khazaria. A three-solidus gold bull. "In the name of the Father and of the Son and of the Holy Spirit, our one and sole true God. Constantine and Romanos, Emperors of the Romans, faithful to God, to . . . [insert Name] the most noble *(eugenestatos)* most renowned *(periphanestatos)* Qagan of Khazaria."

Centered in the lower Volga region on the northeast shores of the Caspian Sea but expansive enough during its epoch of power from the eighth to the tenth century to collide at times with Byzantine interests in both the Crimea and the Caucasus, the qaganate of the Khazars (Hebrew: *Kuzarim*) was more often the prime strategic ally of the empire.[9]

What made the Khazars very valuable as allies was that they were the direct neighbors, not of the empire but rather of its greatest enemy at the time, the invading Muslim Arabs in the Caucasus and the Levant. The Khazar qaganate was well placed to blunt their power by flanking attacks when the Muslim Arabs were themselves attacking the eastern Anatolian borders of the empire, from what is now the Kurdistan of Iran. The Khazars could also help indirectly, by attacking Arab rear areas when the Muslim Arabs were threatening Constantinople by sea from their Syrian or more forward bases.

The Khazars emerged from the disintegration of the great Türk qaganate circa 640–650—it was finally defeated by the Chinese in 659—but fragmentary evidence suggests that they were not subjected tribes reverting to independence, but themselves the core elite of the Türk qaganate: that is, the ruling Ashina ("blue" in eastern Iranian) clan mentioned in Chinese sources. That could perhaps explain why the Turkic allies who provided vital help to emperor Herakleios (610–641) against Sasanian Persia in the supreme crisis of 626–628, when the empire came very close to extinction, are described as "eastern Turks, who are called Chazars" in the *Chronicle* of Theophanes,[10] the best extant source. It is more likely, however, that it was a simple anachronism: the Khazar qaganate was prominent in his time, whereas the Türk qaganate had disappeared. In any case their leader, "Ziebel" in Theophanes, is undoubtedly Tong Yabghu, head of the western Türk qaganate as such, or conceivably an emerging Khazar—because there is no doubt that the latter did come out of the former.

It has recently been suggested that the exceptionally rapid collapse of the Türk qaganate—immense and expanding circa 625, already disintegrating circa 640—was due to a climatic event: a dramatic cooling caused by a volcanic eruption whereby, in 627–629, the eastern Turkic empire experienced severe cold, snow, and frost. Many horses and sheep died, causing a great famine that ruined the pastoral core of the Türk qaganate.[11]

What is certain is that the empire continued to rely greatly on its Khazar alliance, notably against the Umayyad Arab caliphate at the height of its power in the seventh and eighth centuries, when Constantinople itself was attacked. Arab sources report the annihilation of the army of Abd al-Rahman ibn Rabiah in 650 when it invaded Khazar territory in the north Caucasus region, and Khazar forces penetrating as far as Mosul in northern Iraq in 731. Herakleios may have offered his sixteen-year-old daughter Eudokia in marriage to Tong Yabghu, and Justinian II certainly married a qagan's sister, while Constantine V married a qagan's daughter, birthing Leo IV "The Khazar."

Long after the Sasanians and then the Umayyads had left the scene, the Khazars periodically reemerged as useful allies, even as the power of their qaganate waned from the late ninth century. The final collapse came in 969, when the Khazar capital of Itil or Atil on the Volga was destroyed by the forces of Svyatoslav, son of Igor of Kievan Rus'.

They came next in the list of salutations in the *Book of Ceremonies* (book XXX):

To the prince of Rhosia. A two-solidus gold bull.
"Letter *(grammata)* of Constantine and Romanos, Christ-loving emperor of the Romans, to the archon of Rhosia."

That was a Scandinavian, or Slav, or mixed ruler—according to the period, or the national sentiments of the historians who describe them (but the derivation of *Rus* from the Old Swedish *roper,* via Old Finnish *rotsi,* has been much contested to little avail).[12] For the Byzantines it did not matter if the archon in question was post-Scandinavian or proto-Russian; he ruled the state known as Rhosia or Kievan Rus' in today's Ukraine, whence came our Russia or their own later Rossiya—a power almost unknown to the empire until the year 860, when a great number of boats bearing warriors suddenly arrived from the Black Sea to attack Constantinople. The existence of the Rhos themselves as people was certainly known already, because when first attested, by Prudentius of Troyes in the *Annals* of St. Bertinian in 839, they had come from Constantinople:

In 839, an embassy from Emperor Theophilus [830–839] arrived in the court of Louis the Pious at Ingelheim, accompanied by some men who claimed that they belonged to the people called Rhos *(Rhos vocari dicebant)* and who asked Louis' permission to pass through his empire on their way back home. This matter was thoroughly investigated at the Carolingian court, and the Frankish emperor came to the conclusion that they belong to the gens of Swedes.[13]

The attack of 860 was a typical Viking (= "robber" in Old Norse) long-range raid, a concentrated onrush of violence designed to paralyze resistance, in a manner sadly familiar to the coastal populations of western Europe but an utter surprise to the Byzantines. We have an eyewitness reaction in the homily of the patriarch and future saint Photios, whose shock is palpable in spite of the mellifluous style of a literary critic:

> Do you recollect that unbearable and bitter hour when the barbarians' boats came sailing down at you, wafting a breath of cruelty, savagery and murder? When the sea spread out its serene and unruffled surface, granting them gentle and agreeable sailing, while, waxing wild, it stirred against us the waves of war? When the boats went past the city [the Bosporan current at work] showing their crews with swords raised, as if threatening the city with death by the sword? . . . When quaking and darkness held our minds, and our ears would hear nothing but, "the barbarians have penetrated within the walls, and the city has been taken by the enemy? For the unexpectedness of the event and the unlooked-for attack induced, so to speak, everybody to imagine and hear such things—a symptom that is indeed common among men in such cases: for what they fear excessively they will believe without verification."[14]

The Rhos raiders did not penetrate the walls but ravaged the suburbs, thus opening a long chapter of threats, alliance, more raids, alliance, conversion to Christianity, and outright wars.

For the first and so far only time, by 880 a powerful state had arisen in the steppe lands north of the Black Sea in what is now the Ukraine, with its capital in Kiev, and a sometimes vast but variable domain round about. Known as Kievan Rus' to historians, its strength was based on the skills of its boatmen, who could navigate difficult rivers and brave the open sea, and on the fighting power of sturdy foot warriors. They were very different from the mounted archers of the Turkic peoples all around them. Instead of clan, tribal chiefs, and qans who might give their allegiance first to one qagan and then another, its leaders were navigating warrior-merchants. By the year 907 they had a stable ruler, Oleg, who tried to attack Constantinople and was dubbed a prince *(archon)* when he signed a treaty with the empire in 911. The power of Kievan Rus' culminated under Prince Vladimir (from 980), who started its conversion to Christianity in 988, and Prince Yaroslav (1019–1054), after whom there was increasing fragmentation in rival principalities, eventually dozens of them, often at war with each other.

Intrepid long-range traders, the Rus' exported amber, furs, honey, and slaves purchased in the Russian and Baltic lands or collected as tribute from the Slavs in their own territories. Their main trade route started in the far north, in the western Baltic Sea marts of Birka, Hedeby, and Gotland; it then crossed over to the eastern Baltic and down the Neva River where St. Petersburg now stands to Lake Ladoga; it then followed the Volkhov River to Russia's most ancient city, Novgorod, crossed Lake Ilmen, and went upstream on the Lovat River, whence the boats had to be portaged to the Dnieper River at Gnezdovo, where Byzantine and Arab coins have been found. From there, it was another thousand miles southwest to Constantinople via Kiev, by navigating down the Dnieper River—with vulnerable portage intervals over the rapids—and then around or straight across the Black Sea. From Kiev, goods and slaves were also carried southeast overland to the Volga estuary, the terminus of the trade route to Baghdad via the Caspian shore, across the western Zagros Mountains and down to the Mesopotamian plain.

The warrior-merchants of Rus' were even more intrepid as expeditionary and amphibian fighters, brothers-in-arms of bold Norsemen like Harald son of Sigurd nicknamed "Hardråde" (of stern counsel), once a Byzantine guardsman who sailed across the North Sea from his Norwegian kingdom to conquer England, only to be defeated and killed at Stamford Bridge in 1066, shortly before the long-settled, more civilized, but still hard-fighting Norsemen or Normans conquered England in that same year 1066—while still other Norman conquerors took southernmost Italy from Byzantium, and then Sicily from the Muslim Arabs, with few men, much skill, and more courage.

Yet for all the heroic energies of the Kievan Rus', they did not prevent the arrival of new riders from the steppe who would one day extinguish their power. Their interests were focused on the river routes down to the Black Sea along the Don (Greek, Tanais), the Dnieper most importantly, the Bug (Hypanis), and the Dniester (Danastris)—and not on the vast and featureless steppe between the rivers, where successive waves of steppe nomads continued to arrive to pasture their herds and fight whoever stood in their path.

The Avars were long gone by then. Having moved north into what is now Hungary after their failure before Constantinople in 626, they had finally been destroyed by the Franks at the end of the eighth century—the looting of their chief encampment, "the Avar Ring," by Charlemagne in 796 yielded so much gold and silver that it was the richest of all his victories by far.

The bitter enemies of the Avars, the great Türk qaganate, had disappeared much earlier, while the Onogur, Kutrigur, and Utrigur tribes that did not remain east of the Volga were long since established south of the Danube in their Bulgaria. Thus the steppe lands from the Volga to the Danube had become the domain of the qaganate of the Khazars, of the advancing Magyars to their west, and of a much more powerful Turkic nation: the Pechenegs or Patzinaks, who were driving the Magyars before them, and who alternatively coexisted with or fiercely fought Kievan Rus', often for Byzantium.

> To the archon [prince] of the Tourkoi [= Magyars = Hungarians].
> A two-solidus golden bull. "Letter of Constantine and Romanos the Christ-loving emperors of the Romans to the archon of the Tourkoi."

As befits their character, the origins of the Hungarians, or Magyars as they call themselves, are uniquely complicated. The Magyars originally were their leading tribe until all adopted that identity, while *Tourkoi* is only in part a Byzantine misnomer. Byzantine authors disliked outlandish new barbarian names, and just as the Avars were invariably called Huns because that once exotic name had acquired a proper patina through usage in earlier texts, they invariably called the Hungarians after the departed Turks of the great steppe empire. But in this case they had a point, because the Hungarians lived an entirely Turkic way of life as nomadic herdsmen and mounted archers, even though their Finno-Ungric language proves that they started out as forest dwellers well north of the steppe, perhaps in what is now the republic of Bashkortostan of the Russian Federation.[15]

By the mid-eighth century they were formed into a nation inasmuch as they had a common language, but they had many different names: the Onogur that became Ungar and Hungarus; the Slav Ungri, the Khazar Majgar—they called themselves Majier.[16] When the Byzantines first encountered them around 830, the Magyars and the other tribes were living in what is now eastern Ukraine, under Khazar influence if not outright suzerainty—they certainly never had a qaganate of their own. The more numerous Pechenegs were advancing into their pasture lands by 850, driving some of them westward across the Ukraine, eventually to enter what is now Romania. Others remained between the southern Ural Mountains and the Volga River, the area now known as Bashkiria or Bashkortostan, which may have been the name of the entire nation then (in Romanian slang, *bozgori, bozghiori,* and *boangi* are still pejoratives for "Hungarians").

In 894 the Magyars, and the tribes that followed them, raided across the Danube at Byzantine request to attack the Bulgaria of Symeon I, forcing him to call off his offensive against the outmatched forces of Leo VI. Bulgaria remained a threat but the Byzantines did not succeed in keeping their new allies in place—or more likely did not even try, preferring to rely on Kievan Rus' and Pechenegs to control the Bulgarians.

Having crossed the mountains of Transylvania under pressure from the Bulghars or the Pechenegs—steppe herdsmen would not have voluntarily risked their livestock in the mountains—by 900 the Magyars and the tribes with them reached the plains of Pannonia, what is now Hungary, Magyarország to themselves, land of the Magyars. As with all such population movements, some were left behind, and eventually units of Magyar light cavalry *(Vardariotai)* were raised in the Vardar Valley of what is now Macedonia. The flattest part of almost flat Hungary, the Puszta, is the westernmost extension of the great Eurasian steppe, eminently suitable for mounted herdsmen who were indeed still to be found there as late as the 1930s.

As the mounted archers they had become, the Magyars were natural raiders.[17] But like the more formidable Avars before them, they also knew how to construct siege engines. For more than fifty years, the Magyars raided, looted, and burned westward into the German lands, sometimes reaching what is now France—they are mentioned repeatedly in the *Chanson de Roland* (including CCXXXIII.3248 to CXXXIV.3254), where they are listed as one of the tribes that do not serve God and who in battle are felonious murderers: "One are the Huns and the other the Hongres." Otto I, already king of Germany and destined to be called emperor, defeated a very large raiding party of Magyars on August 10, 955, at Lechfeld near the walled city of Augsburg; Otto's armored heavy cavalry massacred their light cavalry of archers, and after that Magyar raiding quickly came to an end. The Avars, their predecessors in Hungary and the Puszta, had also raided westward and continued to do so until they were annihilated, but the Magyars underwent a swift transformation. Within five years of their decisive defeat at Lechfeld, they had a Christian king, the later St. Stephen crowned by Pope Sylvester in the year 1000, and from then on the Magyars waged Christian wars instead of mounting Turkic raids.

The Pechenegs appeared as a new Turkic nation in the steppes during the ninth century, and they replaced the Khazars as the most useful allies of the empire during the tenth century.[18] This was not a strategic alliance as with the Khazars because there were no shared enemies to permanently compel cooperation. But if paid sufficiently, the Pechenegs could be very effective against Kievan Rus' and also against the migrating Magyars and the Bulgarians. If not paid, or not paid enough, they could and did attack the empire themselves, or join others who did.

Like all the semi-nomadic Turkic nations before them, they came across the Volga from Central Asia under pressure from other Turkic nations coming up behind them, chiefly the Oğuz, who would have their turn as allies and enemies of Byzantium. But the Khazars also pushed them westward. By the time the *Book of Ceremonies* was redacted, their power was centered between the Don and the Danube.

> To the archons of the Patzinakitai [Pechenegs]. A two-solidus gold bull.
> "Letter of Constantine and Romanos the Christ-loving emperors of the Romans to the archons of the Patzinakitai."

The Constantine VII Porphyrogennetos of the *Book of Ceremonies* is thus very brief but the Constantine VII Porphyrogennetos of *De Administrando Imperio* has much to write on the Pechenegs, who are indeed the very first subject that is discussed in the text, under a revealing title:

> Of the Pechenegs, and how many advantages accrue from their being at peace with the emperor of the Romans.
> . . . it is always greatly to the advantage of the emperor of the Romans to be minded to keep the peace with the nation of the Pechenegs and to conclude conventions and treaties of friendship with them and to send every year to them . . . [an envoy] with presents befitting and suitable to that nation, and to take from their side sureties, that is hostages and [an envoy] who shall . . . enjoy all imperial benefits and gifts suitable for the emperor to bestow.[19]

The first aim, negative but still very important, was to discourage Pecheneg attacks against imperial territory:

> This nation of the Pechenegs is neighbour to the [Byzantine] district of Cherson, and if they are not friendly disposed towards us, they may make excursions and plundering raids against Cherson.

The positive aim was to employ the Pechenegs as a deterrent against all the powers that they could reach, from the Khazars on the lower Volga all the way to the Bulgarians across the Danube. Under the heading "Of the Pechenegs and the Russians" the text first explains in general terms how deterrence worked:

> The Russians are quite unable to set out for wars beyond their borders unless they are at peace with the Pechenegs, because while they are away from their homes, these may come upon them and destroy and outrage their property.[20]

Then there is the specific mechanism of deterrence, which reflected the very peculiar distribution of power in this case—Kievan Rus' with its boats could control the Dnieper down to the Black Sea, but not the vast steppe on either side:

> Nor can the Russians come at [Constantinople], either for war or for trade, unless they are at peace with the Pechenegs, because when the Russians come with their [boats] to the barrages of the river [Dnieper] and cannot pass through unless they lift their [boats] off the river and carry them past by portaging them on their shoulders, then . . . the Pechenegs set upon them, and, . . . they are easily routed.[21]

The Pechenegs could deter the Magyars—consistently called "Turks" in the text—just as effectively, even though they had no boats and no rapids to cross, simply because they were a much smaller nation that the Pechenegs could always overpower:

> The tribe of the [Magyars], too . . . greatly fears the . . . Pechenegs, because they have often be defeated by them. . . . Therefore the [Magyars] . . . look on the Pechenegs with dread, and are held in check by them.[22]

The author recounts what happened when a Byzantine envoy asked the Magyars to attack the Pechenegs:

> The chief men *(archontes)* of the [Magyars] cried aloud with one voice "We are not putting ourselves on the track of the Pechenegs, for we cannot fight them, because their country is great and their people numerous and they are the devil's brats; and do not say this to us again; for we do not like it!"[23]

The text proposes the same Pecheneg remedy to deal with the Bulgarians, and indicates that at the time they were in direct contact along the Danube:

> The said Pechenegs are neighbours to these Bulgarians also, and when they wish, either for private gain or to do a favor to the emperor of the Romans, they can easily march against Bulgaria, and with their preponderating multitude and their strength overwhelm and defeat them.[24]

On the other hand, the Pechenegs are not identified as useful allies against the Khazars, as are the Turkic Oğuz and the Caucasian Alans. The reason is probably that they feared the Khazars, who along with the Oğuz had originally seized their pasture lands, driving them westward across the Volga and then to the Don.

Obviously the Pechenegs had to be paid to serve imperial purposes, and in kind rather than gold:

> [They are] ravenous and keenly covetous of articles rare among them, are shameless in their demands for generous gifts, . . . when the imperial agent enters their country, they first ask for the emperor's gifts, and then again, when these have glutted the menfolk, they ask for the presents for their wives and parents.[25]

It made sense for the Pechenegs to want goods instead of gold, which would have had to be taken far to be spent. The crafts of the steppe were necessarily limited to a narrow range of leather, wool, bone, gold, and silver articles, with iron less common; its food supplies were mostly of meat, cheese, and kumiss or kymis, fermented mare's milk, an acquired taste. There were many more choices in Constantinople, including spices from afar and wines, while Byzantine artisans produced everything known to antiquity out of all of its materials, including alloys, ceramics, and glass. The specific items identified in the text are rather prosaic but obviously scarce on the steppe: "pieces of purple cloth, ribbons, loosely woven [= lightweight] cloths, gold brocade, pepper, scarlet or 'Parthian'' leather."[26]

What clearly irritated the author and provoked his harsh adjectives was that all concerned asked for payoffs, over and above the emperor's "gift," including the Pecheneg notables held as honored-guest hostages while Byzantine envoys were at risk in Pecheneg territory, and the teamsters who transported both hostages and envoys back and forth:

> The hostages demanding this for themselves and that for their wives, and the escort something for their own trouble and some more for the wear and tear of their cattle.[27]

The same demand was made by the teamsters who transported the imperial envoy. That seemed petty greed to the author but reflected an important political reality. There was no Pecheneg qaganate with an all-powerful head who could punish and also reward by distributing Byzantine tribute. There was only a gathering of tribes loosely governed by different chiefs who might at most meet in council to plan common actions. The text reads:

> The whole of Patzinacia ["Pechenegia" as it were] is divided into eight provinces [*diaireitai* = divisions] with the same number of great princes. . . . The eight provinces are divided into forty districts, and these have minor princelings over them.[28]

That is a bit too schematic to be plausible but explains the absence of an all-powerful central leader served by loyal retainers rewarded by himself alone. Therefore everyone expected to be rewarded individually for any services rendered. That was a novel and disturbing thought for a Byzantine emperor, but still, no Greek could be contemptuous of freedom: "Pechenegs are free men and, so to say, independent [*autonomoi*], and never perform any service without remuneration." The Pechenegs obviously had their price, but so did all the other steppe powers, and evidently the Pechenegs were cheaper:

> So long as the emperor of the Romans is at peace with the Pechenegs, neither Russians nor Turks [= Magyars] can come upon the Roman dominions by force of arms, nor can they exact from the Romans large and inflated sums of money and goods as the price of peace.[29]

But even the valiant Pechenegs could not be useful for ever. To serve Byzantium, an ally had to be both strong enough to be effective against the enemies of the empire, yet not a threat themselves. From 1027 the Pechenegs began to fail on both counts. In that year they started to raid across the Danube, and in 1036 they were defeated by the forces of Kievan Rus' under Yaroslav I—that is, by the very power they were supposed to control. The Byzantines needed a new Turkic ally in the great steppe and duly found it in the Cumans or Qipchaqs in their own Turkic language, known as Polovtsy to Kievan Rus', who were destined for a long season of success under various regimes.[30]

They too were highly mobile and lethal mounted archers like the Huns, Avars, Bulghars, Khazars, Magyars, and Pechenegs before them, and they too became masters of the steppes by outnumbering the previous incumbents.

On April 29, 1091, the Byzantines with their new Cuman allies fought a large number of Pechenegs in the battle of Levounion, in the Maritsa River Valley of what is now southern Bulgaria. Evidently the Pechenegs had been driven into imperial territory by Cuman seizures of their pasture lands, because they came not as a raiding force but in a vast nomadic mass of livestock, men, women, and children.

At the time, the overall strategic situation of the empire was exceedingly unfavorable. Twenty years earlier in August 1071, Romanos IV Diogenes (1067–1071) had led a vast army of regular territorial forces, complemented by Frankish knights, Turkic mercenary mounted archers, and elite palace guard units, to confront the rising power of the Seljuk Turk empire. In what is known as the battle of Mantzikert (the modern Malazgirt), though it was fought in a broader area west of Lake Van in what is still eastern Turkey, the knights fled, some of the mercenaries defected, and Romanos IV Diogenes himself was captured after he was abandoned in the field by forces commanded by a dynastic rival.[31] Though traditionally considered a decisive battle in itself, it was not a catastrophic military defeat. Most Byzantine forces fought quite well until the emperor's capture, and then withdrew in good order to fight another day.

The Seljuk sultan Muhammad bin Da'ud Chaghri better known by his soubriquet Alp Arslan ("valiant lion") treated his captive with respect because he was a refined character, but the mild terms he demanded reflected the continued strength of the Byzantine armies, which had earlier defeated the Seljuks in Cilicia. The emperor and sultan had been negotiating for some time, right up to the eve of battle, and they quickly reached agreement, so that Romanos IV Diogenes was on his way back to Constantinople within a week.

The catastrophe came in the aftermath. The emperor was deposed and blinded, to be replaced by the ineffectual Michael VII Doukas (1071–1078), whose advisers refused to honor the peace treaty yet did not mobilize the armies to defend the frontiers that were then being penetrated by thousands of Turkoman (Turcoman, Turkmen) tribesmen—that being the contemporary term for any Turkic converts to Islam, though most were Oğuz.[32]

Anatolia was the core of the empire, and the loss of any part of it reduced in proportion its power resources of taxable crops and recruitable manpower—and most of it was lost during the twenty years after Mantzikert, both to Oğuz irregulars and to Seljuk *beys*, warlords. Word of the Byzantine defeat incited other enemies as well, chiefly the Normans, who had already seized the last Byzantine enclaves in southeast Italy by 1071, and the Serbs in the Balkans, among others, but it was mainly civil wars that ravaged the empire, under the subsequent emperor Nikephoros III (1078–1081) as well.

What next ensued, however, was a spectacular restoration of the empire under Alexios I Komnenos (1081–1118) that would include the recovery of much of Anatolia. For ten years, Alexios had fought the Normans, the Seljuk beys, and militant Paulician heretics, among others, while restoring the currency, fiscal collection, and territorial governance of what remained of the empire—Greece with its islands, a coastal strip in western Anatolia, and the southern Balkans. All the revenues and manpower of the empire had to come from that diminished realm, hence the Pecheneg invasion of 1091, which threatened its largest remaining part and prejudiced its entire future. That is why the total defeat of the Pechenegs at Levounion had strategic consequences for the rising fortunes of Alexios I Komnenos and the empire—from then on, there was a cumulative recovery of territory in Anatolia that would be greatly aided by the First Crusade, for all its threats and travails. In describing the results of the battle in her *Alexiad,* Anna Komnene, the highly literate daughter of the victor, reveals an appropriate sensibility:

> That day a new spectacle was seen, for a whole nation, not of ten thousand men only, but surpassing all number together with their wives and children was completely wiped out. It was the third day of the week, the twenty-ninth of April; hence the Byzantines made a little burlesque song, "just by one day the Scythian [= Pecheneg] missed seeing the month of May." By the time that the sun was creeping to the West, and practically all the Scythians [Pechenegs] had fallen to the sword, and I repeat the children and the women too, and many also had been taken alive, the Emperor bade them sound the recall, and returned to his camp.[33]

Levounion was a great victory and a massacre—in which the Cumans refused to join—but there were still Pechenegs left in the steppe for the Cumans to attack in 1094, and to themselves invade across the Danube one final time, until the forces of John II Komnenos, son of Alexios I, defeated them completely in 1122, at Beroia, the Stara Zagora of modern Bulgaria.

The dynamics of ethnogenesis worked both ways: just as successful tribal groupings attracted more tribes and individuals, becoming yet more numerous and more powerful as functioning nations, or even imperial qaganates, unsuccessful ones lost individuals, clans, and entire tribes to more fortunate rivals. Some surviving Pechenegs became Bulgarians, others Hungarians, still others Cumans.

"To the archon [prince] of Chrovatia [Croatia]." When the Avars attacked the empire in the early seventh century in a series of offensives that culminated with the siege of Constantinople in 626, they did so with large numbers of Slavs *(Sklabenoi, Sklauenoi, Sklabinoi)* who were fighting under their orders or simply acting as camp followers in the hope of loot—they added the weight of numbers to the elaborate equipment and advanced skills of the Avar warriors. According to *De Administrando Imperio,* emperor Herakleios had successfully separated these Slavs, starting with the Croats: "And so, by command of the emperor Heraclius these same Croats defeated and expelled the Avars from these parts [Dalmatia]."[34]

When the Avars failed to conquer Constantinople and retreated north toward Dalmatia and the Hungarian plain, some of these Slavs retreated with them, gradually differentiating into Croats (*Hvrati* in modern Croat) and the more numerous Serbs (Srbi), for a long time only politically—for they were identical in speech, as they still are today, and presumably in their pagan religion.

By the middle of the ninth century the Croats were undergoing Christianization and had the makings of rudimentary states in both coastal Dalmatia, marked off by the Dinaric alps, and the plains behind them, part of what had been Roman Pannonia.

In the port city of Ladera (Zadar), even after the extinction of the exarchate of Ravenna by the Lombards in 751, there remained the headquarters of the Byzantine theme of Dalmatia under a strategos, which at times had to fight off a Croat ruler Trpimir during the period 845–864. Soon thereafter a more substantial power must have emerged, because in 879, Pope John VIII wrote to *a dux Chroatorum,* Branimir, son of Trpimir, in flattering terms and successfully intrigued to secure the allegiance of the Croat church to Rome. There had long been a bishop in Split, the Roman Spalatum, then as now the largest city of Dalmatia, who was under the authority of the patriarch of Constantinople, as were all the Christian Croats—they used the Slavonic liturgy of Cyril and Methodius written in the glacolitic alphabet, rather than the Latin liturgy.

It was only much later that the final consequences of John VIII's jurisdictional ambitions—there was no doctrinal difference at point—would be realized in the murderous hatred between Catholic Croats and Orthodox Serbs, vehemently encouraged by their respective prelates even in the late twentieth century. By 925, within the time of Constantine VII, the two Croat entities were united under a rex of their own, Tomislav.

"To the Prince of the Servloi [Serbs, Srbi]." Once again it is claimed in

De Administrando Imperio that the emperor Herakleios was present at the creation of their political identity, by granting them "a place in the province of Thessalonica to settle in."[35] That sounds like a good way of separating them from the Avars.

A powerful Serbian state would emerge in the twelfth century under Stephan Nemanja (1109–1199), a dangerous enemy of Byzantium until he was captured and befriended by emperor Manuel I Komnenos (1143–1180).

But when the *Book of Ceremonies* was compiled, there were only petty chiefdoms ruled by "Zupans" (from *zupania* = county), of which the largest was Rascia (Raska). In *De Administrando Imperio,* under the heading "Story of the province of Dalmatia," several of these Zupanates or chiefdoms are described just sufficiently to determine their approximate location, mostly along the Adriatic coast of what is now Croatia, and the Herzegovina province of the contemporary Bosnian Federation:

> From Ragusa [Dubrovnik] begins the domain of the Zachlumi and stretches along as far as the river Orontius; and on the side of the coast it is neighbour to the Pagani, but on the side of the mountain country it is neighbour to the Croats on the north and to Serbia at the front.[36]

There follow separate mentions of the zupanates of Kanali, *Travuni* (Terbounia), *Duklja* (Diocleia), and a very different entity, *Moravia.*

A "Great Moravia" appears in *De Administrando Imperio*—an appropriate name for a large but ill-defined territory that may have included parts of modern Slovakia, Austria, and Hungary, as well as the Czech republic—whose present Moravia is an echo, not a geographical survival. Its first king, Mojmir I (830–846), was the neighbor of the vast Francia that Charlemagne had created and vassal of his son, the emperor Louis le Pieux or Ludwig der Fromme (814–840).[37] When Mojmir was succeeded by his son Ratislav (846–870), the new Moravian ruler tried to free himself from Frankish influence, no doubt because the empire of Charlemagne had been partitioned and his neighbor was now the diminished East Francia of Louis or Ludwig the German. Under that policy, Ratislav sent envoys to the Byzantine emperor Michael III (842–867) to ask for a bishop and teachers who would bring the Gospel to the Slavic peoples in their own language, replacing the Frankish missionaries who were preaching in Latin, and winning Christian souls not only for the deity but also for the pope in Rome, and Francia too.

Michael III responded with the momentous mission of the brothers and future saints Cyril and Methodius. Instead of imposing the Greek liturgy on Slav ears, as the Franks were imposing the Latin liturgy, they created the splendid Old Slavonic liturgy in Macedonian Slav written in the Glagolitic alphabet that Cyril invented (he did not invent Cyrillic). Byzantine missionaries would succeed elsewhere on the grandest scale, but their embryonic Moravian orthodox church was soon extinguished. Louis the German, who had already tried to subject Ratislav in 855, was more successful with a second punitive expedition in 864, and six years later Ratislav was brutally blinded—he soon died—and replaced by his nephew Sventopluk, or Svatopluk in modern Czech. Not coincidentally, the new ruler preferred the Latin of the Frankish priests to the Slavonic liturgy and did nothing to prevent papal legates from expelling the Moravian disciples of Saint Methodius, who came under the jurisdiction of the patriarch of Constantinople.

That determined the religious fate of much of central Europe till this day. Under Svatopulk (870–894), Great Moravia encompassed parts of eastern Germany with its large Slav population—in Cottbus in Brandenburg many still speak the Slavic Sorb language—and Slav western Poland, as well as Bohemia, Moravia, and Slovakia, all lands where the impressive and melodious Slavonic liturgy of the Orthodox Church would have naturally prevailed, unless forcibly excluded, as it was. The popes of the time, and especially Formosus (891–896), were exceptionally experienced in waging ecclesiastical war against the patriarchate of Constantinople. Their ruthless energy often overcame their enormous handicap: their lack of a protective emperor of their own. It was a persistent disparity that added an element of bitter resentment to the competition between the churchmen of Rome and Constantinople, at a time when there was as yet no doctrinal difference to justify all the animosity. Pope Formosus had himself served as legate to Boris I, or Bogoris ruler of Bulgaria (852–889), who in 867 petitioned Pope Nicholas to appoint Formosus archbishop of Bulgaria, a very deliberate attempt to transfer the emerging Bulgarian Church from the jurisdiction of the patriarch to that of the pope.

Four years earlier in 863, Boris had become the first Bulghar ruler to convert to Christianity. Just as Ratislav had wanted his Christianity to come from safely distant Constantinople instead of the excessively powerful Francia of Louis the German right at his borders, so Boris had invited that same Louis to send him missionaries to effect his conversion, instead of inviting clergymen from nearby but overbearing Constantinople.

Both Ratislav and Boris were trying to avoid adding religious subordination to their strategic inferiority. Neither succeeded. The same Michael III who had solicitously sent Cyril and Methodius to support Ratislav's bid for religious independence from the Frankish church sent an army into Bulgaria to force Boris to reconvert to Christianity under the Orthodox rite, which he duly did with his family and retainers in his capital of Pliska in 864, taking the name of his godfather to become Boris-Mihail in history, and just Michael in documents, as in the later seal inscribed "Michael the Monk, who is archon of the Bulgarians."[38]

The two conversions of Boris I were clearly political acts, and one was imposed on him by main force. Yet he was undoubtedly committed to Christianity: when Bulghars still attached to the old religion revolted against the new faith in 865, Boris responded with mass violence, executing fifty-two tribal chiefs *(boyars)* together with their families. Again, having abdicated in 889 to retire to a monastery as a monk—surely a proof of his religious sincerity—in 893 Boris came out of his cell to rally an army to depose and blind his own son, Vladimir, and give the throne to his third son, Symeon I; according to the almost contemporary chronicle of the admittedly distant Regino of Prum, Boris overthrew and mutilated Vladimir because he had wanted to restore the old religion. It would be Symeon who reconciled religion with independence by gaining Byzantine recognition for the autocephaly of the Bulgarian church, whose patriarch he could appoint, just as the Byzantine emperor appointed the patriarch of Constantinople.

In the *Book of Ceremonies,* India is provided for:

> To the hyperechon kyrios [most senior lord] of India.
> "Constantine and Romanos, faithful to Christ the Lord, Great Autocrats and Emperors of the Romans, to . . . [Name] the most senior lord of India, our beloved friend."

The import of spices from India had no strategic significance.[39] But up to the seventh century there was much scope for an alliance, because the Sasanian empire of Persia also threatened the Gupta rulers of India.

With their common enemy between them, the Byzantines and the Guptas could have concerted their military operations advantageously. The mountains of the Hindu Kush, the convergence of the westernmost Pamirs and Himalayas, were an impossible obstacle to overland journeys to India by way of Central Asia, but ships habitually sailed from Byzantine Egypt to Indian ports, which meant incidentally that much was known about the country. The *Indika* of Ctesias of Cnidus (flourished ca. 400 BCE) is full of tall tales, judging by its surviving summary, but the *Indika* of Megasthenes (ca. 350–290 BCE), himself an envoy of Seleucus I, a successor of Alexander the Great to Chandragupta, founder of the Maurya empire, contains accurate information, including a description of caste distinctions. The anonymous Greek text of the second century CE usually known by its Latin title *Periplus Maris Erythraei* has detailed information on trade, while the sixth-century *Christian Topography* of Cosmas Indicopleuste ("India-traveler"), a long-distance merchant who became a monk, described Taprobane (now Sri Lanka), among other things. As compared to the three-year journey of Zemarchos to the Türk qagan in the Altai mountains and back, the passages of envoys to and from India by sea would have been less perilous, more comfortable, and altogether faster.

That potential alliance was never constructed. Given the geographic barriers, there was no possibility of combining forces for unified action—even with transformed logistics, during the Second World War Germany and Japan could deploy no greater joint forces than the brief encounters of their submarines at Penang, Malaya. Coordinated offensives would have been possible, but there is no sign of any such initiative. By the time the *Book of Ceremonies* was redacted, the nearest thing to a "senior lord of India" was the ruler of the Chavda dynasty centered in Gujrat, the last of whom, Samantsinh Chavda, was overthrown in 942 by his adopted son Mulraj, founder of the eponymous dynasty.

Of China the Byzantines knew a little from the Turkic powers in between under the name Taugast (= Turkic Tabghach) for the China of the Wei dynasty. But mostly China was known as the original source of silk—indispensable for the vestments of court officials and high prelates. Silk was also of strategic importance, because of the frequent fights over the control of the way stations of the silk route across Central Asia. We have seen how the Sogdians of the silk-road cities on either side of Samarkand had adapted to the arrival of the Türk qaganate by mediating its alliance with Byzantium. Until the time of Justinian, moreover, the Byzantines were forced to import their silk by way of the Sasanian Persian empire, increasing its gold revenues. Prokopios of Caesarea relates the marginally credible tale of how "certain monks, coming from India" appeared before emperor Justinian (527–565), explained that silk was made in Serinda, north of India, by worms (moths, actually) fed on mulberry leaves, and offered to smuggle in their eggs—their motive being to deny the Sasanians the rich profits of the silk trade.[40]

What is certain is that silk production did start within the empire under Justinian, but without ending imports because Chinese quality and variety could not be matched. A gold solidus of Justin II (565–578) was excavated in a Sui dynasty tomb in Shanxi province in 1953, and with the post–Cultural Revolution revival of Chinese archaeology, an increasing number of Byzantine coins have been found.[41]

The Chinese and the Byzantines often faced the same threats because the largest steppe qaganates, both Turkic and Mongol, spanned the entire distance between them—indeed, we know the early history of the Turks from the Chinese dynastic chronicles.[42] But even the loosest strategic coordination was logistically impossible, so that diplomacy would have been an empty exchange of courtesies. That could not interest the Byzantines; their apparent proclivity for pointless formalities was usually rather purposeful.

Kublai or Khubilai Qan (Khan), grandson of Cinggis Qan, did send two Nestorian messengers, who were eventually received by Andronikos II (1282–1328). Two of his half-sisters were married to great-grandsons of Cinggis Qan, hence Andronikos II was receiving mail from his in-law in Beijing.

Another communication also had no strategic content, but interests for its own sake: in 1372 the first Ming emperor, Hongwu, sent an announcement of his accession to the emperor of Byzantium. He would not have bothered had he known that the domain of John V Palaiologos (1341–1376) was by then reduced to a shrunken and impoverished Constantinople with a few insular and peninsular remnants, and that the emperor himself had suffered the triple humiliation of a stint in the debtor's prison in Venice when trying to summon help from the west, usurpation by his own son Andronikos IV Palaiologos, and subordination to the Ottoman sultan Murad I, who restored him in office—indeed, the end of it all seemed at hand, and was only delayed for another eighty years till 1453 by extraordinary good fortune, as well as thin residues of statecraft.

The text of the announcement explains why the Ming emperor Hongwu (Hung-woo T'i in the old Wade-Giles transliteration), the former starveling peasant, monastery servant, and rebel Zhu Yuanzhang, was impelled to announce his accession as widely as possible. He presents himself as China's first Chinese ruler after the Yuan dynasty of the Mongols of Kublai Qan, and seeks legitimization for his new Ming dynasty by appealing to national sentiments that seem startlingly modern:

Since the . . . Yuan [Mongol] dynasty had risen from the [Gobi] desert to enter and rule over . . . [China] for more than a hundred years, when

Heaven, wearied of their misgovernment and debauchery, thought also fit to turn their fate to ruin, . . . the affairs of [China] were in a state of disorder for eighteen years.

But when the nation began to arouse itself, We, as a simple peasant of Huai-yu, conceived the patriotic idea to save the people. . . . We have then been engaged in war for fourteen years. . . . We have [now] established peace in the Empire, and restored the old boundaries of [China]. . . . We have sent officers to all the foreign kingdoms . . . except to you, Fu-lin [Rome, Byzantium], who, being separated from us by the western sea, have not as yet received the announcement. We now send a native of your country, Nieh-ku-lun, to hand you this Manifesto. Although We are not equal in wisdom to our ancient rulers whose virtue was recognized all over the universe, We cannot but let the world know Our intention to maintain peace within the four seas. It is on this ground alone that We have issued this Manifesto.[43]

Incidentally, the messenger Nieh-ku-lun may have been the Franciscan Nicolaus de Bentra, bishop of Cambaluc, Latinization of *Qanbaliq,* Mongol for the qagan's residence that is the Yuan capital, now known as the "northern capital," Beijing.

CHAPTER 8

Bulghars and Bulgarians

Byzantine dealings with the Caucasian states were certainly complicated but entailed no existential threat to the empire. That was also true of the Muslim Arabs after the failure of their second siege of Constantinople in 718, in spite of periodic scares as in 824, when Arabs in flight from Umayyad Spain conquered Crete. Otherwise there was chronic border warfare with marauding by both sides and occasional gatherings for jihad offensives that remained regional in scope. As for the renascent powers of the west, they threatened Byzantine possessions in southern Italy and Dalmatia from the eighth century, but another four centuries would pass before they could attack Constantinople itself, for that required a fleet stronger than the Byzantine navy.

Bulgaria was different. Because it was so close to Constantinople, its power was a deadly threat whenever crisis on another front, insurrection, or civil war denuded the city's garrison.

The very existence of a Bulgarian state south of the Danube River was necessarily a threat to the survival of the empire, regardless of its strength or weakness, even regardless of its intentions.[1] For across the Danube there was the vastness of the Eurasian steppe, from which the Huns had arrived to exhaust the power of the western empire, then the Avars who almost conquered Constantinople in 626, and then the Bulghars themselves, who would be followed over the next four centuries by the Pechenegs, Magyars, Cumans, and Mongols. Only a Danubian frontier, defended by Byzantine troops backed up by a Byzantine river fleet, could provide due warning and a properly overwatched obstacle to invasions from the steppe.

The Bulgarians could not do it, axiomatically: if they were strong enough to defend the Danube frontier themselves, they would necessarily be a threat to Constantinople as well; if they were too weak, not only they but Constantinople would be in danger. Only a Bulgaria both strong and slavishly obedient could have been a desirable neighbor for Byzantium, but that improbable coincidence only occurred briefly in times of transition, when the Bulgarian state was becoming but had not yet become too weak to defend the Danube, or was becoming but had not yet become strong enough to threaten Constantinople.

The great irony is that Bulgaria was in large degree a Byzantine creation. Having emerged west of the Volga River during the seventh century as distinct Onogur-Bulghar tribes (also Ogur, Onogundur, or Vununtur in the Hebrew of the Khazars), the future Bulgarians acquired a common, still entirely Turkic identity under the suzerainty of the Avars, when they mostly appear as Onogurs in our sources.[2] Under Herakleios (610–641) the empire urgently needed allies able and willing to fight the especially dangerous Avars, who besieged Constantinople in the supreme crisis of 626. At the time there was no newly arrived nation from the steppes already approaching the Danube that could be enlisted in the struggle against the Avars. So the Byzantines found an ally much farther east: Kuvrat (Kubratos, Kurt, Qubrat), paramount chief of the Onogurs who had founded a steppe power that the Byzantines would later describe as Old Great Bulgaria. In 619 Herakleios had received in Constantinople the Onogur chief Organa (Turkic: Orhan), baptized him with his retinue, and sent him off with the title *patrikios,* as well as gifts no doubt. He may have been Kuvrat's uncle.

Around the year 635 Kuvrat cast off Avar overlordship and sent his men all the way west to the Danube and across it to attack the Avars; he was given the title *patrikios* by Herakleios—at least for that there is the perfect archeological evidence of three gold rings marked "Hourvat Patrikios" found in 1912 with twenty kilos of gold and fifty kilos of silver objects, all finely made, near the village of Maloe Pereshchepino, in the Poltava region of the Ukraine—evidently Kuvrat's grave. According to the *Chronicle of John Bishop of Nikiu,* an almost contemporary source but which survives only in an Ethiopian translation of the Arabic translation of the original Greek text, Kuvrat was Orhan's nephew, baptized as a child, and brought up at court along with his lifelong friend Herakleios:

> Kubratos, chief of the Huns [by then a generic term for steppe peoples],
> the nephew of Organa, who was baptized in the city of Constantinople,

and received into the Christian community in his childhood and had grown up in the imperial palace. . . . And between him and the elder Herakleios great affection and peace had prevailed, and after Herakleios' death he had shown his affection to his sons and his wife Martina because of the kindness he had shown him.[3]

This seems to be a "human interest" rationalization of a strategic partnership, for Herakleios apparently spent his youth in Carthage where his father was exarch of *Africa*. Kuvrat's attack was at least one reason for the Avar retreat from Thrace, whence they had directly menaced Constantinople. Until his death in 642, the *patricius* Kuvrat apparently remained a loyal ally of the empire.

At the time, the new Khazar qaganate was expanding westward, squeezing out the Onogurs, or Bulghars as they begin to be named. One of Kuvrat's sons, the Asparuch (Asparux, Isperih) now celebrated as the founder of Bulgaria, forcibly crossed the Danube around 679 to occupy imperial territory in Moesia after defeating the forces of Constantine IV (668–685). The event is recorded in the preserved text of a Hebrew letter of a Khazar qagan, who wrote that the Vununtur (= Onogurs = Bulghars) had fled across the Duna, the Danube.[4] Even if numerous for the steppe, Asparuch's pastoralist warriors and their families were of necessity relatively few as compared to the agricultural Slav population that lived south of the Danube, and thus the Turkic-speaking Bulghars were assimilated linguistically by the Slav majority to form the medieval and modern Bulgarians. This particular ethnogenesis occurred gradually over a period of more than two centuries: there was the Turkic qan (or khan) Krum (803–814), Qan Omurtag (814–831), Qan Perssian (836–852), then the qan who converted Boris I (852–889); then came Tsar Symeon (893–923), Tsar Peter I (927–970), and so on. But this transformation of Turkic shamanists into Slavic Christians did nothing to diminish the warlike character of the empire's new neighbors.

Because even warlike neighbors can be useful at times, the relations between the empire and the new Bulghar qaganate encompassed every possible variation, from intimate alliance to all-out war, as exemplified by the career of the Bulghar qan or khan Tervel (or Tarvel—Terbelis in our Greek sources), the successor and probably son of Asparukh who ruled for some twenty-one years within the period 695–721, extant chronologies being inconsistent.

Tervel is first mentioned by Theophanes when he agreed to help the deposed Byzantine emperor Justinian II Rhinometos ("nose-cut") regain his throne. Overthrown by insurrection in 695 and replaced by Leontios, *strategos* of the theme (see below) of Hellas, Justinian II was exiled to Cherson near Sevastopol in the Crimea, now Ukraine, the most remote of Byzantine cities. Leontios was himself overthrown by the German Apsimarus, commander of the maritime theme of the Kibyrrhaiotai, who ruled as Tiberios III (698–705).

By 703, Justinian II had fled Cherson to seek refuge with the qagan of the Khazars, from whom he received regal hospitality and his sister (baptized as Theodora) in marriage. But when Tiberios III sent envoys to the Khazars to ask for his repatriation, Justinian II fled westward and sent a message to

> Terbelis, the lord of Bulgaria, so as to obtain help to regain his ancestral empire, and promised to give him many gifts and his own daughter [Anastasia] as wife. The latter [Tervel] promised under oath to obey and co-operate in all respects and, after receiving . . . [Justinian II] . . . with honour roused up the entire host of Bulghar and Slavs that were subject to him. The following year (in 705) they armed themselves and came to the Imperial city.[5]

There was neither assault nor siege: with a few men Justinian II infiltrated into the city through one of the aqueducts, rallied supporters, and seized power—he was not just a mutilated ex-emperor, he had an army at his command—eventually executing Apsimarus/Tiberios III along with the deposed Leontios. Justinian II duly rewarded Tervel "with many gifts and imperial vessels."[6] According to Patriarch Nikephoros there was more:

> He showed many favors to the Bulgarian chief Terbelis, who was encamped outside the Blachernai wall, and finally sent for him, invested him with an imperial mantle and proclaimed him *kaisar*.[7]

Caesar, the second-highest rank, had never before been awarded to a foreign ruler. Justinian II may also have conceded some territory in northeast Thrace to Tervel, but nothing more is known of his promised bride Anastasia.

Three years is a long time in international politics: by 708, amity forgotten and gratitude spent, Justinian II "broke the peace . . . and, after ferrying the cavalry *themata* across to Thrace and fitting out a fleet, set out against Terbelis and his Bulghars."[8] It is probable that the fight was over the territory supposedly promised that Justinian II would not hand over. Ingratitude was duly punished: Tervel routed Justinian's troops in the Battle of Anchialos, now Pomorie, southeast Bulgaria. So after intimate alliance, complete with a promised dynastic marriage, there was outright war. Nevertheless, according to Nikephoros, only three years

later, in 711, Tervel sent three thousand of his men to help Justinian II
fight a revolt in Asia Minor—not enough, by intent or otherwise, because he was defeated and executed.[9]

Exploiting the turmoil, in 712 Tervel raided Thrace for plunder,
reaching the vicinity of Constantinople:

> The Bulgarians . . . made great slaughter. They raided as far as the City,
> and surprised many people who had gone across the water (from the Asian
> side) to celebrate opulent weddings and lavish luncheons. . . . They ad-
> vanced as far as the Golden Gate and, after devastating all of Thrace, re-
> turned home with innumerable cattle.[10]

But once again there was a reversal: under Tervel or his immediate successor, the Bulghars were the valiant allies of the empire in defeating the
second Arab siege of Constantinople, in 717–718. Maslama bin Abdul-
Malik, brother of the Umayyad caliph Suleiman bin Abd al-Malik (ca.
674–717) and an enthusiastic jihadi, had brought a vast number of men
across the Bosporos to the European side to invest the Theodosian Wall,
while Arab vessels were blockading Constantinople and attempting to
attack the seawalls. That is, when, according to Theophanes "the Bulgarian nation made war on them and, as well-informed persons affirm,
massacred 22,000 Arabs."[11] Under the fighting emperor Leo III (717–
741), Byzantine forces resisted the Arab offensive at sea and on land,
beyond Constantinople as well—the Umayyad caliph himself,
Sulayman bin Abd al-Malik was killed on the Syrian border in 717, presumably while leading a diversionary attack.

The importance of the Bulgarian contribution to the Arab defeat
emerges very clearly in the *Secular History of the Pseudo-Dionysius of
Tel Mahre,* written in Syriac, that is, late eastern Aramaic, in a passage
that has come down to us by way of its incorporation in the *Chronicle
of AD 1234,* itself extant in a single copy.

The first consequence of the Bulghar intervention was the damage
done to Maslama's personal battle force—it was no mere escort:

> Maslama's army crossed to a point about six miles below the City [down
> the Marmara coast], but Maslama himself with his escort of 4,000 horse-
> men landed after the rest at a distance of about ten miles from the camp
> of those who had preceded him. That night the Bulgarian allies of the
> Romans fell upon him unsuspecting and slaughtered most of the force
> which was with him. Maslama escaped by a hair's brea[d]th.[12]

When landing on the European shore to bring Constantinople under
direct attack from its landward side—essential to conquer the city—the
Arabs exposed their own rear, which they proceeded to leave unguarded

on the normally sound assumption that any Byzantine troops would be inside the city to defend it, and not wondering uselessly in the country-side. Perhaps they knew nothing of the Bulghars, or else they acted on the reasonable assumption that the Bulghars would either seek to join them in attacking the city or at least would do nothing to help the Byzantines to defend it—given their own wars against the empire.

But Byzantine diplomacy had once again been at work—we do not know how or when—and Maslama suffered the consequences.

> A further force of 20,000 under the command of Sharah b. Ubayda was sent out to guard the (landward) approaches of the camp against the Bulghars, and the seaward approaches against the roman ships. . . . One day the Bulghars gathered against Sharah I and his army, did battle with them and killed a large number of them, so that the Arabs came to fear the Bulghars more than the Romans. Then their supplies were cut off and all the animals they had with them perished for want of fodder.[13]

The sequence of the Bulghar attacks on the Arabs in Thrace, and the Byzantine blockade of the Muslim Arabs camped on the shores of the Sea of Marmara, is preserved in the text now known as the "Chronicle of AD 819." Its compiler counted the Seleucid years from 312 BCE, or "years of Alexander the Great," and it was under the year 1028 that he recorded:

> Once again Sulayman [bin Abd al-Malik] mustered his armies . . . and sent a great army with Ubayda as its general to the Roman empire. They invaded Thrace. . . . Ubayda invaded the country of Bulgaria, but most of his army was destroyed by the Bulghars . . . those who were left were oppressed by Leo [III], the sly king of the Romans, to the point of having to eat the flesh and the dung of their horses.[14]

The War of 811, *Themata* and *Tagmata*

Gratitude is not a virtue in strategy. That Bulghars could arrive too swiftly to be detected to fight against the Arabs for the empire, also meant that they could swiftly arrive to fight against the empire. Given that neither a weak nor a strong Bulgaria was compatible with the security of Constantinople, its total destruction was a perfectly rational aim for Byzantine strategy. There was no spare strength for such a venture so long as the Muslim Arabs waged jihad every day, periodically launching larger attacks. It was not until the start of the ninth century that the empire became stronger and the Muslim Arabs much weaker, allowing action on other fronts. The outcome was the large-scale offensive launched in 811 by emperor Nikephoros I (802–811) against the dangerously expansionist Bulgaria of Qan Krum, Kroummos to the Greeks.[15]

Both contenders had recently been freed of dangerous enemies that had kept them preoccupied elsewhere. The Avars had suffered their devastating defeat at the hands of Charlemagne's forces, allowing Krum to invade their territories in modern Croatia and Hungary to finish them off. Yet more recently, the vastly powerful and much celebrated fourth Abbasid caliph Hārūn al-Rashīd (the "rightly-guided") had died in 809, unleashing a succession struggle that paralyzed the dynasty. That allowed the Byzantines to focus their attention on Krum's Bulghars, who had doubled their territory since 800 and were expanding into Thrace toward Constantinople itself. It made eminent strategic sense for Nikephoros to exploit the sudden respite on his eastern front, to contend with the imminent threat on his northern front.

The normal Byzantine strategy would have been to prepare for war by finding allies in the Eurasian steppe willing to attack the Bulghars in their rear while their own forces advanced against them—perhaps not very fast, to allow the brave warriors of the steppes the fullest opportunity to fight gloriously. At the time, the strong Pechenegs driving the Magyars before them were drifting westward from the Volga region, as the Bulghars had once done. Although both were still far away, it would have been in the tradition of Byzantine diplomacy to accelerate the arrival of the Pechenegs with gifts and promises, just as the Bulghars of Kuvrat had once been induced to come westward to fight the Avars. Instead, Nikephoros decided to rely entirely on his own military strength.

Our best source for what happened next is Theophanes Confessor, a loyal churchman who hated Nikephoros, whom he accused of multiple, inconsistent heresies (Manichean *and* Paulician *and* Judaizing), black witchcraft complete with a sacrificed ox, homosexual fornication, and the most outrageous sin of them all: increased taxation of the clergy. "The new Ahab, who was more insatiable than Phalaris or Midas, took up arms against the Bulgarians [in 811]. . . . As he was departing from the Imperial City, he ordered the patrician Niketas, the logothete of the *genikon* [chief tax collector], to raise the taxes of churches and monasteries."[16] Theophanes was hopelessly biased, but subsequent events prove the essential accuracy of his account; moreover a second source, the anonymous fragment published as "The Byzantine Chronicle of the Year 811,"[17] confirms the principal facts of the dismal tale.

Theophanes starts with:

> So having gathered his troops, not only from Thrace, but also from the
> Asiatic *Themata,* as well as many poor men armed at their own expense
> with slings and clubs who were cursing him as did the soldiers, he ad-
> vanced against the Bulgarians.

Here we encounter the principal military institution of the middle centuries of the Byzantine empire: the *themata,* plural of *thema,* Englished
as "theme," were both administrative districts and territorial military
commands; their *strategoi* also had wide civil powers.

Because it ended the clear division between the powers of civil official
and military officers of the later Roman system, the central importance
of the thematic reorganization of the empire is uncontroversial, but
much else is still debated.[18]

Thematic units were manned by part-time farmer-soldiers who were
called for duty as needed, and it was a key responsibility of the *strategoi*
to see to the training and arming of those reservists.

The Asiatic themes mobilized in 811 were presumably the Optimatoi,
Opsikion, and Boukellarion, assuming that the three themes guarding
against Arab raids, the Armeniakon, Anatolikon, and the coastal
Kibyrrhaiotai, would not have been stripped of all their mobile forces.

But for the polemic, the *Chronicle* concurs—Nikephoros took with
him "all the patricians and commanders *(archontes)* and dignitaries, all
the *tagmata* [elite cavalry formations] and also the sons of the archontes
who were aged fifteen or above, whom he formed into a retinue for his
son which he gave the name worthies *(hikanatoi)*"—an unfortunate experiment in a corps of cadets.[19] At the end, when enumerating the casualties, Theophanes adds six patricians, including Romanos, the
strategos of the Anatolikon theme; the strategos of Thrace, whence the
expedition departed; "many" *protospatharoi* and *spatharoi,* midlevel
field officers; the commanders of the *tagmata;* and "an infinite number"
of soldiers.

Of these, the most mobile and presumptively the most valuable forces
were the tagmata manned by full-time soldiers. They were originally
formed by Constantine V in 743 to break up the strength of the then
very large Opsikion theme, which was temptingly close to Constantinople—its chief, Artavasdos (or Artabasdos or Artabasdus), Constantine's
brother-in-law, had just tried to usurp the throne.[20]

Six tagmata were formed out of the Opsikion troops. Each tagma had
an establishment of four thousand men, on paper at least, divided into

two *mere* or *turmae* of two thousand, each in turn divided into two *drungi* of one thousand, each made of five *bandae* of two hundred, in two centuries.

Of the six tagmata, at least two accompanied the emperor into Bulgaria, for the listed casualties included the chief *(domestikos)* of the *Exkoubitoi,* literally the outside-the-bedchamber guards, and the *drungarios* of the *Vigla,* the imperial watch, a contraction of the *Vigiles* of Rome.

In the absence of allies, Nikephoros had evidently mobilized imperial forces on the largest possible scale to defeat Krum as the Romans might have done, with overwhelming force. To add more mass to the trained, drilled, and organized thematic forces, he had also recruited untrained irregulars fighting for cash ("many poor men"). Mass worked. The *Chronicle:*

> When . . . the Bulghars learned of the size of the army he brought with him, and since apparently they were unable to resist, they abandoned everything they had with them and fled into the mountains.

In an exemplary tale of the downfall of the wicked, there must be spurned opportunities for salvation:

> Frightened by this multitude . . . Kroummos asked for peace. The emperor, however, . . . refused. After making many detours through impassable country [a misunderstanding or misrepresentation of maneuver warfare] the rash coward recklessly entered Bulgaria on 20 July. For three days after the first encounters the emperor appeared to be successful, but did not ascribe his victory to God.[21]

The *Chronicle* adds numbers, even if too large given the earlier assertion that the Bulghars had fled into the mountains:

> [Nikephoros found] there an army of hand-picked and armed Bulghars who had been left behind to guard the place, up to 12,000, he engaged battle with them and killed them all. Next in similar fashion he faced another 50,000 in battle, and having clashed with them, destroyed them all.

Subsequent events do indicate that the casualties of Krum's palace guards and elite forces were indeed heavy.

Next came the looting of Krum's palace, made of wood but more than the rustic hall of a barbarian chief—the *Chronicle* says of Nikephoros that while "strolling up the paths of the palace . . . and walking on the terraces of the houses, he exalted and exclaimed 'Behold, God has given me all this.'" Moreover, the palace was filled with the accumulated riches of past depredations. Unwilling to give Nikephoros any credit for having conquered Krum's capital and treasury, Theophanes instead stresses his avarice: "He placed locks and seals on the treasury of Kroummos and secured it as if it was his own."

The *Chronicle* to the contrary presents a generous Nikephoros:

> [He] found great spoils which he commanded be distributed among his army as per the troop roster. . . . When he opened the storehouses of [Krum's] wine he distributed it so that everyone could drink his fill.

What ensued was pillage and destruction. The *Chronicle:*

> [Nikephoros] left impious Krum's palace, and on his departure burnt all the buildings and the surrounding wall, which were built of wood. Next, not concerned with a swift departure, he marched through the midst of Bulgaria. . . .
>
> The army . . . plundered unsparingly, burning fields that were not harvested. They hamstrung cows and ripped the tendons from their loins as the animals wailed loudly and struggled convulsively. They slaughtered sheep and pigs, and committed impermissible acts [rape].

Theophanes inserts another missed opportunity to avert disaster:

> [Krum] . . . was greatly humbled and declared: "Behold you have won. Take, therefore, anything you desire and depart in peace." But the enemy of peace would not approve of peace; whereupon, the other [Krum] became vexed and gave instructions to secure the entrances and exits of his country with wooden barriers.

Evidently Krum was able to rally the Bulghar warriors who had fled into the mountains, and others too from farther afield. In the *Chronicle*, Nikephoros proceeds from hubris to lethargy, conceding the initiative to Krum:

> After he had spent fifteen days entirely neglecting his affairs, and his wits and judgment had departed him, he was no longer himself, but was completely confused. Seized by the torpor of false pretension, he no longer left his tent nor gave anyone an instruction or order. . . . Therefore, the Bulghars seized their opportunity. . . . They hired the Avars [a remnant by then] and neighboring Slav tribes [Sklavinias].

Krum's forces converging on the leaderless Byzantines, who had scattered to loot, employed a characteristic and unique Bulghar technique: the rapid assembly and emplacement of wooden palisades of logs bound with twine across the full width of narrow valleys, erecting "a fearsome and impenetrable fence out of tree trunks, in the manner of a wall" according to the *Chronicle*. These palisades were not fortifications that could resist a siege, but they could protect troops launching missiles from behind them, essentially negating the archery of the Byzantines while allowing the Bulghars to use their own bows through slits in the palisades—as former steppe nomads, many Bulghars must have retained both composite reflex bows and the skill to use them. Fighting barriers like expedient obstacles are efficient insofar as they are not easily circumvented. But according to the *Chronicle,* the Bulghars did not wait for the Byzantines to run into their palisade ambushes on their way home; instead they attacked, achieving complete surprise that induced a panic flight that in turn ended in massacre:

> They fell on [Byzantine soldiers] still half asleep, who arose and, arming themselves, in haste, joined battle. But since [the forces] were encamped a great distance from one another, they did not know immediately what was happening. For they [the Bulghars] fell only upon the imperial encampment, which began to be cut to pieces. When few resisted, and none strongly, but many were slaughtered, the rest who saw it gave themselves to flight. At this same place there was also a river that was very swampy and difficult to cross. When they did not immediately find a ford to cross the river, . . . they threw themselves into the river. Entering with their horses and not being able to get out, they sank into the swamp, and were trampled by those coming from behind. And some men fell on the others, so that the river was so full of men and horses that the enemies crossed on top of them unharmed and pursued the rest.

According to the *Chronicle,* there was but one palisade, which only intercepted fleeing remnants and was unmanned, rather than a fighting barrier:

> Those who thought they had escaped from the carnage of the river came up against the fence that the Bulghars had constructed, which was strong and exceedingly difficult to cross. . . . They abandoned their horses and, having climbed up with their own hands and feet, hurled themselves headlong on the other side. But there was a deep excavated trench on the other side, so that those who hurled themselves from the top broke their limbs. Some of them died immediately, while the others progressed a short distance, but did not have the strength to walk. . . . In other places, men set fire to the fence, and when the bonds [which held the logs together] burned through and the fence collapsed above the trench, those fleeing were unexpectedly thrown down and fell into the pit of the trench of fire, both themselves and their horses. . . .
>
> On that same day the Emperor Nikephoros was killed during the first assault, and nobody is able to relate the manner of his death. Injured also

was his son Staurakios, who suffered a mortal wound to the spinal verte-brae from which he died after having ruled the Romans for two months.

Nikephoros was the first Roman emperor to die in battle since the Goths killed Valens on August 9, 378, at Adrianople, but the catastro-phe of July 811 was even more dangerous because there was no spare emperor ready to exercise control, as the western emperor Gratian did in 378 until he appointed Theodosius as Augustus of the east in January 379. Moreover, the victorious Bulghars were within two hundred miles of Constantinople, unlike the Goths, who were a very long way from Rome when they won their victory.

As Krum offered barbarian toasts from the skull of Nikephoros, lined with silver in the usual manner, all seemed lost. Nikephoros had gath-ered every mobile force to overwhelm the Bulghars, so there was noth-ing left to stop them from seizing Constantinople after his ruinous defeat.

But there is a lot of ruination in an empire. In the east, the Abbasid ca-liphate, the greatest threat of all under the formidable Harūn al-Rashīd till his death in 809, was paralyzed by the war of his son Abu Jafar al-Ma'mun ("Belief") ibn Harun against his other son, the reigning caliph Muhammad al-Amin ("Faith") ibn Harun, whom he beheaded in 813. Hence field forces from the Armeniakon and Anatolikon themes could be summoned to help defend against the Bulghars. To lead them, there was at first only the badly wounded and unpopular Staurakios, son of Nikephoros, hastily proclaimed emperor in Adrianople on July 26; but on October 2, 811, he was forced to abdicate in favor of his brother-in-law Michael I Rangabe, chief palace official *(kouropalates)*, who gained the favor of Theophanes by repudiating Nikephoros to embrace Ortho-dox piety, by gifts of fifty pounds of gold to the patriarch and twenty-five pounds to the clergy, and by ordering the execution of heretics.

Michael readily went to fight, but unsuccessfully, and on July 11, 813, he abdicated in favor of the wily and battle-experienced Leon V (813–820), former strategos of the Anatolikon theme, who allowed Michael and his family to live peacefully as monks and nuns, after castrating his sons. So by the time Krum tried to attack Constantinople in earnest, there was a fighting emperor ready to defend the city.

One reason why Krum delayed so long was that he had lost many or most of the troops who had guarded his palace—probably his only veri-table soldiers, as opposed to Bulghar warriors who might be summoned to war and fight very well, but who were not "hand-picked and armed" and readily commanded. A second reason is that Krum could not attack Constantinople effectively without a fleet to blockade the city and starve it out eventually, or siege engines and the siegecraft to use them to breach the Theodosian Wall. The Bulghars were former mounted warriors of the plains who had also learned to fight very well on foot and in the mountains, but ships, shipping, and naval warfare remained outside their ken. Byzantine defectors were duly found and hired to provide the necessary siegecraft—Theophanes mentions a converted Arab expert, antagonized by the avarice of Nikephoros, of course—but it all took much time, and the necessary machinery was not constructed and ready until April 814, too late for Krum, who died on April 13, leaving an ineffectual successor. By then he had won another major battle at Versinikia on June 22, 813, overrun much Byzantine territory in what is now again Bulgaria, and Thrace, conquering its largest city of Adrianople and many smaller places; but the empire survived, and would one day recover all its lost territories.

The defeat of 811 was not caused by a lack of training or equipment, nor by tactical incompetence or even operational-level shortcomings. It was a fundamental error at the higher level of *theater strategy* that placed the Byzantine forces at a very great disadvantage, which only prompt and fully successful operation-level actions could have compensated and overcome. Carl von Clausewitz explains in his *On War* why no defense against a serious enemy should ever be conducted in mountains, if it is at all possible to defend in front of them instead or even behind them, if necessary conceding the intervening territory to temporary enemy occupation.[22]

It is true that mountain terrain offers many opportunities to establish easily defended strongholds, and narrow valleys offer many opportunities for ambushes. Both strongholds and ambushes can magnify the tactical strength of defending forces, allowing the few to prevail against the many at any one place. But if the army is thus fragmented by mountain terrain into many separate holding units and ambush teams, even if each one of them is tactically very strong, the overall defense is bound to be very weak against enemy forces that remain concentrated in one or two vectors of advance. The few defenders holding each place would then confront massed enemy attackers who can break through ambushes and overrun strongholds to advance right through the mountains, leaving most defending forces on either side marooned in their separate strongholds and ambush positions that were not attacked at all.

When the field army of Nikephoros advanced irresistibly all the way to Krum's capital at Pliska, it left Bulghar forces impotently scattered in mountains and valleys. In their tactically strong but strategically useless positions, they could not resist the Byzantine advance nor defend Krum's rustic palace. But they also remained unmolested by the Byzantine advance, and could therefore rally into action once they were summoned for Krum's counteroffensive against the Byzantines, now cut off a long way from home by the Bulghars in between. None of this could have happened if Nikephoros had read his Clausewitz, therefore concentrating all his efforts against Krum's army instead of Krum's palace. With Bulghar strength destroyed, Nikephoros could have had the palace and everything else without fear of a counteroffensive. Having mobilized the tagmata, thematic field forces, and irregulars and led them into Thrace, Nikephoros should have slowed down his advance or even stopped altogether for long enough to allow Krum to assemble his own forces. The resulting frontal battle of attrition would have been hard, no doubt, with heavy casualties, but given their numerical superiority if nothing else, the Byzantines would have won. Then Nikephoros could have settled down to reorganize the newly regained lands into taxpaying territories, confident that no significant Bulghar forces remained behind to attack him.

Alternatively, if Krum refused combat, Nikephoros could have advanced on Pliska to seize the palace just as he did, but then he should have swiftly retreated back into imperial territory, before the Bulghars could gather to interpose themselves between the Byzantine army and its home bases. That retreat, moreover, would have had to be conducted as carefully as if it were an advance, with scouts ahead and flanking forces to counter ambushes, and battle groups ready to break through Bulghar palisades.

The only way of remaining in Pliska and the conquered lands even though most Bulghar forces remained undefeated would have been to keep the Byzantine army concentrated and ready for combat at all times, to fight off any and all Bulghar attacks. But it is hard for occupation troops tempted by easy looting to retain their combat readiness, and such a choice would have been very dangerous strategically in any case, given that the empire had other enemies besides the Bulghars, starting with the Muslim Arabs whenever they were not divided by civil war.

Because in the event Nikephoros did not redeem his fundamental error of theater strategy, the untrained "poor men" with their clubs and slings were just as good or just as bad as the finest tagma in the field: both were equally cut off strategically and outmaneuvered operationally by Krum's Bulghars.

There would be no further attempt to extinguish Bulgaria until two centuries later. By then, the Turkic-speaking shamanistic Bulghars had been assimilated and refashioned as Slavic-speaking Christian Bulgarians, after the conversion of Qan or Khan Bogoris or Boris I in 865—Michael after his Byzantine godfather emperor Michael III. But not even Christianity could erase Bulgaria's original sin—its proximity to the Theodosian Wall.

That was the strategic context of Byzantine relations with the Bulghars and the Bulgarians. There was also a political context, which was even more unfavorable at times—not in spite of, but because of the adoption of the Byzantine religion and culture by Bulgarian rulers: it meant that they could dream of becoming emperors of all Christians, once they were recognized as emperors in the first place. That they achieved. The *Book of Ceremonies* records the transition from "the God-appointed archon [prince] of Bulgaria" to the later protocol: "Constantine and Romanos, pious Autocrats, Emperors of the Romans in Christ who is God, to our desired spiritual son, the lord . . . [Name] Basileus [= emperor] of Bulgaria."

It was not a willing promotion. In 913, after years of successful military expansion, the descendant of Turkic Bulghar qagans and first-generation Christian Symeon I (893–927)—until then just another "archon" or prince for the Byzantines—was crowned as a *basileus* by the patriarch Nicholas I Mystikos in the Blachernae imperial palace. Our eleventh-century source Scylitzes, a very high official himself as kouropalates, provides a highly official version:

> Symeon, ruler of the Bulghars, invaded Roman territory with heavy forces and, reaching the capital, entrenched himself on a line between Blachernae and the Golden Gate [the main ceremonial entrance on the via Egnatia]. His hopes soared that he would now easily take [the city]. But when he realized how strong the walls were, the number of men defending them and the abundant supply of stone-throwing and arrow-discharging devices they had to hand, he abandoned his hopes and withdrew . . . requesting a peace-treaty. . . . There were lengthy discussions when he came, then the Patriarch and the regents, taking the Emperor with them, came to the palace of Blachernae.
>
> When suitable hostages had been given [to the Bulgarians], Symeon was brought into the palace where he dined with the Emperor. He then bowed

his head before the Patriarch who said a prayer over him and placed his own monastic cowl [they say] on the barbaric brow instead of a crown. After the meal, although no peace-treaty had been concluded, Symeon and his children returned to their own land, loaded with gifts.[23]

Not quite. Symeon was not fobbed off with a cowl but was properly crowned as an emperor. The patriarch had become the residual legatee of imperial power as regent, and was in sole charge of diplomacy with Symeon, who refused to correspond with the emperor Romanos I Lekapenos (920–944). As befits a pious prelate, Nicholas I was first of all generous, materially as well: at one point he wrote to offer "gold, or raiment or even the grant of a portion of territory, such as may be of advantage to the Bulgarians while not causing intolerable loss to the Romans."[24] We only have the Nicholas side of their diplomatic correspondence, but it can be inferred that Symeon was not above accusing the patriarch of serving Byzantine interests instead of God—a shocking accusation.

It was once confidently asserted that for Symeon even crown and title were only half a loaf, that his higher ambition was to be enthroned as the emperor of Byzantium, including Bulgaria.[25] That is now disputed. What is certain is that for the empire it was a great humiliation to award the highest of titles, *basileus*. It was a time of disunity and weakness: the emperor Constantine VII Porphyrogennetos (945–959) was an eight-year-old child, his strong-willed mother Zoe Karbonopsina had been expelled from the palace, his uncle and co-emperor Alexander (912–913) died in June 913 as Symeon was approaching, a popular pretender to the throne, Constantine Doukas, was also advancing on the city, there was turmoil in the Byzantine enclaves in southern Italy, and an Arab invasion threatened Anatolia. This conjunction of threats precluded reinforcements for the capital. Scylitzes depicts a fully manned garrison that could defend the walls, but does not claim that there was enough force to drive away Symeon from Thrace.

Moderns who see only an obsession with empty titles in Byzantine distress over the concession of 913 miss the point: once there was another basileus who also protected the Orthodox Church, the emperor in Constantinople was no longer the only guarantor of the continued existence of the only true church that offered the only path to salvation for all mankind. The loss of that monopoly certainly eroded the authority of the Byzantine emperor over his Christian subjects, and diminished his prestige among Christians everywhere, including the followers of the pope of Rome, for no final schism had yet occurred.

The erosion of Byzantine religious authority would become more intense in 927 when, to reach a peace agreement, the emperor also had to recognize the Bulgarian Orthodox Church as autocephalous, with its own ecclesiastically independent patriarchate. There was no ideological objection to another patriarchate as such—after all, the autocephalous patriarchs of Alexandria, Antioch, and Jerusalem had long existed, but they were all Greek-speaking and with identical Greek liturgies; there was a cultural and professional objection to a non-Greek patriarchate and church that would no longer offer positions to Greek-speaking clerics.

Paradoxically but inevitably, given their missionary vocation, the independence of the Bulgarian church was made possible by Byzantine churchmen. In 886, disciples of the brothers and future saints Cyril and Methodius arrived in Bulgaria to instruct its aspiring clergy. These Greek-speaking missionaries did their work so well that by 893 the Bulgarians had their own priests and monks, and therefore felt free to expel all Greek clergy from Bulgaria. In spite of the perfect doctrinal unity of the two churches, their odium was remarkably persistent: a thousand years later, in 1912–1913, many died in Macedonia in fights over the control of local churches that were vehemently encouraged by their respective Bulgarian, Macedonian, and Greek priests.

The emperors of Byzantium were concerned not with the ownership of village churches but with the preservation of their own legitimacy. If a Bulgarian ruler could nominate his own patriarch, who in turn could anoint him as basileus, the emperor in Constantinople could no longer claim that there was but one legitimate emperor in the *oikoumene,* the Christian world.

In that context, labeling Symeon as "our desired spiritual son" in the salutation was a very deliberate devaluation of his rank, for previously he had been recognized as "our spiritual brother" in letters sent by Romanos I Lekapenos (920–944) collected by his chief of correspondence Theodore Daphnopates.[26]

Better than downgrading the spiritual brother to a subordinate son was to finish him off altogether—that is what the returned empress, mother and regent Zoe, attempted to do four years later, in 917, according to the chronicler Theophanes Continuatus:

> Aware of the elevation of [the Bulghar] Symeon [to imperial rank], and his attempts to gain control over [all] the Christians, the Empress Zoe [regent for Constantine VII Porphyrogennetos, 912–959] determined in council to

effect an exchange of prisoners and a peace treaty with the Agarenes [Arabs], and to transfer the entire Anatolian army to make war on and destroy Symeon. The patrikios John Rodinos and Michael Toxaras, therefore, set out for Syria to arrange the prisoner exchange. And having made the customary cash payments . . . the . . . armies were transported to Thrace. . . . [The commanders] swore together to die for each other, and they set forth in full array against the Bulgarians.

On August 20, in the fifth indiction, the battle between the Romans and Bulgarians was fought by the river Acheloos [near the Black Sea coast of Bulgaria]. And because the judgements of God are unfathomable and inscrutable, the Romans were completely routed. Their headlong flight was punctuated by fearful cries as some men were trampled by comrades and others were killed by the enemy; there was such a letting of blood as had not been seen for very many years.[27]

The destruction of the Bulgarian state had to await the next century.

A War of State Destruction: Basil II, 1014–1018

It was standard Byzantine practice to attack the Bulgarians whenever there was no active campaigning against the Arabs. John I Tzimiskes (969–976), who also defeated the Muslim Arabs and Svyatoslav of Kievan Rus', was notably successful. His victory was total, and extinguished the Bulgarian state in 971. All its territories were annexed and the autocephalous Bulgarian patriarchate was abolished. What ensued, however, was not a tranquil subjection but an almost immediate uprising in what is now Macedonia, led by the *Kometopouloi*, the four sons of a *comes* (local commander, our "count"), the youngest of whom, Samuil or Samuel, outlived his bothers eventually to claim the title of tsar or basileus.[28]

Byzantine armies could not be concentrated against the *Kometopouloi* because it was a time of acute internal strife. In 976, upon the death of emperor John I, the landed magnate and commander *(domestikos)* of the eastern armies, Bardas Skleros, advanced his candidacy to the guardianship—and effective overlordship—of the young co-emperors Basil II (976–1025), then eighteen, and Constantine VIII (1025–1028), then sixteen. On being denied it, he proclaimed himself emperor, starting a large-scale civil war that lasted until 979 when he fled to the Abbasids in Baghdad—leaving a damaged, divided, and demoralized army and empire in his wake. Only the intervention of Bardas Phokas, another wealthy magnate and experienced field commander, allowed Basil and Constantine to retain their thrones.

Samuel was therefore able to extend his sphere of control beyond his Macedonian beginnings eastward into the territory of what is now Bulgaria, while supplying his forces by vigorous raiding in northern Greece and Thrace.

In 986, with order seemingly reestablished within the empire and the Muslim Arabs still absorbed in struggle between the Shi'a Ismaili Fatimids and the Sunnis under nominal Abbasid leadership, Basil II, then fully an adult at twenty-eight but not yet proven as a field commander, set out to attack Samuel. He drove back the Bulgarians while marching to Sardika, modern Sofia, to reduce it by siege. By the tenth century, the techniques of undermining walls and countermining tunnels were dominant, although inherently slower than the mobile towers and battering rams favored earlier. In the event, Sardika held out and Basil II did not persist—his supplies might have been wanting, or perhaps potential instability at home called for an early return to Constantinople.

Basil had advanced on Sardika along the Roman "imperial highway," the *Basilike odos* that once ran all the way from Constantinople to northeast Italy and indeed beyond it to the North Sea. From Adrianople, now Edirne in European Turkey, the highway followed the Maritsa River valley between the high and rugged Haimos or Balkan mountain range to the north after which the entire region is now named, and the equally rugged Rhodope mountain range to the south, then continuing through Sardika or Sofia itself to Singidunum, now Belgrade in modern Serbia, and up to Aquileia across the Italian border east of Venice. Having abandoned the siege, Basil retreated the same way he came.

An emperor who is leading his army between two mountain ranges is anything but unpredictable in his movements—and Samuel's men, accomplished long-range raiders, were anything but lacking in mobility. They also had the specific Bulgarian ambush technique in mountain passes: they could quickly erect palisades in front of an approaching enemy, to better resist breakout attempts with a blocking force, so as to give more time to the ambushing forces to attack downslope the enemy immobilized below. Subsequent events show that Samuel's forces either were not in or around Sardika to begin with, or if they were, they somehow succeeded in outpacing the retreating Byzantine troops, even though they were marching along a road, while any outpacing would have had to be done on the adjacent slopes.

The outcome was a double disaster for Basil. His troops were ambushed in the narrow pass known as Trajan's Gate near Soukeis in modern Bulgaria, suffering very heavy losses. Basil ingloriously fled the scene, saving himself but dangerously undermining his authority. There is the great rarity of a credible eyewitness account by Leo the Deacon:

> The army was traversing a wooden defile, which was full of caves, and as soon as they passed through it they came to steep terrain, filled with gullies. Here the Mysians [= Bulgarians] attacked the Romans, killing huge numbers of men and seizing the imperial headquarters and riches and plundering all the army's baggage. I myself, who tell this sad tale, was present at that time, to my misfortune, attending the emperor and performing the services of a deacon . . . and the remains of the army, going through [nearly] impassable mountains, barely escaped the Mysian attack, losing almost all their horses, and the baggage they were carrying, and returned to Roman territory.[29]

Fana-Khusrau dubbed Adud al-Dawla ("Aid of the Dynasty"), the Persian and Twelver Shi'a emir of Baghdad nominally subordinate to the Abbasid caliph, reacted to Basil's defeat by releasing Bardas Skleros, whom he had both hosted and detained. When Bardas Skleros entered eastern Anatolia to claim the empire once again, Basil in turn had to summon the well-connected Bardas Phokas to help him once more. But this time Bardas Phokas decided to turn against Basil and the Constantinople bureaucratic elite to instead divide the empire with Bardas Skleros.

Three years of civil war ensued, and Basil was only able to restore order in 989 with the help of the six thousand warriors sent by Kievan Rus'.

After that, in spite of a reported operation in Bulgaria in 995, Basil's first priority could not be Samuel because once more he had to restore imperial control and prestige in the east, now threatened not only by the disorder left by civil war but also and to a far greater extent by the rising power of the Fatimids. "Sevener" Ismaili Shi'a, they had established their own rival caliphate in Egypt, and vigorously pursued expansion across the Sinai desert into Syria. Having long since driven back Arab inroads in Cilicia and Anatolia itself, the Byzantines were by then the overlords of both Christian and Muslim potentates as well as Bedouin tribes in Syria and beyond, and their lieges held the important cities of Antioch and Aleppo.

It is in those years that Basil II began to emerge as the most successful fighting emperor in Byzantine history. The restoration of the empire in eastern Anatolia and the rebuilding of the eastern armies was evidently a success. Reacting to the siege of Aleppo by a Fatimid army, in 995 Basil quickly arrived and promptly broke up the siege, then probing an advance toward Tripoli, now in modern Lebanon. Within a few years the expansionist drive of the Fatimids was convincingly stopped: in 1001 the Fatimid caliph al-Hakim negotiated a ten-year truce, which would be renewed for another ten years in 1011, and renewed again in 1023.

Meanwhile, farther north, Basil acquired a considerable territory in what is now southern Georgia, eastern Turkey, and western Iran. Having earlier persuaded or coerced the Armenian ruler David of Tao or Taron or Tayk to make the empire his inheritor, upon David's death in 1000 the transaction was completed. With that, the empire achieved its greatest expansion ever directly to the east, in excess of all Roman conquests, and later also expanded north into the Caucasus through the acquisition of other princely domains.

In the meantime, after his victory of 986, Samuel successfully expanded his own domain westward to the Adriatic Sea, north into what is now Kosovo and south into Greece. He also set the stage for claiming the imperial title by reviving the Bulgarian patriarchate in Ohrid, on the shores of the celebrated lake. When a Byzantine force under the noted general and putative author of the *Taktika* military manual, Nikephoros Ouranos, destroyed a Bulgarian force in 997, his very victory shows how badly the situation had deteriorated for the empire: the battle was fought very deep inside Greece on the river Sperchios, near modern Lamia, much closer to Athens than to Ochrid, Samuel's closest approximation to a capital.

Basil's second major attempt to dispose of Samuel did not start until 1001, with the Fatimid truce safely arranged. This time Basil did not try to drive back the Bulgarians, as Byzantine armies had often done before with inconclusive or disastrous results. Nor did he attempt to attack Samuel in his Macedonian home grounds in the western part of his expanded Bulgaria.

Instead Basil prepared for that more decisive confrontation by depriving Samuel of the most fertile and most populated territories of the original and present Bulgaria, the broad river valley south of the Danube.

There the Bulghar qagans had established their first encampment and then their capital at Pliska, then moving it to Veliki (Great) Preslav, both in modern northwest Bulgaria. To reach them, Basil's forces could advance along the Black Sea coast, well clear of the high Haimos (Balkan) Mountains with their dangerous passes. Or else, if they did cross them—the only nearly contemporary source, the *Synopsis Historion* of

Ioannes Scylitzes is inconclusive—Basil's men had evidently mastered adequate counterambush tactics. The military manuals then in circulation offered the correct remedies, based on the twin principles that low-lying passes and defiles must be overwatched by patrols advancing on ridge lines ahead of the main-force movements below, and that in mountain terrain especially, time and forces devoted to reconnaissance are seldom wasted. At any rate, the bitter experience of 986 was not repeated:

> In the year 6508 [since the creation in 5509 BCE = 1001] thirteenth year of the indiction, . . . the emperor sent a large and powerful force against the Bulghar *kastra* [strongholds] beyond the Haemus [Haimos] range, under the command of the Patrician [*patrikios,* field commander] Theodorokanos and [Nikephoros] Xiphias, [*protospatharios,* force commander]. Great Preslav, Lesser Preslav were taken; Pliska too; then the Roman army returned, triumphant and intact.[30]

After that, Basil's systematic campaign to reduce Samuel's territory—and prestige—continued year after year, to cut away at the political and logistic bases of his power.[31] Having started in the old Bulghar lands, he then directed his annual incursions into Samuel's Macedonian heartland as well. Bulgarian fighters could undoubtedly live off the land to some, or a large, extent—they were not a Byzantine or indeed a modern army that cannot long survive if cut off from its homeland bases and their supplies. But Samuel could not or would not give up his Macedonian base to persist in a war of pure movement. Thus when Basil II set out to attack Macedonia once again, the stage was set for a major battle, which turned out to be decisive.

It was fought in July 1014 in the Kleidion Pass through the Belasica (or Belasitsa) Mountains between the Struma and Vardar river valleys, near the point where the modern borders of Macedonia, Greece, and Bulgaria all meet.

Samuel relied on the usual Bulgarian operational method: ahead of Basil's advancing army, he blocked the pass with ditches and palisades, to set the stage for another successful large-scale ambush. By repeating his operational method, he allowed the Byzantines to study it, to identify its vulnerabilities, and to devise their own *relational* response. Samuel's method required the massing of his own forces behind the obstacles to fight Basil's advance, which meant that they too had to be in low-lying terrain overlooked by heights on either side.

That was the vulnerability that the Byzantines were able to exploit by sending a force to climb and then descend the heights, to fall on the Bulgarians. Surprised, and shocked, the Bulgarians could no longer defend the palisades to hold back Basil's main force, nor could they retreat under attack from the outmaneuvering Byzantine force. The result was the slaughter which—very much later, in the sadly diminished circumstances of the thirteenth century—gained for Basil the soubriquet *Boulgaroktonos*, the "Bulgar-Slayer."[32] It cost Samuel his army, his realm, and his life.

The account in Scylitzes may be a literary construct in part, but it is coherent and precise enough:

> The emperor continued to invade Bulgaria every year without interruption, laying waste everything . . . . Samuel could do nothing in open country nor could he oppose the emperor in formal battle. He was shattered on all fronts and his own forces were declining so he decided to close the way into Bulgaria with ditches and fences. He knew that the emperor was always in the habit of coming by way of what is called Kiava Longus [Campu Lungu] and the mountain pass known as Kleidion ["the key"], so he decided to block this pass. . . .
>
> He constructed a very wide fortification, stationed an adequate guard there and waited for the emperor who duly arrived and attempted to force a way in but the guards stoutly resisted. . . .
>
> The Emperor had already abandoned the attempt to pass when [Nikephoros] Xphias, then [Strategos] of Philippopolis [he had been promoted since 1001], agreed with the Emperor that he would stay there and make repeated attacks on the enemy's line while Xiphias would . . . go and see if he could do anything profitable. . . . He led his men back the way they had come. Then, trekking, around the very high mountain which lies to the south of Kleidion and which is called Valasitza [Macedonian Belasica], passing by goat-paths and through trackless wastes, on the 29th of July, twelfth year of the indiction [= 1014], he suddenly appeared above the Bulgarians and came down on their backs with great cries and thundering tread. Completely taken aback by the unexpected nature of this attack, they turned and fled. The Emperor dismantled the abandoned defence-work and gave chase; many fell and even more were taken prisoner. Samuel was only just able to escape from danger, by the cooperation of his own son who stoutly resisted those who attacked, got his father onto a horse and led him to the fortress called Prilapos [Prilep, Macedonia]. They say that the Emperor blinded the prisoners, about fifteen thousand in number, with orders that one man for each hundred be left one eye so he could be their guide, then sent them back to Samuel, who died two days later on 6 October.[33]

The much-cited story of the fifteen thousand blinded captives sent back in batches of one hundred guided by a one-eyed man sounds like a tall tale, and it probably is, though blinding was much used at the time as

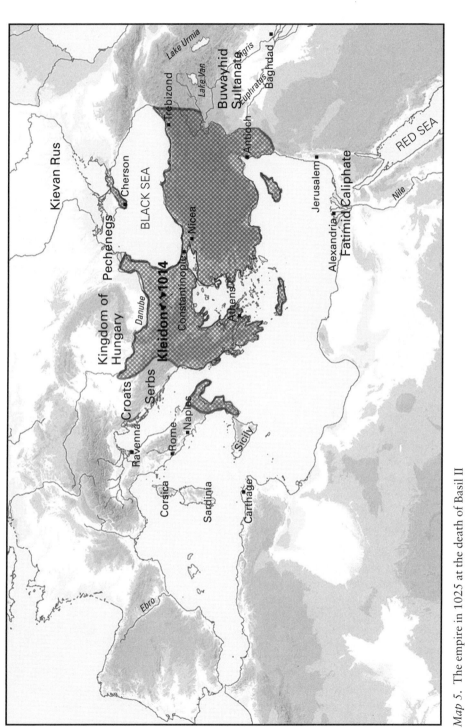

Map 5. The empire in 1025 at the death of Basil II

the more Christian penalty because God-given life was not taken away. But the firm fact that Bulgarian resistance lasted for four more years until 1018 argues against the loss of fifteen thousand fighters, a great number given the size of the population. Even then the final submission of the remaining Bulgarian leaders was not unconditional: they were given lands in eastern Bulgaria.[34] Perhaps only a few prisoners were blinded and sent to Prilep to demoralize Samuel.

More important is that after Kleidion, Byzantine rule was restored from the Adriatic Sea to the Danube for the first time in three centuries. Theater-level *relational* maneuver is the highest form of the art of war.

"Byzantine" Diplomacy in Byzantium

In 896 Leo Choerosphactes was sent by emperor Leo VI (886–912) as an envoy to Symeon of Bulgaria (893–927) to secure the release of Byzantine captives.[35] At that point, Bulgaria was more powerful than the Byzantine empire in the Balkans, and Symeon was striving to be recognized as an emperor also, within a common Orthodox cultural and religious and sphere.[36] In that spirit, Symeon jocularly asked Leo to predict whether the prisoners would be released—for emperor Leo VI had tried to impress the Bulgarians by predicting the recent eclipse of the sun.

Within his letter sent in reply, Leo Choerosphactes responded to the specific question with a sentence whose word order was complicated and which contained no punctuation, so that the meaning remained deliberately ambiguous—though in the simplest reading the answer was no, they would not be released. Symeon replied sardonically that if Leo Choerosphactes had been able to predict the outcome correctly (yes they would be released) he would have released the prisoners, but because did not do so, he refused to release them.

Leo replied by claiming that his letter did predict the outcome, but that Symenon's secretary had failed to interpret the letter correctly, because he did not insert the appropriate punctuation.

Symeon replied, "I did not make a promise about the prisoners; I did not say anything to you; I will not send them back to you."

Leo replied in turn by keeping the same words but twisting their meaning by inserting his own punctuation: "I did not fail to make you a promise about the prisoners—using two negatives to make a positive statement—I spoke to you about it; what is there that I shall not send back to you?" The Bulgarians did eventually release the prisoners.

Leo had failed to advance a cogent argument, and his epistolary attempt to manipulate the text to have Symeon mean what plainly he did not intend, was more childish than cunning. But evidently Symeon wanted to be in communion with Byzantium, and the prisoners were released.

CHAPTER 9

The Muslim Arabs and Turks

In the Book of Ceremonies we find:

> To the kyrios [lord] of blessed Arabia. A gold bull. "Constantine and Romanos, faithful to Christ the Lord, Great Autocrats and Emperors of the Romans, to . . . [Name] ruler of Arabia.

But in the tenth century there was no lord of Arabia, in the sense of the old Roman province of *Arabia Petrae,* "stony Arabia," within today's kingdom of Jordan. The Ghassanid assemblage of Christian Bedouin tribes, which had served the empire well in guarding the desert approaches to the Levant from both Sasanian outflanking maneuvers and Bedouin raids, had been extinguished by the Muslim conquest. Besides, their client-ruler would have borne the title *phylarch*—tribal chief—or more rightly *megaphylarch,* paramount chief, rather than *Kyrios.*

Nor was there a lord of the Arabian Peninsula, even though it had been unified in the seventh century by Muhammad's charismatic leadership and his militant new religion that combined unadulterated Jewish monotheism, redemptive missionary conquest, legitimized plunder, and the promise of superiority over unbelievers in all things. Paradoxically, the very success of the Muslim Arabs in conquering in all directions left Arabia itself without a center of power as Damascus, Baghdad, Aleppo, and Fustat (now in Cairo), among others, became Muslim Arab power centers.

Within a year of Muhammad's death in 632, under the leadership of his erstwhile companions and self-appointed successors, Abu Bakr,

'Umar ibn al-Khaṭṭāb, and 'Uthmān ibn 'Affān, and their field commander Khālid ibn al-Walīd, his followers among the Muslim Arabs went on plunder raids into Byzantine Syria and Sasanian Mesopotamia. The raids were so successful that they were directly followed by conquering and missionary expeditions.

Jihad, the holy struggle against unbelievers, is not an essential "pillar" *(arkan)* of Islam.[1] One reason the Kharijites were marginalized as Islam's first extremists was that they did elevate war against the infidel as a fundamental precept, as do the Alawites of Syria still and all contemporary jihadis—who must now be described as *ultra*-extremists because the eighteenth-century extremism of Muhammad ibn 'Abd al-Wahhab, which prohibits all amity with non-Muslims, is the state religion of Saudi Arabia.

Though not an absolute obligation upon all believers, jihad is a religious duty that all passably orthodox Muslim jurists place immediately after the *arkan,* because of orders from God himself in the Qur'an, notably II:193: "Fight them [the unbelievers] until there is no dissension and the religion is entirely Allah's." Hence jihad is a temporary condition that ends when all humans have become Muslim; until then it is a duty for the Muslims as a whole, though not for every individual Muslim, as extremists would have it.[2] These days much is made of *al-Jihad-al Akbar,* "the great struggle" against one's own carnal desires, which would downgrade war against the infidel to *al-Jihad al-Asghar,* the small struggle. But that is the heterodox interpretation of some Sufis and liberal clerics, largely ignored by mainstream Muslims, including most Sufi movements. Mild, humanistic, tolerant versions of Islam dominate the teaching of Islam in Western universities but remain unknown or at best marginal in Muslim lands, except for such minorities as the Bektashi Alevis of Turkey and former Ottoman lands, whose humanism is both ancient and authentic.[3]

Muhammad's religion promised victory, and the advancing Muslim Arabs saw that promise triumphantly validated by the seemingly miraculous defeat of the vast, ancient, and till then all-powerful Roman and Sasanian empires, which between them had long dominated all the lands of the Middle East fertile enough to be worth ruling.

The two empires had just finished the longest and most destructive of all their wars—almost thirty years of wide-ranging reciprocal invasions that had ruined many of their cities, destroyed commerce, emptied their treasuries, exhausted their manpower, and wrecked frontier defenses and field armies alike, while bitterly antagonizing provincial populations on each side, left undefended to be despoiled by enemy looters yet harshly taxed before and after. A few years of tranquility might have restored the strength of both empires beyond any challenge by Arab raiders, no matter how enthusiastic, but instead both were invaded and each suffered a catastrophic battle defeat.[4]

In 632 when Muhammad died, no reasonable person could have foreseen that the Roman empire that had possessed Syria, Egypt, and all the lands between them for six centuries would lose every part of them by 646. Most had been lost even earlier, after the army sent by the emperor and erstwhile great conqueror Herakleios was utterly defeated at the river Yarmuk in August 636.

Arabs of any faith had never been formidable before. Their new ideological cohesion was probably underestimated, as their ability to mobilize almost certainly was.[5] But battles unfold as tactical and operational phenomena subject to their own circumstances, each side can decide and execute in ways less or more effective, and it seems that the Byzantine commanders Vahan and Theodore Trithurios made identifiable tactical errors.[6]

In this case also, broader factors were more important than tactics, because in the same year the Muslim Arabs also attacked the Sasanian empire of Persia, whose power had very recently stretched from the Mediterranean to the Indus Valley. It too was decisively defeated in 636, at al-Qādisiyyah in Mesopotamia, losing its treasury and capital city Ctesiphon. After a last attempt to defend the Persian hinterland at the battle of Nihawand in 642 commanded by the king of kings Yazdegerd III himself, resistance and the Sasanian empire with it waned, ending by 651.

What the Muslim Arab conquerors themselves humbly saw as a divine victory, *Nasr Allah,* can be recognized in retrospect as something even better, a political victory over both empires that won not merely vast territories but also the consent of many of their inhabitants.

The impetuous Arab advances could have been nothing more than ephemeral raids, destined to be nullified by nativist resistance, had the invaders not offered two very great and immediate advantages with their arrival.

One was a drastic reduction in taxes that had become ruinously onerous. The other was truly paradoxical: by imposing discriminatory rules on all non-Muslims, the Muslim Arabs ended the arbitrary religious persecutions that had recently oppressed a majority of the inhabitants of Syria and Egypt.

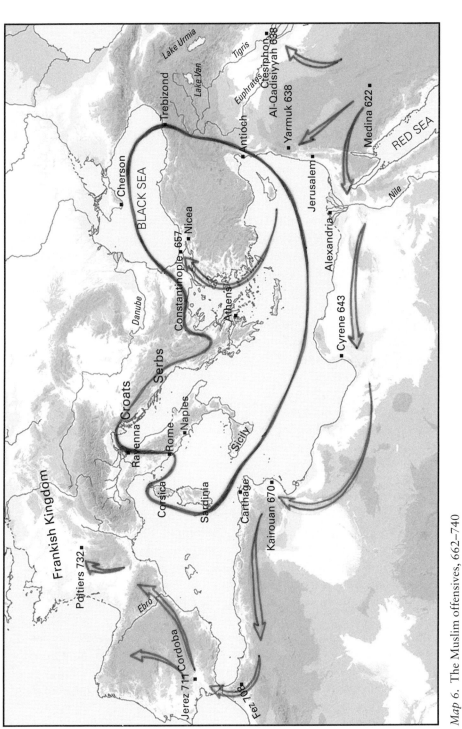

Map 6. The Muslim offensives, 662–740

The Muslim Conquest and Tax Reduction

Muslim taxes could be low because the cost of Muslim rule was very low at first. The conquerors had neither a vast imperial overhead of bureaucrats and courtiers in the austerity of Mecca and Medina, nor were they trying to rapidly rebuild wrecked imperial armies as both the Byzantines and Sasanians were doing in those years. The taxes imposed by the Muslim authorities were both harshly discriminatory, because only non-Muslims had to pay most of them, and blessedly lower than the relatively well-documented Byzantine taxes, and known Sasanian taxes.[7]

While nobody has ever been able to prove—as many have tried to prove—that the Roman empire "fell" because of excessive taxation, it was and remained until the mid-seventh century a top-down system, whereby the total amount of imperial expenditure for the coming year was determined first, the revenue needed was then calculated province by province, and that total was in turn allocated within each province among its registered taxpayers, mostly payers of the land tax, according to periodic assessments of the agricultural yield of each tract *(jugatio)* and the available manpower *(capitatio).*[8]

It was a uniquely sophisticated and very effective system of collection, which was indeed the central advantage of the Roman and Byzantine empire over all other contemporary powers. It did mean, however, that the taxpayer had to pay a precalculated amount regardless of good or bad harvests, droughts or floods, destructive foreign raids, or even outright invasions. An especially dramatic disaster that attracted much attention might persuade the imperial authorities to reduce the revenue obligation of the affected province, but no allowance could be made for ordinary harvest or market fluctuations, because there was no way of offsetting lost revenues: the concept of the public debt and its sale in the form of interest-bearing bonds had not yet been invented.

The purchase of remunerated government positions, which swapped a single capital payment for a revenue stream, was the functional equivalent of selling bonds to the public, but it could not be widely practiced. Hence current expenditures had to be paid for by current taxes in a strict pay-as-you-go sequence—a tolerable burden in good years but harsh in bad years, and sometimes reason enough to flee homes and lands ahead of the tax collectors.

Fundamentally, Byzantine tax collection was simply too effective. Emperor Anastasios (491–518) had his share of foreign incursions to confront with costly military operations, and four years of more costly full-scale war with ever-aggressive Sasanian Persia from 506, and he also spent vast sums on public works, among other things substantially rebuilding and fortifying the Long Wall and building the fortress city of Dara (near Oğuz, Turkey), "fortifying it with a strong circuit wall and bestowing on it . . . not only churches and other sacred buildings but colonnades, and public baths."[9]

Anastasios spent much, yet he was able to abolish the *collatio lustralis,* a top-down capital levy on every form of wealth: buildings, animals, tools, and the slave-value of artisans, merchants, and professionals, excluding teachers but including prostitutes and catamites. It was originally collected every five years *(lustrum),* which became every four years in the normal way of taxes by the time of Anastasios, but either way it was very hard for artisans and small merchants to come up with the gold payment all at once (in spite of its Greek name *chrysargyron,* "gold-silver," only gold was accepted by the tax collectors). The text known as *A Historical Narrative of the Period of Distress Which Occurred in Edessa, Amid and All Mesopotamia,* also known as *The Chronicle of Joshua the Stylite,* describes the ecstatic reaction to the levy's abolition in the town of Edessa, whose assessment was 140 pounds of gold, 10,080 solidi, evidently a crushing burden:

> The edict of the emperor Anastasios arrived this year, remitting the gold which tradesmen paid every four years and freeing them from the tax. This edict did not go only to Edessa but to all the cities of the Roman domain . . . and the whole city rejoiced, and they all dressed up in white, from the greatest to the least, and carrying lighted candles and burning censers, to the accompaniment of psalms and hymns, they went out . . . thanking God and praising the emperor . . . they extended the feast of joy and pleasure for a whole week. . . . All the artisans sat around and had a good time, [bathing and] relaxing in the courtyard of the City church and all the city's colonnades.[10]

Having both spent much and given up much revenue—but he also increased the efficiency and probity of tax collection—Anastasios left 3,200 *centenaria* of gold, that is, 320,000 Roman pounds, in the treasury at his death.[11] As of this writing, the price of gold is roughly US$903 per ounce or 31.1 grams, so the surplus left by Anastasios came to roughly US$3,039,496,257—not much these days, but gold was much more valuable then, in terms of bread, for example.

At the time of the Arab invasions there was no budget surplus to hoard. Thirty years of war had increased expenditures while greatly reducing revenues, leaving the treasury empty or near enough. Hidden reserves—such as ecclesiastical ornaments in gold and silver that could be confiscated in a crisis—were also exhausted. Already in 622 emperor Herakleios "took the candelabra and other vessels of the holy ministry of the Great Church [the Hagia Sophia], which he minted into a great quantity of gold and silver coin."[12] The result was that tax revenues had to be collected from Syria and Egypt as soon as they were reconquered after years of Sasanian occupation—and these were lands that had been taxed by the Byzantines, invaded and taxed by the Sasanians, fought over repeatedly and often looted, before being regained to be taxed again. The empire was rebuilding its strength, and its subjects had to raise the necessary gold, or else face expropriation or worse. It was too much. They welcomed the Muslim Arabs instead, discriminatory poll tax and all.

There is less information on Sasanian taxes, but there was certainly a land tax, *tasqa* in the Talmud's Aramaic, and a head tax, *karga*. The tasqa was set high, at least for buildings, and was inflexible. A passage in the tractate *Nedarim* of the Babylonian Talmud illustrates the first point by noting that the specific transaction being discussed is ethically allowed if the lessee rents from the owner in exchange for payment of the tasqa—implying that the tax could well consume all the rental income derivable from a property.[13]

As to inflexibility, there is a chilling anecdote in the best source we have on Sasanian taxes and much else: "The History of Prophets and Kings" (Ta'rikh al Rusul wa'l-Muluk) of the singularly instructive Islamic historian Abu Ja'far Muhammad bin Jarir al-Tabari (839–923), who wrote a universal history of the lands of Islam replete with accurate information and timeless insights.

In explaining how the Sasanian fiscal system was drastically reorganized, al-Tabari comes to the cadastral survey of agricultural production and yields—evidently copied from Byzantine practice.[14] This had been ordered by Kavad I, who died in 531:

> When his son Kisra [Khusrau I Anushirvan, 531–579, Chrosoes to the Greeks] succeeded to power, he gave orders for it to be carried out . . . and for an enumeration to be made of the date palms, olive trees and heads [of workers, the Byzantine *capitatio*]. He then ordered his secretaries to calculate the grand total of that, and he issued a general summons to the people. He commanded the secretary responsible for the land tax to read out to them the total tax liabilities from the land and the numbers of date palms, olive trees, and heads . . . after which Kisra said to them . . . "We ordain

that the taxation should be paid in installments spread over the year, in three installments. In this way, sums of money will be stored in our treasury so that, should any emergency arise along one of our vulnerable frontiers . . . or anything else untoward, and we have a need to . . . nip it in the bud, involving the expenditure of money . . . we shall have money stored up here, ready and to hand, since we do not wish to have to levy a fresh installment of taxation for that emergency. So what do you think about the procedure we have envisaged and agreed upon?"

Evidently Khusrao I was proud of his innovation, which was actually his father's, and which was actually a copy of the Roman and Byzantine *jugation-capitatio* system. But the assembled multitude was wise in the ways of absolute monarchs:

None of those present . . . uttered a single word. Kisra repeated [his request for comments] three times. Then a man stood up from out of the expanse of persons present and said to Kisra: "O King—may God grant you long life! you are establishing a perpetual basis for this land tax on transient foundations: a vine that may die, land sown with corn that may wither, water channel that may dry up, and a spring or qanat [underground channel] whose water supply may be cut off?"

It was the wrong thing to say.

Kisra replied: "O troublesome, ill-omened fellow, what class of people do you come from?" The man said, "I am one of the secretaries." Kisra gave orders, "Have him beaten with ink holders until he dies." Hence the secretaries in particular beat him with their ink holders, seeking to dissociate themselves, in Kisra's eyes, from the man's views and utterance, until they killed him."

At that point all knew what was expected of them:

The people said, "O King we are in full agreement with the land tax which you are imposing on us."[15]

All states ultimately derive their material power from their ability to extract revenue from their populations, whether by customary obedience or the fear of punishment. Khusrao's system was new, so it could not be sustained by habitual obedience. But he was fortunate in war so that his tribute revenue reduced his need to extract taxes—which were also moderated in important ways. Taxpayers could ask administrative judges to intervene if tax collectors demanded sums in excess of the amount laid down in the master copy of the assessment in Khusrao's chancery, of which they had a copy. Because only designated crops—

wheat, barley, rice, grapes, clover, date palms, and olive trees—were taxed, at a minimum the population was supposed to have enough to live on from the farm animals and vegetables that were tax-exempt. As for the poll tax *(capitatio)*, it was not levied below the age of twenty or over the age of fifty, and it was progressive from 4 to 12 dirhams, that being the drachma of 3.4 grams of gold, less than the weekly wage of a laborer.

Indeed, the system was so moderate in principle that the conquering caliph ʻUmar ibn al-Khattāb added a tax on uncultivated land without meeting known resistance, probably because he too "excluded from liability to taxation the people's means of daily sustenance." But Khusrau II (591–628), who reigned in the generation just before the Arab conquests, needed far more revenue to pay for large-scale warfare. Only terror could raise all he needed from territories increasingly depleted of manpower of military age. It was just the same on the Byzantine side of the border.

Christians, Jews, and the Muslim Conquest

The second advantage of Muslim rule was that its religious discrimination was better than Byzantine persecution. Pagans who refused to convert were to be killed, but in the former Byzantine and Sasanian lands they were few, long since outlawed, and well hidden. By contrast, the "peoples of the book" identified in the Qur'an, the Christians and Jews, to whom Zoroastrians, Sikhs, and Hindus would later be added out of sheer necessity, were allowed to live in safety as disarmed inferiors under the "pact of protection," the *ahl-al-dhimma*.

Exempted from military duties, all dhimmis, "that is protected persons," had to pay the *jizya* poll tax and moreover do so under humiliating conditions. The Qur'an, the very word of God according to its devotees, is explicit: "Fight those who believe not in God nor the Last Day, nor hold forbidden that which hath been forbidden by God and His Messenger, nor acknowledge the religion of Truth, [even if they are] of the People of the Book, until they pay the *Jizyah* with willing submission, and feel themselves subdued."[16] Procedures varied and could be lax, but for those who believe that conversion to Islam is the only path to salvation, there is ample moral justification to vex the dhimmis until they see the light. In later centuries eminent jurists offered varied procedures to implement Sura 9.29: holding taxpaying unbelievers by the beard and striking both cheeks being something of a favorite.[17]

At first Muslim discriminatory rules were largely copied from earlier Byzantine laws against heretics and Jews. It was only later, when Arab and Muslim fortunes declined and the power and the glory were inexplicably vouchsafed to infidels, igniting Islam's crisis of credibility, which still infuriates its devotees, that jurists and local authorities competed in inventing new restrictions and humiliations; and the Shi'a led the way, as the humiliated delighted especially in humiliating, in the usual way of mankind (the post-postmodern Grand Ayatollah Seyyed Ruhollah Musavi Khomeini even revived "purity" restrictions against "unclean" Christians, Jews, and Zoroastrians; in the Islamic Republic of Iran they are forbidden to touch food or drink destined for Muslim consumption).

But in the immediate aftermath of the conquests, with the Muslim Arabs few and mostly ensconced in their garrisons, everyone could live much as they pleased. Muslim discrimination, moreover, had the immense advantage of being nondiscriminatory—all categories of Christians and Jews were treated equally, whether well or badly. That was highly desirable for most of the population in the Byzantine territories that came under Muslim rule, starting with a majority of the Christians themselves: the Monophysites of Syria and Egypt.

They had been harshly persecuted by the Byzantine authorities to persuade them to accept the christology of the Council of Chalcedon of 351, still now upheld by most Christian denominations, whereby both divine and human natures coexist within the single essence of Christ. But most native Christians of Syria and Egypt were and remain Monophysites, adhering to the one-nature doctrine of their Coptic and Syriac Orthodox churches, while only a Greek-speaking and elite minority was Chalcedonian and therefore unpersecuted by the Byzantine authorities.[18]

It was a very damaging breach in the cohesion of the empire. The Monophysite author of the text known as the *Chronicle of the Pseudo-Dionysius of Tel-Mahre* lists the names of the bishops who were "chased out of their sees," fifty-four in all; and the greater figure of Severus, patriarch of Antioch, also had to leave his post. The author then described the newly installed Chalcedonian patriarch of Antioch as "Paul the Jew. . . . The instrument of perdition was chosen and sent here—Paul [also] called Eutyches, that is a jew if it be allowed to say so . . . , it was he who introduced [the doctrine] of the despicable Council of Chalcedon."[19] There was the scandal of Chalcedonian and non-Chalcedonian monks wounding and killing each other in fighting over churches and monasteries. More important politically were the sanguinary tumults that erupted whenever Byzantine authorities tried to confiscate churches and patriarchical installations, expelling or arresting Monophysite prelates, who were supported by most of the population precisely where the Sasanian armies and later the Muslim Arabs invaded the empire, from Antioch in Syria to Alexandria in Egypt.

Such was the intensity of doctrinal hatreds that the two sides had different definitions of the enemy: for the Monophysites, it was the Chalcedonians, not the Muslim Arabs. The *Chronicle of 1234,* written by a Monophysite, recounts how Theodoric, brother of the emperor Herakleios, was marching his forces in Syria to fight the invading Muslim Arabs:

> When they reached the village of al-Jusiya, Theodoric approached a stylite [a very public hermit] standing on his pillar: the man was a Chalcedonian. At the end of the long conversation which ensued between them, the stylite said to Theodoric: "if you will only promise that on your safe and victorious return from the war you will wipe out the followers of Severus [expelled monophysite Patriarch of Antioch] and crush them with excruciating punishments." . . . Theodoric replied, "I had already decided to persecute the Severans without having heard your advice." Then the author gleefully recounts how the Byzantines were defeated by the Muslim Arabs.[20]

Herakleios (610–641) tried to unify his subjects in extremis by offering a neat christological compromise, or at any rate allowed his patriarch Sergius I to do so in the 638 *Ekthesis.* It proclaimed the monothelite ("one will") doctrine, whereby Christ has two natures, human and divine, but in perfect teleological union within a single will.[21] That was a product-improved version of the emperor's first try, monoenergism, whose great virtue was that the "single energy" of Christ was left undefined to accommodate everyone.

At first well-received locally and willingly accepted in Rome by Pope Honorius I (610–638), monotheletism was rejected by its most important target audience, the Monophysites themselves: their Semitic monotheism would not be softened by Greek sophistry.[22] At the same time, firm Chalcedonians opposed any compromise; at their insistence, monotheletism was condemned as heretical by the Sixth Ecumenical Council of 680.

By then almost all Monophysites were under Muslim rule anyway. We have the perfectly contemporary word of John, Monophysite bishop of Nikiu in Egypt, that the Muslim conquest was divine punishment for the persecution of his faith, and a relief for the persecuted:

> [The Byzantine] troops and officers . . . abandoned the city of Alexandria. And thereupon 'Amr the chief of the Moslems made his entry without effort into the city of Alexandria. And the inhabitants received him with respect; for they were in great tribulation and affliction.
>
> And Abba Benjamin, the [Monophysite] patriarch of the Egyptians, returned to the city of Alexandria in the thirteenth year after his flight from the Romans, and he went to the Churches, and inspected all of them. And every one said: "This expulsion of the Romans and victory of the Moslem is due to the wickedness of the emperor Heraclius and his persecution of the Orthodox through the [Chalcedonian] patriarch Cyrus." This was the cause of the ruin of the Romans and the subjugation of Egypt by the Moslem.[23]

Things were simpler for the Jews, who remained numerous in their homeland, in Egypt and in Mesopotamia, where the Babylonian Talmud was redacted from transcripts of rabbinical debates at the schools of Pumbedita (now Iraq's al-Fallujah) Sura, Nisibis (the Nusaybin of modern Turkey), and Mahoza—the Aramaic name of the Sasanian capital of Ctesiphon, near modern Baghdad.

Muhammad had enriched his followers by robbing the Jewish oasis of Khaybar north of Medina, had expelled the Jewish Banu Nadir tribe of ironworkers from Medina among other exactions, and the Qur'an expresses his bitter resentment of the Jewish refusal to accept his improvements to their ancient faith—even though he had paid the ultimate compliment of incorporating much of Judaism in his new religion.[24]

In spite of this, the Jews still welcomed the Arab conquests, as did the Monophysite majority among the Christians, and for exactly the same reason: there was perfect equality under Arab discrimination, with Jews accorded the same limited but stable rights as other dhimmis, including the previously privileged "Christians of the king," the Chalcedonians.

That was a huge improvement, because Byzantine emperors periodically decreed increasingly restrictive laws against the Jews, none more so than Herakleios, who seemingly ordered their forcible conversion, according to the contemporary "Jacob the recently baptized."[25] That would have been in retaliation for the help that local Jews had supposedly given to the Sasanian Persians in conquering Jerusalem in 614—one of the larger disasters of the last and most disastrous war between the two empires. For some time, the Sasanian revival of Zoroastrianism had entailed the persecution of other faiths. Already under Khusrau's predecessor Hormizd IV (579–590), Christians and Jews were driven to flee—among them the entire Talmudic school of Pumbedita, according to the *iggeret Rav Sherira Gaon*—the epistle of Rabbi Sherira, head *(gaon)* of Pumbedita three centuries later.[26]

Unless they were remarkably ill-informed, the Jews of Jerusalem were unlikely to have taken many risks to help Khusrau II replace Byzantium's intolerance with his own. But in a time of defeat and demoralization anything discreditable to the Jews is easily believed—and willingly embellished or rather uglified, in this case by a claimed eyewitness: Antiochus Strategos, monk of the still extant monastery of Mar (saint) Saba, whose text, but for an original fragment, survives only in Old Georgian translated from the original Greek, or possibly from an Arabic translation of the Greek:

> The vile Jews, enemies of the truth and haters of Christ, . . . rejoiced exceedingly [at the fall of the city] because they detested the Christians. . . . In the eyes of the Persians their importance was great, because they were the betrayers of the Christians. . . . As of old they bought the Lord from the Jews with silver, so they purchased Christians out of the reservoir [where they had been imprisoned]; . . . they gave the Persians silver, and they bought a Christian and slew him like a sheep. The Christians however rejoiced because they were being slain for Christ's sake. . . . When the people were carried into Persia, and the Jews were left in Jerusalem, they began with their own hands to demolish and burn such of the holy churches as were left standing.[27]

That Jews would ransom Christians just to kill them for the pleasure of it sounds like malevolent fantasy; Antiochus Strategos was not the first or last churchman to vent his hatred of the Jews in frustration over their sheer persistence—which the Church itself allowed by excluding the Jews alone from the outlawing of all other non-Christian religions. Long before 614 all known non-Christians within the empire had been forced to convert on pain of death, or simply massacred. Only the Jews were allowed to live as non-Christians, but not to live well, or securely.

A cascade of legislation that was continue for two centuries imposed both religious restrictions on proselytization and expression ("mockery") and civil disabilities. Most importantly, a law of March 10, 418 (Theodosian Code XVI,8,24) barred Jews from imperial employment—a huge deprivation because there was no other employment even remotely comparable:

> The entrance to the State Service shall be closed from now on to those living in the Jewish superstition. . . . We concede therefore to all those who took the oath of the service, either among the [*agentes in rebus* = junior administrators] or among the [*Palatini* = palace accountants] the opportu-

nity to terminate their service on its statutory term, suffering the deed rather than encouraging it, though what we wish to be alleviated at present to a few shall not be permitted in the future. As for those, however, who are subject to the perversity of this nation and are proven to have entered the Military Service, we decree that their *cingulum* [the cincture, the military belt, the symbol of a Roman soldier] shall be undone without any hesitation, and that they shall not derive any help or protection from their former merits. Nevertheless, we do not exclude Jews educated in the liberal studies from the freedom of practicing as advocates, and we permit them to enjoy the honor of the curial liturgies [compulsory municipal duties], which they posses by right of their birth's prerogative and their family's splendor. Since they ought to be satisfied with these, they should not consider the interdiction concerning the State Service as a mark of infamy.[28]

That was perhaps unintended irony, because nobody wanted the costly and uncompensated role of a decurion. Nevertheless, until the sixth century the Jews still enjoyed legal protection against violence—including mob actions instigated by priests alarmed by the proliferation of "heaven fearers" *(caelicolae)* who followed Jewish rites without formally converting. A law of August 6, 420, collated into the Theodosian Code (XVI, 8, 21) and copied into Justinian's Code (I, 9, 14) prescribed:

No one shall be destroyed for being a Jew, though innocent of crime. . . . Their synagogues and habitations shall not be indiscriminately burnt up, nor wrongfully damaged without any reason

But the Jews were then warned that they must remain humble:

But, just as we wish to provide in this law for all the Jews, we order, that this warning too shall be given, lest the Jews grow perchance insolent, and elated by their own security *(ne iudaei forsitan insolescant elatique sui securitate)*, commit something rash against the reverence of the Christian cult.[29]

At that point, the legal status of the Jews in the Roman empire was in an average condition: worse than before, better than it would become. On January 31, 438, Theodosios II with Valentinian III promulgated a new law, possibly instigated by monks in Jerusalem, whereby "Jews, Samaritans, Pagans and Heretics" were excluded from all offices and dignities, including municipal ones—except for those officeholders *(curiales)* who were forced to use their own money to carry out their duties. The law also prohibited the building of new synagogues and stated that any Jew who converted anyone to Judaism was to be executed and his property confiscated. Under Justinian, eleven major new laws from 527 to 553 added more civic and legal restrictions and decreed harsher punishments, while offering an incentive for conversion to Christianity—any convert in a group of Jewish inheritors was to have the entire inheritance.[30] In sum, the Jews were allowed to live while all other non-Christians, entire populations of them, were exterminated, but they were not given reasons to be loyal to Byzantium either: when the Muslim Arabs invaded Mesopotamia circa 634, Rabbi Isaac, head *(gaon)* of the Pumpedita school, willingly greeted the conqueror ʿAlī ibn Abī Ṭālib, husband of Muhammad's daughter Fatimah, fourth caliph of Islam.

The Caliphate and Constantinople

Unable to assume Muhammad's prophetic role, his successors Abu Bakr, ʿUmar ibn al-Khattāb, and ʿUthmān ibn ʿAffān had devised the title of *khalifa,* placeholder or deputy, Englished as caliph, for their non-hereditary leader chosen by council. Muhammad's charismatic leadership had tamed the tribes of Arabia, but their allegiance was given only to his person and not to his religious movement, so that upon his death tribalism emerged again, in natural opposition to any centralized government.

The first caliph, Abu Bakr as-Siddīq (632–634), had to fight throughout his short reign to impose his rule. The second, ʿUmar ibn al-Khattāb (634–644) was contested by the partisans of Muhammad's family, though he was assassinated by a Persian slave for unrelated reasons. The third caliph, ʿUthmān ibn ʿAffān (644–656), under whose authority the written text of the Qurʾan was redacted, faced riots and rebellion, and was finally killed by victorious rebels in his own house in Medina. The fourth caliph, ʿAlī ibn Abī Ṭālib (656–661), Muhammad's son-in-law, was outmaneuvered by Muʿāwīyah ibn ʿAbī Sufyān, war leader in Syria and founder of the Umayyad dynasty, though it was an extremist of the Kharijite sect who assassinated ʿAlī (like modern jihadis, the Kharijites demanded unending war against all non-Muslims, denounced all who disagreed as apostates, and opposed all dynasts).

The modern Muslims who wax lyrical about the caliphate of the first four "rightly guided" caliphs *(al-Khulafaʾ ur-Rashidun)*—many of their successors are condemned as tyrants—disregard the violent instability of the institution, no doubt because they celebrate its spectacular victories over the infidels who torment them still. Certainly dissension and even civil war hardly slowed the momentum of Arab conquests. They continued westward right through North Africa to overrun Byzantine *Africa* (centered on modern Tunisia) by 690, reaching Spain by 711; northward, by way of eastern Anatolia and Armenia right through the Caucasus, only beyond it meeting firm resistance from the Khazars; and eastward all the way across Afghanistan to reach Sindh at the western edge of historic India by 664.

The Qur'an is hostile to pharaohs and kings, and its spirit of equality among all believers is inimical to hereditary succession. But within thirty years of Muhammad's death, the fifth caliph, Muʿāwīyah ibn ʿAbī Sufyān (661–680), arranged the succession of his son Yazid I, thereby starting what would become the Umayyad dynasty, condemned by many Sunni jurists and by all Shi'a—short for *Shi'at Ali'* "the party of ʿAli," that being the fourth caliph ʿAlī ibn Abī Tālib, Muhammad's son-in-law, who should have been his hereditary successor according to the Shi'a.

It was the same Muʿāwīyah who had earlier defeated Ali, and as it happens the troops of his son Yazid I killed Ali's son Hussein in the lunar month of Muharram 680—the event mourned ever since by the Shi'a as the greatest crime in history, annually commemorated with tearful lamentations and bloody scenes of cutting and scouring on the Ashura, the tenth day of Muharram. (Sunnis particularly deplore the cutting of the scalps of babies to display their bleeding foreheads as evidence of their family's intense devotion.)

With Muʿāwīyah's caliphate bitterly contested, only further conquests could assuage opposition with the plunder they gained, and by showing evidence of continued divine favor. The Sasanian empire was already destroyed, but the Byzantine empire still stood, even if greatly diminished, and its final conquest was the obvious and compelling priority. In practice that meant the conquest of Constantinople. Arab raids had long since penetrated Anatolia, and it was with larger and deeper raids that the final attack on the city was prepared. By 674 if not before, Arab raiders had reached even westernmost Anatolia, while in the port cities of Syria many ship crews had been converted or simply hired.

With them, the forces of Muʿāwīyah invested Constantinople by land and by sea. There was no continuous siege nor an effective blockade of the city, but rather a series of intermittent attacks by landed forces and sea engagements that lasted into 678. Their momentum had seemed irresistible, but the outcome after five years of sporadic fighting was to be the first Muslim defeat of strategic importance, the first break in the sequence of conquests.[31]

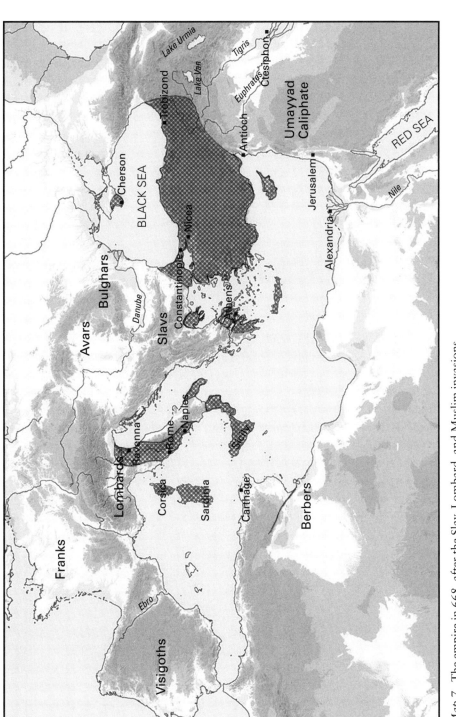

Map 7. The empire in 668, after the Slav, Lombard, and Muslim invasions

By the time Theophanes died in 818, the Muslim Arabs were still dangerous enemies of the empire and border warfare was endemic, but their defeat in the second offensive against Constantinople of 717 had made them seem less threatening than the Bulghars. That is reflected in Theophanes' entry on the first offensive (the year 6165 since the creation):

> In this year the . . . fleet of God's enemies set sail and came to anchor in the region of Thrace . . . every day there was a military engagement from morning till evening . . . with thrust and counter-thrust. The enemy kept this up from the month of April until September. Then, turning back, they went to Kyzikos, which they captured and wintered there. And in the spring they set out and, in similar fashion, made war on sea against the Christians. After doing the same for seven years and being put to shame with the help of God and His Mother, having, furthermore lost a multitude of warriors and had a great many wounded, they turned back with much sorrow. And as this fleet (which was to be sunk by God) put out to sea, it was overtaken by a wintry storm and the squalls of a hurricane . . . it was dashed to pieces and perished entirely.[32]

Prior to that, there had been fighting at sea in which the Byzantine navy for the first time employed siphons that projected *hugron pur*—liquid fire, or "Greek fire," of which more in Chapter 13.

Battle is the great contingency. Outcomes can be determined at the tactical level, or even at the operational level, by chance events, such as great storms. But in this case there was the Theodosian Wall, a garrison to man it, and a superior navy. Then came the storm that scattered and sunk the ships employed by the Muslim Arabs. The results of a battle, even if large, can also remain limited to tactical or operational repercussions. But this time the consequences were strategic.

Caliph Muʿāwīyah ibn ʿAbī Sufyān had evidently made a maximum effort for the maximum goal of taking Constantinople, mobilizing all his battle forces and all the ships he could recruit in the ports of the Levant at very great cost. Having failed, he was in a drastically weakened position. There was fighting in southeast Anatolia also, in which the Arabs were worsted, and the Mardaites, claimed ancestors of today's pugnacious Maronites but of disputed identity, took over the Amanus (Nur) mountain range that runs inland from Antioch and Mount Lebanon, attracting many fugitive slaves and runaways.

Hence the Arab jihadis fighting against the Byzantines in Cilicia had enemies on both sides of them. Under year 6169 since the creation, Theophanes lists the consequences: Muʿāwīyah had to sue for peace as the Byzantines understood it ("a written treaty of peace"), though it was only a *hudna,* an Islamically allowed truce and hence had a time limit.

There was no doubt about which side had won: Muʾāwīyah agreed to pay an annual tribute of three thousand pieces of gold, fifty thorough-bred horses, and fifty prisoners in exchange for a thirty-year truce.

Cilicia was far away from the west, but the critical struggle had been fought in Constantinople: "When the inhabitants of the West had learned of this, namely the Chagan of the Avars as well as the kings, chieftains, and *castaldi* [*gastaldi,* Lombard chiefs] . . . and the princes of the western nations, they sent envoys and gifts to the emperor, requesting that peace and friendship should be confirmed with them."[33] Much of Italy was still Byzantine, the Lombards had more of it, and their deterrence by victory over the seemingly irresistible Muslim Arabs was therefore of strategic importance, because with Bulghars to fight and an Arab frontier to guard, no large Byzantine forces could be sent to Italy to replace deterrence with defense.

The second Arab attack on Constantinople was not attempted until 717 when the Umayyad caliph Sulayman bin Abd al-Malik (715–717) mobilized for jihad to send a naval expedition to Constantinople, under his brother Maslama bin Abdul-Malik, which he followed with his own advance overland, through Cilicia presumably. As already noted, the forces that Maslama landed in Thrace to invest the Theodosian Wall were attacked from the rear and defeated by the Bulghars, while the forces landed on the Marmara shore were blockaded and starved; the caliph himself was killed in 717 so no help could reach Maslama.

The new caliph, ʿUmar ibn Abd al-Aziz, is described as a pietist indifferent to both the famed Umayyad elegance and the Umayyad striving to capture Constantinople. According to the Syriac Chronicle of AD 1234:

> As soon as he became king [caliph], he put all his energies into rescuing the Arab people who were trapped in the Roman empire. Seeing that news of them was unobtainable, he appointed a trustworthy man, gave him a sufficient escort and sent him into the Roman empire. . . . This man found his way into the Arab camp and learned all about the situation of the army; then Maslama gave him a letter full of lies to take to Umar saying, "The army is in excellent condition and the City is about to fall."

It was not until the winter of 717 had passed and navigation was again possible that Umar could order Maslama home, but that meant break-

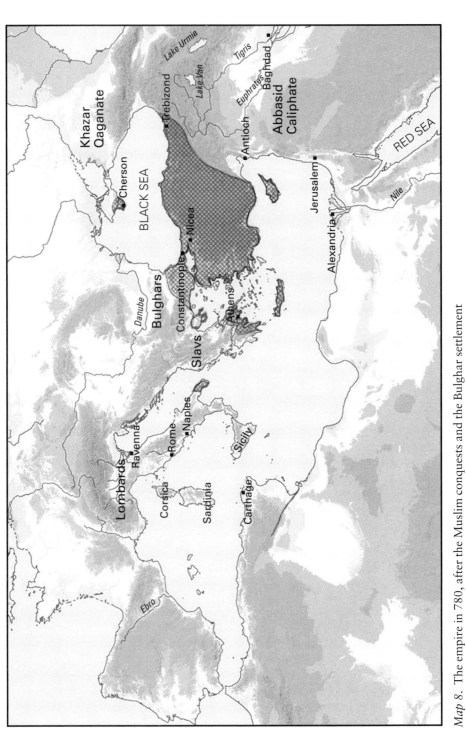

Map 8. The empire in 780, after the Muslim conquests and the Bulghar settlement

ing through the naval blockade: "They embarked on their ships and set sail on the sea and the Romans did battle with them there and burned many of their ships. The survivors were caught at sea by a storm and most of their ships went down."[34]

Characteristically, defeat was followed by the persecution of Christians and attempts at forcible conversion by order of Caliph Umar, who combined pietism with extremism.

Until the tenth century, the Muslim Arab power remained vigorous, raiding all around the Mediterranean and periodically attacking the empire's land borders as well. Such was the damage inflicted on the empire that many cities were reduced to villages; during the eighth century, even Constantinople shrank to less than fifty thousand inhabitants, who lived amidst abandoned houses without even one functioning aqueduct until 768.[35] During the ninth century the empire was recovering vigorously from earlier territorial losses and depredations, but Arab land attacks remained costly and naval invasions and booty raids still continued. In August 902 the last Byzantine stronghold in Sicily, on the mountain of Taormina, was lost; by then almost all the islands of the Mediterranean were occupied or raided, and coastal cities large and small were also attacked.

In July 904 the convert Leo of Tripoli led the most destructive raid of all: after entering the Sea of Marmara seemingly headed for Constantinople, Leo's large fleet fled before the massed sortie of Byzantine warships, only to attack the second city of the empire, Thessalonike. The city was unwarned, its defenses unready. Many of its inhabitants were killed, a huge number of captives was taken away in slavery.

Jihadi mobilization still worked, the twin incentives of plunder and slaves in victory or a luxurious afterlife in martial death could still enlist many volunteers, but politically Arab Muslim power was fatally undermined by chronic disunity. The Byzantine empire had its mutinies, insurrections, usurpations, and civil wars. But until 1204 there was only one empire, not two or three, or four. To be sure, Arab Muslim expansion, even before it became a multinational Muslim expansion, gained territories far larger than the Byzantine empire, reaching as far as the outer edge of China in the Talas River battle of 751 against Tang dynasty troops, and at the same time entering the Indus Valley of India (now Pakistan).

With this enormous expansion came both political and sectarian fragmentation—with the two often intertwined—as well as ethnic tensions, first of all between Arabs and Persians. At first overwhelmed and silenced by the Islamic conquest, the ancient and attractive culture of Persia with its Zoroastrian rites and customs found ways of reemerging within Persian Islam, as it still does today: not even the fanatical founders of the Islamic Republic tried to stop the purely Zoroastrian Nowruz fire ritual and subsequent festivity; and it is indicative that since the sixteenth century, Persian Muslims are assertively Shi'a rather than Sunni, as are most Arabs and Muslims at large.

Byzantine victories were often predicated on Muslim disunity—never more so than during the later years of the indefatigable and ultimately victorious Basil II (d. 1025), under whom the empire expanded in all directions. By then the single caliphate that was supposed to command all Muslims as one nation *(Umma)* was no more. Instead there were multiple powers often at war with each other. The most important for Byzantium at the time was the heterodox caliphate of the Fatimids, *al-Fātimiyyūn,* named for Fatima, Muhammad's daughter from whom the founder Abdullāh al-Mahdī Billah claimed descent. Started in what is now Tunisia and centered in Egypt, its maximum domain reached westward to the Atlantic coast of Morocco, south into the Sudan, east over Syria to the edge of Mesopotamia, and down to Mecca and Medina in western Arabia.

The Fatimids were therefore adjacent to Byzantium during the eleventh and twelve centuries, with much warfare and even more peace, because they were tolerant in religion and prudent in statecraft, presiding over economic expansion and wide-ranging trade. The Fatimids were Ismaili "Sevener" Shi'a, who like all Shi'a believe that 'Alī ibn Abī Tālib, husband of Fatima, should have been Muhammad's successor by dynastic right, and that his line is perpetuated by infallible Imams, the last of whom is still alive in hiding or "occulted"; but unlike the Twelver Shi'a of contemporary Iran and Iraq, for whom the last Imam is Muhammad al-Mahdi born in 868 (and still alive), the Ismailis only recognize the succession up to the sixth Imam, Ja'far ibn Muhammad, who died in 765, then inserting their own final and immortal Imam Muhammad ibn Ismail born in 721 (there are also "fiver" Shi'a).

The Abbasid caliphs were Sunni and recognized no occulted Imams, but their caliphate, though originally established with the force of the frontier Arabs of Khurasan, was chiefly supported by Persians, displacing the purely Arab elite of the previous Umayyad caliphate. After its destruction in Syria circa 750 by the Abbasids, the Umayyad line was revived by a lineal descendant in *al-Andalus,* Muslim Spain, as an emirate, implying at least a tacit acceptance of the Abbasid caliphate. But in 929 a restored Umayyad caliphate was proclaimed in Cordova, so that the Fatimids were challenged doctrinally and politically by a Sunni caliphate in Spain to their west, and by the Sunni caliphate of the Abbasids in the east. The latter had no strength of its own by the tenth century, but it was protected, dominated, and in that way empowered first by the Persian revivalists and Fiver Shi'a Buyids or Buwaihids *(āl-i Būya)*—and then by the Sunni Seljuk Turks who reconquered Baghdad for the Abbasids and ruled in their name. In between, there was even a revenance of Zoroastrian Persia, albeit in Muslim garb, when the Qarmatians, *Qarāmita,* emerged in Bahrain in 899 as a specifically Persian version of Shi'ism, and challenged the Fatimids then ruling Mecca by raiding the city in 928, removing the Black Stone, reestablishing Zoroastrian fire worship, and proclaiming the abolition of *Shari'a,* Muslim law.

With the Arabs in decline, and the Persians chronically unable to reconcile their ancient national culture with Islam—a dilemma that persists—the time had arrived for the primacy of the Turkic converts to Islam.

The Seljuk Turks and the Decline of the Empire

At the death of Basil II in 1025, the Byzantine empire was at the peak of its second expansion. Although it included less territory than in the first expansion of half a millennium earlier under Justinian, its possessions were not perilously scattered across the three-thousand-kilometer width of the Mediterranean, its Christianity was far more cohesive, and its more compact frontiers were not threatened by vigorous new enemies, except in the remaining enclave of southeast Italy. For the rest, with the Bulgarian state extinguished, there were only pliant Serbs, the minor power of the Christianized Magyars of the new kingdom of Hungary, the Pechenegs in decline before the approaching Cumans or Qipchaqs, and the Kievan Rus' of Yaroslav I. That power was at the peak of its geographic expansion, yet was not a sustained strategic threat either, as it wavered between hostility and deference—a Russian constant it seems: in 1043 a fleet arrived to attack Constantinople, but after it was defeated and burned by the Byzantine navy, Yaroslav I gratefully accepted the illegitimate daughter of Constantine IX Monomakhos (1042–1055) for his son Vsevolod, the future prince of Kiev.

As for the normally more dangerous eastern front, it was quiescent by 1025 because the Sunni Arab Abbasid caliphate was powerless, and its awkward protectors, the very Persian Buwayhid Viziers—who used the pre-Islamic "king of kings" *Shahanshah* as their title—were increasingly weakened by both internal dissension and rival powers.

All seemed propitious, but the Byzantines existed in a chronically unstable strategic environment. When Basil II acquired control of the Armenian lands east of Lake Van (now in eastern Turkey and western Iran) in the year 1000, he could not have known anything of Toğrül, then a child perhaps seven years old, grandson of Seljuk, the first of his Turkic Oğuz clan to convert to (Sunni) Islam. Yet by the time this same Toğrül died on September 4, 1063, the Seljuk, or Seljuq or Selçuk, had evolved under his leadership from a clan of nomad warriors into a Great Power.[36] Much more contributed to their success, but there was also a tactical factor: as newcomers from Central Asia, their archery—a perishable skill—was of the highest quality.[37] Upon entering Baghdad in 1055, Toğrül was graced with the title Sultan ("power holder") by the Abbasid caliph, by then a spiritual authority at best, beset by internal dissension, by the Fatimids—they had even seized Mecca and Medina—and by the declining Ghaznavids in the east.[38]

The Seljuks had not even existed as a power in 1025 when Basil II died, yet within thirty years they had become rulers of a vast domain that included the territories of modern Iraq, Iran, and Uzbekistan. As such they were a strategic threat to Byzantium, but they were also involuntary allies because they too resisted the expansionism of the Fatimid caliphs of Egypt. Empowered by Egypt's ample tax revenues, the Fatimids had both an effective fleet and capable Turkic mercenary troops.

The Seljuks were therefore strategic allies of Byzantium whether they wished it or not; but at the same time, they threatened the eastern border zones from northern Iraq to northwest Iran, and in the Caucasian lands of the Armenians and Georgians by then under Byzantine control. With the arrival of increasing numbers of hungry, landless, newly converted Oğuz tribesmen, border raids and deeper incursions became more frequent, and outright invasion was becoming a definite threat. In 1064 the important Armenian cathedral city and religious capital of Ani was sacked.[39]

As new converts, the Seljuks and their Oğuz, or more broadly Turkoman or Turkmen (= any Turkic Muslim), followers were strongly motivated to fulfill the religious duty of jihad, to expand the Dar el-Islam by invading the Dar el-Harb, the "land of war" of the unbelievers. But for the Ghazis—the border warriors of jihad—Islamic duty and personal profit were intertwined: they could have their loot, their captives for sale or for service as slaves, or in a fighting death the promise of a richly furnished and well-watered heaven *(jannah)* of unending joys with black-eyed virgins and handsome boys.[40] That had been true for their Arab predecessors as well, but the Arab conquering impulse that had transformed North Africa and western Asia and beyond from the mid-seventh century, was entirely spent by then.

Already under Toğrül the looting raids of Oğuz and other Turkoman horsemen greatly afflicted eastern Anatolia—and they intensified under his very able successor Alp Arslan (1063–1072). Turkoman tribesmen under their Ghazis, like the Bedouin and Kurdish marauders before them, functioned as the advanced echelon of Seljuk expansionism—and by all accounts they were considerably more skilled in fighting, as befitted mounted archers from Central Asia.

There were no organized frontier defenses to hold them, no chains of forts connected by patrols, but only the point defense of walled towns, fortress-monasteries, and the fortified mansions of local magnates. They sustained the *akritai,* the mostly Armenian border warriors, much celebrated in song and romance, less useful for local defense than for vigorous counter-raiding across the border. By such means the Anatolian eastern border had been held for three centuries against Arabs from the area of Trebizond on the Black Sea all the way to Cilicia on the Mediterranean coast, as explained in the manual *De Velitatione* examined below. But counter-raiding is futile against nomads and could not contain Turkoman raids, and neither could the ambushes and pursuits of whatever imperial forces were present.

Only a classic Roman frontier defense at its most elaborate could have protected eastern Anatolia, by combining fortified watchtowers within sight of each other, forts with garrisons of hundreds in every valley on the frontier, and large formations in the rear to reinforce them—a Hadrian's Wall extended for hundreds of miles, impossibly costly to build, garrison, or supply. There was also the cheaper alternative that the Romans had applied in arid zones of the Middle East and North Africa, where there was no province-wide agriculture to protect but only occasional oases large or small: light-cavalry units patrolled the frontier and beyond to detect marauders or outright invasions, which were then to be intercepted by five-hundred- or thousand-man auxiliary cavalry, infantry, or mixed units kept in forts at some depth behind the frontier, which could in turn be reinforced by the nearest legionary and auxiliary field formations.

An immediate reaction was impossible because messages first had to reach the forts, then the auxiliary forces would need time to ride out ready for combat; and finally the intruders had to be found, to be engaged or just frightened across the frontier once more. That would have allowed ample time for the marauders to loot and enslave, but all significant oases and villages also had their own "point" defenses—either walls or simply outer rings of stone houses built very close to each other, leaving only narrow passages that raiding horsemen would not find welcoming. Undefended farms or hamlets could not exist in any arid frontier zone near nomads wondering with their flocks. Pastoralists do not normally allow undefended agriculturalists to survive within their reach—there is no incentive for moderation in looting harvests because what one raiding band would leave to ensure next year's crop, another would seize—it is the pastoral raiders' version of the "tragedy of the commons."

The Roman arid-zone solution would not have sufficed to provide adequate protection for the predominantly Armenian farmers and shepherds who inhabited the valleys and watered plateaus of eastern Anatolia. Nor could their security needs be ignored—they provided the empire with some taxes, many recruits, and all its part-time frontier forces. Moreover, a patrol-and-intercept solution would have collided with an elementary military fact: no cavalry properly equipped for combat could expect to outpace the Turkoman, who mostly rode without helmet, corselet, shield, sword, mace, or lance, but only with the compound bow and a scimitar, or just a dagger—a much lighter load that obviously made for a faster ride.

That much had already been proven in a long series of frustrating encounters with elusive Turkoman horsemen, when in the summer of 1071 the emperor Romanos IV Diogenes (1068–1071) assembled an exceptionally large army—of forty thousand, it has been estimated—to attack the problem at its source.[41] His aim was to dislodge the Seljuks from their newly acquired strongholds in northeast Anatolia, which served as bases for Oğuz raids, and their own more directed inroads into imperial territory. Each place could not be strong on its own, certainly not against an army of forty thousand, so Romanos could have continued to progress from one to the other, to demolish the Seljuk infrastructure of Turkoman terror, in modern parlance. One of these strongholds was Mantzikert, modern Malâzgird, north of Lake Van in far eastern Turkey. It duly surrendered to the Byzantines.

What happened next is a perfect illustration of the contradiction between strategy and tactics that often occurs—and can ruin the best of plans. There is nothing to prevent such contradictions except forethought and talented command, because while strategy and tactics are governed by exactly the same logic, the level of the action is very different and is subject to different influences, including divergent human proclivities.

To begin with, Romanos was there to safeguard the inhabitants and imperial subjects from Turkoman raiding, to prevent the abandonment of more cultivated and taxable land—much of it was already deserted.

That was his strategic purpose. Yet though the forty thousand troops were supposed to bring sixty days of food with them, they themselves pillaged the long-suffering population of the area, many or most of them Armenians with their own ethnic identity and agenda, and even the emperor's Nemitzoi (Slav for "German") bodyguards joined in, to his displeasure—he reportedly sent them away, leaving himself with that much less personal protection, a mistake as it turned out. Instead of bringing reassurance and security to imperial taxpayers with the strength of its forty thousand, the expedition apparently intensified local disaffection from the empire—the large Christian population around Lake Van would later remain docile under Seljuk rule, showing no nostalgia for Byzantine government.

The only fit objective for a costly army of forty thousand, perhaps half of them foreign mercenaries—Oğuz and Pecheneg mounted archers, Norman heavy cavalry, Varangian guards, and Armenian infantry—was a strategic offensive to conquer Iran, but no source suggests that such an ambition was even contemplated by Romanos. As for the limited objectives he did have, perhaps four thousand good troops would have sufficed—unless of course the Seljuk sultan Alp Arslan foolishly chose to concentrate his major forces in that backwater, just to fight off a limited attack. It was another matter altogether once forty thousand troops started marching—they could not be ignored. It seems that Alp Arslan was preparing a large-scale offensive against the Fatimids when word reached him that a huge Byzantine army was on the march in the mountains of northeast Turkey.

That was not even the major theater of war between the two sides, which would more naturally fight each other over far more valuable territories exactly where Sasanians and Romans had once fought, in northwest Mesopotamia (now southwest Turkey) with its often besieged cities of Amida, Dara, Edessa, and Nisibis. In any case, Alp Arslan's strategic priority was not to fight the Byzantines at all, but rather to fight the Fatimids of Egypt, the only really important competitors for a ruler of Baghdad, as he then was. Given that his political power as sultan could legitimately extend as far as the religious authority of the Abbasid caliph, who had authorized him to rule in his place, if the Fatimids with their heterodox Ismai'li faith were destroyed, then the caliph's religious writ would embrace Egypt once more, and there his sultan Alp Arslan would rule over fertile lands with especially abundant tax revenues. For that was a further advantage of Egypt, for a Muslim ruler: its population was still largely or at least abundantly Christian, and was therefore subject to the head tax, as Muslims were not.

In the event, Alp Arslan chose not to ignore the Byzantine counterattack to pursue his strategic offensive against Egypt—no doubt it would have been politically damaging for a new dynasty of newly converted Muslims to be attacking other Muslims, howsoever heterodox, rather than defend Muslim conquests against the supreme Christian power. Or perhaps it was politically damaging in quite another way to let Romanos have his way uncontested: there were still many Christians and Zoroastrians in nearby parts of Iran, and more Christians in the Caucasus who might be emboldened by the uncontested advance of a large Christian army, which could win over newly converted Muslims as well.

Once Alp Arslan abandoned his plans for Egypt to stop Romanos with his own forces and a much larger number of Turkoman volunteers, the stage was set for the accidental encounter at Mantzikert. Romanos, for his part, thought that he was engaged in little more than a police operation, and therefore had spread out his strength to cover as many localities as possible: a substantial force under the Norman mercenary Oursel or Roussel de Bailleul had been sent to seize the stronghold at Chliat (now Akhlat) on the northwest shore of Lake Van. Then a second force under the Armenian commander Joseph Tarchaneiotes was sent to reinforce Roussel de Bailleul, while the emperor's German bodyguards, as noted, had been sent to the rear. Yet another force, of cavalry under the Armenian commander Nikephoros Basilakes, was badly defeated two days before the battle, when it impetuously chased a band of horsemen hurriedly fleeing—who led Basilakes into a well-prepared ambush.

The enemy had faithfully followed the standard tactic of the mounted archers of the steppe, while Nikephoros Basilakes had failed to heed the clear anti-ambush instructions of Byzantine field manuals. Four hundred years before, the feigned retreat tactic of the nomads had been correctly analyzed in the *Strategikon of Maurikios* with very definite results: if they are really fleeing in panic, you have already won the battle and there is no need to pursue them; and that way you are also safeguarded if they are simulating flight to lure you into an ambush. It is difficult to tell one kind of flight from another, but fortunately there is no need to do so, for the same sovereign remedy applies: do not chase fleeing nomads; they are faster, so you will not catch them anyway, but they might lead you into an ambush—so no pursuit is ever justified. Evidently Basilakes was uninstructed, impulsive, or both—he would end as a defeated rebel in the Balkans.

For these reasons, when the battle at Mantzikert started on the morning of Friday August 26, 1071, Romanos IV Diogenes did not have forty thousand men concentrated with him, or half that number. When he suddenly discovered that Alp Arslan had gathered his own fresher forces to attack him on Friday, August 26, 1071, much of his strength was elsewhere and beyond quick recall. That preordained defeat, barring undeserved tactical fortune.

But instead of a difficult tactical redemption from operational error, there were further tactical errors, though of course after a defeat any and all tactical dispositions and movements can be shown to be gross errors, just as they may be judged brilliant in the event of victory.[42]

The sources also report treachery. That is a commonplace explanation of unexpected defeats, but is fully credible in this case because Romanos was surrounded by political enemies in his own court, notably the Doukas in-laws of the previous marriage of his wife Eudokia Makrembolitissa.[43] Most incautiously but possibly unavoidably, Romanos relied on Andronikos Doukas, son and executive of John Doukas, his brother-in-law and most obvious political rival, to command the rearguard element of his army.[44]

A key advantage of the learned ways of fighting of sophisticated armies like the Byzantine over the onrush—or retreat—of fighting mobs, is that specific forces can be kept separate to perform synergistic attacks if the battle unfolds favorably, or for defensive insurance if it does not.

Dispositions varied infinitely in accordance with circumstances, but they almost always included flank guards and a rear guard—the manuals insist on the need for both even at the cost of weakening the main battle force. The rear guard could be summoned forward to reinforce success, or remain in place to backstop the first-line forces if they were falling back under enemy pressure. In the event of enemy breakthroughs, only the rear guard could stabilize the situation by plugging the breach in the first line, just as it could contain sudden panics by simply standing where it was in good order. It was also a function of the rear guard to block outflanking attempts by spreading out behind the first line to intercept them—often a much better way than to loosen the order of the first line to widen its frontage. Finally, the rear guard normally allowed the overall commander to make his second move. By standing between the first line and the rear guard, he could actually command the latter and direct its action, when the first line was already totally caught up in fighting and hard to control.

But Romanos did not so place himself to control both echelons. Instead he played warrior rather than general and fought at the front. As soon as he saw that the hated emperor was in trouble and needed help, Andronikos Doukas simply led his forces away—all the way to Constantinople, to participate in the deposition of Romanos and the elevation of Eudokia Makrembolitissa's son by her Doukas first husband, Michael VII.

The outcome was a catastrophic strategic defeat for the Byzantine empire, not merely a dislodgment from some forward slice of territory, or the loss of many troops—neither necessarily decisive in the long run, and certainly not for an empire that held all the sub-Danubian territories of the Balkan Peninsula regained by Basil II by 1025, as well as Anatolia and Greece. The catastrophe was that Anatolia was the core of the empire, and much of it would never be recovered.

Byzantine losses were not especially heavy at Mantzikert, and perhaps they were not heavy at all.[45] The light cavalry of Oğuz warriors was excellent for raiding and surveillance patrols but not for pinning down more heavily armed enemies, and still less for killing them in bulk—a job for the heavy infantry or perhaps the heavy cavalry of the time, whose armored men with maces could bodily break enemy forces.

The Seljuks had won the field, but that was mainly because the larger part of the Byzantine army was not there to begin with, or had retreated safely if treasonably. But the sensational result of the battle was that the lightly wounded Romanos IV Diogenes was captured. He was found the day after the battle by men plundering his reportedly lavish expeditionary tented camp and convoy of supply carts. They brought Romanos IV Diogenes to Alp Arslan.

This was not an encounter with a savage: the Seljuks had been in communication with the empire for years, from the time that Toğrül was forming his state. The latest exchange between the two sides had occurred as recently as the day before the battle, when Romanos imprudently turned away envoys offering a settlement. Characteristically—

and cleverly—these came in the name of the caliph in distant Baghdad, not of Arslan, who was probably just out of sight on the other side of the hill with his main forces.

Alp Arslan did not humiliate or torture the captive emperor, instead offering honorable hospitality while politely negotiating with him. Evidently knowing that his enemies at court, the Doukas relatives of his wife, had no use for Romanos, Alp Arslan did not even try to extort a ransom. Instead after a week he simply released Romanos to go home with an escort, in exchange for his personal promise that he would pay a ransom, the cession of a slice of territory in eastern Anatolia, and a generic promise of friendship. Chivalrous generosity of spirit aside—it started a cycle of courteous reciprocation amidst intermittent warfare that would last for two centuries—Alp Arslan was reaffirming his strategic priorities, which were not to destroy the Byzantine empire but to widen Seljuk control within the Muslim sphere, against the Fatimids in the name of Sunni Islam and the caliph in Baghdad, and against Sunni rivals in the name of the Seljuks.

The accord was not honored in Constantinople, where Romanos had already been deposed in the name of his stepson Michael VII Doukas (1071–1078). In the ensuing civil war, Turkoman bands and organized Seljuk forces had ample opportunities to advance deep into Anatolia and indeed all the way to Nicea, modern Iznik, and Cyzicus on the Sea of Marmara, within a day's ride from Constantinople.

It could have been the end for the empire even then, as rival contenders for the imperial title competed for Seljuk support against each other by conceding more territory, while expending dwindling imperial revenues to fight each other's forces. But three unconnected forces were to change the balance of power between Byzantines and Seljuks in unexpected ways.

First, the Seljuk offensive against the Fatimids gave them Jerusalem by 1071, but in the ensuing chaos the Holy Land became insecure for western pilgrims evoking, along with any other set of causes one wishes to assert, the crusading movement in western Europe. Twenty-six years after Mantzikert in 1097, the fighters of the First Crusade arrived, just as lusty for war as any Turkoman raider or Ghazi holy warrior. They would conquer western Anatolia on their way to distant Antioch and the Holy Land.

Second, civil war in Byzantium was an exercise in the survival of the fittest, and Alexios I Komnenos (1081–1118), winner in the decade-long contest that followed the deposition of Romanos, was certainly tal-

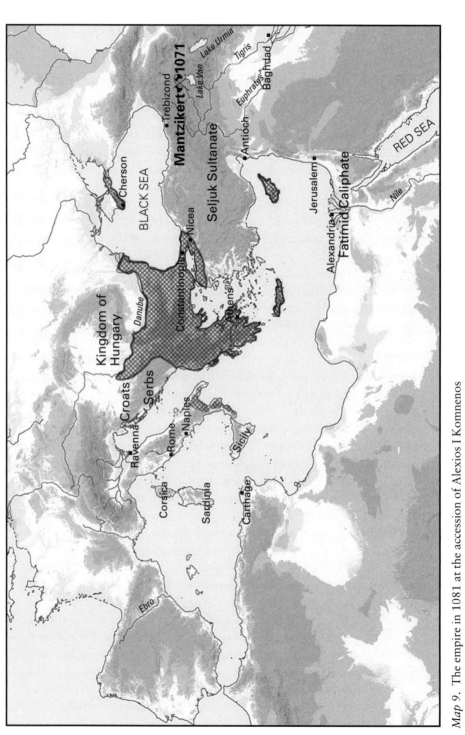

Map 9. The empire in 1081 at the accession of Alexios I Komnenos

ented and fit to rebuild a devastated empire; he also had time to do so, ruling for thirty-seven years.[46]

Third, the core of the Seljuk empire was Iran, and Alp Arslan's priority was evidently to control of the adjacent region of Central Asia—it was on the Oxus River (Amu Darya) between modern Turkmenistan and Uzbekistan that Alp Arslan was killed in 1072, outliving his Mantzikert victory by only one year. Moreover, the Seljuk exposure to the chronic instability of the great steppe would have disastrous consequences: in the Qatwan steppe near Samarkand, the Seljuk sultan Sinjar lost an army on September 9, 1141, at the hands of the Qara Xitay.[47]

The Seljuks could not therefore exploit the victory at Mantzikert—and the ensuing ten years of civil war in which they were even invited to participate—by conquering the whole of Anatolia. Had they done so, the empire could not have lasted long, because Anatolia was its indispensable demographic and tributary base. But having reached very close to Constantinople, under Kilij Arslan I—the beneficiary of gentlemanly reciprocity from Alexios I Komnenos (who returned his captured family without ransom)—the Seljuks were driven back to central Anatolia, establishing their court at Iconium (Konya), which became the capital of their Sultanate of Rûm (= the Roman empire = Anatolia), which would last until the end of the thirteenth century, albeit under Mongol suzerainty from 1243.

Manuel I Komnenos (1143–1180), intrepid, irreverent in religion, uniquely multicultural in favoring both Latin and Turkic subordinates and customs, was also talented both in diplomacy and war. At different times, he intervened effectively in Italian politics though he had to abandon an attempted invasion; he fought alliances of Normans, Serbs, Hungarians, and Kievan Rus', gaining territory in the process, both in the Balkans after defeating the Hungarians at Semlin (in modern Kosovo) in 1167 and by reestablishing a Byzantine presence in the Crimea. Most importantly, he increased Byzantine control of all the coastal plains of Anatolia, reducing the territory of the Sultanate of Rûm to its interior, and he strengthened the Byzantine hold on Cilicia and western Syria.

It was in this context that Manuel I attempted a theater-level offensive to finish off the Sultanate of Rûm and reestablish imperial rule over all of Anatolia. He had already succeeded in regaining territory from the sultanate in a series of small operations. They did not ruin the traditionally amicable personal relations between sultans and emperors in between bouts of ferocious fighting; thus there occurred the extraordinary

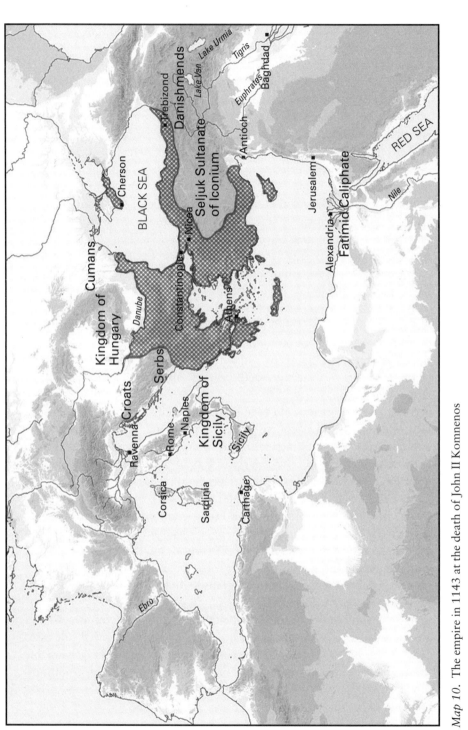

Map 10. The empire in 1143 at the death of John II Komnenos

episode of the 1162 visit of Kilij Arslan to Constantinople. It was not a businesslike official visit and it was not brief; he was a cultured man and as open-minded as his host—who even dared to toy with a restructured theism that would suit both religions.

The reaction in the city was enthusiastic:

> [It was] something tremendous and wonderfully extraordinary, such as I know never happened to the Romans before. Of the many magnificent emperors, who is not outdone, that a man who rules so much land and lords it over so many tribes should appear at the [court of the Roman Emperor] in the guise of a servant?[48]

There was a magnificent reception ceremony, followed by festivities and banquets. Only a procession to the Hagia Sophia was prohibited by the patriarch Louka Chrysoberges, whose authority was undoubtedly enhanced by the coincidence of a serious earthquake.

During the 1162 visit a pact of peace was added to a merely personal amity, but the entente broke down, and in 1176 Manuel decisively abandoned the path of gradualism recommended by Byzantine strategic manuals to mount a deep-penetration offensive to conquer the Seljuk capital of Iconium, the modern Konya. Elaborate preparations assembled stone-throwers and engineering equipment for the siege of Iconium, a reported three thousand carts of supplies, from extra arrows to food, and at least ten thousand, and possibly twice as many, infantrymen both light and heavy, and cavalry including *kataphraktoi*, the armored cavalry trained to charge with the lance, and for close combat with mace and sword, that could scatter any number of light horsemen.[49]

The usual risks of deep-penetration maneuver were present: there was difficult terrain—the Phrygian mountains—which had to be crossed swiftly to achieve surprise, through narrow defiles and passes suitable for Seljuk ambushes, not suitable for a rapid advance. After that, however, Manuel's forces would be able to spread out in the more open terrain leading down to Iconium, and nearer to the city the kataphraktoi would find flat ground suitable for their devastating charges.

Either because the Byzantine advance was too slow, or because the Seljuks moved too fast, it was not in the plain of Iconium that the two armies met on September 17, 1176, but still in the mountains—the place that gave its name to the battle, Myriocephalon, means "a thousand mountain peaks."

The terrain was unfavorable for the Byzantine forces, which lacked the room they needed to deploy out from long marching columns to broad fighting lines. The Seljuks, moreover, had reached the Tzibritze Pass that would be the battleground ahead of time, positioning bowmen on the slopes on both sides, ready to release their arrows against the enemy below, or charge down to attack weaker elements.

The outcome was a perfect ambush on an operational scale, in which tactical advantages added up to more than the sum of their parts: archers had the advantage of gravity against any archers below them; even the most powerful heavy cavalry was negated because it could not ride up the slopes, and Seljuk forces on high ground could select when to stay there and when to descend to attack the enemy below—it was the supply train of carts that was most thoroughly destroyed. The Seljuks had averted the immediate threat to their capital but lacked the strength to fight off Manuel's army. It mostly did survive to retreat, but the offensive momentum of the empire was no more.

Defeat in the Tzibritze Pass did not lead to any immediately momentous consequences. Manuel was not overthrown as Romanos IV Diogenes had been after his defeat at Mantzikert in 1071, Seljuk armies did not advance on Constantinople, and the Crusaders did not turn on their Byzantine patrons in their moment of weakness. But in subsequent years the empire could not reconstitute its military strength to regain the initiative. That required, first of all, political unity under effective emperors, administrative efficiency in collecting taxes, and more efficiency in raising armed forces. Instead of political cohesion within the ruling elite, indeed within the court, there was murderous factionalism that drove the losing faction to seek help from the forces of the Fourth Crusade, a transnational gathering of quarrelsome, hungry, predatory knights and hapless pilgrims brilliantly manipulated by the Venetian doge Enrico Dandolo, who succeeded in extracting real gains for his city from the chaotic violence of the Crusaders.

It was by no means the first time that foreign forces summoned by contenders for the throne determined who would rule Byzantium. Khazars, Bulghars, and Russians had all served in that capacity without lasting consequence, as the strong Byzantine sense of identity, resilient morale, and enduring administrative abilities each time achieved an ample restoration. But in 1204 the outcome of foreign intervention was fatal, in part because Catholics no longer accepted the legitimacy of Orthodox rule. The year before, the forces of the Fourth Crusade had restored the deposed Isaac II (1185–1195) of the Angeloi family and his son Alexios IV as co-emperor. When a disaffected courtier, Alexios V

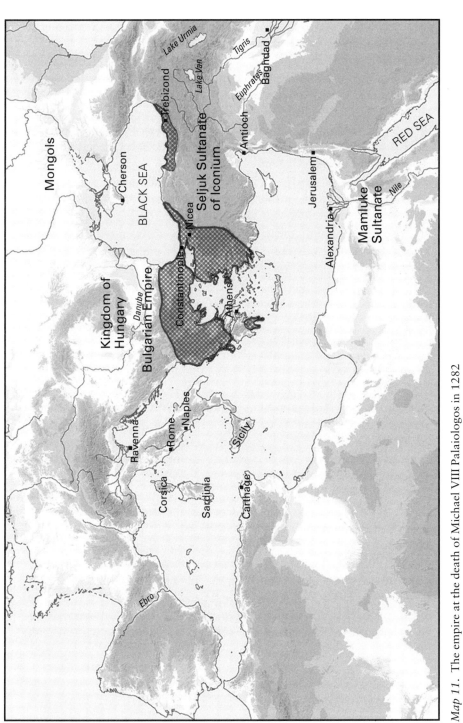

Map 11. The empire at the death of Michael VIII Palaiologos in 1282

Murtzuphlus, overthrew them, the Venetians and the Crusaders reacted on April 13, 1204, by storming, looting, and seizing Constantinople for themselves, installing a Catholic emperor of their own. The colossal resilience of the Roman empire of the east had finally been defeated not by pagan steppe nomads from Central Asia or inflamed Muslim jihadists but by fellow Christians, rival claimants to the same Roman tradition.

The extreme fluidity of the strategic environment that the Byzantines had to contend with was again exemplified by what happened in 1204. When the Crusaders broke into Constantinople to rob it of its accumulated treasures—some to be seen in Venice to this day—there were many in the city who could recall that in the days of their youth emperor Manuel I Komnenos seemed on the verge of regaining Italy, just as much of Anatolia and northern Syria had already been regained, while Byzantine influence was projected deeper into Europe than ever before.

The empire had come very close to destruction several times before, only to recover very quickly, but there was no real recovery from the downfall of 1204. When Michael VIII Palaiologos seized Constantinople in 1261, it was a Greek kingdom that he ruled, not an empire.

A few years later, Osman, a talented warrior-chief, started to gather and lead followers as one more Ghazi, albeit in dubious standing as a jihadist: he had Christians riding with him. A sultan of Konya lingered until 1308, but by the time Osman died in 1326, his Osmanli ("Ottoman") followers had started building a powerful state that accommodated the increasing sedentarization of the Oğuz and other Turkic migrants, and had a definite capacity for important military innovation. None was more important than the invention of uniformed, regimented, "new soldiers," *yeniçeri* ("janissary"), the ancestors of all modern armies, marching bands included. The territory controlled by the increasingly misnamed emperors in Constantinople on both sides of the straits kept shrinking amidst endemic dynastic struggles, as the cumulative loss of tax revenues enfeebled the remnant. Surrender to Sultan Bayezid dubbed "Yildirim" (thunderbolt) seemed imminent by 1402, when the irruption of Timur-i-lenk, the Tamerlane of Western memories, claimant to both Cinggisid Mongol and Turkic ancestry, destroyed the army of Bayezid at Ankara on July 28, 1402. That allowed an emperor to linger in Constantinople until 1453, to then fight and die with the utmost heroism.

PART THREE

The Byzantine Art of War

In organizing and training their forces, in devising their tactics and operational methods, in evaluating their strategic choices, the Byzantines were informed by an entire military culture rooted in ancient Greece and the earlier Roman empire, but increasingly of their own making, and sharply different.

As successive layers were added from the fifth century onward, this distinctive culture was preserved and transmitted, as cultures always are, in all sorts of ways, by institutions, by customs, by norms, and by word of mouth, but most durably by the written word. Ancient Greek military texts were duly honored, and there were some Roman writings, but the Byzantines increasingly relied on their own growing body of military literature, which included detailed handbooks. We do not have any veritable Roman field manuals, that is, guidebooks written by experienced soldiers for the use of soldiers, but we do have several Byzantine manuals of evident practical value; each is examined in what follows, and not all their recommendations are entirely obsolete.

The most direct benefit of this accumulated military culture was to broaden the repertoire of Byzantine armies and navies, endowing them with a greater variety of tactics, operational schemes, and practiced stratagems than any of their opponents could command. Sometimes this enabled Byzantine forces to surprise and overwhelm their enemies by employing tactics or methods or stratagems or weapons entirely unknown to them. More often the benefit derived from this military culture was of a more subtle order, adding an advantage that was marginal rather than overwhelming—but then it is also true that it was by small margins that the empire survived its worst crises.

More important in its consequences than any number of cunning stratagems was the distinctive Byzantine concept of war and peace, which evolved by the end of the sixth century into the veritable "operational code" defined in this book's Conclusion. Its starting point was the impossibility of decisive victory—the very aim of warfare for the earlier Romans as for Napoleon, Clausewitz, and their emulators till this day, though with waning conviction, perhaps. The Byzantine concept was thus a revolutionary reversal. Its powerful implications are manifest in what the Byzantines did, in what actually happened, and sometimes in what Byzantine voices reportedly said, but they emerge more clearly and more fully in the varied texts of their military literature.

Byzantine military commanders were not intellectuals. On the whole, they were probably less educated than the ordinary soldiers of the Roman army in its better years, judging by its voluminous record keeping and the personal letters and varied writings that have survived on papyrus and bark. In the later sixth century, at any rate, we can presume from the finest of Byzantine military handbooks, the *Strategikon* attributed to the emperor Maurikios, that illiteracy was the norm even in fairly senior field ranks, because the author writes that *merarchs* should be "prudent, practical, experienced, and, if possible, able to read and write. This is especially important for the commander of the center meros, . . . who has to, if it becomes necessary, take over all the duties of the [*strategos*, the commander]."[1] A merarch could command as many as seven thousand cavalrymen—one-third of the entire field army envisaged by the author, the equivalent of a modern brigadier general in charge of a small division or large brigade battle group. And of three envisaged merarchs, one would be the *hypostrategos*, the under-general, or very literally the lieutenant general ("placeholder") of the commander of the entire field army. Yet the author does not even insist on literacy but merely recommends it: "if possible." Literacy must have been rare indeed among cavalry officers.

One likely reason was that the late sixth-century Byzantine cavalry described by the author, which owed so much to the methods of the steppe nomads, fought alongside mercenary mounted archers, the "Huns" much mentioned by Prokopios. They were probably Onogurs or other Turkic warriors rather than descendants of Attila's few Huns, and they were certainly recruited into regular Byzantine units as well. During the endless wars of Justinian, the illiterate ways of the steppe warriors are likely to have shaped the army's camp culture and the army itself, from which cavalry officers were necessarily promoted—for young gentlemen sent from literate Constantinople were unlikely to be successful in commanding semi-wild horsemen.

Ostensibly, their Roman predecessors had done just that in the rank of *praefectus alae* of the auxiliary cavalry, the first stage of a public career in the equestrian class; but the officers actually recorded in that rank were not young gentlemen mostly, but rather veteran centurions or native chiefs.[2] It may be noted parenthetically that in European armies until 1914, cavalry officers, and especially Hussars and other light cavalry, were generally less educated than their colleagues of the infantry and certainly of the artillery, and that may have been true in the sixth century also.

The prevailing illiteracy would explain very well why the author of the *Strategikon* is so meticulous in listing the nomenclature of units and ranks, and in specifying the different command phrases required by the tactics he explains—many of them still in Latin rather than Greek. When illiterates repeat words they hear from other illiterates, especially in a language they do not know, over time most of those words are transformed beyond recognition, retaining their operative meaning only within the in-group but not beyond it, with the lively possibility of disastrous misunderstandings when officers transfer from unit to unit.

Illiteracy among officers also explains why the author justifies his book by writing: "Those who assume the command of the troops do not understand even the most obvious matters and run into all sorts of difficulties."

Yet if the dating of the book in the late part of the reign of Justinian is correct, the ignorant officers the author deplores had just reconquered North Africa, the southeast edge of the Iberian Peninsula, Sicily, and much of Italy. They must have been better fighters than readers.

In any case war is a collective enterprise. If one literate commander remembered a clever stratagem, or a training procedure, he had once encountered in his reading, it could be applied by an entire army of illiterates.

CHAPTER 10

The Classical Inheritance

Illiteracy among cavalry officers did not prevent the study, dissemination, and retention of entire repertoires of tactics originally learned from books. That indeed was an important comparative advantage of the Byzantines, whose own military literature was more useful than the earlier Roman, as far as we know, including lost texts by Cato, Celsus, Frontinus—whose *Strategemata* survives—and Paternus. Successive Byzantine military manuals followed one another, some mere recapitulations of earlier works going all the way back to Greek antiquity, starting with Aeneas Tacticus, who wrote before 346 BCE, but others that were undoubtedly original works.[1]

By contrast, the only surviving Roman military textbook, "The Summary of Military Matters" by Vegetius, was written by a scholar of antiquarian bent with no military experience at the very end of the fourth century or in the early fifth century—when not much of the Roman army was left.[2] Unlike his *Mulomedicina*, a veterinary manual full of practical advice, Vegetius's military epitome offers exhortations and noble examples of ancient glories alongside tactical prescriptions and instructions that are sometimes impractical and often inconsistent, because the Roman army presented in the book is a collage of earlier realities duly identified as such, some current realities, and what the author wished were true of the contemporary Roman army. Sometimes Vegetius copied a text too far.

On archery training, for example, Vegetius first offers rather useless generic advice, incidentally revealing no awareness of the significance of the composite reflex bow, by then already widely introduced:

> About a third or a quarter of recruits, who prove to have more aptitude, should be trained constantly . . . using wooden bows and mock arrows. Instructors should be chosen for this training who are experts.[3]

Clearly Vegetius was not an expert, because it is foolish to train with weak wooden bows for combat with powerfully resistant composite bows. On the contrary, it was a fundamental Roman rule to use extra-heavy shields, swords, and javelins for training, to ease at least the physical effort of combat. If anything, training bows would have to be even more resistant to prepare men for combat.

Other surviving Latin texts on military matters are not useless, but they are not systematic military manuals either. The *Strategemata* of Sextus Julius Frontinus, as he himself explains, is not a work of strategy—he enjoins the reader to differentiate between "strategy" (*strategikon,* in Greek in the text) and "stratagem" *(strategematon);* his own work is a compilation of exemplary episodes of tenacious, courageous, innovative, clever, cunning, and deceptive leadership in war.[4] Divided into four books, on stratagems (starting with "On Concealing One's Plans"), on the conduct of battle, on siege operations, and a final fourth book on principles of war rather than stratagems, the examples are well chosen and well presented—one does see how modern military commanders might still benefit from reading the text. Book II on battle leadership offers a number of interesting stratagems under its headings: "On Choosing the Time for Battle," "On Choosing the Place of Battle," "On the Disposition of Troops for Battle," "On Creating Panic in the Enemy's Ranks," "On Ambushes," "On Letting the Enemy Escape, lest, Brought to Bay, He Renew the Battle in Desperation" (a much-valued principle of eighteenth-century warfare with its "golden bridges," easy escape routes deliberately left unguarded)—and eight more, ending with "On Retreating."

Of particular interest for the light they shed on the Roman, and in this case also the Byzantine, military mentality are the quotations that Frontinus chose for the seventh and last section of Book IV dedicated to military maxims. Some were taken from the "memorable deeds and sayings" of Valerius Maximus from which the very word derives. They show that there was no desire to emulate the compulsive boldness of Alexander the Great, enormously admired though he was. Julius Caesar is plausibly quoted as saying that "he followed the same counsel towards the enemy as did many doctors when dealing with physical ailments, namely, that of conquering the foe by hunger (through sieges) rather than by steel."[5]

The successful first-century commander Domitius Corbulo is likewise quoted as saying that the *dolabra* (a combination pick-axe tool) "was the weapon with which to beat the enemy." The next maxim reinforces the point:

> Lucius Paulus [Lucius Aemilius Paulus Macedonicus, 229–160 BCE] used to say that a general ought to be an old man in character, meaning thereby that moderate counsels should be followed.

As does the fourth:

> When people said of Scipio Africanus that he lacked aggressiveness, he is reported to have answered: "My mother gave birth to a general *(imperatorem)*, not to a warrior *(bellatorem)*."

Bellatorem being the word for a wild fighter as opposed to a soldier, *miles.*

And the fifth:

> When a Teuton challenged Gaius Marius [the consul and military reformer, 157–86 BCE] and called upon him to come forth, Marius answered that, if the man was desirous of death, he could end his life with a halter.

Not coincidentally, Frontinus was himself successful in war as legionary commander and military governor *(legatus)* in warlike Britain from the year 74 CE, where he subdued the dangerous Silures of Wales and constructed the Via Julia highway, whose traces may still be seen in Monmouthshire. Much later, in 97, the emperor placed him in charge of all the aqueducts of Rome, and his very precise description of how they worked *(De Aquis Urbis Romae)* is wonderfully instructive. Unfortunately his tactical manual, or *Art of War,* has not survived, and Frontinus himself indicates that there were no other comparable works in the Rome of his day, a most revealing absence: "since I alone of those interested in military science *(militaris scientiam)* have undertaken to reduce its rules to [a] system."[6]

The second-century literary lawyer Polyaenus from Bithynia in western Anatolia dedicated his *Strategika* in Greek—about stratagems and not strategy, in spite of the title—to the emperors Marcus Aurelius and Lucius Verus on the occasion of their war against Arsacid Persia or Parthia that started in 161. He was currying favor—probably hoping for a well-paid sinecure such as the one that Hadrian had granted to prolific Plutarch. To that end, Polyaenus claimed Macedonian ancestry: "I, a Macedonian who has inherited the ability to conquer the Persians in war, want to do my part at the present critical time."[7]

The examples selected by Polyaenus are drawn in part from proper classical texts about early times and the petty warfare of the Greek cities of the classic age, already several centuries old by then, in part from the Hellenistic age preserving some historical data and detail otherwise unrecorded, and in part from Roman history down to Julius Caesar, all with a definite emphasis on tricks rather than other forms of ingenuity.

It is not an inspiring work. It seems certain that Polyaenus had no military experience—there are none of its characteristic signs—even if he did write a lost work on tactics, as the tenth-century Byzantine encyclopedia *Suda* indicates.[8] But some Byzantines valued Polyaenus very highly. He was commended by the learned Constantine Porphyrogennetos as a valuable source of historical information and by the successful general Nikephoros Ouranos for his stratagems; he was repeatedly paraphrased with or without emendations, and also excerpted. In the new edition of Polyaenus, two of these efforts are translated, the ninth-century *Excerta Polyaeni* and the tenth-century *Strategemata*, wrongly attributed to the emperor Leo, which forms the latter part (sections 76–102) of the work published as *Sylloge Tacticorum*.

It can even be said that both are more useful than the original work, in part because the selections do favor the better material, and in part because the anecdotes are classified by subjects, including "tactics," under which we find in the excerpts only three rather tame examples: how the Athenians amazed the Lacedaemonians by standing still with spears extended when they were supposed to charge, how the Spartan Cleandridas outmaneuvered the Leucanians by first thickening the phalanx to allow himself to be outflanked and then extending it to trap them, and how Alexander the Great fought Porus, Raja Puru of Punjab, by adopting a novel tactical deployment, with the cavalry projecting at an angle from the right of the battle line and both phalanx and light troops on the right—it was supposedly the hardest battle he ever fought. As for the *Strategemata*, it chooses rather different subjects, starting with how secret messages should be sent: a most complicated contrivance is suggested, which would indeed function:

[Lucius Cornelius] Sulla [Felix] took the urinary bladder of a pig and after stoutly inflating it and binding it up until it dried out, he wrote on it with encaustic ink about something he wanted. Then he opened it, folded it together, and inserted it in an oil jar; after . . . filling it with oil, he gave the jar to one of his most trustworthy men and sent him off to [the recipient] with orders to tell him to break open the jar in private.

Other subjects include the matched pair on how to make a small army seem large (add men on donkeys and mules to the scarce cavalry kept well in front, etc.), and how to make a large army seem small (light few fires in the camp at night, etc.). That Polyaenus can entertain and divert the reader there is no doubt, but his is certainly not a work of systematic instruction.

The same is true, and much less excusably, of the work of another, better educated, and much grander Roman citizen of Greek Bithynia, Lucius Flavius Arrianus, Englished as Arrian, friend of Hadrian before he became emperor, and his appointee afterward to the high position of governor of Cappadocia in northeast Anatolia. He too did not write usefully, but by perverse choice in his case because he had ample and most interesting military experience, having served on several fronts in rising command positions. But he preferred to strike an antiquarian pose, no doubt to please his patron, the Hellenophile Hadrian.

This prolific writer, best known for of his account of the offensive of Alexander the Great all the way to India *(Anabasis)*, also wrote a *Techne Takhtike*.[9] The title is promising, for *techne* indicates practical knowledge, but the contents are disappointing because instead of writing of the Roman tactics he knew very well firsthand, Arrianus chose instead to emulate another, slightly earlier second-century Greek litterateur living in Rome, Aelianus "Tacticus," whose *Taktike Theoria* is a very detailed description of the drills and basic tactics of the long-defunct Macedonian phalanx; it was much consulted in Europe from the sixteenth century, as medieval levies fit only for the melee gave way to fixed formations meant to execute tactical orders.[10] Arrian interrupts his rendition of that text only to describe, not Roman tactics, but rather a parade-ground cavalry exercise he evidently commanded or just witnessed alongside his patron Hadrian (117–138), in which again he perversely provides no description of what was actually done, but only the orders given—and with mawkish apologies for the exotic words that disfigure his Greek text, inasmuch as the Romans took up Iberian and Celtic terms as they take "the thrones of rulers."

In 136 Arrian led a large field army of two legions with strong auxiliary forces into Roman Armenia to repel an attack by Alan horsemen from the North Caucasus and the steppe beyond. Only part of his account of the deployment of the forces, *Ektasis kata Alanon*, survives, and again it is deformed by Arrian's antiquarian compulsion—he has the defunct phalanx instead of *legio,* properly classical Scyths instead of outlandish Alans (though they appear in the title) and others such, even identifying the army's commander—himself—as Xenophon, evidently his admired predecessor of half a millennium before, as both writer and man of action—in fact he used Xenophon's most famous title for his own work on Alexander.

In spite of his avoidance of any such inelegant precision, exact unit identifications have been recovered for the *Ektasis* (also known by the Latin *Acies contra Alanos*), and the result is interesting indeed, both in itself as an example of the composition of an actual Roman field army ready for combat, and as a base of comparison with subsequent Byzantine field forces—because some of the standard Roman formations reappear as prescribed troop combinations in Byzantine tactics. The base force of two legions was purely Roman: XII *Fulminata* and XV *Apollinaris,* each with stone-throwing and bolt-launching artillery, 120 light horse for liaison and scouting, various specialists, and ten infantry cohorts, for a total of some five thousand heavy infantry at full strength.

It is usually asserted that in the Roman army of the time, the legionary forces of heavy infantry, engineers, and artillery, were matched by a roughly equal number of mostly noncitizen auxiliary troops, recruited on the fringes of the empire and beyond. They certainly complemented the tactically dominant but slow and stolid heavy infantry with a wide variety of light-infantry units, and both light and heavy cavalry, adding mobility, versatility, agility, and the missile capacity of slingers and archers lacking in the legions. But if it was indeed true that the proportions were roughly equal in the Roman army as a whole—there is no definitive evidence—they would not necessarily be equal in every field army, for it would have made no sense to assemble expeditionary field armies without trying to select auxiliary units sized and suited to the nature of the terrain and the enemy.

In this case, Arrian was fighting the Alans, who came from the steppe with their horses just as the Huns and Avars would do from farther afield, but they were not armed with the powerful composite reflex bow any more than the Roman auxiliary units of bowmen. The mounted Huns and Avars, moreover, combined forces with warriors from their subject nations who fought on foot, the Huns with their Goths and Gepids, the Avars with their Slavs. By contrast, the Alans whom Arrian confronted were seemingly all horsemen with no fighters on foot. (The

Alani horsemen reported in western Europe circa 400 had been driven westward by the Huns. The Caucasian Alani survived in the medieval kingdom of Alania, and survive still as the Ossetians of Georgia and Russia.)

That explains the high proportion of mounted troops among the auxiliary units of Arrian's expeditionary army: first a unit of mounted scouts that precedes the main force, a *numerus exploratorum* with perhaps 300 horsemen; then the main cavalry contingent, four *alae* with 500 horsemen each at full strength;[11] then came that peculiar type of Roman auxiliary unit *cohors equitata* that mixed horsemen with infantry within the same formation, to provide both patrols and stronghold guards on the defense, and light cavalry and infantry on the offense: three double-sized *(miliaria)* units of 240 mounted and 760 foot soldiers at full strength, including one of bowmen, and five standard-size mixed units of 120 mounted and 380 foot soldiers at full strength.[12]

All these cavalry and mixed units were accompanied by a single cohort of pure infantry, and an unusual one, having originally been raised from Roman citizens: *Cohors I Italica voluntariorum civum Romanorum,* with 500 men at full strength, for a total of 4,680 auxiliary infantry as opposed to 3,620 horsemen. If all auxiliary units present were fully manned, a thing most unlikely in any army at any time, their total would have come to 8,300, considerably less than the 10,000 well-armored infantrymen in the two legions—but they of course were even less likely to be fully manned, in part because legionary bases could not be left without any troops at all, given the presence of the men's families, convalescents, and new trainees, and in part because legionary soldiers who could usually read, write, and count were forever being detailed to all sorts of assignments in support of the civil administration.

As for the proportion of horsemen to foot soldiers in Arrian's force, it amounts to one-quarter if absentees were evenly distributed, very much less than in all the Byzantine field armies for which we have numerical information. More cavalry than infantry forces, and especially more elite, multipurpose cavalry, were a much better fit for the less decisive and more flexible Byzantine style of war.

To invent a new style of war was precisely the aim of the anonymous *De Rebus Bellicis* (Of Military Matters), the final Roman military text—for the empire was reaching its end in the west when it was written.[13]

This pamphlet, which has reached us in a single manuscript complete with its essential illustrations, is not an exercise in nostalgic antiquarianism in the manner of Vegetius, nor a literary exercise. It was written in deadly earnest by an anxious author who rightly feared for the survival of his civilization.

Having established that the empire was being ruined by the enormously costly upkeep of vast military forces that nevertheless failed to prevent ruinous barbarian incursions, and recognizing that to simply reduce troop levels would open the door to more destruction, the author offers a solution of startling modernity: mechanization. He even presents a sort of cost–benefit analysis to prove that Roman military strength could be increased while reducing troop levels, by increasing fighting power per soldier. That was to be done by investing in better armor and personal weapons—including the lead-weighted darts that would later indeed find much favor with Byzantine forces—as well as in new military machines, including cart-borne multi-arrow projectors, war chariots with rotating blades, and an ox-propelled ship, whose illustration has often been reproduced. Some of the machines are quite practical, and others fantastical, but if the writer was not a practiced engineer he was certainly a coherent analyst of his own tragic times.

Half a millennium earlier, decidedly more practical machines of war were described by Marcus Pollio Vitruvius, a first-century BCE combat engineer with Julius Caesar, author of the widely read *De Architectura,* a strong influence on the sixteenth-century Andrea Palladio and not unknown in Byzantium because the twelfth-century freelance writer and poet John Tzezes refers to a specific passage.[14]

Most of *De Architectura* is dedicated to detailed explanations of building methods and designs, Greek as well as Roman and the history thereof, with such admixtures as the inspection of the livers of local sheep to determine the suitability of a site for habitation: if the livers were consistently livid in color and damaged *(livida et vitiosa),* the site was dangerous (Book I. 4, 11). It is characteristic of Vitruvius that this method would actually work to detect chemically contaminated soils. Book X, is mostly about machines, windlasses, pulleys, water pumps, vehicles, levers, hoists, mechanical mills, siphons, a water-powered organ, and a distance-measuring carriage—all of which can be built following his directions, and all of which would work.

The same is true of the descriptions of arrow-launching catapults—all the dimensions are determined by the length of the intended arrow—in the tenth chapter of the book. Vitruvius explains how to design catapults powered by the torsion of ropes and/or twisted human hair or dried tendons, giving the exact dimensions of each component; replicas have been successfully built in modern times by following his directions. Next, in chapter XI, there are equally complete instructions for stone-throwing *ballistae,* whose detailed design is a function of the weight of the intended projectiles.[15] They are meant for people ignorant of geometry who cannot be delayed by calculations *(cogitationibus detineantur)* amid the perils of war, that is, soldiers in the field; only the obscurities of notation would complicate a modern reconstruction by perfect amateurs. In chapter XII, directions follow on how to prepare both designs for combat.

Next Vitruvius writes of the history and design concepts of swinging-pole battering rams, evidently on the basis of lost Greek texts, of movable towers—discussed below—boring machines, elevating machines *(ascendentem machinam)* that lift assault troops to wall height (see the discussion of *sambucae* below), and more such. He then proceeds to give detailed construction directions for a mobile swinging-pole, protected, battering ram *(testudinis arietariae)* or "tortoise-ram." Chapter XIV begins with the description of a mobile, ditch-filling machine protected from gravity projectiles by stout timbers and from fiery incendiaries by two raw ox hides, sewn together with a filling of seaweed or straw soaked in vinegar. The conceptual designs of more machines of ancient Greek design follow, including defensive devices from chapter XVI, such as a crane to lift enemy machines that reach the wall to hoist them into the city; and the method employed by the freelance consultant Diognetus of Rhodes to defeat the gigantic mobile assault tower built for "King Demetrius" (Demetrius I of Macedon, 337–283 BCE, dubbed *Poliorcetes,* "the besieger"): 125 feet high, 60 feet wide, protected to resist stones of 360 pounds, and weighing 360,000 pounds. Diognetus had volunteers pour water, sewage, and mud in front of the walls, the mobile tower was stuck in the softened ground, and Diognetus received his requested fee: the mobile tower itself. More such accounts follow, good stories and sound engineering. Later, more Hellenic Byzantines could have dismissed it all as derivative of ancient Greek engineering; so it was, but for the Roman brevity, precision, and practicality.

Those are also the virtues, to an even greater extent, of the third-century work *De Munitionibus Castrorum,* on the fortification of military camps, which was once attributed to the second-century Hyginus Gromaticus ("The Surveyor"); accordingly, the author is now listed as pseudo-Hyginus. The extant part—"fragment" would imply that the complete text was many times longer, which seems unlikely—of this treatise runs to fifty-eight sections, mostly of a few lines each, making

up ten pages or so in print. Their extreme brevity and clarity are enough
for exact unit-by-unit instructions on the layout of a Roman marching
camp—a *"marschlager"* for the editor, translator, and commentator, the
eminent Alfred von Domaszewski.[16]

The Roman *castrum* was certainly one of the secrets of Roman military success—a secret not lost in Byzantium: the tenth-century work
known as *De Re Militari,* newly edited as "Campaign Organization and
Tactics," begins with the detailed layout of a marching camp.

By constructing an entrenched and palisaded camp for themselves,
if necessary each and every night when marching through insecure territories, the Romans and the Byzantines after them not only guarded
against dangerous night assaults, but also ensured a calm sleep undis-
turbed by harassment raids or infiltrators. When thousands of soldiers
and horses are crowded inside a fortified perimeter, which must be as
short as possible to be well guarded, a tightly defined layout of the tents,
baggage, and horses, unit by unit, with clear passages between them,
leading to broad "streets," is the only alternative to chaos, congestion,
and confusion in the event of a enemy attack, or simply an urgent exit
from the camp. Moreover, it is the only way to keep latrines well separated and downhill from streams or wells. In the fundamental Byzantine military manual known as the *Strategikon* of (emperor) Maurikios,
night attacks on their camps are suggested (Book XI, 1, 31) when fighting the Sasanian Persians; otherwise highly competent, the Sasanians
were lacking in their camps. Although they too entrenched and guarded
a perimeter, they did not enforce a disciplined internal layout unit by
unit—the troops camped where it suited them.[17]

The camp described in *De Munitionibus Castrorum* is very large indeed—too large, most would have been far smaller—for it assigns
places for three complete legions, four cavalry *alae miliariae* of 1,000
men each with more than 1,000 horses, five *alae quingenariae* of 500
men each, and thirty-three more legionary detachments and auxiliary
units, with a broad panoply of unit types represented, including 1,300
marines or assault-boatmen (500 *classici misenates* and 800 *classici
ravvenates*), 200 scouts *(exploratores),* 600 Moorish and 800 Pannonian
light cavalry, and many more, for an impossible total of more than
40,000 troops and 10,000 horses. Evidently this was a design exercise,
and there are specific places for each unit in the layout: the cohorts of legionary heavy infantry are tented in the outer perimeter, which they
would be the first to defend, and the usual twin headquarters the
Quaestorium and the Praetorium are in a spacious central segment. In

its small compass, the work is highly instructive, and it may well have sustained the marching-camp concept that we know was studied and practiced for at least another seven hundred years.

For the Byzantines, *Roman* military literature, whether in Latin or Greek, could not be classical—only the texts of ancient Greece could aspire to that status, starting with the impeccably antique fourth-century BCE Aeneas, usually known as Tacticus, on the defense of fortified positions.[18] The surviving text is only part of a longer work cited and quoted by Polybius in Book X, 44, with faint praise for the method of signaling suggested by Aeneas.

The contents necessarily were original for the times, but although they are practical enough they show no particular ingenuity and did not perceptibly influence Byzantine practice, though the work was remembered, cited, and excerpted, by Julius Africanus among others—of whom more below—whose fragments add to the available text. The same is true of several lost military texts of that era of which we only know the authors' names.[19] One of these we have much reason to miss: that by Polybius, who mentions his notes on tactics in passing in *Histories*, Book IX, 20; in that detailed and reliable work, the treatment of military matters shows insight and expertise, which would have been consolidated in the lost work on tactics.

From the third century BCE, we have a very interesting technical treatise by a Biton, otherwise unknown, dedicated to Attalus I Soter ("The Savior") of Pergamon—giving us a date because Attalus was enthroned in 239 BCE.[20]

Biton describes six artillery weapons: a small stone-thrower designed by Charon of Magnesia in Rhodes on the crossbow principle; a large stone-thrower by Isidorus of Abydos constructed at Thessalonike on the same principle; a mobile siege tower *(helepolis)* man-powered by an internal capstan that turned the wheels with assault drawbridges, designed for Alexander the Great by Posidonius the Macedonian—and a practical design, unlike larger specimens outfitted with catapults that proved too heavy to be driven within reach of enemy walls; a *sambuca*—the name borrowed from a triangular harp—that is an assault ladder hinged on a pedestal to swing down onto the assaulted wall, designed by Damis of Colophon; a medium artillery *gastraphetes*, in effect a large crossbow built by Zopyrus of Tarentum at Miletus, and a lighter, "mountain" *gastraphetes* built by the same Zopyrus at Cumae.

Biton's text is so precise in its descriptions and fully consistent measurements that all six war engines could have been reconstructed quite easily even without the drawings that came with the surviving manuscript. Its survival is evidence in itself of continued interest in his work, but he is also cited by the better-known technologist Heron of Byzantium.

In his account of the Roman siege of Syracusa, a splendid story with real military and technical content, Polybius described a naval sambuca in action:

> A ladder was made four feet broad and of a height equal to that of the wall. . . . Each side was furnished with a breast work, and it was covered in by a screen at a considerable height. It was then laid flat upon . . . the ships . . . protruding a considerable distance beyond the prow. At the top of the masts there are pulleys with ropes, and when they are about to use it, they attach the ropes to the top of the ladder, and men standing at the stern pull them by means of the pulleys, while others stand on the prow, and supporting the engine with props, assure its being safely raised. After this the rowers on both the outer sides of the ships bring them close to the shore, and they now endeavor to set the engine . . . against the wall. At the summit of the ladder there is a platform protected on three sides by wicker screens, on which four men mount and face the enemy resisting the efforts of those who from the battlements try to prevent the *sambuca* from being set against the wall. As soon as they have set it up and are on a higher level than the wall, those men pull down the wicker screens on each side of the platform and mount the battlements.[21]

That is how seemingly unassailable seawalls could be assaulted effectively from ships below them, and the Roman commander Marcus Claudius Marcellus duly relied on sambucae to quickly conquer Syracuse from the sea in 214 BCE during the second Punic war—except that the chief engineer on the other side happened to be Archimedes, who had his powerful anti-sambuca hooked levers ready to upend them—along with the ships on which they were mounted. The siege lasted more than two years. Marcellus incidentally deserves our respect for his reported reaction to the debacle, which had evoked the garrison's derision: "Archimedes uses my ships to ladle sea water into his wine cups, but my sambuca band is flogged out of the banquet in disgrace."

Philon of Byzantium was another technologist of the third century BCE but with a far broader scope. His many writings included volumes on the mathematics of engineering, on levers, on constructing harbors, on constructing artillery projection weapons *(Belopoeica),* and other subjects, including a volume on pneumatics that survives only in an Arabic translation. He himself gathered his writings into a comprehensive survey *(syntaxis)* of mechanics, but most are lost, though not the *Belopoeica*.[22] In the preface, Philon recognizes the difficulty of explaining how to build the machines he proposes with sufficient precision to ensure consistent results:

> Many who have undertaken the building of engines of the same size, using the same construction, similar wood, and identical metal, without even changing its weight, have made some with long range and powerful impact and others which fall short of these.

His remedy is very careful calculation—an emphasis on mathematics as the basis of engineering, even though Philon himself, along with others, misread Aristotle, who did not write that large weights fall more quickly than light ones—an enduring error that could have been refuted experimentally but was not until much later (though before Galileo, contrary to legend). It was a Byzantine weakness to accept authority too readily, but they certainly understood the importance of mathematics whether in architectural or military engineering.

Philon's *Belopoeica* comprises a description of torsion-powered arrow- and bolt-firing catapults, which employed hair or tendon for elasticity; of an arrow-firing engine with a stretching device, accompanied by criticisms of ordinary catapults whose operation, he writes, is compromised by flimsy components; of the *chalcentonon* of Ctesibius (the third-century BCE mathematician who may have headed the library of Alexandria) powered by bronze springs in a double-sided torsion box, and his *aerotonos katapaltes lithobolos,* a totally original pneumatic stone-thrower powered by bronze cylinders and pistons; and of the repeating bolt-firing catapult *polibolos katapaltes* of the otherwise unknown Dionysius of Alexandria—a wondrous machine that could launch bolts nineteen inches (twenty-five dactyls) long in quick succession from a topside magazine that fed the bolts down by gravity, but regulated by a rotary placement device. This was no fantasy—it is clear from the text that Philon examined such a machine, and his description is sufficiently precise to have allowed his ingenious modern editor E. W. Marsden to draw it in full detail, the original drawings having being lost; incidentally Marsden notes that in the 1894–1895 war, the Chinese used repeating crossbows most unsuccessfully against Japanese infantry armed with bolt-action magazine rifles, and that their bamboo-powered weapons were much less powerful than the *polibolos katapaltes* of Dynosius of Alexandria that was probably used in the siege of Rhodes in 304 BCE.[23]

To determine what earlier writings influenced Byzantium, the indispensable guide remains the celebrated survey by Alphonse Dain revised by J. A. de Foucault, a culmination of Dain's high scholarly achievement in editing Byzantine military texts.[24]

Dain's survey cites the first-century BCE treatise on war machines by Atheneus the mathematician, before proceeding to the less derivative and more precise Heron of Alexandria, variously known as "The Ancient" and "The Mechanical"—an intellectual disciple of the celebrated Ctesibios. Dain dates Heron at the cusp of the second and first centuries BCE, while his more recent editor places him two hundred years later, but given the technological stasis of the age, that hardly mattered to his Byzantine readers.

Two of Heron's works survive, a *Belopoeica* that describes projection artillery, enhanced by seventy-six illustrations, mostly of components and characterized by extreme precision; and a fragmentary *Cheiro ballista,* a title that implies the description of a mobile arrow-firing catapult—and therein hangs a tale. A French architect, Victor Prou, actually built a model of a bronze-spring catapult in the 1870s; a new examination of the surviving text by a combative German editor showed, however, that it does not describe a weapon but only a number of mechanical components whose Greek names all start with the letter *K (kanones, kleisis kamvestria, . . .),* inducing that editor to argue that the fragment was just the K section of a mechanical lexicon. But the most recent and most authoritative editor, E. W. Marsden, has concluded that the components do assemble into a weapon, specifically the sinew-powered catapult depicted on Trajan's column.[25] The *Belopoeica* itself, which undoubtedly does describe complete weapons, is prefaced by a bold argument for the virtues of studying war that would be much imitated:

> The largest and most essential part of philosophical study deals with tranquillity [that was before linguistics conquered philosophy, alas] . . . and I think the search for tranquillity will never reach a definite conclusion through the argumentative method. But Mechanics, by means of one of its smallest branches—I mean, of course, the one dealing with what is called artillery-construction—has surpassed the argumentative training on this score and taught mankind how to live the tranquil life. With its aid men will never be disturbed in time of peace by the onslaughts of enemies.

In other words, *si vis pacem para bellum:* if you want peace, prepare war.[26]

Of the first-century BCE Asclepiodotus, "The Philosopher," another

Greek living in Roman imperial times, we have a "tactics" *(Techne Takhtike)*, whose promising title is deceptive.[27] There is nothing on contemporary Roman tactics or on tactics of any age properly defined. Instead the text consists of an exceedingly detailed, indeed obsessive, description of the ranks, titles, structures, hierarchies, dispositions, and drilled movements of a Macedonian phalanx, and as such is valued for its lexical contributions—many of the words are otherwise unattested. In the manuscripts there are also many graphics of the drills, with the use of countless symbols to keep track of the prescribed positions at each remove. But it would have all been quite useless to Byzantine soldiers, because nothing could undo the obsolescence of the phalanx (the Swiss pikemen and halberdiers who dominated European battlefields in the fourteenth and fifteenth centuries fought in deep columns, as did their Byzantine predecessors with their pikes).

The antiquarian nature of the work is further affirmed by its section VIII on war chariots ("two chariots are called a pair *zygarchia*, two pairs a double-pair *syzygia*"), which were entirely archaic even then, and section IX on war elephants ("the leader of a single elephant is called an animal-commander *zoarchos*, the commander of two a beast-commander *therarchos*"), last used—unsuccessfully—against Julius Caesar in 46 BCE at Thapsus in modern Tunisia, where the legionnaires of the V *Alaudae* earned their elephant symbol by using their axes to hack away at the elephants' legs.

A work much better known, the *Strategikos* (The General) of Onasander or Onesandros of the first century CE is much disliked by Dain. He ridicules the author as a *graeculus*, the Roman insult for a little Greek on the make, of which there were many in early imperial Rome, and denigrates him as a flatterer of his Roman masters entirely lacking in useful originality; his book is filled, Dain writes, with obscure counsels of prudence and hollow exhortations.[28]

The Byzantines disagreed, for the work is cited by John the Lydian (book I.47.1) in the sixth century, was entirely annexed by the imperial anthologist Leo VI into his *Taktika*, and is mentioned by the celebrated general and writer Nikephoros Ouranos in the tenth century. We know that many others in post-renaissance Europe also disagreed, because Dain himself lists numerous editions and translations from 1494 onward, and mentions the praises of that successful and thoughtful soldier Marechal de Saxe, cited in several French editions. Modern readers will find the work neither heaven nor hell. There is much good sense on the choice of generals and the characteristics of good generals—one can readily think of cases to whom the exclusions should have applied—and more (in book III) on the general's council:

> It is not safe that the opinions of one single man, on his sole judgment, should be adopted. . . . However, the general must neither be so undecided that he entirely distrusts himself, nor so obstinate as not to think that anyone can have a better idea than his own.[29]

Also innovative for the times is the advice (book IV) that war should be plausibly justified:

> It should be evident to all that one fights on the side of justice . . . for with the knowledge that [the soldiers] are not fighting an aggressive but a defensive war, with consciences free from evil designs, thy contribute a courage that is complete.

We are a long way from Homeric war fought for personal honor, or unabashedly fought for booty or imperial expansion—this is definitely a functional "just-war" argument before its time, exactly what the Marechal de Saxe would have appreciated.

There is nothing much wrong with the tactical advice, either—it is useful even if neither original nor detailed. It proceeds from the advancing formation to the passage of dangerous defiles (book VII), the making and changing of camps (for hygienic reasons), foraging with care, captured spies—to be killed if the army is weak, to be released if it is impressively strong—and more such, including some things that one does not expect from an author that Dain dismisses as militarily inept *("nullement versé dans l'art militaire")* such as the holding of a reserve force in battle, kept aside to intervene at critical moments; that is what allows the general to shape the battle, the uncommitted reserve without which he is a mere spectator. On the whole, the reader is likely to agree with the Byzantines and Marechal de Saxe rather than Dain, for all his splendid scholarship.

The multilingual, multitalented, multinational Sextus Julius Africanus, or more properly Sextos Ioulios Aphrikanos, inasmuch as he wrote in Greek, born in Jerusalem by 180 and probably a Jewish convert to Christianity, dedicated a collection of writings—*kestoi* (literally embroideries, but possibly amulets)—on the most disparate subjects to the emperor Alexander Severus (222–235).[30] Only parts of it survive, including some on military matters, but his real talent was in logic and mathematics, which he applied to all manner of things, becoming the founding chronographer of Christianity, much cited as such by Eusebius and other Church fathers, and most curiously, to a study of the flight of arrows.

He begins by asserting that an arrow is capable of flying over a distance of 25,000 stadii (4,675 kilometers) in a day and a night, that is, twenty-four hours. He then explains how to prove this with an elaborately clever experiment in which distance is replaced by timing, and the calculation is ultimately based on the premise that no more than six thousand arrows can be released in *immediate sequence* within an hour. The experiment is actually valid, and so are the derived calculations: disregarding the falloff in initial velocity, 4,675 kilometers in twenty-four hours corresponds to 194.8 kilometers an hour, just about the measured initial velocity of modern, medium-power bows in the 50–55 pound class, and only slightly more than the velocity of modern reproductions of ancient compound bows.[31]

The first great epoch of Byzantine military literature begins in the sixth century, but at first very unpromisingly, with the *Taktikon* of Urbicius or Orbicios dedicated to Anastasios I (491–518)—a mere summary of Arrian on the phalanx, mostly just terminology. Another even slighter work by the same author, the *Epitedeuma*, is dedicated to his own wondrous invention designed to defeat the extraordinary impetuosity of barbarian cavalry: *kanones*. These are assemblies of portable tripods with sharp points that can be placed end to end to protect the infantry—which is a light-missile infantry in his text, without the sturdy pikes and large shields that can fend off cavalry charges.[32]

Syrianos Magister on Naval Tactics

Syrianos Magister is not impressive either, but it is now believed that the works of his that survive—the recently attributed "Anonymous Byzantine Treatise on Strategy" previously known as *Peri Strategikes* or *De Re Strategica*, and the *Rhetorica Militaris*, both discussed below, as well as the fragment on naval warfare next examined—are all unrepresentative of the lost whole.

The naval fragment, derived from a much older source by the ninth-century Syrianos, starts in section IV in the middle of a phrase with the recommendation that crews immediately line up in battle ranks when they land on the shore, until they are entirely sure that there is no enemy nearby.

That is a reminder that with galleys all naval warfare was amphibious warfare, if only because fresh water had to be taken aboard very frequently, and the men could be properly rested only ashore—it little mattered if the oarsmen were citizens, professionals, or slaves (not used by the Byzantine navy), none could do much rowing without better sleep than they could have at their rowing bench.

Next, the author advises that the oarsmen should also be swimmers, underwater as well. Combat swimmers can be very effective. They can surreptitiously cut the anchor ropes of enemy vessels, causing them to drift onto rocks propelled by currents, and by swimming underwater they can evade pursuit even if seen.[33]

In section V the author notes that the *strategos*—a military leader but not necessarily of the seafaring kind—must have with him real experts in winds, currents, shallows, ports, and all marine things, and each separate vessel must have one such, for the fleet can easily be separated by the winds. In addition, at least two oarsmen in each vessel must be skilled in hull repairs, and all of them must know how to plug holes with clothing and blankets.

Intelligence comes next in section VI, with useful advice. The lightest and fastest ships, with oarsmen selected for stamina rather than bravery—for their task is not to fight but to observe and report—are to look for enemy vessels hidden behind promontories or islands, in port, or in river mouths. Four fast vessels are to be sent out from the fleet, two remaining within six miles and the other two venturing out, so that the information can sent back by prearranged signals from ship to ship. On land, able scouts are needed as well. In section VII the signaling methods are specified: white sheets at close range, smoke signals farther out and in sunlight, mirrors or a well-polished sword. A code for each is implicit.

Tactics are explained in the much longer section IX. For the fleet at sea, station-keeping in tight formation is as essential as well-ordered ranks and files on land; to learn the skill, the fleet should not wait for a battle but always move in formation. The strategos goes ahead in one of the bigger ships; they are slow, so he must have two fast vessels with him to send orders around the fleet. The other bigger ships should form the front, as the most heavily armed infantry does on land.

Admiral Nelson would not have approved of the entire concept of naval warfare offered by the text: instead of always seeking battle, instead of always being on the offensive, as he was, the strategos is urged to be cautious before entering combat. First he is urged to ask himself and his most trusted subordinates if it is really necessary to fight. Next, he is urged to assess the balance of strength once again, trusting deserters if several say the same thing. If his fleet is stronger, he is reminded that even the outnumbered can win. If the two sides are equal in strength and the enemy does not attack, neither should the strategos.

This caution is not tactical but strategic, and derives from the very essence of the Byzantine style of war. To seek out the main enemy fleet and attack it with every vessel at hand to achieve a "decisive" victory was the only proper purpose of naval warfare for Nelson, as was its equivalent on land for Napoleon or Clausewitz, and all those who followed them till the present day. The fundamental premise was shared by the Romans of the united empire as well: the aim of war is decisive victory because it will destroy the enemy's power and bring peace—on favorable terms of course. That is where matters end, happily and finally.

But the Byzantines knew that there was no end, that new enemies would emerge if the old were utterly defeated, and there was an even chance that they might be as dangerous or more dangerous. In that case, the enemy fleet left undestroyed might be useful, for the newcomers might be as threatening to the old enemy as to the empire. It was the same on land, where to destroy one enemy pressing against the imperial frontiers would merely leave a vacant space for the next enemy to threaten imperial borders.

Therefore the purpose of war for the Byzantines could not be to seek battle for decisive victory, for there was no such thing, but only to contain immediate threats by weakening the enemy with expedients, ruses of war, and ambushes, leaving any form of two-sided large-scale combat as only a last resort. That is the strategic context of this, as of every piece of serious Byzantine military writing.

The hardest case is also plainly the most important for the writer: what to do if the enemy is much stronger but battle cannot just be avoided for it would expose cities to attack. Then the fleet must fight outnumbered and win by tactical maneuver and by stratagems. To use the winds to contrive to fight in narrow waters between islands or in straits is very advantageous for the outnumbered, because the enemy will not be able to deploy his full strength, as many of his ships will be too far behind to engage in combat. To divide the enemy fleet is another way, either by arriving first before all ships have assembled, or by stringing out the enemy formation, so that the outnumbered in total can outnumber in each engagement.

The author again recommends avoiding battle, and suggests that if it results in depredations, they can be balanced by raiding enemy coasts.

When battle is actually joined, the strategos must be ready to encourage the brave and threaten the fearful; incipient desertion or worse may happen even in the picked first line of warships, with oarsmen stopping or rowing slowly. If exhortations fail, fast light vessels are to be sent to kill deserters who jump into the water. Medium-size vessels, which have nothing to fear from small vessels and can outrun the big ships, should be sent behind the enemy fleet formation to attack it when the frontal battle is already under way. Medium ships are also suitable as flank guards.

The detailed tactics recommended by the author are realistic, and so is his envoi to the strategos in section X: remain prudent if victorious, do not despair if defeated, but gather the surviving ships to fight another day.

Peri Strategikes / De Re Strategica

Translated as "The Anonymous Byzantine Treatise on Strategy" by its most recent editor, George T. Dennis, but conventionally known as *Peri Strategikes* or more commonly *De Re Strategica,* the text was dated to the later sixth century by Dain and Foucault—mostly because the author mentions Belisarios in a manner that suggests that he was a near contemporary.[34] Recently, however, the work has been persuasively reinterpreted as a surviving part—or rather as a series of parts—of a much more comprehensive treatise on military matters attributed to the elusive Syrianos Magistros, whose dating might be as late as the late ninth century, though in the usual way it contains much older materials.[35] The hypothesis of a lost major work has the virtue of explaining why Syrianos is praised so highly by the praiseworthy Nikephoros Ouranos, as well as by the uncritical Constantine Porphyrogennetos, for nothing of his that survives is all that impressive.

The first page is missing, but otherwise the treatise is complete, and a substantive work it is, even if Dain opines that the author was not a practical soldier or experienced general but rather a *"stratège en chambre,"* an armchair strategist. Dennis disagrees and is impressed by the realism of the sections on military engineering, visualizing the author as staff officer or indeed a combat engineer if not a field commander;[36] but the author's familiarity with ancient authors, and his preference in places for thoroughly obsolete texts rather than sixth- or seventh-century expertise, suggest that Dain was right.

The work begins with three sections on society and government before turning in the fourth section to strategy itself, "the most important branch of the entire science of government." His definition is succinct: "Strategy is the means by which a commander may defend his own lands and defeat his enemies. The general is the one who practices strategy." Interestingly enough, before starting on the subject, the author felt the need to justify his interest. The delegitimization of war as a valid human undertaking has progressed greatly in the aftermath the prolonged mass wars of the twentieth century. But the process must already have started in the sixth century, owing to the influence of Christianity (and is still largely confined within it), because the author writes:

> I know well that war is evil and the worst of all evils. But since our enemies clearly look upon the shedding of our blood as one of their basic duties and the height of virtue, and since each one must stand up for his own country and his own people with word, pen and deed, we have decided to write about strategy.[37]

What follows is in part theoretical, in part antiquarian—there is the Macedonian phalanx, forever recalled by Greek writers still unreconciled to Roman predominance—and in part far more practical, realistic, and certainly useful for subsequent Byzantine commanders. The sections on "a good defense," on guards and guard posts, on operating signal fires, and on forts, belong to the first category, sensible but unoriginal and too general in any case to be really instructive. Three more sections, on studying terrain to determine if it is suitable for a new city, on choosing its site, and on how to build the new city, all apply exclusively to a fortified frontier city built for strategic reasons; accordingly, the section on countermeasures against mining and siege engines contains much technical detail. Among other things, the author gives specific instructions on how to detect and oppose attempts to mine under the walls, including effective and ingenious countermining techniques.

The building of fortified cities could in fact be a tool of theater strategy for the Romans and Byzantines, albeit necessarily of rare application, given the cost. Most notably, Dara or Anastasiopolis, now Oğuz in southeast Turkey, was built in 505–507 under Anastasios I (491–518) as a fortress city, evidently to consolidate imperial control over that sector of the frequently threatened border with the Sasanian empire.[38] Its location proves as much: sixty Roman miles—four days' march for equipped troops—south of the greater and much-contested fortress city of Amida, lost to the Sasanians in 359, retaken in 363, lost again to the Sasanians in 502, regained in 504, lost once more in 602, regained in 628, and lost to the Muslim Arabs in 640; and fifteen roman miles west of Nisibis, Nusaybin of modern Turkey, which had been the impregnable bulwark of the frontier until given to the Sasanians under the 363 peace treaty. As its belated replacement, Dara was duly exposed to attack, falling under Sasanian control between 573 and 591, and again from 604 to 628, during the longest and last of the Persian wars.

In the contemporary account of the building of the city, by Marcellinus Comes, there are clear parallels with the instructions in sections 10–12 of the *Anonymous Treatise*:

> Anastasios . . . despatched outstanding workmen and ordered building to begin. He put Calliopius . . . in charge of the work. Indeed with wonderful perception, he marked out with a hoe a furrow for locating the foundations on a hill ending up on a level ground; and he surrounded it on all sides to the edge of its boundaries with the erection of very strong stone walls. So too he enclosed the stream called Cordissus . . . which winds its way as it roars along, and at the fifth milestone it divides . . . the new city, falling into a concealed entrance [under the walls] at each end. . . . The so-called Herculean tower, the city's huge lookout, was built on higher ground and connected to the walls. It looked up to [toward] Nisibis [across the Sasanian border] to the east and looked back to Amida to its north.[39]

An antiquarian strain emerges when the author of the *Anonymous Treatise* turns to tactics, in spite of his own military experience; he repeatedly refers to "ancient writers."[40] He does not name them, but they have been identified in the aforementioned Aelian and Asclepiodotus. His own account of the infantry is mercifully much simpler but not more useful, as it discusses the obsolete phalanx, its equipment—"the spears should be as long as possible" he writes (16, 31) (calling them *dorata*, a more classical word than the Macedonian *sarissa*)—the cavalry phalanx, and the phalanx on the move. At the time, Justinian's infantry was fighting in much simpler and certainly more fluid formations—and in case it was the cavalry that was the dominant arm, and very different it was from the Macedonian and indeed Roman cavalry, with the missile power of its composite bows added to the shock of the charge, and also a readiness for close combat with javelin and sword. The author must have known these things full well, but in a time of perceived decline from greater glories, the splendidly victorious Macedonians of Alexander the Great evidently retained their irresistible appeal, even to a writer elsewhere dourly practical.

A radically different tone pervades the next section on "crossing rivers" against enemy resistance—one can sense that the author has done it, and certainly he no longer echoes "ancient writers." He starts by noting that missile superiority is needed to effect an opposed crossing, and suggests the use of decked boats with "artillery to fire missiles and stones, while the men below deck carry on their part by shooting through the portholes." It is with disenchanted military expertise that he discusses the ingenious idea of an Apollodoros—presumably Apollodoros of Damascus, architect of Trajan's bridge across the Danube described by Cassius Dio (LXVIII.13), whose highly imaginative model is displayed in Turnu Severin, Romania, the Roman Drobeta. It was to build a raft along the riverbank as long as the river's width. At the upstream end, a turret is to be added, to be manned by archers. When all is ready, the downstream end is secured to the riverbank while the upstream end is pushed off into the current to swing round to the enemy's side of the river. Under the cover of the arrows launched from the turret, the troops land.

In effect, this was a variant of the pontoon bridges, some very long, built by both the Byzantines and the Persians. In December 627, a desperate Khusrau II fleeing from Herakleios crossed the lower Tigris River "and cut the cables of the pontoon bridge" behind him.[41] But that was a bridge constructed in the usual way by securing current-wise boats with long logs affixed to them, on which planks were laid to form the roadway, and of course firmly attached at both ends. To swing a long bridge in the current is a far different thing. This is our author's disenchanted comment on the brilliant suggestion of Apollodoros:

> In theory this operation may seem reasonable, but in practice I do not think that it will work out so well. Look at it more closely. If the river is narrow, I am certain that enemy arrows will easily prevent the construction of the raft. Even if there should be no worry on that score, it would be impossible to build such huge rafts or to maneuver them. The width of the raft should obviously be proportionate to its length; otherwise when both ends are secured to each bank, the current will bend it like a bow and eventually break it in two. Moreover, the depth has to be proportionate to its width, especially since the raft also has to provide support for a tower, parapets, and a large fighting force. If it is to be deep enough for this, then the whole construction becomes impossible. . . . In my opinion it is much safer to make use of boats.[42]

To transport boats to the crossing site would be difficult, of course, unless they were small, and if so the soldiers would have to cross to the enemy side of the river in small boatloads, a few at a time, and without stable archery platforms, hardly the right thing for opposed landings. But the author has a perfectly practical solution: he suggests that larger boats be safely constructed away from the site in separate components, with "each piece marked to indicate where it belongs in assembling the vessel." In this way, fairly large assault boats could be transported in disassembled form by wagons and pack animals: "when we reach the river, the timbers of the ships may again be fitted together and reassembled, their joints caulked with pitch, wax."[43]

After this sound advice there is more antiquarianism in what follows, on the drilled movements of the phalanx, through sections on turning, wheeling about, countermarches by files or by ranks. ("These are the kinds of countermarches named by ancient writers. They add that there are three ways of executing each of them.") Here again we glimpse the obsessive lexigraphy of Asclepiodotus. But after that there is more advice of practical value, on setting up camps, on allocating space within them—the cavalry should be in the middle to keep horses calm and safe from arrows—on fortifying camps, on how generals can best pass their orders in combat (by voice, by trumpet, by signals), and on the handling of the battle. The author continues to use the term *phalanx* but now evidently as a generic term for an infantry formation rather than the Macedonian phalanx of shock infantry with thrusting spears.

More follows on when to engage and when to avoid battle: "If we are facing the risk of defeat, it is wise not to join battle with the enemy until it is getting towards sunset . . . [so] . . . that we do not suffer serious harm, for, I know well, the gathering darkness will prevent pursuit at night."[44] This suggests combat experience, though not necessarily on the losing side. Just before that the author cites the remedy of Belisarios when faced with a superior force that he could not resist: he would retreat, destroying provisions beforehand so as to compel the enemy to separate his forces to feed and forage them, and would then fight each in turn with superiority on his side.

The author is again practical when giving advice on tactical dispositions in the next few sections of his treatise, and we can be quite certain that his "phalanx" is a contemporary force and not an antique remembrance, because his infantrymen are armed with the bow in facing the cavalry and not the long Macedonian *sarissa:*

> The men stationed in the first and second ranks should keep up continuous [volleys] with the bow, aiming at the feet of the enemy's horses. All the rest of the men should shoot at a higher angle, so that when their arrows drop down from above, they will cause all the more injury, since the horsemen cannot use their shields to protect themselves and their horses.[45]

On what would now be called "battle management," on how to engage, how to keep reserves ready to offset reverses, and on night combat, the author certainly writes with the authority of combat experience:

The average person thinks that fighting at night is a simple matter. . . . On the contrary, very careful organization is needed. . . . If the sky is cloudy and we cannot see the stars, we should have men who are very well acquainted with the road and with the enemy's camp go on ahead of us. They should have lanterns suspended from their spears. These lanterns should have four sides covered with hides; on three sides the hides should be black, but the fourth white, through which the small lamp can light the way. . . . These men . . . should have iron soles under their feet as a protection against caltrops [multiple spikes, one of which must points upward, against horses and men].[46]

The author makes other detailed suggestions for effective attacks at night, making it very clear why many talk of night combat and few attempt it—everything conspires to impede smooth action in daylight, twice so at night, even with the latest night-vision devices, incidentally.

This passage also shows that the anonymous author was writing for interested fellow citizens rather than to instruct generals. And that is true also of the subsequent section on ambushes and the reception of deserters, and the management of spies. He recommends that spies be sent to operate in pairs, so that the one "in place" to spy on the enemy can remain there instead of traveling back and forth, transmitting his information by meeting the other, who does report back, in a public market on the pretext of trading. "In this way they should be able to escape the notice of the enemy. One offers our goods for sale or barter, and the other gives foreign goods in exchange and informs us of the enemy's plans."[47]

A further section on envoys has already been cited, but it is the last three sections on archery training that are especially practical—they may indeed have been written by a different author, who was most certainly a very expert archer. He sets out to explain how to train bowmen to achieve the three goals of military archery: to shoot accurately, to shoot powerfully, to shoot rapidly.

Both the "Roman" thumb-and-forefinger and the "Persian" three-finger draw should be practiced, so that when the fingers used become tired "he may use the others." The bow is drawn to the ear for maximum power, with the breast draw mentioned only to get in the obligatory reference to Amazons burning off the right breast to lengthen their draw a bit.[48]

By way of combat technique, the author recommends that when the enemy is drawn up in ranks directly in front of them, bowmen should aim not straight ahead but at an angle to bypass the shields.

But the main challenge is to achieve accuracy. Targets should be broad and high, for if the trainees "keep shooting and missing, they may lose heart." The width of the targets should be reduced "gradually until they become quite narrow." The author points out that while the men might miss laterally, they should "not be off very much as far as height is concerned"—and indeed it is easier to learn to correct for the fall of shot than to get the azimuth right. Eventually, the height of the targets is also to be reduced, to remain with a round target. As skills improve, targets with holes of different sizes can be used, with the aim-point moving gradually from the larger to the smaller holes. Moving targets are next, birds, animals, or such artificial targets as balls pulled by cords.

To train for power, he suggests using a bow "not too easy to draw or a long arrow," which requires a proportionately long draw—the opposite of the wooden bows and mock arrows of Vegetius. The construction of a simple and effective machine is recommended to stimulate competitive training for power: a wooden disk segmented by 360 lines to mark off as many degrees is to be affixed horizontally to a rod inserted into the ground; then another wooden disk is to be attached vertically to slip rings around the axis of that same rod so that it can rotate—but not too easily. Blunt arrows are to be used in aiming at the vertical disk: "The lines inscribed in the circle on the [flat] disk indicate the . . . impact of the shot. A weaker shot will turn the [vertical] circle, say, one degree, the stronger blow two or more degrees."[49] The archer-scholar Giovanni Amatuccio, whose illustration fully clarifies the workings of the device, remarks that the device could still be of use for the training of bow hunters and in archery competitions.[50]

Powerful shooting was the characteristic of sixth-century Roman archers, according to Prokopios, who writes that the Persians

> are almost all bowmen and they learn to make their shots much more rapidly than other men. But their bows were weak and not very tightly strung, so that their missiles, hitting a corselet, perhaps, or helmet or shield of a Roman warrior, . . . had no power to hurt the man. The Roman bowmen are always slower indeed, . . . but inasmuch as their bows are extremely stiff and very tightly strung, and one might add that they are handled by stronger men, they easily slay much greater numbers for no armour proves an obstacle to the force of their arrows.[51]

Finally, the author notes that training for rapidity requires much practice, which cannot be replaced by sheer technique. He suggests that all mark their arrows with their names or a symbol before starting to launch them in continuous volleys until the signal is given to desist. At that point the arrows of each man can be counted to determine his rate of release.

The final training technique the author recommends is to have a quick-launching bowman move out in a straight line while he keeps releasing arrows to his side, until they are picked up and a marker left in each spot; opposite this line at a distance of 56 meters (30 *orgyai*; each *orgya*—the opening of arms—is some 1.87 meters) another set of markers is to be placed.[52] The men are to go rapidly from marker to marker along the second line aiming at the first line of markers—to simulate the movement of the bowmen in actual combat. Nothing comparably useful would be written until the *Toxophilus* of Robert Ascham dedicated to Henry VIII in 1545.

Dain is coldly dismissive of another text that he attributes to the author of the *Anonymous Treatise,* known under the title *Rhetorica Militaris,* a compilation of suitable hortatory speeches for generals in forty-eight chapters: "armchair strategists are followed by armchair orators" (*"Aux stratèges en chambre font ici pendant des orateurs en chambre"*).[53] But since then the authorship has been attributed to the Syrianos Magister cited above, while the work "was more important and influential than Dain judged it to be."[54]

The work became influential when the empire had to confront the Arab, and then more generally Muslim, jihad with its strong ideological component. When a Muslim frontier emir could raid imperial territory successfully, he gained more than loot and captives—he also validated the Islamic promise of victory, and enhanced his personal reputation as a conquering hero while eroding the standing of the emperor as defender of the faith and of the faithful.[55] The ideological threat evoked vigorous Byzantine responses.

CHAPTER II

The *Strategikon* of Maurikios

Vegetius was much admired and often cited by practitioners of Renaissance warfare and long after. By contrast, the infinitely superior *Strategikon* attributed to the emperor Maurikios (ca. 582–602) remained largely unknown until recent times. This fundamental field manual and military handbook, much copied, paraphrased, emulated by subsequent Byzantine military writers, and much used by warring emperors and their commanders over the centuries, was simply not available when the classics of ancient warfare were rediscovered and mined for useful ideas by Europe's military innovators from the fifteenth century onward. There were numerous medieval manuscripts of Vegetius, and the text was already printed by 1487 in the first of many editions, both in the original Latin and in translation, some lavishly illustrated.[1] At the time, however, Greek was still an unknown language even for the most learned scholars of western Europe.

When the language and its writings were rediscovered very soon after that, it was the archaic and classical Greek authors from Homer to Aristotle that evoked passionate interest, not the later, presumptively decadent, and certainly schismatic Byzantine writings. So it was that the text of the *Strategikon* was not printed until 1664, in the back of the antiquarian and decorative *Techne Taktike* of Arrianus, a Roman officer, albeit writing in Greek, and therefore the more prestigious.[2] Even after 1664, neglect long persisted, for with the Enlightenment came the black legend of Byzantine minds paralyzed by obscurantist religiosity, and so it was that the *Strategikon* was not rediscovered until the eve of the twentieth century, eventually attracting the interest of strategic theorists and even practitioners, who could best recognize the real expertise it contains.

The author modestly claimed only a limited combat experience, but he was evidently a highly competent military officer. In the preface, he promises to write succinctly and simply, "with an eye more to practical utility than to fine words," and keeps his promise.[3] The work was written at the end of the sixth century or very soon thereafter—the modern editor of the text has convincingly shown that it was completed after 592 and before 610.[4]

The *Strategikon* depicts an army radically different in structure from the classic Roman model, most obviously because of a fundamental shift from infantry to cavalry as the primary combat arm. That was no mere tactical change; it was caused by a veritable strategic revolution in the very purpose of waging war, which compelled the adoption of new operational methods and new tactics.

Before proceeding, it is interesting to note that there was no such radical change in the language of the army, which had been partly Latin-speaking even in the eastern half of the Roman empire. From the time of Justinian, there was instead a very gradual transition from Latin to Greek, though many of the Greek terms in the *Strategikon* are still Latin words with Greek endings added and pronounced in a Greek way. Its *strategos* (general) and *hypostrategos* (lieutenant general, or "under-general") are pure Greek, but below them Latin persists in the field grades: a *dux* (which became our "duke") commands the (purely Greek) *moira* of one thousand troops, while a *comes* (which became our count) or *tribunus* commands one of the three units that form a *moira,* for which the author records three different names of three different linguistic origins that all mean exactly the same thing—a unit of some three hundred men: the purely Greek *tagma* which just means "formation," *arithmos* which is a straight translation of the Latin *numerus,* and *bandon* from the same Germanic word (which actually meant "flag") as in our "war band." We encounter *koursores* (later *prokoursatores*), our "skirmishers" or "reconnaissance troops" in modern parlance, taken from the original Latin *cursores,* and while there is no change in *defensores,* "defenders"—troops armed and trained to fight in close order to hold the battle line—the *deputatoi,* "paramedics," are the Latin *deputati,* as pronounced by Greeks.

Armies are anxiously conservative, especially about the fragile certitudes of battle, and so we find that the combat orders of the glorious legions of the victorious Romans are preserved perfectly unchanged in the original Latin: *exi,* "out," for "march out," when the width of the battle line is to be doubled by halving the depth of the file from eight men to four; *dirige frontem,* to "straighten out the battle line"—"when some men . . . step out in front and the whole line becomes uneven"; *junge,* "join up" or "close ranks." A scholar who tabulated fifty such terms to show how they were translated into Greek in the subsequent paraphrase incorporated in Leo's *Taktika* gave a contemporary example of linguistic conservatism, the retention of German by an elite Swedish regiment of Hussars.[5]

In modern wars, hand-to-hand fighting is exceedingly rare and combat mostly begins very suddenly, with the impact of munitions fired, projected, or launched by invisible enemies from afar. In ancient combat there were no long-range weapons, so that except in ambushes the last moments before the fight were fully experienced, in the deliberate approach of the enemy or toward the enemy, until the first clash of arms. The prescribed sequence of orders for those last minutes of maximum tension, as horribly intense for veterans who know what to fear as for first-timers who know not, amounted to a rather subtle step-by-step process of psychological preparation:

> *silentium* (silence) *mandata captate* (understand your orders);
> *non vos turbatis* (do not be anxious);
> *ordinem servate* (maintain your position in the rank and the file);
> *bando sequute* (follow the flag, the unit standard);
> *nemo demittat bandum et inimicos seque* (do not outrun the
>   standard to pursue the enemy).

When the fight is about to begin as the troops approach within bowshot range of the enemy, "the command is *parati,* be ready. Right after this another officer shouts: *Adiuta,* help us! In unison everyone responds loudly and clearly: *Deus,* 'O God.'"[6] At that point the bowmen are to release their first volley of arrows and the better-protected heavy infantry is to advance in close order, shield touching shield across the front rank.

To retain Latin orders in a Greek-speaking army was not mindless conservatism, it was a way of maintaining continuity with what was then, and still is, the most enduringly successful military institution in human history, the most important inheritance that the empire of new Rome received from the old.

The *Strategikon* of Maurikios is the most complete Byzantine field manual in spite of its brevity. To describe the training and tactics that could allow one man to defeat three, the author himself used one word where other writers might have used three. It was certainly the most useful of all books for Byzantine military chiefs down the centuries, and is not entirely useless even now. Behind a veil of sometimes sanctimonious Christian ceremoniousness, the Byzantines were very Roman in their sheer practicality, nowhere more so than in the *Strategikon,* which begins by invoking "our Lady, the immaculate, ever-virgin, Mother of God, Mary" before immediately turning to the training of the individual soldier, the right starting point for any serious field manual then as now.

Military history is often written without even mentioning how the soldiers on each side were trained. Yet that is routinely the decisive factor in the strength of armies. It is not only historians who disregard the essential importance of training in general, starting with a thorough course of initial individual or "basic" training. If new recruits do not acquire their weapon and field-craft skills while still in basic training, before they are posted to their units, the latter cannot practice their unit tactics and are instead forced to remedy the lack of elementary skills of each new intake. Most units in most armies do just that, essentially because their officers have better things to do with their time than to supervise the individual training of new recruits, with many early-morning starts that preclude late-night fun, long hours of numbingly repetitive instruction, and a great deal of marching, running, and crawling in all climates. So in most of the world's armies, recruits reach their units after a couple of weeks of parade-ground drill and ceremony, having fired ten or twenty rounds on a range, if that, with unsurprising results in combat.

Only a small proportion of all contemporary armies train their soldiers seriously, and therefore enjoy a decisive tactical superiority over the ill-trained majority of military mankind.

That was the aim in the *Strategikon,* whose primary type of soldier was neither an infantryman nor a cavalryman but rather both, and a bowman first of all. He therefore required training in both foot and mounted archery with powerful bows, in using the lance for thrusting and stabbing while mounted—with unit training for the charge—and in wielding the sword in close combat. The old term "mounted infantry" does not apply, because in most cases it was nothing more than infantry with cheap horses that could not fight on horseback, let alone with the bow; the even older term "dragoon" is suggestive insofar as the better class of dragoons were equipped with rifles for accuracy and range rather than muskets. Under the heading "The Training and Drilling of the Individual Soldier" we read:

> He should be trained to shoot [the bow] rapidly on foot, either in the Roman [thumb and forefinger] or the Persian [three middle finger] manner. Speed is important in shaking the arrow loose [from the quiver] and discharging it with force. This is essential and should also be practiced while mounted. In fact, even when the arrow is well aimed, firing slowly is useless.

The tactical effectiveness of bowmen is obviously a function of their rate of fire, accuracy, and lethality, but there is no homogeneous trade-off between the three, because enemies will normally either withdraw beyond the useful range of accurate and lethal arrows, or else to the contrary seek to charge and overrun the bowmen, either way making the rate of fire the dominant variable. "He should also shoot rapidly mounted on his horse at a run [galloping], to the front, the rear, the right, the left."[7]

Most riders are content if they remain secure on their horse at full gallop, and would hesitate to rely on their gripping knees alone while using both hands to discharge arrows straight ahead. It is that much more difficult to turn away from the galloper's forward tilt to aim sideways, and yet more so to release a Parthian shot by turning right around in the saddle to shoot directly rearward. But given some initial aptitude and much training, even these virtuoso archery techniques can be mastered to an adequate level. Originally learned by the Byzantines from the Huns, whose own training started in childhood, these techniques are still a side-show attraction at *Eeriin Gurvan Naadam* festive competitions in Mongolia, where one may admire the accurate bowshots of local champions at a full gallop "to the front, the rear, the right, the left" just as the *Strategikon* prescribes. According to Prokopios, that was an established skill for the Byzantine horsemen he saw in action not long before the *Strategikon* was written:

> They are expert horsemen, and are able without difficulty to direct their bows to either side while riding at full speed, and to shoot an opponent whether in pursuit or in flight [the rearward "Parthian shot"]. They draw the bowstring along by the forehead about opposite the right ear, thereby charging the arrow with such impetus as to kill whoever stands in the way, shield and corselet alike having no power to check its force.[8]

Mounted and dismounted archery had its specific roles in every stage of battle, from initial sniping at long range to the rapid volleys of all-out engagements, to the pursuit of retreating enemies with forward bow-shots, or defensively, to provide rearguard covering shots against advancing enemies.

The weapon of the Byzantine cavalryman described by the *Strategikon* was certainly not the simple bow of wood and string drawn to the breast that was left to auxiliaries in the Roman army, and repeatedly ridiculed in Homer, in spite of Apollo's godly archery: "Argives, you arrow-fighters, have you no shame?" (*Iliad* 4. 242); "You archer, foul fighter" (Iliad 11.386); "[the bow] is the . . . weapon of a useless man, no fighter" (*Iliad* 11.390),[9] and most contemptuously by the mighty hero Diomedes, whose foot Paris, the lover of Helen, had just pierced with an arrow:

> Bowman and braggart, with your pretty lovelocks and your glad eye for the girls; if you faced me man to man with real weapons, you would find your bow and quiverful a poor defense. . . . All you have done is to scratch the sole of my foot . . . a shot from a coward and milksop does no harm. But my weapons [throwing spears] have a better edge. One touch of them, and a man is dead.[10]

By the sixth century, Byzantine archers were armed with the composite reflex bow, the most powerful personal weapon of antiquity. Well before the *Strategikon* was written, when the Byzantines were fighting the Goths in Italy in the mid-sixth century they were already doing so with the tactical edge of mounted archery. The *Strategikon* provides the specifics of the required training:

> On horseback at a run (gallop) he should fire one or two arrows rapidly and put the strung bow in its case, if it is wide enough, or in a half-case designed for the purpose, and then he should grab the [*kontarion* = lance] which he has been carrying on his back. With the strung bow in its case, he should hold the [lance] in his hand, then quickly replace it on his back, and grab the bow. It is a good idea for the soldiers to practice all this while mounted.[11]

Compound bows, held together by animal-bone glues and powered mostly by dried tendons, had to be protected from the rain by special cases, broad enough to hold the bow when already strung for battle and not just when unstrung. Ottoman bow cases, waterproofed with thinned leather, survive to be examined, unlike their Byzantine counterparts.

In addition, the *Strategikon* recommends "an extra-large cloak or hooded mantle of felt . . . large enough to wear over . . . [body armor and] the bow" to protect it "in case it should rain or be damp from the dew."[12] The second notable point is the recommendation to quickly alternate from shooting one or two arrows, to pulling out the spear from its back strap, to replacing the spear, to taking out the bow again.

That is how useful weapon training must be conducted in any epoch. In this instance, after the initial stage of mastering the weapon as such, by shooting at targets as many times as necessary to be able to aim accurately (hitting is another matter), the next stage is to learn how to use the weapon in combat, when it is no longer alone but rather accompanied by shields, swords, lances, or throwing spears. At that point, the aim was to acquire facility in using the equipment fluidly, in switching back and forth among thrusting, cutting, and missile weapons.

It was an art already much cultivated in the Roman army. There is evidence of compelling authenticity in a famous inscription that records a blunt speech of the year 128 by the touring emperor Hadrian (117–138) before the men of the mixed cavalry-infantry *Cohors VI Commagenorum* (Sixth Cohort of Commagene, now southeast Turkey). They had just executed a weapon-switching combat drill, unfortunately for them immediately after the virtuoso performance of a distinctly more elite unit of pure cavalry, *Ala I Pannoniorum* (First Cavalry of Pannonians):

> It is difficult for [a mixed unit] to please even by themselves, and more difficult not to displease after the exercises given by the men of an *Ala*. The areas of the parade ground are different, the number of javelin-throwers is different, [there are] the right wheelings in quick succession, the Cantabrian gallop in close order [virtuoso acts that the *Ala* had just performed] while the appearance and quality of [your] horses and the training in and elegance of the weapons is in keeping with [your lower] level of pay. But my aversion you have avoided by [your] ardor . . . in carrying out your duties briskly. Moreover, you have fired stones from slings and fought with missiles. . . . The outstanding care . . . [of your commanding officer] Catullinus . . . is clear from the fact that such men as you [are] under his command.[13]

For the Romans, who believed in destroying enemies not wise enough to recognize the advantages of submission, the cutting and thrusting and besieging heavy infantry was the most important arm, because it could best achieve decisive results.

By contrast, most of the time, and certainly when the *Strategikon* was written, the Byzantines believed in containing but not destroying their enemies—potentially tomorrow's allies. Therefore for them the cavalry was the more important arm because its engagements did not have to be decisive, but could instead end with a quick withdrawal, or a cautious pursuit that would leave both sides not too badly damaged. Still, even at the height of the cavalry era there was a need for some infantry, both light and heavy. The *Strategikon* accordingly offers its advice for the training of both while admitting that the subject had long been neglected.

Under the heading "Training of the Individual Heavy-Armed Infantryman" there are only a few words:

> They should be trained in single combat against each other, armed with shield and staff [a real shield and a simulated spear], also, in throwing the short javelin and the lead-pointed dart at a long distance.[14]

There was more on "Training of the Light-Armed Infantryman or Archer":

> They should be trained in rapid shooting with a bow . . . in either the Roman or the Persian manner. They should be trained in shooting rapidly while carrying a shield, in throwing the small javelin a long distance, in using the sling, and in jumping and running.[15]

The equipment specified in the *Strategikon* for each type of infantry clarifies its character, with armored coats for at least the first two men in the file of heavy infantry, so that the front rank and the one behind it were both protected against enemy arrows, as well as cutting weapons, if not maces and such; helmets with cheek plates for all, greaves of iron or wood to protect the legs below the knees, and shields of unspecified type but of full size—elsewhere small shields or "targets" are mentioned. An exhaustive if not excessively insightful modern study contains a long list of different equipment types or perhaps of equipment names, and although there are illustrations, they are insecurely related to the names.[16]

What is certain is that the function of the heavy infantry at the time and for centuries later, indeed until the introduction of firearms, was to seize and hold ground. No great agility could be expected from it, nor missile strength beyond the modest impact of pebble-throwing slings and the launch of throwing spears, javelins, and lead-pointed darts.

The long weapons in the *Strategikon* begin with the stabbing lance (Latin *contus*, Greek *kontos*) of the horsemen, which infantry could also use to keep charging cavalry at bay. Then there were many names for light throwing spears or javelins of different origins and designs: *monokontia, zibynnoi, missibilia* or the classical *akontia,*[17] particularly important, under whatever name, for light-infantry men who for whatever reason could not use the bow.

In the *Strategikon,* as in all other Byzantine texts, the light infantry is chiefly a missile force, equipped with quivers holding up to forty arrows for its composite reflex bows, though it is specified that for "men who might not have bows or are not experienced archers" small javelins, Slavic [light] spears, lead-pointed darts, and slings were to be provided.

There was also a more recondite and much misunderstood item of equipment, the *solenarion,* not a small crossbow with short arrows as was once believed, but rather "tubes" or more fully—translating the original *solenaria xylina meta mikron sagitton*—wooden launch tubes for small arrows. They are the overdraw or "extender" devices still sometimes used by modern archers.[18]

With it, short arrows that can fly farther than full-size arrows are inserted in a tube with a central slit; in that way, the bowstring can still be drawn back fully even though the arrow is only, say, 40 instead of 140 centimeters long. Known as *myas,* flies, these short arrows were useful for harassing volleys against the enemy when still out of range of full-length arrows, which were of course more lethal because they could penetrate thick coverings and armor as the *myas* could not.[19]

In the *Strategikon* the primary type of soldier is undoubtedly the mounted lancer-archer, and naturally there is more detail about its equipment. (The impossibility of training everyone to use the composite reflex bow, and the normal prevalence of dismounted archery, may have misled the distinguished scholar who opined that "the Byzantine composite lancer/horse archer is probably something of a myth.")[20]

The author recommends hooded coats of sewn-on scale armor *(lorica squamata),* or interlinked lamellar armor, or chain mail *(lorica hamata),* down to the ankles—the latter prized equipment for fighting men for another eight hundred years or so, until muskets became common. There were also carrying cases for them covered in water-resistant leather, for armor was expensive and it would rust; it was further specified that light wicker cases for body armor should also be carried behind the saddle over the loins, because "in the case of a reversal, if the [servants] with the spare horses [and ancillary equipment] are missing for a day, the coats of armor will not be left unprotected and ruined."

Helmets, swords, iron breastplates, and head armor for horses are mentioned, but special attention is devoted to the primary weapon: "Bows suited to the strength of each man, and not above it, more in fact on the weaker side."[21]

The composite reflex bow was effective because it accumulated much energy but was equally resistant, so it was a good idea to choose a bow whose string could be pulled back quickly and confidently even on the thirtieth arrow, and not just the first. Cases wide enough for combat-ready strung bows are specified, as mentioned above, as are spare bow-strings in the soldier's own saddlebag and not just in unit stores, quivers with rain covers for thirty or forty arrows—more were in unit stores—and small files and awls for field repairs.

The author specifies that cavalry lances with leather thongs and pennons, round neck pieces, breast and neck coverings, broad tunics, and tents (round leather yurts) are to be of the "Avar type." The Byzantine mounted archers that featured so largely in Prokopios a half century before were patterned on the Huns, but by the time the *Strategikon* was written, the Byzantines had been repeatedly attacked by the Avars, the first of the Turkic mounted archers to reach the west, who had the same composite reflex bow as the Huns but also much other equipment picked up along the way from the other two advanced civilizations of the time—China, from where they originated, and Iran, whose culture they encountered on their way westward as soon as they reached the trading cities of central Asia.

Unlike the Huns, the Avars from the start could build and operate elaborate siege equipment, possibly including the traction trebuchet, which with its powerful simplicity made all previous stone-throwers obsolete, and they probably introduced in the west a most famous item of equipment first mentioned in the *Strategikon*: the *skala*.[22] Literally "stair," the term is used to mean "stirrup"—"attached to the saddles should be two iron [stirrups]"—the Avar term is unknown. That is unfortunate, because it might have clarified the origin of the stirrup, that being just one of the controversies that the subject has generated.

Contrary to the myth propagated by nonriding historians (e.g., "Without the stirrup, the shock charge with couched lance could not have been a possible maneuver"),[23] the stirrup is not essential to allow horsemen to charge with the lance without being thrown off by the impact. If pre-stirrup lancers tumbled off their horses, it was not for the want of stirrups, because it is tightly gripped thighs that hold the rider, not loosely hanging stirrups. Especially valuable in that regard is the testimony of a modern jouster, whose "examination of the mechanics of shock combat and the development of shock tactics" proves the point experimentally and finally.[24]

Moreover, recent research has allowed the reconstruction of Roman cavalry saddles, and their particular design would allow competent riders to remain well horsed while absorbing the shock of a thrusting lance, or half-turning to swing a sword. Consisting of a sturdy wooden frame covered in leather, they had a built-up pommel at each corner to help anchor the cavalryman in all four directions.[25] Reenactors have tried the design, demonstrating its functionality without stirrups. There is no evidence that the Byzantine cavalry had the same saddles, but it seems unlikely that this successful design was forgotten.

Both the Romans and their enemies, most notably the Sarmatian cavalry in its characteristic scale armor, as well as the successive great powers of Iran, Arsacid Parthia, and Sasanian Persia, had heavy cavalry, that is, cavalry trained for the charge. Indeed, they had charging armored cavalry long before the stirrup arrived, with thick boiled leather, chain mail, lamellar or even plate armor in superficial resemblance to medieval knights—or rather to images of renaissance jousters impersonating medieval knights, for the latter rarely had a full suit of armor. To be sure, the weapon used for the charge by Romans, Sarmatians, Parthians, and Persians—and indeed all others who ever charged their enemies in real combat—was the handheld stabbing lance *(kontos)* that also equipped the lancer units of eighteenth- and nineteenth-century European cavalry, not the very heavy thrusting pole of late-medieval tournaments and their cinematic evocations.

When they first encountered them in the searing summer heat of Mesopotamia, the Romans mocked the Persian cavalry in plate armor as *clibanarii,* from *cliba,* "bread oven." Yet they still imitated this heaviest form of armored cavalry, expensive and easily exhausted as it was (especially in hot weather), for the very good reason that in suitable terrain it could offer "escalation dominance" in short, sharp, charging actions. The fifth-century *Notitia Dignitatum* lists ten units, including several whose name proclaimed their eastern origins: *equites primi clibanarii Parthi* (first Parthian armored cavalry), referring to Arsacid Parthia; *Equites secundi clibanarii Parthi* and *Equites quarti clibanarii Parthi,* for the second and fourth units; *Equites Persae clibanarii* (Persian armored cavalry); *Cuneus equitum secundorum clibanariorum Palmirenorum* (second squadron of Palmyrian armored cavalry).[26] Other units were identified by their specialty alone: *Equites clibanarii,* plate-armored cavalry; *Equites promoti clibanarii,* selected plate-armored cavalry; and *Equites sagittarii clibanarii,* plate-armored mounted archers.

Again, as noted, the *Notitia Dignitatum* cannot have been an accurate order-of-battle inventory—at any moment of time it probably included defunct units that were still carried on outdated payrolls (profitably for paymasters) while excluding new ones not yet registered in the capital. Moreover, military formations tend to retain traditional titles even when their actual character changes—the armored cavalry regiments of the contemporary U.S. Army have no horses, while by contrast the infantry divisions have many tanks. So there is no saying what was the actual nature of the listed *clibanarii* units when that part of the *Notitia* was redacted, but they would hardly have been named as they were if they did not originally have distinctive plate armor.

There was also another category of heavy cavalry listed in the *Notitia* that was destined to endure much longer, the *catafractarii* (Greek *kataphraktoi*, from *kataphrasso*, "cover up"). They too were well protected to confront close combat, and they too were trained to charge with the lance, but originally at any rate they were not as heavily armored as the *clibanarii*.[27] Instead of heavier plate or lamellar armor, they had sewn-on scale armor or chain mail coats as mentioned in the *Strategikon*, or body armor of boiled leather or thick, dense cloth—which, if tightly woven to begin with, could be sewn and knotted in multiple layers to function as a sort of proto-Kevlar.

In the *Notitia Dignitatum*, nine units are listed, including one that probably dates back to the third century, the *Ala prima Iovia catafractariorum* (first Jupeterian armored cavalry "wing") of the Thebaid in southern Egypt, while the others are listed as simply *equites*, "cavalry unit" or "squadron" in modern parlance, except for a *cuneus* ("wedge") *equitum catafractariorum*, and a unit identified by its commander, *Praefectus equitum catafractariorum, Morbio*, in Britain. It is likely that in the usual way of military formations over long periods of time, the distinction between the two kinds of armored cavalry faded away, even as their ancient names persisted.

It seems evident that the historical importance of the arrival of the stirrup has been greatly exaggerated, notably by Lynn White Jr., who tried to erect an entire explanation of social change on that narrow footing, if the pun be allowed.[28] But it is certainly true that the stirrup increased the relative combat value of all forms of cavalry, just as it still eases all forms of riding. Men wearing armor, who could not easily leap onto their horse as Roman training prescribed, could instead swing up to them by stepping on a stirrup. In combat, stirrups enhanced lateral stability whether wielding sword, mace, or the charging lance.

Most important, stirrups allowed mounted archers of sufficient skill, whether trotting, cantering, or even galloping, to stand upright and level over their heaving horses while loosening their arrows, greatly improving their aim.

In the *Strategikon* there is no mention of plate-armored *clibanarii*, while the *catafractarii* had evolved into the primary lancer-archer with scale or chain-mail armor. In the *Notitia Dignitatum* there is no mention of *catafractarii sagittarii*, which would have been their exact predecessors.

Along with the light missile infantry and the ground-holding and ground-seizing heavy infantry, three other categories of soldiers are mentioned in the *Strategikon*. The first are the *bucellarii*, "biscuit-eaters," named for the twice-baked dehydrated bread issued to ship crews and soldiers on campaign; originally they were raised and paid privately by field commanders as their personal guard and assault force, but evidently they evolved into a state-paid elite force, for we find that special attention is devoted to their appearance:

> It is not a bad idea for the [*bucellarii*] to make use of iron gauntlets and small tassels hanging from the back straps and the breast straps of their horses, as well as small pennons hanging from their own shoulders over coats of mail. For the more handsome the soldier is in his armament, the more confidence he gains in himself and the more fear he inspires in the enemy.[29]

That would have been just as true of other categories of troops, but it is revealing of their status that the point is made about the *bucellarii* specifically. The latter, incidentally, would soon evolve further into a territorial army corps that was in turn given a fixed military district, or *theme*, to both govern and defend, when that emergency response to defeat and retreat became an administrative system in the later seventh century. The theme *Boukellarion* duly appears in the tenth-century survey of Constantine Porphyrogennetos known as *De Thematibus*.[30]

The second category of troops mentioned as such or simply as "foreigners" were the *federati*, originally "treaty" *(foedus)* troops supplied to the empire as complete units under their own chiefs by tribes too poor to pay taxes, or too strong to be taxed; later they could simply be units serving under contract.[31] Unlike today's mercenaries provided by security contractors, who often cost much more than even well-paid soldiers, units of federati were much cheaper than an equivalent number of legionary troops, because the citizen-soldiers of the legions received good salaries, well-built barracks, careful medical care, and substantial retirement allowances. Roughly half the army of the Principate was cheaper because it consisted of lower-pay, noncitizen auxiliary troops serving under Roman officers—they provided almost all the cavalry of what was still an infantry-centered army; but because they did not have expensive Roman officers, the federati were even cheaper. That is the reason, no doubt, why they continued to serve in the Byzantine forces till the end in one form or another, most often as more expendable light troops, as in the "javeliners, whether *Rhos* (early Russians) or any other foreigners" of the *Praecepta Militaria,* a tenth-century work.[32]

They sometimes distinguished themselves for their skill and valor, as with the Onogurs ("Huns") who fought for Belisarios in Italy; less often, they were blamed for defeats or even accused of battlefield betrayals, especially if the enemy was of the same ethnicity. That was supposedly one of the causes of the important strategic defeat suffered by Romanos IV Diogenes at Mantzikert on Friday, August 26, 1071: some of his mercenaries were of the same Oğuz Turkic ethnicity as his Seljuk enemies and reportedly changed sides. In the *Strategikon,* a specific precaution is recommended under the heading "Peoples Akin to the Enemy":

> Long before the battle, troops of the same race as the enemy should be separated from the army and sent elsewhere to avoid their going over to the enemy at a critical moment.[33]

Finally, the *Strategikon* refers to some kind of citizen militia, or at least to a general preparedness to serve in that capacity:

> All the younger Romans up to the age of forty must definitely be required to possess bow and quiver, whether they be expert archers or just average. They should posses two [spears] so as to have a spare at hand in case the first one misses. Unskilled men should use lighter bows. Given enough time, even those who do not know how to shoot will learn, for it is essential that they do.[34]

Given all the incursions that penetrated right through imperial territory to reach Constantinople itself, one can understand why the author of the *Strategikon* would favor universal military training, so that all of the able-bodied could help defend their own localities, supplementing professional imperial forces. The same recommendation was made in later military texts as well. We hear, for example, of the valiant role of the population of Edessa (Şanlıurfa, Urfa) in fighting off the Sasanian Persians in 544:

> Now those who were of military age together with the soldiers were repel-
> ling the enemy most vigorously, and many of the rustics [*akgroikon polloi*]
> made a remarkable show of valorous deeds against the barbarians.[35]

But Roman and Byzantine law prohibited private weapons, while organized militias were rarely sanctioned by the Byzantine authorities.[36]
That is not surprising. Their potential and episodic military contribution, in the event of enemy incursions that reached their particular part
of the empire, was outweighed by their actual and continuing political
threat to the imperial authorities in place, and indeed the stability of the
empire. The empire's governance was not arbitrary, for it was regulated
by laws, but it was not consensual either. The political premise of a militia is that its citizen-soldiers must necessarily serve the government loyally, for it represents them as citizen-voters, or else soon will upon the
conclusion of the next elections.

Obviously that could not apply to an imperial autocracy, however
benign—and none was more so than that of Trajan (98–117), at least
according to his admiring appointee Plinius Caecilius Secundus, our
Pliny the Younger, imperial governor *(legatus propraetore consulari
potestate)* of the important province of Bithynia-Pontus, in westernmost
Anatolia. In a letter to Trajan, Pliny reported on a

> widespread fire in Nicomedia (the modern Izmit) which destroyed many
> private houses and . . . two public buildings. It was fanned by the strong
> breeze . . . but . . . would not have spread so far but for the apathy of the
> populace. . . . Will you, Sir, consider whether you think a company of [vol-
> unteer] firemen might be formed, limited to 150 members. I will see that
> no one shall be admitted who is not genuinely a fireman. . . . It will not be
> difficult to keep such small numbers under observation.

As befits an experienced imperial official, Pliny was being properly cautious, though 150 men could hardly threaten the empire. But for Trajan
he was not cautious enough:

> You may very well have had the idea that it should be possible to form a
> company of firemen at Nicomedia . . . but we must remember that is socie-
> ties like these which have been responsible for the political disturbances in
> your province. If people assemble for a common purpose, whatever name
> we give them and for whatever reason, they soon turn into a *hetaeriae* [po-
> litical gathering]. It is a better policy then to provide equipment and to in-
> struct property owners to make use of it.[37]

With this exception, the *Strategikon* is consistently realistic in its recommendations about individual training, just as in its recommended tactics
and operational methods.

Attrition and Maneuver

For the strong, who can win by outmatching the enemy in straight-forward force-on-force combat, tactics can remain simple procedures to convey troops and their weapons against the enemy. The resulting "attrition', an almost mechanical process, has to be paid for in casual-ties—when still a commercial nation, the English called it the "butcher's bill"—but it can reliably grind down the enemy, avoiding all the risks of more ingenuous and complicated maneuvers.

Even in the absence of the remote firepower that can nowadays make attrition entirely one-sided, in antiquity also there was no equal-ity: with superior individual training, armor, and weapons, the butcher's bill could be reduced in like measure. That is how it was for the Romans in their better days. They could rely on the powerful frontal attacks of their well-armored and well-drilled legionary infantry to win their bat-tles by cutting up the enemy—attrition to be sure, but rather economi-cal. Auxiliary noncitizen units of cavalry (*alae*, "wings") could guard the flanks and rear of the infantry formation and fight off enemy horse-men, while auxiliary light-infantry units *(cohortes)* variously armed with javelins, bows, and slings, could hit and harass the enemy with their missiles, as could the arrow-firing and stone-throwing field artil-lery. But usually it was the meat grinder of the legionary infantry in close order that decided the fight.

Enemies who stood their ground before an advancing legion—un-wisely in most cases—were first hit by two successive volleys of *pila,* javelins with heavy metal heads that penetrated shields and could pierce helmets. Next, the advancing wall of legionnaires behind their heavily bossed shields was upon them, pressing forward with lethal thrusts of their short swords. With helmets and grieves to protect them above and below their large shields, with more shields held overhead from the sec-ond rank in the *testudo,* or tortoise, formation, the legionary infantry pressing forward in cadenced steps had relentless armored momentum. Death or flight were the usual choices for those who faced them, but flight had better start early—preferably before the battle: although the ponderous legionary infantry could hardly chase anyone far or fast, the auxiliary cavalry and light infantry were ready to pursue and cut down fugitives.

The Byzantines admired the glorious antiquity of the Macedonian phalanx and even more the power of the Roman legions in their prime. But they rejected their style of war. They never tried to replicate those infantry killing machines, because they did not want to sustain the inevitable casualties of fighting decisively in the Roman manner. Instead they persistently favored less decisive tactics with more mobile and, if need be, more elusive cavalry forces. In the *Strategikon* the author summarizes the tactical argument for avoiding attrition whenever possible:

> Warfare is like hunting. Wild animals are taken by scouting, by nets, by lying in wait, by stalking, by circling around, and by other such stratagems rather than by sheer force. In waging war we should proceed in the same way, whether the enemy be many or few. To try to simply overpower the enemy in the open, hand in hand and face to face, even though you may appear to win, is an enterprise which is very risky and can result in serious harm. Apart from extreme emergency, it is ridiculous to try to gain a victory which is so costly and brings only empty glory.[38]

Although they constantly trained for their typical battle of annihilation, the Roman themselves usually tried to avoid having to fight it. They much preferred to let their enemies retreat into strongholds, which could then be defeated by hunger in slow, systematic, relentless sieges. Not for nothing were the legions cross-trained and equipped as combat engineers, as handy in demolishing fortifications as in building them, along with highways, bridges, viaducts, storehouses, even theaters. Julius Caesar finally defeated Vercingetorix in his Gallic war in the siege of Alesia, just as Vespasian and his son Titus ended their Judean war in the sieges of Jerusalem and Masada. By trapping and starving the enemy until surrender, the Romans avoided the casualties inevitable in any two-sided combat and the vagaries of fortune of meeting engagements and set-piece battles, and the outcome was the same, and more certain.

The Byzantines too favored the slow and safe siege when they thought that they could afford to wait unmolested by other enemies. But those were rare occasions, while the Byzantines were very consistent in their reluctance to consume expensively trained, expensively equipped, and always scarce forces in battles of attrition.

Operational schemes designed to achieve the objective with a minimum of attrition could be as complex as combined actions by multiple infantry, cavalry, and riverboat forces converging on a single objective; or as simple as a plain sequence of actions. But synergistic actions: as in the standard three-step operational scheme for the Byzantine cavalry: first, threaten a charge to induce the enemy to close up in dense ranks; but do not charge, and instead launch volleys of arrows into the closely packed mass; then charge in earnest, but only if the enemy is depleted and visibly wavering, to provoke a rout.

Less simple in theory, but quite practical if exercised enough, were different operational-level schemes that combined the tactical actions of the light and heavy infantry along with those of the light and heavy cavalry.

To fight at all, the heavy infantry had to be within a spear's length of the enemy; to use its missiles, whether darts, javelins, slings-stones, or arrows, the light infantry had to be within their range of the enemy—yet needed protection from the enemy's spears and swords. That was a very ancient problem by Byzantine times, and many solutions had been tried; all were variations on linear and nonlinear arrays. In the first, both kinds of infantry are mixed in the same dispositions: "Sometimes the archers are posted to the rear of each file in proportion to the number of men, that is, four for the sixteen heavy foot soldiers. . . . Sometimes they are placed within the files alternating one heavy-infantryman with one archer."[39]

That had the virtue of simplicity—a very important virtue in the confusion that attends combat—but it would mean that archers and other missile troops had to either project their weapons over the heads of the ranks of heavy-infantry men in front of them, or else do without their protection by being interspersed on the front line itself, thereby also making it weaker in resisting a charge or assault. Without light-infantry men between them, the heavy infantry could form a shield wall in the first rank to repel cavalry, while those of the second rank behind them could elevate their shields to cover their heads from plunging missiles— the *testudo*. Likewise, when attacking, a block of heavy infantry would have more mass and capacity for momentum than a mixed force that contains light infantry with its small shields and without armor.

In nonlinear arrays, units of light infantry and units of heavy infantry were kept separate, so that they could each perform to maximum effect. The problem was instead overcome in one of two ways. Very easily, if there was high ground on which light-infantry units could be posted safely, to rain down their missiles on the enemy below—the ideal part of an ideal ambush in the ideal terrain of a narrow pass or defile through the mountains; then terrain could wholly nullify the light infantry's lack of holding strength, and the same was true of walls and towers in sieges. Otherwise, much more effort was needed to overcome the ancient problem: units of light infantry had to move back and forth between the front and the rear, passing through corridors between heavy-infantry units. The latter could lengthen their files to narrow their front ranks, thus opening the corridors, or shorten their files to widen their frontage, thus closing up the corridors and forming a continuous front.

It sounds complicated, but the Byzantines did it all the time—and much of the drilling that so greatly occupied their troops was needed precisely to quickly change the depth of unit dispositions, and thus the width of their frontages. Unit frontages could be narrowed in order to narrow the entire line of battle to make it coincide with that of the enemy, or to make the formation deeper and more resilient, or to open corridors for the light infantry or cavalry.

That too was an ancient problem: how to combine the cavalry and infantry to achieve operational-level synergies. It can even be said that combinations of different kinds of infantry alone are just tactics, even if complicated.

The striving to minimize attrition was not just a question of conserving scarce resources. There was also a strategic reason to avoid attrition even if its costs were low. The Byzantines always confronted a multiplicity of actual or potential enemies. There was never just one enemy whose destruction could be imagined as tantamount to the end of conflict—as the demise of the Soviet Union was once mistaken by some as the end of history. Ever since the arrival of the Huns, the Byzantines knew that behind the enemies already at their frontiers there were others waiting their turn to attack—so that to destroy one enemy totally might simply open the way forward for another, who might be even more dangerous. Besides, yesterday's enemy could become today's valuable ally. To court potential enemies, to recruit them as allies, was hardly an original Byzantine invention, but it did become their specialty. They therefore learned to view their enemies of the moment with a distinct element of ambivalence, evaluating them not only as immediate threats that had to be countered and possibly fought very hard, but also as possible future allies. That made attrition tactics inappropriate at the strategic level, as well as costly.

There were occasions when the empire did pursue maximalist aims, notably under Justinian (527–567) when the power of the Vandals in North Africa and the Ostrogoths of Italy was entirely destroyed, and repeatedly in fighting the Bulgarians, most famously under Basil II (976–1025). In those cases—so rare in Byzantine history that the examples given are separated by half a millennium—attrition tactics would have been perfectly congruent with the strategic aim, yet still remained inappropriate for the Byzantines because they would have required proportionately large forces, and would have cost commensurate casualties. The Byzantines down the centuries lacked the former and could not afford the latter, so that instead of frontal offensives and swiftly decisive battles of attrition, even Justinian's war in Italy and Basil's in the Balkans were fought mostly by prolonged campaigns of maneuver and by sieges, which could hardly cost that many casualties because the total number of Byzantine troops was not large to begin with.

An acute shortage of combat-ready troops was indeed the perpetual condition of Byzantine warfare.[40] Not withstanding the catastrophic demographic collapse caused by the bubonic plague from 541, it was not a lack of healthy males of military age that caused the shortage. Most enemies of the empire (except for the dispersed Bedouin of the desert) were diminished in like measure, and besides the empire could always recruit beyond its borders and often did. Not even the high cost of maintaining military forces explains the shortage, for the empire often bribed foreign rulers with gold that could have paid for more troops.

The critical constraint was neither manpower nor money but training—or rather the time needed to train soldiers thoroughly. Given its style of war, troops with only rudimentary training were of little use to the Byzantine army. It needed versatile craftsmen of war, integrated into cohesive and well-drilled units ready to execute different tactics on command. The troops had to practice their way through the tactical repertoire again and again to achieve such competence, and that required much time. While in modern armies, including the U.S. army and marine corps, troops may be sent into combat within six months of recruitment or less, Byzantine soldiers with more than a year of military service were not yet considered ready for combat. That is how Prokopios explained an episode in which heavy casualties were suffered in fighting the Persians:

> Eight hundred others perished after showing themselves brave men . . . and almost all the Isaurians fell . . . without even daring to lift their weapons against the enemy. For they were thoroughly inexperienced in this business, since they had recently left off farming and entered the perils of warfare.[41]

Prokopios also records that four thousand men recruited in Thrace by Belisarios for his second Italian campaign were considered insufficiently trained for battle a year later—no doubt they had lost much time in transit, but even the initial training implied by the *Strategikon* certainly required at least six months (it is less than four weeks in the contemporary U.S. army).[42]

That created an insurmountable strategic problem.

The taxes that could be collected from the modest surplus of a primitive agriculture could not pay for the permanent upkeep of a sufficient number of trained troops; but neither could young men be recruited only when they were needed to fight the enemy, because it would take too long to train them for combat. Therefore the empire had to make do with a chronically insufficient number of trained troops, and we find that the tactics in the *Strategikon* are all characterized by the avoidance of attrition.

This tactical orientation is emphasized in a series of maxims:

> When a populous city is taken, it is important to leave the gates open, so that the inhabitants may escape and not be driven to utter desperation. The same holds when an enemy's fortified camp is taken.[43]

And again,

> When an enemy is surrounded, it is well to leave a gap in our lines to give them an opportunity to flee.[44]

Frontinus offered analogous advice, but his fellow Romans normally wanted their sieges to end in the total destruction of the enemy and the enslavement of survivors, while it was standard operating procedure for the Byzantines to leave the enemy a way out. Finally, there is the recommendation that encapsulates the principle: "A wise commander will not engage the enemy in pitched battle unless a truly exceptional opportunity or advantage presents itself."[45]

In other words, even if a numerical and qualitative superiority is certain, even if victory is certain, it is not enough reason to start a battle.

With attrition rejected there must be alternative ways of fighting, and indeed the *Strategikon* largely consists of a reasoned catalog of the two alternatives: stratagems (or ruses of war), and "relational maneuver" made of tactics and operational schemes specifically designed to circumvent the peculiar strengths of a given enemy and to exploit his peculiar weaknesses.

Stratagems and relational maneuver both are described in some detail, but are also summarized as maxims. Under the heading of what would now be called "cover and deception," we find: "It is very important to spread rumors among the enemy that you are planning one thing; then go and do something else,"[46] and "If an enemy spy is captured while observing our forces, then it may be well to release him unharmed if all our forces are strong and in good shape."[47]

Deception provides "the bodyguard of lies" for the truth that must be kept secret. Security measures meant to deny information to the enemy are necessary, of course, but may not suffice: if there are leaks of bits of information, only a cover story already in place can misdirect the enemy as to their meaning. In addition, deception can be a weapon in itself:

> When a delegation comes from the enemy, inquire about the leaders of the group, and on their arrival treat them very friendly [*sic*], so their own people will come to suspect them.[48]

And:

> A way of arousing discord and suspicion among the enemy is to refrain from burning or plundering the estates of certain prominent men on their side and of them alone.[49]

The Byzantine Style of War

Under the heading of matters to be considered before the battle, the *Strategikon* offers this excellent advice in book VII: "That general is wise who before entering into war carefully studies the enemy, and can guard against his strong points and take advantage of his weaknesses."[50]

This is the precondition of "relational maneuver," an entire style of war that is of a different order than any number of mere stratagems, and is one of the characterizing differences between Roman and Byzantine warfare, as has been pointed out: "un tournant important qui conduit de la guerre de conquête romaine vers la guerilla byzantine"[51]—except that it is reductive to use the term *guerilla* in this case, because for the Byzantines it was just one mode of warfare among several.

When relational maneuver is successful, it changes the effective military balance by circumventing the enemy's strengths and exploiting his weaknesses. If in a straight contest of attrition, 3,000 equal-quality soldiers must prevail over 1,000, barring extraordinary circumstances, with relational operational methods or tactics, it can easily happen that 1,000 can defeat 3,000. Or if the numbers are even, 1,000 can defeat 1,000 but with many fewer casualties, or with the expenditure of fewer resources, or both.

So why would anyone ever fight in any other way?

The first reason is that to uncover the enemy strengths to be avoided and weaknesses that can be exploited, the enemy itself must be understood, and that requires an intellectual effort, and also an emotional effort to overcome hatred, for there can be no deep understanding without empathy.[52]

The Byzantines too may have hated their enemies but apparently not enough to prevent them from understanding their characteristics. Even if not blinded by hatred, the powerful and the ignorant may simply lack the elemental curiosity to study the enemy. Indeed, greatly superior power habitually induces ignorance by making it seem unnecessary to study the despised inferior. That alone explains many a nasty surprise in war—from the year 9 CE destruction of the legions XVII, XVIII, and XIX of Publius Quinctilius Varus by Arminius, his Cherusci and other Germanic tribesmen all deemed loyal, or thoroughly subjected, or at least incapable of defying the immense conjoined strength of three entire legions, to many a modern debacle. Napoleon's fatal underestimation of the Russian army of illiterate serfs and drunken officers in 1812, which ruined irremediably his magnificent sequence of victories, had disregarded the debacle of his predecessor in magnificent victory, Charles XII of Sweden, defeated by clumsy Russian recruits at Poltava in 1709, and Napoleon's error was in turn repeated on a larger scale by Hitler, when he set out in 1941 to destroy Stalin's armies of racially inferior subhumans.

The Romans were not racists—they were very properly culturalist, if the word be allowed—but they were simply too powerful to be much interested in the trivial lives of non-Romans. With the signal exception of the *Germania* of Cornelius Tacitus, which was little known anyway, the Romans did not follow the ethnographic tradition of the Greeks that started with Herodotus, and it was on the *Geographia* of the Greek Ptolemaios, Ptolemeus, Ptolemy (83–161), that they relied on, if interested at all.

The Byzantines were quite different: their writings show an intense curiosity about the culture and lives of foreign peoples, and not just in utilitarian intelligence regarding their governance and military characteristics—the early history and culture of many nations, including the Bulgarians, Croats, Czechs or Moravians, Hungarians and Serbs, can only be reconstructed from their texts.[53]

As evidence of the centrality of relational maneuver in its recommended style of war, the entirety of Book XI of the *Strategikon* is dedicated to the military ethnography of diverse nations—the essential prerequisite for the design enemy-specific operational methods. The author moreover is not content with technical detail—there is psychology and sociology in addition to the specifics of enemy weapons, tactics, and battlefield habits.[54]

At a time when the protracted struggle with Sasanian Persia—a series of fierce wars interrupted by amicable relations secured by formal peace treaties—was mounting toward its cataclysmic seventh-century climax, the Persians are naturally the first to be considered. To begin with, the text explains that they are especially dangerous antagonists because they alone are highly organized, much like the Byzantines, and unlike less civilized and more individualistic enemies:

> They prefer to achieve their results by planning and generalship; they stress an orderly approach rather than a brave and impulsive one. . . . They are formidable when laying siege but even more formidable when besieged.[55]

That was a clear warning against engaging in positional warfare with them. As an advanced civilization, Sasanian Persia could organize the supply of large field armies, keeping them in food, fodder, and water even in arid areas. Unlike barbarians, who would soon be forced to withdraw by starvation, the Persians could therefore sustain long sieges against fortified cities, and they also had the equipment and training to undermine and breach defended city walls.

That much is clear from the detailed account of the siege of Amida in 359 by the soldierly eyewitness Ammianus Marcellinus; among other machines he describes mobile iron-clad towers incorporating stone-throwers that were brought up to overlook the walls.[56]

Another feature the Sasanians had in common with the Byzantines, another sign of a high degree of organization, is that on campaign, they encamped within fortifications, and when expecting battle, they constructed a perimeter with a ditch and a sharpened palisade. On the other hand, as we saw, they did not have the Roman tradition of erecting a tent city in ordered streets within the four squares formed by the intersection of the *via principalis* and the *via praetoria;* instead the Persians pitched their tents anyhow within the fortification, and that made them vulnerable to surprise attacks.

It is noted that the Persians wear both body armor (presumably a corselet, that is, a breastplate and back piece) and mail, and are armed with swords and bows. It is not specified if that refers to cavalry or infantry or both, but only cavalrymen are likely to be so heavily armored.

As the *Strategikon* continues to describe the equipment and tactics of the Persians, the weaknesses to be exploited begin to emerge: "They are more practiced in rapid, although not powerful archery." That can only mean that the Persians use smaller or otherwise less resistant bows that can be outranged by Byzantine archery. "They are really bothered by cold weather, rain and the south wind, all of which loosen their bow strings"—that useful information is frequently repeated in military manuals. Byzantine composite bows also had an outer edge of dried sinew that would lose its elasticity with humidity, but as we saw, special waterproof carrying cases were to be issued, and in the meantime to launch missiles in the rain the troops could rely on their slings—every archer was to carry more than one in his belt.

"They are also disturbed by very carefully drawn-up formations of infantry." That must reflect the prevalence of Sasanian combat with less organized enemies, the nomads and semi-nomads encountered across the Oxus, the Bedouin Arabs to the west, and the mountaineers and no-mad tribes of Afghanistan and Baluchistan. For them too, Byzantium was the only enemy with an advanced civilization, and Persians unused to seeing enemies in well-ordered ranks would be understandably wary. Persians also fear "an even [battle] field with no obstacles to the charge of lancers," because Sasanian infantrymen were all archers, and were not trained and equipped as heavy infantry to stand its ground against the charge: "they themselves do not make use of lances and shields."

As for the Persian cavalry, "charging against them is effective because they are prompted to rapid flight" and—unlike the horse-warriors of the steppe—"they do not know how to wheel about suddenly" to con-front their pursuers. The unstated reason is that the Persian cavalry was trained to fight in formation, unit by unit, and it is impossible to reverse the direction of an entire formation by wheeling about each one of its units, for that would work only if they all turned at exactly the same time, at exactly the same speed—no large cavalry force can perform vir-tuoso riding with that degree of precision. The steppe nomads did it eas-ily enough by wheeling about their own horses on command, because they always rode in loose order and were quite used to maneuvering with each other to avoid collisions and entanglements—which were not disruptive, anyway, in the absence of formations.

For the same reason, the *Strategikon* advises against attempts to re-verse withdrawals by wheeling or turning around—the men would run headlong into the ordered ranks of the Persian cavalry formation bear-ing down on them, because the Persians strive to remain in tight forma-tion even when in rapid pursuit. The recommended alternative is to de-viate from the direction of the withdrawal to circle halfway back, then ride parallel to the enemy's line advance, until reaching the rear of the Persian formation.

That is their weakness in general: "[the Persians] are vulnerable to attacks and encirclements from an outflanking position . . . because they do not station sufficient flank guards."

After the Sasanian Persians, the *Strategikon* examines the "Scythians," thus indulging in the Byzantine passion for antique classical terms, but then immediately adding, "That is Avars, Turks, and Others whose way of life resembles that of the Hunnish peoples." They were by then a familiar category for the Byzantines: the horse nomads and mounted archers of the steppe, starting with the Huns themselves, who were eventually followed by the more versatile Avars, who arrived just ahead of their ancestral enemies, the first Türk qaganate.[57] In between these arrivals other steppe peoples also came within the purview of the Byzantines, notably the Hephthalites first mentioned by Prokopios. But after writing that they were all the same, the author draws a distinction by noting that only the Türk and Avars have organized military forces, and therefore are stronger than the other steppe peoples in fighting pitched battles.

It is obvious who was the main enemy at the time of writing, the Avars: "scoundrels, devious and very experienced in military matters."

Without specifying that he is referring to the Avars, the author lists their weapons as swords, bows, and lances; with "lances slung over their shoulders and holding bows in their hands, they make use of both as need requires." That is exactly what the *Strategikon* prescribes for the Byzantine cavalry, along with the mail body armor and the frontal iron or felt armor of the horses of Avar chiefs. It is also noted that they emphasize training in mounted archery—which indeed requires much training to be effective at all.

Under the heading of enemy strengths to be avoided, there are a number of warnings. What appears to be a long battle line conceals units of different sizes that have hidden depth, and there will also be a hidden reserve: "Separate from their main formation, they have an additional force which they can send out to ambush a careless adversary."

Even without reckoning the dust raised by thousands of horses' hooves, even in broad daylight, in fighting the steppe riders it was difficult to estimate the strength of enemy forces, to thereby decide whether to attack boldly, defend resolutely in place, or retreat quickly, as outnumbered Byzantine commanders were enjoined to do by all their manuals. The implication is that scouting is necessary all around enemy forces because their frontal appearance understates their real strength.

When pursuing fleeing enemies, they will not stop to plunder but will press on until they achieve total destruction. The implication is that if no orderly retreat with strong rear guards is possible, it is better to stand and fight rather than retreat. The contrary is also true: if they retreat or even flee, there should be no hasty pursuit, because they are practiced in the quick turnaround and counterattack, and also in luring pursuers into ambushes by feigned retreats, as most famously in 484 when the Sasanian Shah Peroz was killed by the Hephthalites.

When it comes to vulnerabilities to be exploited, the first is the reverse of their great horse mobility. Unlike the Byzantine cavalry, the steppe warriors did not have just one mount and a spare at most: they rode with vast herds of horses that provided their basic nourishment of blood and milk and also many remounts, so they could ride back to fetch a fresh horse even in the midst of combat (they kept them hobbled next to their tents).

There is an eyewitness description of the Cumans (or Qipchaks), who replaced the Pechenegs as the steppe warrior ally/enemies of Byzantium in the twelfth century:

> Each of them has at least ten or twelve horses, which must follow them everywhere they wish to lead them; they ride first one and then the other. Each of the horses, when they are on a journey, carries a nosebag containing food, it eats as it follows its master, not ceasing to travel by night or by day. They ride so hard that they cover in a night and a day at least six, seven, or eight days' journey.[58]

All those horses needed pasture, and all those nosebags needed forage, which limited the strategic reach of the steppe warriors, especially in winter. That comes up in the *Problemata* of Leo VI, consisting of extracts from the *Strategikon* in the format of questions and answer:

> What must the general do, if the [enemy] be Scythian or Hunnic?

> He should attack them about the month of February or March, when their horses are weakened by the hardships of winter.[59]

That is taken from Book VII, "Before the Day of Battle," although it has also been attributed to Urbicius.[60] Their dependence on pasture or forage also meant that the warriors of the steppe could be weakened by setting grasses on fire when the weather allowed. But the stronger remedy was to campaign in a way that would maneuver them away from good grazing grounds, and into lands already overgrazed, or lacking in grass to begin with.

A greater weakness to be exploited was structural. They were cavalrymen, not infantry, not formidable when fighting on foot and wholly untrained in fighting in close formation. Therefore their cavalry could be readily halted by infantry in disciplined ranks, so long as there were enough bowmen to prevent the steppe archers from simply standing in front of them to discharge their arrows into the massed ranks. Moreover, while the steppe riders were the best horsemen, they were not heavy cavalry and had no heavy infantry with them, so they could be defeated by Byzantine cavalry charges followed by hand-to-hand fighting. Accordingly, the *Strategikon* points to the necessity of choosing level and unobstructed ground for the battle.

This suggests that for all their efforts, the Byzantines could not count on superior archery against the steppe masters of the art, that they could not outrange the mounted archers as with their other enemies. The right move was therefore to close the distance as soon as possible, to negate archery on both sides and replace it with the contest of sword, dagger, and mace—after the cavalry charge had delivered its impact. It is also remarked that "night attacks are also effective," presumably because the steppe enemy could not fall back on standard drills to overcome confusion.

There is also a political vulnerability: "Composed of so many tribes as they are, they have no sense of kinship or unity." Subversion will therefore be effective: "If a few begin to desert and are well received, many more will follow." That presumes, however, that the tide of battle has already turned: just as victory enlarged the Huns and later the Avars with subjected nations and camp followers, defeat diminished them.

It was the fate of the Byzantines that they had to contend not only with the Sasanian empire in the east and the mounted archers of the steppe to the north, but also with the warriors of northern Europe, collectively dubbed "light-haired peoples" in the *Strategikon*.[61] They are itemized as the Franks, Lombards, "and others like them."

The Franks entered Italy from the northwest in 539, attacking Milan just when the Byzantines were defeating the Goths of Witigis, who was besieged in Ravenna where he surrendered. Prokopios describes how they fought:

> At this time the Franks, hearing that both Goths and Romans had suffered severely by the war . . . they straightway gathered to the number of one hundred thousand under the leadership of Theudibert, and marched into Italy: they had a small body of cavalry about their leader, and these were the only ones armed with spears [*dorata*, not long lances] while all the rest were foot soldiers having neither bows nor spears, but each man carried a sword and shield and one axe. Now the iron head of this weapon [the cele-

brated *francisca*] was thick and exceedingly sharp on both sides, while the wooden handle was very short. And they are accustomed always to throw these axes at one signal in the first charge and thus to shatter the shields of the enemy and kill the men.[62]

The Lombards (*Langobardi* in Latin) entered Italy from the north-east in 568, only twelve years after the final defeat of the Goths in 554, but the Byzantines had encountered many other "light-haired peoples" long before the Franks or Lombards, most recently the migrating Vandals they defeated in their final destination of North Africa before entering Italy in 535, and the Gepids, whose power was centered in Sirmium (Stremska Mitrovica, in Vojvodina, Serbia), whence they threatened Byzantine lands until their crushing defeat by the Lombards and Avars conjointly in 568. When the Lombards under Alboin invaded Italy, seizing lands from Byzantine control all the way down to Benevento near Naples, they came with Gepids, Bavarians, and other Germanic camp followers, and also Bulghars it seems, but assimilation into a common Lombard identity was rapid.

It is with a high compliment that the author of the *Strategikon* begins his comments: "The light-haired races place great value on freedom." Earlier he had described the Persians as a "servile" lot, who "obey their rulers out of fear"—some things never change after millennia—and attributed a monarchical form of government (= *qaganate*) to the steppe peoples, whose rulers "subject them to cruel punishments for their mistakes."

By contrast, the light-haired freely fight for their honor, and that gives them strength but also limits their tactics: "They are bold and undaunted in battle. Daring and impetuous as they are, they consider any timidity and even a short retreat as a disgrace." Not for them, therefore, the feigned retreat or any other such maneuver; heroic rigidity offered opportunities for the Byzantines to exploit.

Their greatest weakness, however, was their lack of missile power, as Belisarios noted of the Goths in Italy, for their weapons are "shields, [spears] and short swords slung from their shoulders" with no mention of the bow—which they no doubt had, but not in large numbers, and not powerful bows either. The Italian campaign began in 535 to last in fits and starts for three decades, yet the Goths, and for that matter the Franks and the Lombards who followed them, did not adopt the composite bow from the Byzantines as the latter had imitated Hun archery.

Why did the "light-haired nations" fail to adopt the superior weapon? It was certainly not because they were too backward to learn how to dry layers of horse sinew onto a wooden core, file down bone plates, and prepare glue to hold the three-part bow together. In the first place, specifically Gothic jewelry has survived that required far greater technical skill; secondly, the "Goths" were another fighting power misnamed as a nation, and they included other ethnic groups, including Romans of course, and even camp followers from the steppe.

Only conjecture can offer an explanation for the mystery, but in this case it does not have to stretch very far: the Goths did not adopt the composite bow, and the archery tactics it allowed, for the same reason that the English longbow was scarcely imitated even after its dramatic victories (and the first handheld firearms, exceedingly clumsy arquebuses, were preferred to bows superior in both accurate range and rate of fire)—endless training is needed to acquire and preserve proficiency in using very powerful bows, whether the longbow or the composite bow of the Byzantines.

The freedom of the light-haired did not yet coexist with discipline, and that created vulnerabilities to exploit: "Either on horseback or on foot they are impetuous and undisciplined in charging, as if they were the only people in the world who are not cowards. They are disobedient to their leaders." They can therefore be lured into incautious advances, where powerful forces await them in concealed ambush. This might work on any scale, and might be battle-winning if a vital component of the battle force is lured away. "They are easily ambushed along the flanks and to the rear of their battle line, for they do not concern themselves at all with scouts and other security measures."

The light-haired were therefore only really formidable when their numbers and impetuosity could overcome their shortcomings, that is, in all-out battle. The resulting advice follows logically enough: "Above all, therefore, in warring against them one must avoid engaging in pitched battles. . . . Instead make use of well-planned ambushes, sneak attacks (to exploit the absence of flank guards and scouts) and stratagems."

There is also important nontactical advice: talk to them. "Pretend to come to agreements with them." Why? Our author says that the aim is to delay battle, to reduce their enthusiasm "by shortage of provisions." For the greatest vulnerability of the less organized must be logistic. Except in sieges and in the worst of times, Byzantine troops were reliably fed by the Byzantine state with its armies of tax collectors, clerks, and storekeepers. But much of the time the light-haired had no state to speak of, only battle leadership. It was mostly at the higher operational level that this greatest of vulnerabilities could be exploited, by containing the enemy in an elastic way with a minimum of actual fighting, to exhaust his provisions through the mere passage of time. Talks also provided opportunities to divide the light-haired, in part because underlying ethnic identities could perhaps be awakened, more likely because, as the author previously claims, "they are easily corrupted by money."

When the *Strategikon* was written, the Danube frontier and the Balkan Peninsula below it, including mainland Greece, had to contend with the incursions and invasions and permanent settlement of the Slavs. As compared to Goths and the rest of the light-haired, the Sasanians or the Huns, the Slavs were much newer enemies. That may be the reason why the chapter on the Slavs and Antes (*Sklavois,* or more commonly *Sklavenoi, Antais, Antes,* Latin pl. *Antae*) "and the like" is much longer than the others. Who were they? The *Sklavenoi* can reasonably be identified with the Slavs, but only because the term is broad indeed, encompassing many peoples whose many languages have many elements in common. But unless the Antes were merely a particularly troublesome segment of the Sklavenoi, they were probably not an ethnic group at all but rather a gathering of warriors of diverse origins, as with the Alans, variously reported as mounted warriors from the Caucasus to what is now France; both would have had only a camp language in common, like the Urdu of Mughal armies in India.

In Prokopios, on the other hand, the Antes and the Sklavenoi, formerly one people, became separated, and his description of their way of fighting seems to refer to the Antes: "The majority of them go against their enemy on foot carrying little shields and javelins *(akontia)* in their hands, but they never wear corselets."[63]

In the *Strategikon,* the opening remarks irresistibly evoke images of twentieth-century Russians at war: "They are . . . hardy, bearing readily heat, cold, rain, nakedness and scarcity of provisions." Their strengths are several: "They make effective use of ambushes, sudden attacks, and raids. . . . Their experience in crossing rivers surpasses that of all other men" (also definitely true of modern Russian armies). The author describes a ruse that was also used during the Second World War:

> [When] caught by surprise and in a tight spot, they dive to the bottom of a body of water. There they take long, hollow reeds . . . and hold them in their mouths, the reeds extending to the surface of the water. Lying on their backs on the bottom they breathe through them and hold out for many hours without anyone suspecting where they are.

Slav vulnerabilities begin with their weapons—the poor weapons of primitive peoples: "They are armed with short javelins, two to each man. Some also have nice-looking . . . but unwieldy shields."

Though often overrun and subjected by mounted archers, they do not have the composite bow; instead they make do with the simple wooden bow, good enough to hunt birds, perhaps, but evidently lacking in penetrating power at any significant range. Instead they have "short arrows smeared with a poisonous drug." That is not a formidable weapon of war. It works in hunting animals that can be patiently tracked till they die, but it is unlikely to have been effective against troops protected by thick clothing or leather, let alone the hooded chain mail prescribed for cavalrymen in the *Strategikon*.

No tactical sophistication can be expected of them: "Owing to their lack of government . . . they are not acquainted with an order of battle. They are not prepared to fight a battle standing in close order." In other words, they cannot execute battle drills by ranks and columns to stop arrows with shield walls, to thrust forward with spear and pike, to cover light-infantry projecting missiles with heavy infantry in front, or fight with the sword side by side for mutual support. The author's recommended way of exploiting this shortcoming is straightforward: "They are hurt by volleys of arrows, sudden attacks launched against them from different directions, hand-to-hand fighting."

But the Slavs are unlikely to be caught, they will "run for the woods, where they have a great advantage because of their skill in fighting in such cramped quarters."

The recommended operational method for fighting the Slavs is to field a combined cavalry-infantry army equipped with a large number of missiles ("not only arrows, but also other throwing weapons") and with materials for building bridges, pontoon bridges if possible. There are numerous and unfordable rivers in the country of the Slavs—on the other side of the lower Danube and its delta ("They live among impenetrable forests, rivers, lakes and marshes"), and the advice is to build assault bridges concurrently in the Scythian manner, some men already laying down the planks while the framework is still being built; ox hide or goatskin rafts are also needed, in part to hold the armor and weapons of soldiers swimming across to launch surprise attacks, in summer of course. But the author actually recommends winter campaigns, when rivers can be swiftly crossed on the ice, the Slavs suffer from cold and hunger, and they cannot hide among the bare trees.

In what would be a contradiction in the coldest months, what is proposed is an amphibious operation, with the warships (the *dromon* type is specified) in apposite locations along the Danube. A *moira* of cavalry is to provide security, with the entire force kept a day away from the

Danube. Rivers can be crossed by sending across a small force of infantry, both archers and heavy, on the night before, to hold a formation with backs against the river: "they provide enough security to put a bridge across the river." When all is ready, there should be a sudden crossing to engage the enemy in force, preferably on clear and level ground. The battle formation should not be too deep, and wooded terrain should be avoided at all times, if only to safeguard horses from rustling.

A standard sequence is recommended for surprise attacks: a detachment is to approach the enemy frontally to provoke them before turning to flee, while a second force is positioned to await the pursuers in ambush.

Actually splitting the force is recommended on the offensive as well, even if there is only one suitable road; the reason is to keep up the advance with a lead force while a second force does the looting and ravaging of Slav settlements, because "they possess an abundance of all sorts of livestock and produce." Even if the Byzantines did not need the food, it was important to deprive the Slavs.

Subversion by gifts and persuasion should be particularly effective, because there are many "kings" among them "always at odds with one another." But the author is resigned to the necessity of fighting, because he provides several operational methods to do so, all very much relational.

Why does the *Strategikon* devote so much attention to such ill-prepared enemies? Or rather, how did the Slavs become such formidable enemies if they were so poorly armed and hardly organized at all? In a word, because they were "populous," as the author writes.

Sasanian forces were large, of course, but only by the standards of highly organized armies—in the thousands for a battle, in the tens of thousands or perhaps a hundred thousand for the lot. The light-haired were more numerous but not by much—we read of all the Visigoths of Alaric on the march, and they could all be fed by requisitions even in a greatly decayed Roman empire. As for the Huns and the Avars, leaving aside all controversies over the number of the Huns or Avars as such—no vast hordes, of course, but only ruling elites—it is beyond argument that the sum total of mounted archers in any given area could not exceed the pasture available for their numerous horses, rigidly limiting their number. Whatever that was, it was a number bound to decline whenever the mounted archers ventured into less flat and less wet lands, as they did in moving south across the Danube to the Balkans, then to Thrace and Greece. As formidable as they were, the mounted archers could not also be numerous when invading those lands. It was otherwise with the Slavs, who were numerous enough to resettle much of Greece, and therefore too numerous to be contained by far smaller Byzantine forces. The author actually writes little of defense against them and much of offensive operations in their own lands, to fight other Slavs beyond the Danube, not the ones that had already entered the empire under Justinian, not be dislodged again.

It is now fashionable to deride such texts as colonialist inventions, designed to denigrate the Other, filled with imagined fears or perhaps secret desires but always motivated by the will to dominate with words as with oppressive deeds. Perhaps so, but it does seem that the author of the *Strategikon* was trying to understand rather than invent, because his aim was to uncover real strengths and weaknesses, not imagined ones.

The information needed to devise relational methods and tactics is an obstacle that can be overcome with enough of the intelligence effort recommended in the manuals. But there is also risk, and that cannot be eliminated so easily. Relational maneuver can succeed wonderfully, but it can also fail catastrophically. To boldly penetrate deep behind enemy lines into the soft rear, to throw him into confusion and disrupt his supplies, is very fine—if the enemy does indeed collapse in disorder. But if the enemy can tolerate confusion and remains calm, the advancing columns can be caught between the remaining enemy forces they encounter in the rear and those returning from the penetrated front to attack them from behind.

The risk that bold maneuver will defeat itself by overextension is the usual reason why relational schemes are mostly avoided. But another reason is that any military action more complicated than a straightforward attack or a stand-fast defense is that much more likely to break down simply because of "friction"—the sum total of many seemingly very minor delays, errors, and misunderstandings that can combine to wreck the best-laid plans. That is true of any military action, but more so of maneuvers that seek to achieve surprise by fast timing, or unexpected approaches through difficult terrain, or long-distance penetrations.

The *Strategikon* accordingly follows a middle course, placing great emphasis on gathering intelligence by scouting and with spies, but recommending prudent rather than bold operations.

The discussion of tactics proper begins with a criticism of the single, long battle line of cavalry, especially of lancers; the author explains that it would be disordered by varied terrain, hard to control at the extremities far from the field commander, and there may even be desertions there. Moreover, the single line has no depth and no resilience, because there is no second line behind it and no operational reserve such as the Avars keep, so there is no remedy if the line is outflanked or breached.[64] But why would anyone favor the single battle line to begin with? The question is not raised or answered in the *Strategikon* because it is too obvious. It is not that the single long line looks more impressive than any deeper formation to enemies that stand directly before it on level or lower ground, with horsemen side by side stretching all the way to the horizon, as it were. That would impress only enemies ignorant of contemporary war, a rare commodity sometimes encountered by the British in their far wanderings but never reported by the Byzantines.

The compelling reason was simply that the single, long line requires no prior drilling to teach everyone how to swiftly assume their place in different formations, and how to move individually on command to change the shape of the entire formation, deeper with longer files and fewer ranks, or shallower and longer; or of a different shape altogether.

That is why the single, long line was the ordinary disposition of both Romans and Persians as the author acknowledges (II, 1, 20), because when the two superpowers of the age fought each other, they did it with their forces as fully mobilized as possible, both the guard units well drilled in every sort of formation, and everyone else they could muster, including part-time militia (thematic) cavalry, barbarian allies and auxiliaries, perhaps very skilled horsemen and fighters but not trained to keep and modify formations.

The *Strategikon* then proceeds to advocate and explain a battlefield array composed of different formations each under its own commander instead of the single formless line. Exactly as in all modern armies, the recommended force structure is triangular: the basic combat unit is the *bandon* of three hundred men or more; three of them along with officers and specialists form a *moira*, and three of the latter form the *meros* of six thousand men or so.

While the author rejects the long, single line, the formation he advocates cannot be very deep—indeed, he favors a depth of four ranks.

In the text (book III) a number of formations are offered and illustrated in detail with symbols for each kind of junior officer and soldier: in the basic layout, at the head of each file there is a *dekarch*, a squad leader commanding ten, with lance and shield (= heavy infantry); behind him in the file is a *pentarch*, a team leader commanding five, with lance and shield; an archer is third in the file without a shield (= light infantry); fourth in the file, another archer is behind him, but with a shield for possible rear-guard combat; a further archer is fifth in the file without a shield, and behind him there is a sixth soldier with his choice of weapons.

Alternating the sequence of heavy- and light-infantry men, opening up or closing the files, different formations are illustrated for a single *tagma,* and then the battle disposition of an entire army is illustrated, complete with flank guards, baggage train, and reserve, as well as the different combinations of heavy infantry to stand, hold, and engage in close combat, with light infantry to reduce and harass the enemy with its arrows.

The most ambitious battle array in the *Strategikon:* "The Order of Battle Called Mixed" depicts a complete army composed of heavy infantry and cavalry with only a handful of light-infantry men in the rear.[65] Both the heavy infantry and the cavalry are in files seven deep, but while each infantry meros has five men across, the cavalry meros is seven across.

In the battle array, there is an infantry meros on each side to secure the flanks, with a cavalry meros on the inside, another infantry meros next to it, and finally a central cavalry meros, for a total of seven. For rear-guard protection, there is a single infantry force on each side, also five across, with heavy infantry five files deep and a final rank of just five light-infantry men on each side, so that the two outer-flank files on each side are actually thirteen deep, but in two separate units. That leaves a large space, five meros deep, in the rear of the formation, but no enemy is likely to venture into it, given the ease with which the cavalry could maneuver against such a move.

One can readily envisage how such a formation would fight. A mounted enemy, counting on momentum and the charge, would collide with the solid holding power of the heavy infantry with its interlocked shields and spears fixed into the ground. An enemy counting on the mass of his infantry would run into the missiles of the cavalry archers— that is why only a few light-infantry men are needed as rear guards. Of course, there is no surprise if the formation is observed in advance. Cavalry is too high off the ground to remain hidden, but the infantry can be concealed in the dust and confusion. "To prevent the formation from being too closely observed by the enemy before the battle, a thin screen of cavalry may be deployed in front of the infantry phalanx until the enemy gets close."[66] Such combinations can achieve operational-level effects greater than the tactical strength of each force on its own, but there are no free gifts in warfare, and the price paid is yet more training, as the *Strategikon* duly notes: "This sort of formation requires constant practice for both men and horses."[67]

Professionalism was also required of the men manning the river fleets, whose task, in addition to surveillance patrols, was to ferry across soldiers, horses, artillery, and supplies, and to effect opposed river crossings against enemies waiting on the far bank. That is the subject of five paragraphs inserted in part B of Book XII of the *Strategikon,* which are conventionally cited as *De fluminibis traiciendis,* "On River Crossings." Their significance has recently been recognized and lucidly explained:[68] *Strategically,* the value of river fleets exceeded that of the sea fleets, because for more than half a millennium there were no seagoing enemies (the third-century Goth raiders were the notable exception), except for scattered pirates and sea robbers, while there were always dangerous barbarians across the Rhine and Danube. *Operationally,* to combine frontal offensives with backstopping, ambushing, or raiding, by forces inserted behind the enemy by river fleets, was an excellent way of exploiting the comparative advantage of the Byzantines in organization and planning—indeed, it became standard operating procedure when fighting the Bulghars and Bulgarians to send forces via the Black Sea into the Danube to attack their rear. *Tactically,* opposed river crossings required specialized training, because to land on the far bank against alert and deployed enemies, it is first necessary to "dominate it by fire," in modern parlance, that is, to scatter the enemy with massed volleys of arrows and stone-throwers embarked on warships while a pontoon bridge is assembled section by section (para 5).

The well-trained and well-exercised military man depicted in the *Strategikon* was not an impressed ragamuffin but a professional, and his social status was commensurate:

> The men, especially those receiving allowances for the purpose, should certainly be required to provide servants for themselves, slave or free, according to the regulations in force (to avoid having to detail soldiers to the baggage train). . . .
> . . . But if, as can easily happen, some of the men are unable to afford servants, then it will be necessary to require that three or four lower-ranking soldiers join in maintaining one servant. A similar arrangement should be followed for pack animals, which may be needed to carry the coats of mail and the tents.[69]

Servants are mentioned again in the section "On Baggage Trains," implying that there were many of them:

Included with the train are the servants needed by the soldiers. . . . We advise that no large number of servants be brought into the area where the main battle is expected . . . [but] there should be enough servants to each squad to take care of the horses. . . . At the time of battle these servants should be left behind in the camp.[70]

It was a very demanding form of warfare that the *Strategikon* recommended, but the reward of intensive individual training, much tactical drilling, and discipline in executing operational schemes was to achieve the objective with a maximum of maneuver and a minimum of attrition. That was the one thing to be avoided at all times, lest tactical victories result in strategic defeat for an empire that always had one more enemy arriving just over the horizon.

CHAPTER 12

After the *Strategikon*

The value of the *Strategikon* of Maurikios was amply recognized by Byzantine military officers and Byzantine writers, who excerpted, paraphrased, summarized, plagiarized, and updated the text in subsequent centuries. The work that Dain labels *De Militari Scientia* is such a paraphrase; it dates at least from the later seventh century, because the Muslim Arabs are studied in it instead of the Sasanian Persians. It counts as further evidence of the vitality of Byzantine military literature, of which other examples noted by Dain but not seen by me include a version of Aelian, an extract of the *Taktikon* of Urbicius, another *De Fluminibus Traiciendis* extracted from the *Strategikon,* as well as a variety of lost texts indicated by their surviving traces and paraphrases in extant works.[1]

The first great age of Byzantine military literature was the sixth century, not coincidentally the age of Justinian's wars and conquests. Then came Justinian's plague, a world-historical event that killed a large part of the population, necessarily wrecking every imperial institution, including the army and navy. Mutiny, usurpation, the catastrophic Persian invasions, the ruinously belated Byzantine victory, and the almost immediate Arab invasions left a Byzantium much diminished and greatly impoverished, in which there was little reading and less writing by the end of the seventh century. But decline was followed not by fall but by a recovery that had clearly started by the end of the eighth century and would progress into a veritable economic, cultural, and military renaissance.

The second great age of Byzantine military literature was one of the products, and in turn perhaps one of the contributing causes, of its military renaissance, starting with the works attributed to Leo VI "The Wise" (886–912).

His first try at a military manual, the *Problemata,* consists of nothing more than extracts from the *Strategikon* of Maurikios arranged as answers to the author's questions. That unimpressive beginning—he may have been in his twenties at the time—was followed by an altogether more substantial work: the *Taktika,* or *Tacticae Constitutiones,* which was written in stages and later edited by Leo's even more studiously literary son Constantine VII Porphyrogennetos.[2] The Roman "constitution" was not that in its modern sense but rather a law, and more specifically an imperial decree in the form of a personal letter containing instructions and orders addressed to a named official or to the generic holder of a post; in the *Taktika* Leo is sending letters to an unnamed *strategos,* a general or admiral. A greatly awaited new edition by George T. Dennis SJ is forthcoming, but in the meantime there is only J. Lami's 1745 Florence revision of the first 1612 Leyde edition by Joannes Meursius, plagiarized in the immense *Patrologia Graeca* published by Jacques-Paul Migne (vol. 107, cols. 669–1120) and variously cited as edited by Meursius, Lami, or Migne, by *droit de seigneur* presumably.

The contents are mostly paraphrases of earlier texts derived from the *Strategikos* of Onasander, the *Taktike Theoria* of Aelianus "Tacticus," and more largely from the *Strategikon,* whose preface Leo also echoed while translating into Greek its Latin words of command.[3] But there are also original parts that are historically valuable. The material is organized not by authors or texts but according to a logical scheme, by subjects. Constitution I: Tactics, or rather drills. II: The qualities of the strategos or general. III: The force structure and ranks of the army. IV: Military councils and decisions. V: Weapons. VI: The weapons of the cavalry and infantry, mostly from the *Strategikon.* VII: Training, in which two-sided simulated combat exercises are recommended, with wooden lances and swords, and arrows without arrowheads or blunt ones; Also to be practiced is the advance of the cavalry against volleys of arrows, to be achieved by keeping very tight formations, with shield touching shield, both horizontally in the first two ranks and overhead from the third rank onward. VIII: Military punishments. IX: Marches. X: Baggage trains. XI: Camps and marching camps. XII: Preparing for combat. XIII: The day before the battle—an important subject when it could be presumed that the day of battle could be determined, because it would in effect be set by mutual agreement. XIV: The day of battle. XV: Siege warfare. XVI: The day after the battle. XVII: Unexpected incursions. XVIII: The customs of different nations, another *ethnika* as in the *Strategikon,* but mostly on the Muslim enemies. XIX: Naval warfare. XXI: Maxims of war. And an epilogue.

Leo VI on Fighting the Muslims

As in much of the *Taktika,* Leo's principal source is the *Strategikon*—in this case, book XI on the ethnic characteristics of the different enemies of the empire—the starting point of relational maneuver with all its potential tactical and operational advantages. But Leo adds original material, adapted to contemporary realities.[4] The principal enemy on the scene in the time of Leo had not existed when the *Strategikon* was redacted: the Muslims, originally Arab but increasingly non-Arab Muslims of Turkic or Iranic origins, notably the Kurds and the Daylami highlanders from the Caspian region.[5] All of them could also be described as *Sarakenoi,* Saracens, originally the name of pre-Islamic Bedouins of the northern Sinai, but later a word used in many languages, including the *Saracini* of my Sicilian childhood, for any and all Muslims.

What made the Muslims dangerous was their ideological commitment, as the text fully recognizes. But matters are not so simple, because while the ideological commitment is genuine, the conditions of jihad also provide opportunities for poor warriors who are in it for the loot: "They are not assembled for military service from a muster list, but they come together, each man of his own free will and with his whole household. The wealthy [consider it] recompense enough to die on behalf of their own nation, the poor for the sake of acquiring booty. Their fellow tribesmen, men and especially women provide them with weapons, as if sharing with them in the expedition." Leo's admiration is evident—he despises the religion but respects the militant altruism it inspires.[6]

Leo diverges from the *Strategikon* almost from the start: the generic injunction to provide weapons according to regulations is followed by: "In particular make sure you have a large number of bows and arrows. For archery is a great and effective weapon against the peoples of the Saracens and the Kurds who place their entire hope of victory in their archery." That was true of the Kurds who fought as mounted archers, and of the increasingly prevalent Turks, but not of the Bedouin irregular cavalry or of the Daylami, who fought on foot with javelins and swords.

What follows by way of tactical advice is sound enough:

> Against the archers themselves, defenseless at the moment of firing the arrow, and against the horses of their cavalry, the arrows shot by our army are extremely effective . . . when the horses so highly prized by them are destroyed by continuous archery, and the result is that the morale of the Saracens, who had been so eager to ride out to battle, is completely cut off.

For the steppe warriors of the great grasslands, who commonly had a dozen horses for their own riding and more with their families, and who kept remounts hobbled nearby when fighting, a dead horse was so much meat for the pot; not so for the riders of arid lands where each horse had to be kept alive by hand-feeding in the driest months, whose appetite for horse meat was that of an average English horse-fancier, and who rarely had remounts at hand, as the Byzantine cavalry often did. That is why it was so useful to target the horses.

There was also an ideological vulnerability, which remains of transcendental importance till this day: because Muslims "do not go on campaign out of servitude and military service" but rather to fight for their creed, when they suffer defeat "they think God has become their enemy and they cannot bear the injury"—hence the deep trauma that has accompanied recent Muslim defeats at the hands of Christians and Jews, and the global mobilization that followed apparent victory over the Soviet foe in Afghanistan.

After this aside, Constitution XVIII reverts to tactics, the necessity of marching camps, and the different modes of pursuit, where the *Turkoi* are mentioned, meaning the newly arrived Magyars at this time. That leads to an interesting digression:

> When the Bulgarians had disregarded the peace treaty and were raiding through the Thracian countryside [c. 894] . . . Justice pursued them for breaking their oath. . . . While our forces were engaged against the Saracens, divine Providence led the Magyars, in place of the Romans, to campaign against the Bulgarians.

In this case Providence was assisted by the Byzantines:

> Our Majesty's fleet of ships . . . ferried them across the Danube . . . and, as though they were public executioners, they decisively defeated [the Bulgarians]. . . . , so that the Christian Romans might not willingly stain themselves with the blood of the Christian Bulgarians [the Magyars were still pagan].

What follows is an evocation of Book XI of the *Strategikon* on the military customs of the "Scythians," that is, the mounted warriors of the steppe, of the Franks and Lombards, and of the Slavs before reverting to the "nation of the Saracens that is presently troubling our Roman commonwealth."

In an encapsulated history, the text recounts that the Arabs were formerly scattered about toward Syria and Palestine; but

> when Mohammed founded their superstition, they took possession of those provinces by force of arms. . . . They took Mesopotamia, Egypt and the other lands at the time when the devastation of the Roman lands by the Persians allowed them to occupy those lands.

Next come the customs relevant to war, and the methods:

> They make use of camels to bear their baggage instead of wagons, and pack animals, asses and mules. They use drums and cymbals in their battle formations, to which their own horses become accustomed. Such great din and noise disturbs the horses of their adversaries, causing them to turn to flight. Moreover, the sight of the camels likewise frightens and confuses horses not used to them.

This is obviously useful information: horses can be trained accordingly and certainly were.

Another useful remark is that they "fear battle at night and all that is connected with it, especially when they are raiding in a country foreign to them." Of course all sensible soldiers fear the added uncertainties of battle at night, because battle is so uncertain even in full daylight; but Leo evidently felt that there was a more-than-normal reluctance to fight at night. That is readily explained by the composition of Muslim armies: their men were volunteers from many lands, much less homogeneous than the Byzantines, and not the products of uniform training, and therefore less likely to be able to coordinate spontaneously when—literally—in the dark.

When the text turns to the specifics of theater strategy, there is no question of echoes or imitations. The author is describing the annual arrival of jihadists responding to the summons of the frontier warlords and militant preachers—once they leave, normal raiding for loot resumes: "They flourish . . . in good weather and in the warmer seasons, mustering their forces, especially in summer, when they join up with the inhabitants of Tarsos in Cilicia and set out in campaign. At other times of the year only the men from Tarsos, Adana and other cities of Cilicia launched raids against the Romans."[7] The sovereign remedy is pre-emption:

> It is necessary to attack them as they are marching out to pillage, especially in winter [i.e., against looting raids]. This can be accomplished if [our] armies remain in a location out of sight. . . . When our men observe them marching out, they can launch an attack against them and so wipe them out. [We can also attack] when all of our troops have come together at the same time in large numbers, fully equipped for battle.

The point is that such a battle, a "meeting engagement" in modern operational parlance, is subject to all possible uncertainties and is therefore to be avoided absent a numerical advantage—and a very large one. That is emphasized in what follows: "It is very dangerous, as we have frequently said, for anyone to take the risk of a pitched battle, even when it seems perfectly clear that [our forces] far outnumber the enemy."

The alternative is the "nonbattle," or more fully an elastic defense that allows the enemy to invade—because it is impossible to defend strongly every segment of a long frontier—and then intercepts the enemy on his way back home:

> If they ever raid inside the Taurus [mountains] for the sake of pillaging, it is necessary for you to deal with them in the narrow passes of that mountainous region, when they are on their return journey and are particularly exhausted, perhaps bearing along some booty of animals and objects. Then you must station archers and slingers on some of the high places to shoot at them and thus have the cavalry charge.

More follows that would be spelled out in full detail in a specific field manual conventionally entitled *De Velitatione,* or "Skirmishing." But the latter is confined to the analysis of ground combat, while Constitution XVIII considers the entire threat:

> The Saracens in Cilicia think it a good thing to fully train all their infantry forces to engage in battle on two fronts, that is, on land along the road leading out from the Taurus mountains and on sea by means of their ships. . . . When they do not go out to sea, they campaign against the Roman towns on land.

The recommended remedy is very interesting, it is to find out what the Arabs intend and then do the opposite:

> You . . . O general, must keep an eye on them by means of trusted spies. . . . When they campaign by sea, you go by land and, if possible, launch an at-

tack against them in their own territory. But if they intend to campaign on land, then you should advise the commander of the Kybyrraiote [thematic] fleet so that, with the dromons under his command he may fall upon the Tarseote and Adanan towns which lie along the coast. For the army of the Cilician barbarians is not very numerous, since the same men are campaigning both on land and on the sea.

The operational logic is evident: it is impossible to defend a long frontier preclusively defending every inch of the frontage. It may be necessary to allow the enemy to invade here and there, but is it is not desirable, of course. So the most effective response is to launch an asymmetrical counteroffensive that can achieve surprise by attacking on land in response to a seaborne attack, and vice versa.

The imperial author is commendably modest in concluding, "We have presented Your Excellency with these regulations. Perhaps they contain nothing new." That was excessively modest, as we have seen, for Constitution XVIII was certainly a new response to a new threat. The ideological dimension of this response is present throughout the text with its frequent invocations to the deities and its condemnation of the false religion. They occur in all Byzantine field manuals of that time but form the core of two hortatory speeches—harangues before the troops—attributed to Leo's son Constantine VII Porphyrogennetos and appended to the *Rhetorica Militaris* already mentioned. That scholarly emperor did not lead his men into battle, nor, presumably, did he harangue them before the battle; but before being retained as model speeches for Byzantine commanders, they may well have been read to the troops at the conclusion of the campaigning season, when the paramount concern was to have them return in good form in the spring.[8]

The first speech follows a victorious defensive campaign, and accordingly starts with praise for the troops and their exploits against a well-equipped enemy with "horses whose speed made them impossible to overtake" the Arabians of the Bedouin and whose weapons were "unmatched in strength, equipment unmatched in craftsmanship."[9] But they could not win because they lacked "the one paramount advantage, by which I mean hope in Christ," who is contrasted to the Beliar or Muhammad and baldly equated to the Hebrew God of Ezekiel and the Psalms at his most martial, which are then quoted: "Who alone is *strong and mighty in battle*," "whose *weapons are drunk with the blood* of his enemies," "who *makes strong cities a heap*"—with more of the same.

Next, the enemy—the Sayf ad-Dawlah, already called by his dynastic name as the Hamdanid—is denigrated as vainglorious and boastful:

> He is afraid . . . and without a reliable force . . . [but] he is trying to put fear in your minds with ruses and deceptions. One moment he proclaims that another force is on its way to him and that allies [are coming] . . . or that . . . a vast sum of money has been sent to him [jihad contributions]. He has exaggerated rumors spread about for the consternation of his listeners.[10]

This passage is interesting in itself. How did Sayf ad-Dawlah spread his propaganda? Did he send agents to disseminate it? That is entirely probable: the frontier was evidently porous, there were large Christian communities in Syria, and there was a steady traffic of merchants and pilgrims.

Next the troops are told that this boasting is itself a proof of weakness:

> Were it possible to look into the mind of the Hamdanid, then you would see how much cowardice, how much fear oppresses it . . . pay no heed to his theatrics, but with confidence in Christ rise up against the foe You know how virtuous it is to fight on behalf of Christians.[11]

That being so, why is the emperor denying himself? He wants to fight but cannot, it is the divine will:

> What great yearning possesses me, what great desire inflames my soul. . . .
> I would much prefer to do my breastplate and put my helmet on my head,
> to brandish my spear in my right hand and to hear the trumpet calling us
> to battle.

But God orders him to wear the "crown and the purple" instead.[12]

After duty to God and to the Christians, there come the rewards for officers and common soldiers: promotions, gifts, land grants, cash donatives, shares in the booty; the promotions are comprehensive: "The *strategoi* in charge of the smaller themes will be transferred to larger ones . . . the commanders of the *tagmata* and other units who fight courageously will be rewarded in proportion to their deeds, some to become *tourmarchs,* others *kleisourarchs* or *topoteretai.*"[13]

How does the emperor know who deserves to be rewarded and in what degree? Before listing the rewards the emperor asks for accurate information: testimony under oath from his commanders, or "better yet, you will keep written records." That is how bureaucracy mediates heroism even now.

The *Sylloge Tacticorum*

Edited by Dain himself, wrongly attributed to Leo VI, the *Sylloge Tacticorum* is to a large extent a paraphrased summary of prior military writings, such as Leo's *Taktika*, though in part derived from different texts.[14] But in addition the work contains some significant original material, which Dain unaccountably failed to note in his survey. Most notably, chapter 47 (pp. 86–93) on tactics for combined infantry and cavalry forces (which follows earlier chapters on each arm separately) formed the basis of the tactical system described in the *Praecepta Militaria* of Nikephoros Phokas. Among other things, the *Sylloge* is "the first text in which a square is prescribed as the standard battle formation for Byzantine infantry"—a device that would have a very long history with armies that possessed well-disciplined troops. A stoutly held infantry square offers a secure base from which cavalry units can sally out—and to which they can return when exhausted by a charge, defeated or in risk of defeat, or simply for rest and recuperation in between strenuous tasks.[15] Moreover, the *Sylloge* includes contemporary information on Byzantine and Magyar shields and weapons.[16]

As for the derivative textual material in the *Sylloge* that does not coincide with the selections in Leo's *Taktika,* according to Dain it was taken from two lost prior compilations, which he dubbed *Tactica Perdita* and *Corpus Perditum.* From the first, the *Sylloge Tacticorum* paraphrases sections on the qualities desirable in generals; metrology; different kinds of combat; deployment of the army in peacetime in cantonments and fortresses; measures to be taken against the enemy; and many more. For Dain, clearly mistaken in this case, the entire *Sylloge* is the product of the library rather than of contemporary military experience, notably once again Onasander, Aelian, the *Strategikon, the Anonymous Treatise,* a work on metrology and the Theodosian code—on how to divide booty. The reconstruction of the eighty-seven sections of the *Corpus Perditum* from its paraphrases in later texts was a great philological coup but added no interesting new materials, being composed of recycled earlier texts otherwise known more fully.

Heron of Byzantium

"Heron of Byzantium"—in the splendid new edition his name is within quotes—is what an earlier editor chose to call the unknown writer of two very interesting tenth-century treatises on siegecraft and on measurement. Dain, whose own book on the manuscripts is another monument to his scholarship (although written in an outrageously popular style by his standards—the text is not even in Latin but in French!)—notes that anonymity is not the same as a lack of personality.[17] Indeed, the writer has original ideas even though most of his material is taken from engineering manuals that dated back some seven centuries or more by his time. The editor of the authoritative new edition rightly refers to the "generally static nature of the methods of fortification,"[18] but in spite of the overall technological immobility, there were some significant innovations, including two mentioned in the text: the traction *trebuchet* unknown in the west before the seventh century, and Greek fire, normally considered a naval weapon but also used in sieges, and indeed in open-field battle as well, both discharged from siphons and in the form of potted projectiles.

The first text, *Paragelmata Poliocertica*, begins in a manner that suggests that the intended reader is not an engineer: "Everything about siege machines is difficult and hard to understand, either because of the intricacy and inscrutability of their depiction, or because of the difficulty of comprehending the concepts."

Drawings are not of much use either, we are told—the explanations of the original inventors are best—but the author promises three-dimensional drawings far more easily understood.[19] He then lists some of his sources: the text of Apollodorus of Damascus, who built the Danube bridge for Trajan and wrote a treatise for Hadrian according to the author (he did not), the book that Athenaeus wrote for Marcellus[20]—that being the nephew of Augustus, not the rueful opponent of Archimedes; and the Biton, already reviewed. He then lists the machines needed for siege operations, including "tortoises," heavily protected mobile assault shelters with rams or without rams, for excavating and for filling trenches; the "recently invented" *laisai*—mobile shelters against arrows made of plaited branches, vines, or reeds; portable wooden towers "which are easy to procure," that is, not self-propelled with capstans to power the wheels; very high scout ladders; tools for undermining; "machines for mounting walls without ladders," of the *sambuca* type; bridges for assault crossings; and more.

After a ritual apology for his "commonplace and flat writing," the author offers his version of *si vis pacem para bellum*, quoted from Heron of Alexandria: to live without fear of domestic or foreign enemies, rely on "artillery construction" and a few more warlike preparations, including the storage of "long-lasting rations." Interestingly, a scholiast described them in detail: boiled, dried, squill (the nutritive bulb, not the flower) cut very thin, with one-fifth part sesame seed (as in Israeli combat rations) and one-fifteenth part poppy seed, with the "best honey kneaded into it"; or alternatively, sesame seed, honey, oil, peeled sweet almonds, roasted, ground, and pounded very smooth with an equal amount of squill—the scholiast described this ration as "sweet, filling and causes no thirst."[21]

That of course is what sieges were mostly about: food, both for the besieged, whose inventory of food must decline each day if they are fully encircled, and the besiegers, who must bring their food from ever-larger distances as local foraging, plundering, requisitioning, and purchasing are exhausted.

As each side tried to starve out the other, sieges acquired a desultory character. The *Chronicle of Pseudo-Joshua the Stylite*, the first historical text in Syriac, describes a telling episode of the siege of Amida in 503— when the Persians were defending the city they had conquered in 502:

> One day, when the whole Roman army was at rest and peace, fighting was provoked in the following manner. A young lad was feeding the camels and asses, and one of the asses walked up to the wall as it was grazing. The boy was too frightened to go in and retrieve it, and when one of the Persians saw it, he came down from the wall by rope, intending to cut it up and take it up for food, for there was absolutely no meat in the city. However, one of the Roman soldiers, a Galilean by birth, drawing his sword and taking his shield in his left hand, rushed towards the Persian to kill him. Because he went right up to the wall, those standing on the wall hurled down a large stone and struck the Galilean and the Persian began to climb up to his place by the rope. When he reached halfway up the wall, one of the Roman officers came near, two shield bearers going in front of him, and from between them he shot an arrow and hit the Persian, bringing him down near the Galilean. Shouting erupted from both sides, and for this reason they became agitated and rose up to do battle.[22]

This accidental transition from a sleepy day around the walls to a sudden outbreak of fighting over a donkey, is far removed from the technical dimension of siege operations but provides a realistic context for all the engineering.

The tactical and technical siege preparations described in *Paragelmata Poliocertica* presume the offensive—appropriately enough, because tenth-century Byzantium was advancing and taking Arab-held cities in southeast Anatolia and what is now Syria.

First comes the scouting of enemy fortifications, then preliminary diversionary actions with simulated preparations against sectors of the wall where no assaults are intended; to press the siege, assault trenches are to be dug diagonally to preclude enfilading missiles. Explanations of how different kinds of tortoises are to be advanced to the enemy wall are brief but coherent: men inside wheeled tortoises well protected frontally are to fill ditches and holes and depressions in the approach to the enemy wall, to allow other machines to be deployed smoothly. It is also necessary to probe the intended path to the wall with iron lances to detect fall-traps concealed by layering earth upon fragile clay pots; the soles of the soldiers' feet must likewise be protected against caltrops; the mining of tunnels to undermine the enemy wall ends with the ignition of the flammable dry sticks and pine torches placed around the wooden props that kept up the wall till then. Alternatively, stone blocks at the base of the enemy wall can be shattered by pouring vinegar or urine onto them after they have been strongly heated with charcoal, itself inserted into place by tubes projecting from the front of the tortoise that protects the entire operation; the chemistry is certainly correct: the reaction between acid and the calcium carbonate of the stone intensifies with heat. Inflatable ladders "like wineskins, and smeared around the stitches with grease," are recommended for surprise attacks (modern commandos have used inflatable scaling poles); and drilling many holes in wall stones with bow-drillers (still used today by Indian craftsmen) can make it much easier to breach walls.[23]

In between perfectly clear explanations of perfectly practical techniques, the author commends the specifications of the giant ram made by the ancient Hegetor of Byzantium: it was 56 meters, 150 feet, long. In other words, it was much too long to be an effective weapon. But that is followed by more practical instruction on how to build and use quite functional siege equipment, such as a shielded scouting ladder that can be instantly elevated by four sets of ropes (modern commandos still use them, but made of light alloy tubes and unshielded).

There is a much in the text on movable siege towers, with credit given to Diades and Charias, who served Alexander the Great, but with fully functional specifications and detailed dimensional data, and a list of available accessories, such as siphons to project water to quench flames set by enemy incendiaries and mattresses filled with chaff soaked in flame-retardant vinegar or marine moss or seaweed, to dull the blows of projectiles hurled by stone throwers. There is no known source for some of these detailed suggestions, as the modern commentator notes, but instead of speculating on a lost text he agrees with Dain in crediting "Heron's own ingenuity."[24]

The author offers a method for mounting walls without a ladder, a kind of *sambuca* but tubular, shielded by hides, with a protective door, and mounted on two vertical beams affixed to a four-wheeled wagon; only one armed man can fit inside the tube to step down to the top of the enemy wall, but when the top is up, the bottom of the tube touches the ground, and other soldiers can climb up inside it to reinforce the first stalwart.[25] The exact dimensions of each component follow, along with the suggestion of a bigger device with a tube of larger diameter—but no numbers are supplied—to allow two armed men to fit inside, side by side, instead of just one.

Then the author seamlessly proceeds to make another suggestion: "The [protective] doors . . . together with the front part of the tube, should have a frightening facade with deep carvings and polychrome painting, depicting a fire-bearing figurehead of a dragon or lion; this leads to terror and fear among the enemy."[26] That reminds us in what era the text was written, or perhaps of whom the Byzantines were fighting—both sophisticated enemies as little frightened of painted dragons as themselves, and also primitives who might be shocked by the sudden elevation of the machines, even before they see that fire-breathing beasts are attacking them. This is for a siege engine, and a siege presumes a fortified city, not an encampment of primitives, but the tenth-century context is the struggle to recover cities from the Muslims, and by then it was mostly Turkic warriors from Central Asian steppes who did their fighting, not urban sophisticates.

The work closes with the famous raft of Apollodorus, a little longer than the width of the river to be bridged so that the current swings it neatly into place on the far bank, and which the *Anonymous Treatise* had long since condemned as unfeasible (19.40 ff). Here it is presented without attribution and with emphasis on a clever accessory, a hinged wooden rampart from behind which the troops are to launch their missiles, until the rampart is dropped flat and they launch their assault onto the far bank.[27]

De Obsidione Toleranda: A Manual on Resisting Sieges

Siegecraft was important in the tenth century, and especially in its earlier part and not only offensively: in July 904 an Arab fleet under the convert Leo of Tripoli, known in Arabic sources as Ghulam ("Slave Soldier") Zurafa or Rashik al-Wardani, captured Thessalonike, second largest city of the empire, after a siege of just three days. It was a stunning surprise and a very major defeat. Obviously the city was unprepared.

This disastrous episode is mentioned in, and may conceivably have inspired the author of, the tenth-century instruction manual on defensive siege warfare conventionally known as *De Obsidione Toleranda,* which is now available in a valuable annotated edition.[28] This didactic text is addressed throughout to an imagined general or strategos, with both political and military authority.

At the outset the general is told that there is no need to capitulate even if he anticipates a lengthy siege—implicitly a siege that outlasts his food and water. The enemy might be riven by quarrels, other powers might intervene, the besieging army might exhaust its "wheat," and pestilence "may occur when large forces remain in one place for a long time," and other fortunate events might happen.[29] In other words, first comes the will to resist and then the required logistics, which are next examined in considerable detail.

Provisions must be accumulated for up to a year for noncombatants as well; if that is impossible because of the lack of state funds with which to purchase them, bad harvests, lack of transport, or enemy pillage, the merchants and the wealthy must participate in the distribution to all of one month's supplies of wheat, barley, and legumes taken from public and private stores. But the greater remedy is an organized evacuation to a safer and presumably well-supplied location of "old men, the ill, children, women, beggars."[30]

Likewise, the destruction of "beasts of burden [donkeys] and horses and mules and whatever is not essential for the army [?]" is prescribed because they become "agents of destruction for besieged cities, [by] using up provisions."[31]

The counterpart to provisioning is to deny provisions to the enemy by reaping the fields "even if they are not ready for reaping, and to remove everything useful . . . not only livestock but also people, . . . and it is necessary to poison the rivers or lakes or local wells. . . . It is necessary to poison the rivers upstream from the camps at the lunch hour, in order that when the heat is burning the bodies of the enemy, worn out with toil, when they [drink], the . . . water . . . will totally destroy them.[32] Poisonous berries, roots, and seeds are common in the Mediterranean lands, as elsewhere, but only a few are sufficiently toxic to be of any use in dilute form to poison water; one such is Pseudaconitine ($C_{36}H_{49}NO$), a powerful alkaloid found in dense concentrations in the roots of common and pretty Aconite flowers.

The author offers a long list of the technicians and craftsmen who are to remain and start producing shields, arrows, swords, helmets, spears, javelins, and siege artillery: *tetrareai, magganika, elakatai, cheiromaggana*—all terms that cannot be interpreted with high confidence but which must refer to the known types of stone-throwers and arrow-launchers.[33] Also needed were projecting-beam stone-droppers and grappling irons, as well as the *epilorika* surcoats already encountered and the thick felt headgear, *kamaleukia,* which were provided as substitutes for more costly metal helmets. All this production will require raw materials: iron, bronze, wet and dry pitch, sulfur, tow, flax, hemp, pine-wood torches, wool, cotton, linen, boards, saplings, cornel trees (essential for sturdy pikes, *menavlia*)—with some of the quantities determined by standard issue criteria: ten javelins per javeliner; fifty arrows per archer (a much smaller number than in field conditions—but in sieges arrows could be aimed much more carefully) and five spears (*kontaria,* not the heavier *menavlia*) for each spearman.[34] A well-wooded region is assumed by the author, because each inhabitant is threatened with the death penalty (!) if he fails to gather and store enough firewood for the duration specified, while there is an injunction to gather brushwood and willow branches to weave the *laisai,* anti-arrow screens.

In what amounts to a checklist, an essential precaution is to search out and secure any tunnels, such as forgotten aqueducts or sewers, which could compromise the entire defense of walled towns under siege—the examples of Caesarea of Cappadocia, Naples (from Prokopios), and ancient Syracusa are mentioned.

By contrast, the walls are to be pierced by many loopholes, not only for arrows but also to allow the defenders to push back ladders with spear shafts.[35]

There is to be a water-filled ditch, that is, a moat, or better two or three of them, each with its own palisade and outwork (made with the upcast), especially useful when there is no cavalry within the city that needs to sally out. But if there is, strong oak bridges will be needed.[36]

Bells are to be provided outside the battlements to sound emergency warnings of stealthy incursions—in the event that guards fail to report them discreetly, because of negligence or treason (!)—a recurring preoccupation in this text; in the event of festivals—evidently not entirely devoted to sober devotions—the general himself must oversee the watches, sound advice.

When it comes to the training of the garrison, the advice is analogous to that found in the *Strategikon,* but for an appropriate emphasis on projectiles: *riktaria* (javelins), hand-thrown stones (effective enough with gravity), slings, and arrow-firing and stone-throwing artillery.

Siege warfare need not be only reactive, to repel attacks: the author recommends that ambush parties be stationed outside the gates, presumably when they are not closely invested by the enemy, and on a larger scale, that outlying infantry and cavalry forces should be established in "suitable places"—mountains, it is later specified, offering concealment and obstacles—from which they can descend out to "harm the enemy and not allow them to prosecute the siege with impunity."[37] These forces can also combine with any incoming allies or attack enemy convoys bringing food for the besiegers. But if things go so very well that the enemy camp can actually be encircled and attacked, then the Byzantine difference emerges: instead of urging a battle of annihilation in the classic Roman style, the author writes, "it is necessary to leave the enemy a place through which escape is easy, lest, completely surrounded and despairing of safety, they resist to the death."[38]

After describing raiding tactics, covert sallies through tunnels, exits through posterns and others such, all of which presume a leisurely siege, after describing stalwart fighting on the walls, the author turns to the grimmer fare of close investment: "And if it happens—and I pray it does not—that the ditches are filled in and they bring up the rams at that point, build an additional wall; for there is nothing which can stand against the momentum of the ram."[39]

That is excellent advice—if a new wall, preferably fronted by a new ditch, can be built fast enough behind the shelter of heavy mats to protect the builders from arrows. But the author then recalls the standard remedies: sacks filled with chaff to absorb the blows, grappling irons to deflect the ram, hooked ropes to pull up the ram beam, heavy stones, siphons for Greek fire.[40]

In the capture of Thessalonike of 904, the seaward walls were attacked directly from the sea by the use of ships fitted with artillery and elevating ladders. For this, the remedies of Archimedes as described by Polybius are recommended: powerful stone-throwers to damage the approaching ships, then heavy stones dropped from projecting beams when they reach the seawall, grappling hoists to lift ships right out of the water, and of course arrows against their marines on decks, all as discussed in Heron's *Parangelmata Poliocertica*.[41]

The author is undeterred by the antiquity of these remedies—on the contrary, he interrupts the long narratives drawn from Polybius (the siege of Syracusa), Arrian (the siege of Tyre, and of Sogdiana), and Josephus (the siege of Jerusalem) to assert that they would work even better than in ancient days because contemporary enemies ("foreign peoples in our time") are much less accomplished than their predecessors of the days of Alexander or Titus, who could mount sieges on a much larger scale.[42] The author then finds reassurance in this for the contemporary defenders of cities: in spite of the strenuous and skillful efforts of the ancient besiegers, the besieged were nevertheless often able to resist.

It is evident that the author was striving to raise morale above all. That in itself indicates a practical purpose. This was not a literary exercise, in spite of its extensive quotations from the revered ancients, nor was it a case of playing soldiers—there is a sense of urgency in this poorly organized text.

A much shorter text on defensive siege warfare, first edited and published by Dain as *Mémorandum Inédit sur la Défense de Places*,[43] was a tenth-century derivation from the same lost source as *De Obsidione Toleranda*, according to Dain. Indeed, this severely practical work reads very much as a series of extracts from a fuller work. With no prologue and no narratives of ancient sieges in between, it amounts to a stark sequence of thirty-two injunctions ("Be aware that . . .") on a great many things, including the training of workmen "who are useful to a besieged city"; the preparation of artillery and of stocks of arrows; raising the height of walls behind the protection of anti-arrow *laisai;* the use of ship masts or large poles bound together to ward off enemy ships from sea-walls;[44] and the need for the general to tour the wall when it is under attack, along with an elite contingent ("valiant soldiers") to serve as his personal operational reserve "to give aid to a section in difficulty."[45]

That is truly one of the eternal verities of warfare worth restating at frequent intervals because it is counterintuitive. Any siege implies that the besiegers are a stronger force than the besieged, otherwise the latter would come out to fight off the attackers. Yet the general is to weaken an already weak garrison by taking away elite soldiers to form his personal mobile reserve. That is illogical and makes sense only in dynamic terms: by arriving with his operational reserve at wall sectors under the strongest attack, the general can counterconcentrate the enemy, improving that sector's balance of forces. There is also a psychological dynamic: just when the enemy thrust against the wall seems to be prevailing, emboldening the attackers and intimidating the defenders—thereby further shifting the local balance of strength—the general arrives with his valiant soldiers to reverse the situation psychologically as well as materially.

Further injunctions concern siege engines, some as simple as well-anchored heavy poles with sharpened points to "ward off machines";[46] familiarization-training for night combat (not to be tried without it); the necessity of offensive action even on a small scale; the usefulness of vertical iron gates that can be dropped down suddenly to strike at the enemy underneath; the necessity of locking up the women in their homes "and not to allow their weeping to weaken the spirit of the fighting men" (yet women are frequently recorded as active participants in ancient sieges, doing everything from digging trenches and dropping stones to exposing themselves to taunt the attackers); the need to guard against enemy tunneling—the dubious device of using thin bronze plates as amplifiers is recommended—"[the general] is to put his ear to these and listen";[47] and there is a final injunction on the need to keep an eye on the strongest as well as the weakest sectors of the perimeter where the fortification seems most formidable, "for many cities have been taken at unexpected positions."[48]

There is only a partial modern edition of *Apparatus Bellicus,* another tenth-century compilation that includes twenty large extracts from Julius Africanus,[49] just one of the texts that were written and presumably read at the time. Most of their contents are extracts, paraphrases, or elaborations of earlier writings, but they are nevertheless evidence of the vitality of Byzantine military culture.

CHAPTER 13

Leo VI and Naval Warfare

Leo complained that for his writing on naval warfare (Constitution XIX), he could find no ancient texts to copy and was therefore forced to rely on the practical knowledge of his naval commanders. One could hardly find a better example of the slavish textualism—if the word be allowed—of the Byzantine mind that coexisted with ample pragmatism, and indeed even with transgression. (Leo himself famously took his concubine Zoe of the coal-black eyes, *karbonopsina,* as his fourth wife, contrary to canon law, in order to legitimize her son, the future Constantine VII, born in the imperial bedchamber, *porphyrogennetos,* but to an unmarried mother.) In a greater sin, perhaps, Leo improperly claimed the invention of the hand grenade, that being Greek fire in a pot, of which more below.

The substance of Constitution XIX begins with an echo of Syrianos Magister: the commander is enjoined to study the theory and practice of navigation, including the forecasting of the winds by observing the movement of celestial bodies—accurate wind forecasting would indeed have been most precious intelligence, but unobtainable by the recommended method.

Next there are vacuous generalities on how warships should be built, not too narrow, not too wide. From the sixth to the tenth century and even later, that would be the *dromon* ("runner") in one of its many versions, but all with a single mast, two decks, propulsion by both oar and sail, and aphract—no top deck over the upper bank of oarsmen.[1]

Standard designs ranged from twenty-five to thirty-six or even as many as fifty rows of oarsmen on each side of each deck, for a total of up to two hundred oarsmen, and a hundred others could also be aboard, mostly sea-trained infantrymen ("marines"), as well as the ship's captain and officers. It seems likely, however, that a smaller vessel, an *ousakios,* with one hundred oarsmen as the name implies and a marine contingent of thirty or forty, was more common, especially because the upper-deck oarsmen could also fight, unlike the lower-deck oarsmen, who could at most thrust lances through their oar slots to damage enemy hulls alongside. There were also distinctly lighter and faster two-deck ships for reconnaissance and raiding, and also small galleys *(galea)* with a single bank of oars.

The side gangways and rowing positions were protected by detachable shields, and the oarsmen worked their oars directly through the hull without an outrigger or the protection of an oar box. Square sails were replaced from the seventh century by the lateen rig. Rams were still present at the time of Leo VI but were gradually replaced by beaks—over which marines could reach enemy vessels—but naval combat was mostly by missiles: the marines could launch their arrows from an elevated *xylokastron* (wood castle) near the mast, there were also one or more stone-throwers, and *hugron pur*—liquid fire, or "Greek fire"—was hurled in ignited flasks or projected by piston-activated or even pump-fed siphons.

Greek Fire

In romance, even in historiography of middling repute, Greek fire is a mysterious and most formidable weapon, the technological secret of the Byzantines alone, that none could ever emulate, perhaps not even now. At least some Byzantines, or perhaps just one, pretended to believe in the myth. In the manual of statecraft *De Administrando Imperio* attributed to the emperor Constantine VII Porphyrogennetos (912–959), the text suggests a pompous and outrageously mendacious reply if any foreigners should ever demand access to the "the liquid fire which is discharged through [siphons]."

> This . . . was revealed and taught by God through an angel to the great and holy Constantine, the first Christian emperor, and concerning this . . . he received great charges from the same angel, as we are assured by the faithful witness of our fathers and grandfathers, that it should be manufactured among the Christians only and in the city ruled by them [= Constantinople], and nowhere else at all, nor should it be sent nor taught to any other

nation whatsoever. And so for the confirmation of this among those who should come after him, this great emperor caused curses to be inscribed on the holy table of the church of God [Hagia Sophia], that he who should dare to give of this fire to another nation should neither be called a Christian, nor be held worthy of any rank or office; and if he should be the holder of any such, he should be expelled therefrom and be anathemized and made an example for ever and ever, whether he were emperor or patriarch. . . . And he adjured all who had the zeal and fear of God to be prompt to make away with him.[2]

It is remarkable to encounter a warrant for regicide penned by an emperor, or by his loyal scribes, which would seem to further confirm the unique importance of Greek fire, and its possession in absolute monopoly by the Byzantines alone. Actually, by the time this warning was written, the secret was out.

The first extant report of Greek fire occurs in the *Chronicle* of Theophanes under the year 6164 since the creation, that is, 671–672. Vast Arab fleets were converging on Constantinople:

The aforesaid Constantine [IV, 668–685], on being informed of so great an expedition of God's enemies against Constantinople, built large biremes bearing cauldrons of fire and *Dromones* equipped with siphons [to project liquid fire].[3]

Under the year 6165, that is, 673–674, Theophanes also writes of the origins of the invention:

Kallinikos an architect from Helioupolis [Baalbek in modern Lebanon, then newly under Arab rule] took refuge with the Romans and manufactured a naval fire with which he kindled the ships of the Arabs and burnt them with their crews. In this way the Romans came back in victory and acquired the naval fire.[4]

But according to the Syriac chronicle of the Jacobite patriarch Michael, Kallinikos—described as a carpenter—first employed his invention the year before in Lycia, southeast Anatolia:

[He] concocted a flaming substance and set fire to the Arab ships. With this fire he destroyed the rest of those which were confidently riding [at anchor] out to sea and everyone on board. Since that time the fire invented by Callinicus, which is called *naft* (petroleum in Arabic) has been constantly in use by the Romans.[5]

Myths aside, including those uncritically repeated in some modern works, five things are reliably known about Greek fire, whose nature has also recently been clarified experimentally by an eminent Byzantinist who successfully set fire to a harmless sailboat.[6]

First, it continued to burn in contact with seawater. That much is known from the credible report of Liutprand of Cremona (*Antapodosis*, cols. 833–834); he wrote that the Kievan Rus' who abandoned their ships in Prince Igor's failed attack on Constantinople in 941 (Liutprand was there eight years later) "burned as they swam on the waves." That requires no magical compounds: crude oil will burn persistently in water if first ignited, and it was certainly available because it seeps to the surface on the Caspian shore well within reach of Byzantine traders even when it was beyond the limits of Byzantine power. The locals dug shallow wells to lift it out more conveniently. In *De Administrando Imperio* there is a list of localities where there are "wells yielding *naphta*"—that is, crude oil (not the light distillate fraction now called naphtha).[7]

Further, it has been suggested that Greek fire ignited spontaneously upon coming into contact with water. That could have been true if it contained rather pure sodium (Na) or sodium peroxide (Na_2O_2), both of which react violently with water to form sodium hydroxide (NaOH) while generating intense heat. Sodium compounds are as common as common salt (NaCl), but there is no evidence that Byzantine chemistry was up to the task of extracting pure sodium metal, or its peroxide.

Another suggestion is that petroleum was mixed with pine resin to make it more viscous and "sticky," thus forming a kind of napalm.[8] In preparing modern napalm—something one may comfortably do at home—palm or other oils are added to much lighter gasoline jelly to make it more sticky, but crude petroleum is already more than viscous enough without the need of resin.

More credibly, if resin was present at all, it served to facilitate ignition, because crude petroleum will burn vigorously but is not as easily ignited as its lighter fractions, such as gasoline. With resin, moreover, the temperature of the flame is higher.

Second, all accounts agree that Greek fire was primarily projected against its targets by siphons—tubes with an internal piston that is rammed forward to eject the liquid through a nozzle. To do that, however, the liquid first had to be warmed, confirming that it consisted entirely or largely of crude petroleum, which is too viscous to be efficiently ejected unless first heated, just as in modern pipelines oil is heated for a better flow if too waxy. Hence, to use Greek fire its containers had to be heated by fires kept going inside the hull not far from the siphons—a tricky proposition in wooden ships.

Third, the combination of the siphons' very short range—it is the technology of a child's water pistol, twenty yards would be much—and the need for internal warming fires, plus the probable need to ignite the fluid, required precise movements to approach enemy ships close enough, while staying out of boarding range—and also *very* calm waters. Again that is documented by Liutprand (*Antapodosis,* cols. 833ff): "God . . . wished . . . to honor with victory those who . . . worshiped him. Therefore, he quieted the winds and calmed the sea. For otherwise it would have been difficult for the Greeks to shoot their fire."

Fourth, it follows that Greek fire was primarily effective in the calmer waters of the Sea of Marmara rather than in open sea, particularly when the Byzantines were too outnumbered to prevail by ramming, by projectiles, or by boarding. Hence Greek fire was primarily useful as a defensive weapon against enemies strong enough to attack the empire at its core, rather than as a strategically offensive weapon on the high seas against weaker enemies. That circumscribes the overall importance of Greek fire for Byzantine naval power, which owed infinitely more to sound Roman traditions.

Fifth, the secret of Greek fire was not preserved for long. Arab sources discuss it soon enough, and it was used in the Arab conquest of Crete circa 824–826.[9] Petroleum, which seeps to the surface in the Caspian shore near Baku and the Kirkuk area of modern northeast Iraq, had always been known, while by the ninth century Abbasid scholars had translated the Hellenistic technical work that explained how to make siphons, the *pneumatica* of Hero of Alexandria. Neither petroleum nor siphons could remain a mystery to the Arabs once they were demonstrated in action. Both Greek fire and siphons are recorded as having been used by the fleet of Leo of Tripoli in the assault on Thessalonike in 904, and they were probably used by Arabs much earlier.[10] Conversely, that the enterprising and innovative Italian seagoing republics of Amalfi, Genoa, Pisa, and Venice never adopted Greek fire reveals its limited military value, a function of the short range of siphons and the difficulty of using it in projectiles.

The Dromon

By the standards of the time, the dromon was a fast and maneuverable ship, but that was due to its shallow draft and light structure. The vessel had a low freeboard, as little as one meter, and therefore poor sea keeping—it could be swamped by two-meter waves, not that rare in the

Mediterranean even in the warmer months. That made any prolonged open-sea crossings dangerous at any time of the year, and virtually ruled out winter navigation. Propulsion under oars could be very fast in short bursts of twenty minutes or so, up to ten knots, that is, 11.15 statute miles or 18.5 kilometers per hour—and this could be very useful in combat. Cruise speeds under oars of up to three knots could be kept up for as long as twenty-four hours by rowing in shifts. Under sail with favorable winds astern, speeds could exceed seven knots, but not much headway could be made by tacking into the wind, given the lack of a proper keel—but in any case the low freeboard and oar slots meant that the dromon could be swamped by 10 percent angles of heel.

Because of its long, thin, shallow-draft design, there was little room aboard for stores, including the water that was usually needed in large amounts. The minimum requirement was half a gallon per man per day, with twice that for hard-rowing oarsmen. Decks had to be kept clear, allowing no extra water stowage on deck in hot weather.[11] Given the uncertainties of winds, currents, and enemy action, no prudent captain of an *ousakios* (a dromon with 108–110 oarsmen, not a distinct type of ship) could leave the shore with less than 650 gallons of water, and preferably twice that. Water stowage was therefore the decisive constraint on the endurance of the ships, limiting them to ten days at sea at most but more often seven, while ranges from point to point were diminished by the strong preference for coastwise routes rather than more direct open-sea crossings.

The text begins with an outfitting checklist (para. 5), as trivial and as essential as checklists always are:[12] "there have to be spare rudders, oars, oar-rings, ropes, wooden planks, fuse rope, pitch, liquid pitch and all needed shipwright tools including axes, drills and saws."

Next Greek fire enters the picture, but, interestingly, not as an essential: the text merely advises that it is opportune to have a bronze siphon at the prow to launch fire on the enemy. Over the siphon there should be a platform with a parapet from which trained men can fight the enemy in hand-to-hand combat in addition to launching arrows or other projectiles (darts, sling rounds). On large ships there should be fighting towers—not just one *xylokastron*—from which the soldiers can hurl big stones, sharp-sided maces, or ignited pots of Greek fire.

In defining a standard dromon for his navy, Leo specifies that there should be at least twenty-five rowing benches on each side on two decks, for a total of one hundred men. Every warship must have its captain, ensign, two helmsmen, and first officers, and also an assistant to the captain. One of the last two oarsmen at the stern is in charge of the pump, and the other of the anchor. There should be an armed officer at the prow to lead the fight there while the captain—who also commands the fighting force—should remain at the stern, visible to all aboard but protected from arrows. From there he can command both the fighting and the maneuvering of the ship.

Larger ships could be built with two hundred men or even more, with 50 oarsmen in the lower deck and 150 armed to fight the enemy—but presumably also oarsmen in part. Smaller, very fast warships with a single bank of oars are used for exploring and generally when speed is needed.

Auxiliary ships must be fitted out to transport cargo and horses. The latter needed specialized techniques—hoists, underbelly slings to avert injuries in rough passages, bandaging, feed with added olive oil— all of which were by then very ancient: specialized horse transports (*hippagogos, hippegos*) are attested from 430 BCE.[13] More generally, transport vessels are to carry all military material so that warships will not be loaded down. They can supply food, weapons (extra arrows especially), and other necessities.

Auxiliary ships need to be equipped not only for navigation but also with bows, arrows, and whatever else is needed for war. The upper-tier oarsmen and everyone who is near the captain will be armed from head to foot with shields, long lances, bows, different kinds of arrows, swords, javelins, helmets, and body armor; they should have metal helmets, arm guards, and chest armor, as if they were on the battlefield. Those who lack iron armor should make their own with doubled boiled leather; taking cover behind the front rank, they should launch their arrows and hurl their sling stones. Fighters should not exhaust themselves but instead rest periodically, because the enemy will attack tired soldiers and defeat them:

> Saracens [Muslim Arabs] at first resist the assault. Then when they see that the enemy has become tired and is short of weapons, arrows, stones, or other things, they become insolent, and in tight formation with swords and longer lances they move to attack with much impetus.

The text enjoins the commander addressed throughout to ensure vigilantly that the men are well supplied—for in a state of deprivation they could rebel or engage in extortion against the cities and populations of the empire. If possible, the commander is to ravage the enemy's land to gather abundant food for his men. Justice for the men is a great concern:

the commander is responsible for the fairness of the chiefs under him. On the other hand, none is to ease his service by giving gifts, not even the most ordinary things. "What can be said of your dignity if you think of gifts?" Leo writes. "Do not accept gifts for any reason from those under your orders, be they rich or poor."

From section 22 of Leo's Constitution XIX we learn that there was an imperial fleet based in Constantinople, whose commanders came under a single commander in chief, and separate thematic fleets. But their commanders—the drungaries of the Kibyrrhaeot and other maritime themes—also served under the orders of the commander of the imperial fleet.

Leo recalls at that point that drungaries were once only in charge of auxiliary ships, but now it is the rank of the commander of an entire theme.

In the best Roman tradition, the author advocates vigorous warlike exercises by the marines with shields and swords, and by the ships that should alternate between battle lines, close formations, and head-on attack among themselves: the ships should train in all the ways that the enemy might want to fight, so that their crews get used to the screams and clamor of combat and will not be unprepared for the real thing.

In arranging the camp—crews had to sleep ashore as noted to get a decent night's rest—the commander is enjoined to ensure that the men rest in orderly fashion, without fearing the enemy, and without touching anything that belongs to the indigenous population.

The next section echoes the advice of every Byzantine manual: the commander is to avoid battle. The enemy must be attacked by raids or incursions rather than by the entire fleet or a big part of it, unless there is impellent necessity. Entanglements that can lead to a major battle should be avoided—fortune is mutable and war is full of unknowns. The commander must not be provoked into combat; when warships are very close, combat can be impossible to avoid, hence the commander must keep his ships away—unless he is certain that he is superior in the number of ships, in their weapons, and in the courage and readiness of his men.

If the course of the battle requires it, the commander is to deploy the warships in open order in scattered locations. If he is convinced that his force is superior and therefore seeks battle, the commander should still not attack in his own territory but rather near enemy territory, so that enemies will prefer to flee to their own land instead of fighting.[14] The commander is warned by Leo that "every soldier is fearful when combat is about to begin, and is tempted to find safety in flight, abandoning his weapons." Leo ruefully writes that few Romans or Barbarians prefer death to a dishonorable and shameful flight.

The day before battle the commander is to decide with his officers the line of action to be followed, and the strategy that seems best; he is then to ensure that his ship captains will faithfully execute his orders. If then, because of enemy action, a different plan is called for, all will look to the commander's ship and must be ready to receive whatever signal is necessary; at the signal, all are to strive to fulfill what orders it entails.

The commander is to have the best ship, superior to the others in size, agility, and robustness; it is to be manned by selected fighters. That selected ship is identified as a *pamphylos*, evidently larger than the ordinary dromon of the time. In the same manner, the subordinate commanders should also choose the best men and keep them on their ships. All will look to the commander in chief's ship during the fight, and will receive their orders to carry out the plan from it.

The signaling gear is to be placed high on the deck, with a flag, a torch or any other device to communicate what needs to be done, so that others can receive word of the movements intended, of the decision to fight, or to withdraw from the fight, of whether the fleet needs to deploy out to look for the enemy, or rush to help a garrison that has been attacked, of whether it is necessary to slow down or to increase speed, set up ambushes or avoid them, so that all the orders signaled from the commander's ship will be carried out. Leo explains that all of the above is necessary because as soon as the fight begins it will not be possible to receive commands by voice or by trumpet because of the cries of the men, the sounds of the sea, and the clashing of boats.

Leo explains that the signal can be shown upright, inclined to the right or to the left, agitated, lifted, lowered, removed or changed in its figures and colors. The commander is to ensure familiarity with those signals so that all his subordinate (flotilla) commanders and all the ship captains have a reliable knowledge of them, and all will understand the same thing at the same time, and will be ready to recognize and execute what is signaled to them.

The author next turned to tactics. The commander is to deploy the fleet in a crescent moon formation with warships on each side as the horns, while the strongest and fastest ships are in the front of the center of the half-moon. The command ship is to monitor everything, issue orders, manage the action, and if reinforcements are needed, to send support to that part of the formation. The crescent moon formation is said to be extraordinarily effective to encircle the enemy. Sometimes the commander will be able to deploy the fleet in a line ahead, to attack the prows of the enemy ships and burn them with the flames of the Greek fire siphons. Sometimes the fleet will be deployed in two or three ranks depending on the number of warships; after the first rank has engaged the enemy, the second rank will attack the now tighter enemy formation from the flanks or rear, so that they will not be able to resist the attack of the first rank.

Naturally, stratagems are to be employed. If enemies attack when they see that the Byzantine fleet is small, fast and agile ships are to simulate flight; the enemy will chase them at maximum speed without catching them, then other warships with fresh crews will assault the enemy and seize them—even if the best-trained and strongest enemy ships escape, they will take the weaker and less trained. Then fighting till night with the enemy in tight formation, other fresh ships, strong and capable, are to join the battle in all its violence. All of this will happen when the commander can overtake the enemy in numbers and in capability.

Next, there is advice on what to do when lacking numerical and qualitative superiority—the normal condition of the Byzantines at sea when the work was written, because the fleets of jihad were amply supported by the taxes and donations of the vast hinterland that had come under Muslim rule.

Sometimes by simulating flight with fast ships the commander will provoke the enemy to pursue his ships once they have turned stern. In the excitement of the pursuit, the enemy will break their formation. Then by inverting course, the commander will attack the strung-out enemy and with two or three ships against each one of theirs, he will win effortlessly.

The commander is told that he should engage in naval battle against the enemy when it has suffered shipwreck, or is weakened by a storm, or when its ships can be set on fire during the night; the commander is to attack when enemy crews have gone ashore or whenever circumstances are especially favorable.

Implicit in the above is that in normal conditions the commander should not engage in battle—the usual Byzantine advice, given the impossibility of truly decisive battles. Techniques, "kill mechanisms" in modern parlance, are the subject of the next passages. Leo writes, "Many are the means of destroying warships and sailors that war experts have invented both in the past and recently. Of the latter kind is fire projected by siphons that burns ships with flames and smoke."

Bowmen at the stern and at the prow of ships can launch small arrows known as mice (or "flies," *myas*). Also mentioned is that some keep in vases and launch into enemy ships poisonous snakes, scorpions, and other dangerous animals that will bite and kill the enemy.

It is improbable that this happened often, but the next device is more practical: throwing vases full of quicklime. When the vases break, a gas is emitted that can choke. Other projectiles mentioned by Leo include iron balls studded with sharp points that when thrown onto enemy ships can become a notable impediment to further fighting. Vases full of Greek fire already aflame are to be hurled onto enemy ships—when they break they will start fire. The commander is also told to use hand siphons that the soldiers can hide under their bronze shields; already filled with Greek fire, they can be hurled against the enemy. A different approach is to use hoists to drop down weights, burning liquid pitch, or other materials onto enemy warships after having rammed them.

The commander is instructed that he can destroy the entire enemy fleet if he brings his own ships next to the enemy's ships, and then has other of his ships arrive to ram the enemy ships from the other side. The first lot of ships should retreat slowly, and then the ramming can sink the enemy ships. The commander is warned to be alert not to have the same thing happen to him. Also, the oarsmen in the lower deck can thrust long lances through the oar slots. In addition, specialized tools and pumps should equip the warships, so that enemy ships can be filled with water by way of the lower bank of oars.

But there are more recondite techniques that Leo does not want to specify because they are too sensitive:

> There are also other war strategies invented by the ancients which because of their complexity can only be partly described; and here it is best not to recall them to avoid their becoming known to the enemy who would use them against us. Once known, these ruses of war can easily be understood and elaborated by the enemy.

The text was indeed translated into Arabic.[15] After discussing larger ships, Leo VI turns to the need for smaller vessels, writing that there should also be smaller and faster warships that can capture enemies that pursue them and that cannot themselves be caught and attacked. These ships should be kept in reserve for particular combat situations. The commander is to prepare large and small warships according to the enemy that is to be fought. The fleets of the Muslim Arabs and of the Kievan Rus' are different: the Arabs use rather large and slow warships, while the Rus' use lightweight ships that are small and agile, for they reach the Black Sea navigating down on rivers, so they cannot use large ships.

Manpower management comes next, especially important because sailors, even complete ship crews, can easily defect—and the Muslim Arabs had both great need of sailors and marines and also the means to reward them. At the end of the war, the commander is to distribute the booty equally, and is to prepare lunches, banquets, and feasts. He is to reward with gifts and honors those who behaved like heroes, and severely punish those who behaved in ways unfitting to military men.

In conclusion there is further emphasis on the importance of the human factor: the commander is warned that a great number of ships will be of no avail if their crews lack courage, even if the enemy are few but brave. He is reminded that war is not measured in the number of men: "How much harm can a few wolves inflict to a numerous herd of sheep?"

Naval Strength in Byzantine Strategy

On land even the best-trained troops with the best tactics could be overwhelmed by a mere mob of warriors, if large enough. Not so at sea, where no warship can function at all without the required minimum of trained teamwork, and where a well-exercised fleet could prevail against any number of incompetently operated or poorly outfitted enemy craft.

The qualitative advantage of the imperial navy was therefore more consequential than that of the army—both could be qualitatively superior, but only in the case of the navy could that relative superiority result in the absolute destruction of the enemy fleet.

This was just as well because the interior land masses of the empire, chiefly Anatolia and the Balkans after the loss of Egypt, were much less important economically and politically than its coastal plains and coastal cities, including Constantinople of course, the large islands of Crete, and Cyprus as well as Sicily, the numerous small islands of the Aegean, and the mountainous promontories of very difficult access except by sea.

Besides, overland travel along the coastal plains was interminably long, either because of all the twists and turns of shorelines with their gulfs, bays, and inlets, or because even linear distances were very great: in the sixth century, when the conquests of Justinian had extended the empire's original portion of the Mediterranean's southern shore beyond Cyrene (eastern Libya today) to reach all the way west to Tingis (Tangier), thus giving it the entire North African coast, it would have taken at least three months to walk the four thousand and more kilometers, and it would have been ruinously expensive or simply impossible to transport goods that far by cart or pack mule. Except for incense and spices, precious stones and others such, any commerce more than very local was likely to be seaborne—and navigation in reasonable safety required a navy.

But safety was a commodity never to be had at sea. In 960, Crete would be conquered from the Muslims by the future emperor Nikephoros Phokas, but two previous expeditions in 911 (probably against Syria first) and 949 were defeated. Their muster lists happen to have survived as appendices to the compilation now known as *De Cerimoniis* by Constantine Porphyrogennetos, and they give us some idea of the empire's expeditionary capacity at the time:[16]

In 911:

The imperial fleet: 12,000 sailors and marines; + 700 Rhos mercenary ("Varangian") guards
To be sent by the strategos of the theme Kibyrrhaiotai: 5,600 sailors and marines + 1,000 reserves
To be sent by the strategos of Samos: 4,000 + 1000 reserves
To be sent by the strategos of the Aegean islands (Aigaion Pelagos): 3,000 + 1,000 reserves
Total of sailors, marines, and reserves: 28,300
Imperial ships: 60 dromons with 230 oarsmen and 70 marines each; 20 larger *pamphyloi* with 160 oarsmen each, 20 smaller *pamphyloi* with 130 oarsmen each[17]
Kibyrrhaiotai thematic ships: 15 dromons as above; 6 larger and 10 smaller *pamphyloi*
Samos thematic ships: 10 dromons as above; 4 larger and 8 smaller *pamphyloi*
Aegean islands' thematic ships: 7 dromons, 3 larger and 4 smaller *pamphyloi*
From the theme of Hellas: 10 dromons as above
Army of the Mardaites: 4,087 officers and men, 1,000 auxiliaries

The strategos of the Kibyrrhaiotai and the *katepano* (one rank below strategos) of the Mardaites are to send scouting ships to observe Syrian ports to determine if any fleet is preparing to sail from there (which could counterattack the expedition, or threaten imperial possessions elsewhere).

The theme Thrakesion is to supply 20,000 modioi of barley (also used as horse feed) 40,000 modioi of wheat and biscuit, 30,000 modioi of wine, and 10,000 animals (sheep?) for slaughter and other supplies.

For the expedition of 949 there is a different list of ships and crews, but there is also detailed information that is missing in the 911 list on the equipment of each dromon:

70 klibania (sleeveless corselets—lamellar breast armor)
12 lorikia (lighter body armor) for helmsmen and Greek fire siphon operators
10 other lorikia
80 helmets (implying 80 marines aboard)
10 helmets with visors (for officers?)
8 pairs of arm guards, tubular—vambraces—(for siphon operators?)
100 swords
70 light shields of cloth
30 metal shields *(skoutaria ludiatikai)*[18]
80 trident lances
20 long, light bladed rigging cutters *(longchodrepana)*
100 pikes *(menavlia)*
100 throwing spears, javelins *(riktaria)*
50 compound "Roman" bows
20 crossbows
10,000 arrows (these are "imperial" arrows in reserve, additional to individual kits; 240,000 arrows were purchased for the entire expedition)
200 short arrows ("mice/flies") (the number is too small—20,000 would make sense—they were used for longer-range harassment)
10,000 caltrops
4 anchors with chains
50 surcoats *(epilorika)* to protect the bows of the bowmen from wet weather
50 signaling flags *(kamelaukia)*
Equipment (bolts, weights, chains . . .) for artillery: 12 *tetrareai, lambdareai, and manganika*

Much more follows in the lists for the 949 expedition, including "as many leather shields as God may guide the holy emperor to provide,"[19] as well as battle axes both double-bladed and single-bladed (for throwing), slings, Greek fire siphons, processed materials: lead sheets, hides, nails, bolts of cloth, and unworked raw materials for expedient equipment: bronze, tin, lead, iron, wax, linen, hemp, cables to be worked with tools: crowbars, sledge-hammers, mattocks, pins and spikes, fasteners, braziers, rings, clamps, shackles, and more, each in specified quantities. The amount of money allocated for each item is also listed; evidently there were administrative offices in the imperial palace with the technical expertise to compile comprehensive inventory lists, and the financial expertise to know what everything should cost, e.g., 88 nomismata (coined at 72 to the pound of gold) for 122 ox hides, or 5 nomismata for the purchase of 385 oars.

The Byzantine navy of galleys and embarked soldiers waxed and waned over the centuries in a familiar cycle: security at sea that made its expensive upkeep seem unnecessary was followed by the disastrous arrival of seaborne enemies, which was followed in turn by frantic efforts to build, arm, and man galleys. But until the political collapse of the later twelfth century, which was followed by the Latin conquest of Constantinople in 1204, the Byzantine navy through its up and down cycles always remained powerful enough when it was most needed. In the great crisis of 626, when the Sasanian armies of Khusrau II (Chosroes) had already conquered the entire Levant and Egypt, and were menacing Constantinople from the Asian shore, the Avars besieging the great Theodosian Wall on the European side sent their Slav subjects with their handy boats into the Golden Horn, to attack the seawall and to cross over to the Asian side in order to ferry Sasanian troops to join in the attack on the Theodosian Wall. According to Theophanes, the *monoxyla*[20] of the Slavs: "filled the gulf of the Horn with an immense multitude [of Slav fighters], beyond all number, whom they had brought from the Danube."[21]

They had numbers on their side but not quality. The boats and their occupants were destroyed by the rams and bowmen of the Byzantine galleys. According to the Armenian history of Sebeos:

> The Persian king . . . commanded his army to cross by ships to Byzantium. Having equipped [ships] he began to prepare for a naval battle with Byzantium. Naval forces came out from Byzantium to oppose him, and there was a battle at sea from which the Persian Army returned in shame. They had lost 4,000 men with their ships.[22]

Sebeos was not a naval expert, and nor were the Persians especially maritime. Any actual ships, as opposed to local boats or Slav *monoxyla*, would have had to have been conscripted in the many Levant and Anatolian ports that the Persians had captured by that point; but it is not clear if any or many were so conscripted. It is unlikely that Persians could have built and operated ships in the Sea of Marmara, off the march, so to speak. The contemporary *Chronicon Paschale* under the year 626 describes the fate of the Slavs:

> They sank them and slew all the Slavs found in the canoes. And the Armenians [infantry] too came out from the wall of [the palace] of Blachernae and threw fire into the portico which is near St. Nicholas. And the Slavs who had escaped by diving from the canoes thought, because of the fire, that those positioned by the sea were Avars, and when they came out [from the water] they were slain by the Armenians.[23]

In the four years from 674 when Arab attacks by land and by sea reached their maximum peak, at a time when the Levant was entirely lost, part of Anatolia was overrun, and greater parts ruinously raided, the navy of Constantine IV (668–685) achieved a colossal victory in 678. According to Theophanes, Constantine had prepared well for combat:

> In this year the deniers of Christ equipped a great fleet. . . . Constantine, on being informed of so great an expedition of God's enemies against Constantinople, built large biremes bearing cauldrons of fire and dromones equipped with siphons, and ordered them to be stationed at the . . . harbor of Caesarius [on the Propontis, Sea of Marmara side].[24]

The resulting tactical superiority of the Byzantine navy did not prevent a long and very damaging siege, but it did contribute very greatly to the ultimate defeat of the Muslim offensive.

From the seventh century to the twelfth, the imperial fleet again and again saved the day. It was the *deus ex machina* that came out from its fortified bases recessed into the seawalls on the Golden Horn and the Propontis (Sea of Marmara) to attack the vessels of the invaders.

Sometimes enemy warships were of comparable individual quality—when the Arabs first attacked Constantinople, their ship crews were mostly Christians from the Levant and Cilicia, including former imperial sailors. But even well-built and well-manned enemy warships were outmatched by fleet maneuvers they could neither defeat nor imitate. Those skills were more important than "Greek fire," useful though it was, and they outlived the Arab acquisition of its secrets.

CHAPTER 14

The Tenth-Century
Military Renaissance

After centuries on the defensive, from the middle of the tenth century the Byzantine empire mounted a series of strategic offensives against both the Muslims to the south and the Bulgarians on the northern front, which resulted in large territorial gains in both the Balkans and the Levant. Even before he became emperor, Nikephoros II Phokas (969–976) was a protagonist of this strategic transformation, which continued under his murderer and successor John Tzimiskes (969–976) and culminated with Basil II (976–1025), who expanded the empire in all directions and entirely defeated the Bulgarians to regain the Danubian frontier.

In one way or another, Nikephoros II Phokas was personally associated with a set of field manuals that complement each other much more than they overlap, and are replete with shrewd advice.[1] What makes them especially interesting is that they incidentally reveal much information on all sorts of things, from Byzantine weapons to everyday life on the contested frontier of jihad.

As I have learned from personal experience, the writing of field manuals may reflect any combination of different aims: to provide a moral context for war in the manner exemplified by Onasander; to tame the chaos and confusion of war with a neatly ordered framework of ranks and files, perfectly arranged camps, and so forth, as exemplified by Hyginus Gromaticus and Arrian; and to provide information on techniques that could actually be used in combat, which seems to be the aim of the tenth-century manuals discussed in what follows.[2]

For all their hard practicality, these writings emerged from a specific cultural tradition whose most important, and possibly most bookish, exemplar was Constantine VII Porphyrogennetos. In what reads very much as a personal text, we find him at work. He has set out to write the memorandum headed "What Should Be Observed When the Great and High Emperor of the Romans Goes on Campaign" for his son Romanos:

> Now this subject has been reported upon in past times and discussed by many up to the present day, but it has not been contained in writing, a fact which we have held to be neither just nor good. . . . Having completed a great deal of research, yet finding no memorandum deposited in the palace, we were at last just able to discover one which dealt with these matters in the monastery called Sigriane, in which Leo the *magistros,* named Katakylas, had embraced the monastic life. For this *magistros* committed these things to writing by order of [Emperor] Leo. . . .
>
> But since the magistros was unaccomplished in Hellenic letters, his book contains many barbarisms and solecisms and lapses of syntax, . . . nevertheless it was praiseworthy and accurate . . . [in its contents]. . . . Since we found this work composed in a negligent fashion . . . setting matters forth indistinctly as though in the footprints of a phantom, so to speak, . . . we have written these things down for you in order to bequeath them as a memorandum and a guide.[3]

The document itself is valuable for the incidental information it conveys, especially on logistics, but what it describes is an imperial progress to certain victory rather than a veritable expedition; as von Moltke might have said, it has no tactical value. But there was immense value in the practice of compiling instructive documents, editing them, and preserving them—the practice that allows us now to read the three manuals, not at all free of the barbarisms and solecisms that agitated Constantine but rich in realistic advice.

De Velitatione (Skirmishing)

The Byzantine-Arab borderlands of eastern Anatolia were the geographic setting of the work most recently published as *Traité sur le guerilla* but traditionally known as *De Velitatione Bellica Nicephori Augusti* and now translated into English as *Skirmishing by the Emperor Lord Nikephoros.*[4]

It is a largely original work on the operational method and tactics of defensive border warfare with the Muslim Arabs that owes nothing to the library and everything to actual combat experience—the author tells us as much, and we can safely believe him.

Doctrinally, under Islam all imperial territory was *Dar al-Harb,* the land of war, in which private raiding, insofar as it weakened the infidel, was just as religiously legitimate as volunteering for the wars of caliphs or local potentates—both were equally jihad and would yield either glory (or at least respect) on earth or joyful martyrdom. At the same time, raiding had become a trade in the borderlands, risky of course but evidently profitable—or at least much less laborious than to grow crops or raise livestock.[5] Actually they too were risky activities because of Byzantine raiding.

The borderland had its own religions: jihad-centered Islam—there was a steady inflow of newly converted Turkomans who may have known little else of their new faith—and what one is tempted to call Byzantine crusading, long before the First Crusade. The borderland also had its own literature, exemplified by the epic poem or poetic romance *Digenis Akritas,* one of many "akritic" (*akra* = border) compositions, which was matched on the other side by Bedouin raiding poems that increasingly gave way to Turkic ballads—some of which may still be heard in Istanbul's Istikal Street music cafes, in the form of the Bolu Bey cycle of folk songs.[6]

As for the military culture of the borderlands, the very purpose declared by the author of *De Velitatione* was to preserve it at a time when fortunately there was no need of it, in case it would be needed in the future.

> Christ . . . has greatly cut back the power and strength of the offspring of Ishmael. . . . Nonetheless, in order that time . . . might not blot out this useful knowledge, . . . we ought to commit it to writing.[7]

There is a specific method, and our author writes that he was instructed by "the very ones . . . who invented it"—the reference is to Bardas Phokas, *domestikos tou scholai,* senior field commander and his three sons: Constantine, who died a captive in 953, Leo, who won several major battles, and the son who rose to become emperor Nikephoros II (963–969), and also the latter's nephew, assassin, and successor, John Tzimiskes (969–976).

In *De Velitatione,* the aim is to do much with little, with raids by relatively small forces that magnify their strength by achieving surprise—that is, the temporary nonreaction of the unprepared enemy. Surprise transforms the balance of strength because so long as and in the degree that it lasts, the enemy is transformed into a mere inanimate object that cannot react: it is then very easy to attack in effective ways. If, moreover, surprise can be used to diminish and dislocate the enemy, there is no return to the prior balance of strength even when surprise ends.

Raids are the usual way of achieving surprise, but the defensive equivalent of the raid is the ambush—with it too there is surprise, and the suspension the entire predicament of war, in which everything is so difficult and dangerous because the enemy does react.

Given that not every tract of the frontier can be secured, the first priority is to overwatch the mountain passes to detect incursions as early as possible. In high and rugged mountains, sentry posts are needed "three or four miles apart."[8] The Byzantines, like the Romans before them, could send signals with fire at night and smoke in daylight, but there is an advantage to keeping warning information secret, so that enemy intruders do not know that they have been detected; hence a relay is recommended instead: "When they observe the enemy . . . they should quickly hurry off to the next station and report. . . . In turn, those men should race off to the next station." The relay continues until it reaches cavalry posts "situated on more level ground," which will inform the general.

We can infer that the sentries are thematic soldiers called up for short tours of fifteen days at a time—officers are enjoined to relieve them on schedule. Nevertheless, these part-timers are expected to serve as clandestine (hidden) scouts and even as covert (disguised) agents: "They should not stay in the same station for a long time but should change and move to another place. . . . Otherwise . . . they will be recognized and might easily be captured by the enemy." In the borderlands, men had to be intrepid to safeguard their livestock and their families from infiltrating marauders, and also cunning enough to do some infiltrating and cattle rustling of their own, for otherwise the livestock would all be going one way and there would be none left to protect. But there were also specialists, *expilatores,* legal Latin for violent robbers as opposed to mere thieves, but here clearly in the meaning of scouts—although the editor wisely notes that "in these border areas the distinction was probably minimal."[9]

The text is harsh with the Armenians, more commonly praised for their valor in Byzantine military texts: "Armenians carry out sentry duty rather poorly and carelessly." Monthly rotations, a regular salary, and monthly allowances are all recommended, but "these men are not very likely to perform the sentry duty very well, for, after all, they are still Armenians."[10]

These mountaineers were not native Greek speakers, not quite Orthodox—the Armenian Apostolic Church rejected both the Chalcedonian Creed and pure Monophysitism—and somewhat exotic with their admixture of Persian habits and tastes—each a sufficient reason for mistrust by the very Greek and very Orthodox author. But the *Praecepta Militaria* discussed below commends them. (Another possible reason would incidentally date the work as post-976: John Tzimiskes, who killed and succeeded Nikephoros II Phokas and died in 976, was of Armenian origin; so was the Phokas family, whom the author greatly admired, but Tzimiskes, Armenian for "shorty," was more recently Armenian.)

In any case, good information is the key, and the author recommends the use of both spies and light-horse scouts—the *trapezites* or *tasinarioi* (elsewhere called *tasinakia*), who can capture prisoners for interrogation while doing their own raiding and ravaging.[11]

When incursions can be anticipated, the army is to move out to hold the passes to dissuade the enemy or to beat him back with combined forces of heavy infantry to block narrow passages along with archers and light infantry with javelins and slings in ambush on higher ground; a second line is to stand behind the first. Special care is needed to block lesser paths that the enemy could use to reach behind those defensive forces, as the Persians memorably did at Thermopylae, and countless others also did—because the seemingly formidable advantage of mountain terrain to the defending side is so often a trap, whenever the enemy is determined enough to outflank positions through supposedly impassable terrain.

The author tells us that if all is properly prepared, the enemy will be defeated, or induced to try another and more circuitous route that will wear out his strength, or else demoralized and forced to retreat.

It was rarely possible to successfully predict the timing and direction of enemy incursions, mobilize the part-time thematic forces, and deploy them into their assigned positions ahead of the enemy's arrival.

The Roman imperial solution, at any rate from the end of the first century to the fourth, was to protect each and every segment of the entire frontier from Britain to Mesopotamia—the *limes* that then acquired a physical form—with manned walls, palisades, guarded river lines, or patrolled highways according to the terrain, all strengthened at frequent intervals by forts containing auxiliary infantry cohorts and cavalry "alae," which could in turn be reinforced by the well-armored heavy infantry and artillery of the legions of each frontier province, which could themselves be strengthened by detachments *(vexillationes)* sent by legions stationed in other provinces near or far. This was the magnificent theater strategy of permanent, preclusive defense that long allowed the empire to prosper by keeping out marauders as well as invaders, and which was altogether too expensive to maintain for the Byzantine empire with its greatly diminished resources.

The next best thing to a permanent preclusive defense manned day and night, year after year, would have been a *reactive* preclusive defense, under which adequate forces to keep out incursions would each time be deployed to garrison the specific segment of the frontier that was threatened, ahead of the enemy's arrival. But even this cheaper variant would still have required many more spies and scouts to anticipate the timing and direction of each enemy incursion, and more full-time troops in addition to thematic farmer-soldiers, to be able to man all the threatened segments of the frontier quickly enough.

Even that, moreover, would still have been a purely defensive strategy that simply waits for the enemy to attack, conceding the initiative without any ability to influence the attack beforehand. Therefore, in this case, even with all the additional spies and troops, Arab raiders could still approach imperial territory, discover from their own scouts and spies that the Byzantines were ready to repel an attack, and then react by calling off the raid, or by journeying elsewhere to raid a quite different segment of the long frontier that extended from the Mediterranean to the Caucasus. Either choice would impose costs on the raiders as well—but over time a reactive theater strategy would impose greater costs on the Byzantines, who could not keep mobilizing their part-time farmer-soldiers and keep them away from their homes, fields, and livestock for as long as Arab raiding forces were gathered near the frontier in readiness to invade.

The author therefore recommends an alternative theater strategy of *elastic* defense with a deterrent purpose: instead of trying to preclude incursions—much too hard to do—enemy columns are to be trapped on their way back home. At the price of exposing imperial territory to destruction and depredation, that circumvented the impossible problem of predicting the timing and direction of enemy incursions: it was altogether easier to predict what routes the raiders would follow to cross back into Muslim territory. It also avoided the problem of mobilizing, assembling, and deploying the thematic forces ahead of time, then perhaps keeping them mobilized away from home while nothing happened. Instead, the thematic troops could be summoned only when they were needed, and with plenty of time to deploy into position to intercept enemy columns as they returned home with their captives and their booty:

> Instead of confronting the enemy as they are on their way to invade Roma-
> nia, it is in many respects more advantageous to get them as they are re-
> turning from our country to our own. They will then be . . . burdened with
> a lot of baggage, captives and [seized] animals. The men and their horses
> will be so tired that they will fall apart in battle.[12]

Byzantine forces by contrast will have had time to mobilize, assemble—
even from farther afield—and deploy properly.

The costs of a theater defense so elastic that it protects nothing could
be mitigated, however: the author earlier noted that when incursion
warnings arrive, the civilian population "may take refuge with their animals in fortified locations." The raiders would normally try to take
fortified cities by assault or siege, for cities could yield enough plunder
and captives to justify their time, efforts, and casualties; but only very
hungry raiders would besiege rustic fortifications in high mountains for
the sake of the cattle they contained—though slaves had their value, of
course.

The best protection for much-raided border populations left undefended, especially when their able-bodied males were mobilized into
thematic units called for duty elsewhere, was situational, so to speak:
the siting of towns and villages in inaccessible locations, even if that
meant twice-daily treks to work fields perhaps far below them; virtual
wall circuits, obtained by building the outer circle of houses side by side
with no alleys between them, especially thick out-facing walls and no
openings at ground level; and dense layouts that left only narrow alleys
between buildings in which horsemen would not dare to venture, and
which could be easily blocked. The traces, remnants, and ruins of Armenian and Greek towns and villages in what is now eastern Turkey abundantly illustrate all three protective features; they are still visible because more recent settlements by Oğuz and other Turkomans, Yörüks,
Tatars, Kurds, and Zaza since the twelfth century are mostly sited in
low-lying terrain near streams, having started as semi-nomadic encampments.[13]

Border populations also had other ways to limit damage, though each
had its implicit costs: a preference for livestock rather than field crops
that raiders could burn; a preference for crops harvested in the spring
instead of the summer and early autumn months favored by raiders; ingenious blinds, not only for possessions but for the villagers themselves;
and where the mountains were high enough, the shift of entire village
populations instead of just their livestock to high-mountain summer
pastures.

Without such damage-limitation measures, the theater strategy of elastic defense would have collapsed for lack of manpower, because the cumulative damage of successive raids would have caused the civil population to abandon the region, leaving thematic units without part-timers to call up. Eventually that is exactly what happened, but when the text was written the strategy could still succeed by a cumulative deterrent effect—which the author explicitly identified as such: "Attacking them as they return . . . will instill in them the fear that each time they want to invade, we will occupy the passes, and after a while they may cut down their constant incursions."[14]

Three defeats of "Ali the Son of Hamdan"—that is, Sayf ad-Dawlah—are given as examples of what could be achieved: in 950 he was ambushed by Leo Phokas when retreating from a successful plunder raid that had reached across the river Halys (now Kizilirmak) deep into what is now central Anatolia; in 958 he was defeated by the future emperor John Tzimiskes, who in the process captured Samosata (Turkey's Samsat, until flooded by the Ataturk dam); and in 960 he was again defeated by Leo Phokas, whose older brother Nikephoros was then successfully campaigning to reconquer Crete.

Leo Phokas had few troops. His brother had mobilized a major expeditionary force for his successful offensive, possibly a reason for Sayf ad-Dawla's decision to invade the empire once again, only two years after his defeat by John Tzimiskes. There is an account by the historian Leo the Deacon, whose understanding of the battle may have been enhanced—or colored—by his own experience at the scene of Basil II's rather similar 986 debacle at the Gates of Trajan in Bulgaria—Basil too was retreating after having successfully invaded, only to be ambushed. Leo the Deacon explains how Leo Phokas deployed his forces:

> [Because] he was leading a small and weak army . . . Leo decided . . . to occupy the most strategic positions on the precipices, to lie there in ambush and guard the escape routes.[15]

One might think that the classicizing Leo the Deacon next depicted the oblivious approach of unweary, downright playful Sayf ad-Dawla as an exemplar of hubris literally riding before the fall, except that there is apparent confirmation from the other side in a poem by the celebrated al-Mutanabbi who was on the scene riding with his patron:

> [Ibn] Hamdan, confident and priding himself on the multitude of his followers . . . and bragging at the quantity of plunder and the number of cap-

tives . . . was rushing this way and that, now he was riding at rear of the army, now he led the way, brandishing his spear, tossing it to the winds, and then retrieving it with a flourish.[16]

What happened next is what the author of *De Velitatione* / "Skirmishing" wishes for, a tactical situation in which one man can defeat ten, whose military gains were to offset the costs of a theater strategy that could not defend imperial territory but only deter further attacks:

> The barbarians had to crowd together in the very narrow and rough places, breaking their formations, and had to cross the steep section each one as best he could. Then the general ordered the trumpets to sound the battle charge to make the troops spring up from ambush, and attacked the barbarians.[17]

The outcome was a slaughter—the Byzantines had the added advantage of being well rested while their enemies were tired by their march. Sayf ad-Dawla lost all his plunder and was almost captured—the usual trick of scattering silver and gold to divert pursuers is mentioned.

The text offers specific recommendations on how to implement the strategy, starting with the need to secure water supplies by controlling each spring in the defiles and passes where the ambush is laid.

Next come the tactics.[18] Against the frequent occurrence of raids by horsemen alone *(monokoursa)*, skilled scouts are essential to estimate their numbers from hoofprints and trampled grass and guess their direction; competent officers and troops with good horses are then needed to catch and attack the enemy on the move. These were usually raids for plunder and slaves under thin jihadist pretensions, if that, but the author also recommends preparations against large-scale jihad by volunteers who were far more likely to be religiously motivated *mujahideen* (= those who struggle):

> In [August] large numbers would come from Egypt, Palestine, Phoenicia, and southern Syria to Cilicia, to the country around Antioch, and to Aleppo, and adding some *Arabas* [Bedouins] to their force . . . they would invade Roman territory in September.[19]

In a momentous shift, the jihadis were increasingly not Arab but rather Turkic warriors, many of them *ghilman* (singular *ghulam*). In the Qur'an (52:24, 56:17, 76:19), *ghilman,* or *wuldan,* refers to the "divine youths, forever young, beautiful as pearls" whose intimate services, as those of female concubines *(houris),* are heavenly rewards for righteous Muslims, dead jihadis, and now suicide bombers; but the term later came to describe not Ganymedes but Turkic warriors, the so-called "slave soldiers"—a peculiar indentured status that could be compatible with wealth and power. The tenth-century Ghaznavid empire was founded by the *ghulam* Abu Mansur Sebük Tigin, born circa 942 and sold in Bokhara as a boy, while the enslaved Qipchaq Turk Baibars rose from bodyguard to commander of the guards circa 1250 and then became sultan of Egypt and Syria in 1260 as al-Malik al-Zahir Rukn al-Din Baibars al-Bunduqdari, celebrated victor over both Crusaders and Mongols.[20]

With jihadis periodically gathering to attack the frontier, advance intelligence was the key; in addition to the spies, scouts, and light-cavalry patrols already mentioned, merchants should be sent across the border to find out what they can.

Subversion comes next: the caliph was far away in Baghdad and by then impotent, and even a very active enemy like Sayf ad-Dawla was far behind the border in Aleppo. So letters and "gift baskets" should be sent to the "local emirs who control the border castles."[21]

Next comes the shadowing of the advancing enemy forces, until it can be determined where they will camp. With special dark surcoats over their body armor, the main forces are to approach the enemy encampment sending out scouts ahead, so that they can continue to monitor the enemy advance when it resumes from the camp.[22] The author evidently assumes that enemy numbers are so large that the camp cannot be usefully attacked—not even by night, with some advantage of surprise.

A special method is to be used to monitor the enemy advance effectively. Three selected teams of scouts are needed: one to remain so near the enemy that its men can hear the murmur of massed voices, the second to keep its distance while keeping the first within sight, and the third doing the same in turn. All three teams are to focus on monitoring the enemy, not on communicating their findings back to the *turmarch*—highest military rank below *strategos*, nominally in charge of a *turma* or *meros* of two thousand or more, here used to mean the deputy commander of the operation. He is to get his information from the three tracking teams by way of four-man units who in turn remain within sight of them, sending back two of their number when there is information to report. That way the strategos can slow, accelerate, or redirect the march of his shadowing force to match the enemy's movements.[23] It is dangerous to simply follow the enemy: they are experienced in leaving concealed rear-guards behind them to ambush pursuers.

All this presumes, as noted, an enemy force too large to be attacked in its totality. But one purpose of shadowing the incursion so closely is to be ready to attack promptly if enough raiders in search of booty venture far enough from the "emir's battle formation" to leave it vulnerable. That calls for night movements beforehand, because otherwise the invaders will see the dust clouds of the shadowing force and refrain from separating to raid.

Even if all is done well, the "emir's battle formation" may still remain too strong; in that case, it is one or more of the raiding parties that peeled off that can be attacked. If the latter have their own protective force to secure their rear while they loot—the word in the text is *foulkon*, Germanic for a Roman-style infantry shield wall but here any detachment not absorbed in looting—the officer in command should divide his force into two, to engage the *foulkon* with one while himself leading the other to attack the looters "with great speed and spirit, shouts and battle cries."[24] That would not work against the troops of the *foulkon* arrayed in battle order, but could panic raiders scattered to seize plunder and captives into disorderly flight, to then be cut down by their pursuers.

The author explains why he offers yet more tactical detail: "we will not be the least bit hesitant in writing down what we actually observed."[25]

What follows are detailed variations on the themes of shadowing, pouncing, ambushing, blocking—everything but the head-on battle of attrition, main force against main force, which, win or lose, must cost heavy losses. The empire had no expendable men, only valuable farmer-soldiers who garrisoned the frontier zone by living there with their families, and even scarcer professional soldiers who could not be replaced by new recruits until they were properly trained. Today's fight will be followed by tomorrow's, so casualties lost today will be missing from the battle line tomorrow. By contrast, mujahideen, who could achieve their aim by being killed in battle, could easily be replaced by fresh volunteers from the depths of Islamic territories where the proportion actually engaged in jihad was small, leaving many potential recruits. Even Sayf ad-Dawla and his battle group could be replaced, because mujahideen seeking glory or martyrdom and freebooters seeking plunder and slaves would soon enough find another leader to follow.

It was therefore necessary to defeat the principal enemy in the field, Sayf ad-Dawla and his battle group, in this instance—but there was no advantage in eliminating him altogether because he would soon enough be replaced by another emir with another battle group. That made attrition unattractive, for the casualties suffered were irreparable in the short run, and damaging even in the long run by causing survivors to leave the frontier zone, while the casualties inflicted on the enemy would be quickly replaced by a new influx of volunteers and predators.

Instead of attrition, the text recommends maneuver, to dislocate and disrupt the enemy instead of destroying his units and his men one by one in direct combat. And this is all *relational maneuver* aimed at a specific enemy with specific strengths to be circumvented and specific weaknesses to be exploited.

For example, when enemy forces become separated because horsemen race ahead to plunder, leaving the foot fighters behind, the latter can be attacked advantageously; when raiding parties can be outpaced as they march or ride to destinations that are easily predictable because there are no other plausible targets in that direction, cavalry forces can get there ahead of time, hide well, and wait until the invaders scatter to loot, when they can be cut down or captured; alternatively, or as complements, ambushes can also be laid on the way to such predictable destinations, or from them to finish off enemy bands as they are fleeing.

When enough damage has been inflicted by lesser attacks and ambushes, the time arrives to engage the enemy "battle line," the main force. For that, cavalry is not enough, infantry is also needed both to launch missiles and engage in close combat. If left behind by the maneuvers of the cavalry, infantry troops should strive to catch up before battle is joined. If that is impossible because the distance is too great, some of the "capable" cavalrymen must be ordered to dismount to fight on foot with bows and slings, as well as spears and shields.

While the troops are being prepared for battle, the strategos prepares by setting up his own tented camp with its baggage within sight of the enemy, to induce "consternation and despair" by this display of assured confidence.

The author also has suggestions for a badly outnumbered force. It cannot possibly attack the enemy's main battle group in open combat, but it can prevail by ambush against enemy columns that must pass through the mountains. Infantry is essential, and must be concealed on both sides of the road. The force is small, and it could be demoralized by the spectacle of the advancing enemy army. The commanding officer can reassure the men by his own calm presence. He should be just behind the infantry, "very, very close behind them . . . his own position should be almost in the rear ranks of the infantry."[26]

Ambushes may fail simply because the enemy goes elsewhere. The dynamic ambush is the remedy: a cavalry force runs away from combat to lure the enemy in pursuit. It heads into the prepared killing ground, which may be a double ambush of hidden infantry in one part of the mountain pass, with concealed cavalry waiting in the next, to cut down enemies retreating from the infantry ambush. If it can be concealed, the cavalry can ambush even if there is no defile or pass to canalize the enemy, because any column is vulnerable to a charge against its flank, and more so if men and horses are exhausted by their pursuit of the decoy force.

The reliable countermeasure is to refrain from pursuing fleeing enemies, but that is exactly when the cavalry can be most effective by cutting down soldiers reduced to fugitives.

To predict the enemy's movements accurately is the highest accomplishment of field intelligence. But it can also fail totally. The enemy's "sudden, concentrated attack" can surprise the strategos with only a few troops in hand, with no time to mobilize, assemble, and deploy his thematic forces. The civil population is also caught unprepared, still in its dwellings instead of the area's evacuation fortresses. Interestingly, civilians at risk are the first priority. The strategos is to send officers "with great speed" to get ahead of the enemy incursion to "evacuate and find refuge for the inhabitants of the villages and their flocks."[27] After that, it is back to counter-raiding, to whittle down the enemy's strength.

But that too may fail. The enemy can prudently remain in battle formation ready to repel attacks, and refrain from sending out raiding parties that excessively weaken his main force. In that case, it still possible to act, by detaching small forces ("three hundred or fewer combat-ready horsemen") to ambush the advancing enemy before prudently withdrawing to the main body, which should be secured by fortifications if possible ("if there is also a fortress in the vicinity"). Foot soldiers who cannot find safety in a quick retreat from mounted pursuers need a fortified place for protection, from which they can come out as needed for combined action with the cavalry.

At the very least, there will be ambush opportunities against small bodies of enemy troops, such as advance parties sent out to choose and measure out campsites for the main army that is marching up. Even for that, however, safeguards are in order: the main body should be in readiness nearby, so that if the enemy advance party is strong enough to counterattack and pursue the ambush force, it will be able to ride to safety, leaving the main force to overpower the pursuers "with a noble, brave charge"; and if that too is not enough, because the enemy's advance party consists of many troops and not just a few surveyors, the infantry can come out of its fortress to join in the fight.

To shadow, outpace, and pounce, the defensive army must be highly mobile in a transportation sense, as well as agile tactically. The latter is a matter of training and leadership, but the former depends on good organization and sound procedures. The baggage train *(touldon)* with its wagons and mules is absolutely necessary—the army cannot long operate without the food, additional arrows, javelins, and spare coats, armor, and shields as well as the varied repair and digging tools they carry. Fodder is too bulky to be transported efficiently, though some of it at least might be essential in extremely arid areas. But the baggage train is too slow and must be separated from the fighting force, to remain within a fortress; some combat-ready horsemen are assigned to it, so that they can escort the wagons and mules that come out to a designated place to resupply the fighting force. Two or three days of fodder can be carried on "fast mules" and in the saddlebags of the cavalry.[28]

There are many tricks that can be used to conceal, disguise, minimize, or maximize forces. Cavalry rides out in the late evening, when the dust clouds it raises are invisible; the terrain is closely studied by commanders and by scouts to find blinds for ambush forces, or any forces whose presence on the battlefield should be concealed from the enemy (not the case when dissuading attack); strong forces can minimize their signature by riding very quietly in close order, while some soldiers can be disguised as farmers with heads uncovered, weapons well hidden, and barefoot, with real farmers and herdsmen mixed in among them to serve as decoys lo lure pursuers into ambushes. Weak forces can maximize their signature by trailing branches to raise more dust.

There are also countermeasures against such tricks, starting with advance units of agile light horsemen to trigger ambushes ahead of the more vulnerable main force, or to the contrary, to force the enemy to reveal his strength; and the much-mentioned *saka,* the rear guard that protects the vulnerable wagons and mules of the *touldon* at the back of the army's column, but which is always needed to intercept raiders trying to attack the less alert rear of a forward-facing force.

Attrition is to be avoided, but if there is both a large enemy army and a large Roman army in the field, the enemy cannot just be allowed to plunder at will—it must be engaged. If the enemy does send out raiding columns, they can be ambushed as before by a combination of a small ambush force, a larger reserve cavalry force, and infantry from nearby fortresses if available. But otherwise battle must be engaged.

If the enemy is too strong, the strategos can seek refuge inside the nearest fortified town *(kastron)* but only after he has seen to the safety of all his troops, and of civilians and their livestock too—anything else would be dishonorable, despicable, and would lead to the destruction of the country, its depopulation.[29]

All of the above presumes troops that are especially well trained, more strictly disciplined, and of higher morale than ordinary line forces, because this kind of warfare was far more demanding of the individual soldier than set-piece battle in ordered ranks. In modern times it has been the light infantry and its more elite derivatives that have maintained the distinction.

It is the first task of the strategos to train his soldiers and exercise his units. Aside from individual and group skills, there has to be toughening up:

> There is no other possible way . . . for you to prepare for warfare except by first exercising and training the army under your command. You must accustom them to, and train them in, the handling of weapons and get them to endure bitter and wearisome tasks and labors.[30]

Border warfare called for inordinately long foot marches and long days in the saddle. Morale is very important in every form of combat, but even more so in frontier warfare, because much of the action must be performed quite independently by soldiers in pairs, foursomes, or very small units far from the commanding gaze of senior officers.

The upkeep of morale begins with discipline to repress laziness, slackness, drunkenness, but there must also be incentives: salaries paid on time and food allowances provided on schedule, but also gifts and bonuses beyond what is "customary or stipulated," so that the soldiers can obtain the best horses and equipment, and serve with a "joyous spirit."[31]

The thematic farmer-soldiers must also be respected, not humiliated or even beaten by

> [tax-collecting] manikins who contribute absolutely nothing to the common good, but whose sole intent is to . . . squeeze dry the poor, and from their injustice and abundant shedding of the blood of the poor they store up many talents of gold.

Thematic soldiers received both salaries and land allotments, but had to pay land taxes; accordingly, they were treated as any other taxpayer by imperial tax collectors, that is, very badly according to the text. Worse still was the delegation of tax collection to large landowners, and village-wide collective responsibility for paying taxes.[32] Thematic judges must also show more respect, the farmer-soldiers must not be "dragged off as prisoners and whipped, bound in chains, and—oh, what a terrible thing—pilloried."

Apparently civilian judges had been judging cases and issuing typically harsh Byzantine sentences on the farmer half of thematic farmer-soldiers, while our author insists that it is their officers who should judge them as soldiers, under the authority of the strategos appointed by the holy emperor himself. But he does not rule out cooperation with civilian judges and civil servants, so long as they defer to military authority "as is clear from the law." This part ends with a peroration on the "enthusiasm, happiness and good cheer" that will obtain once "those elements dragging the men into poverty" are eliminated by the emperors.

After this there is more on operational methods on a theater scale, specifically the very interesting case of a "defensive-offensive," mounted in the hopes of compelling the enemy to withdraw when his army is too strong to be fought directly, no matter how cleverly. The author cites the earlier prescriptions of Leo's *Taktika* or *Tacticae Constitutiones* (XI, 25) before himself describing what happened when around the year 900 a huge jihadist army invaded from Cilicia, besieging Mistheia (Claudiocaesarea) while ravaging far and wide. The strategoi of the Anatolikon and Opsikion themes were left behind to defend "as best they could" while Nikephoros Phokas, the senior field commander and ancestor of the future eponymous emperor, invaded deep into Cilicia with a strong mobile army heading for the walled city of Adanes, the modern Adana. The garrison—what was left of it after the departure of the jihadist army—came out to meet the Byzantine force. They were defeated, fled back to the walls, and the laggards were killed or captured. Nikephoros did not invest Adana but instead acted according to the doctrine of *De Velitatione* by destroying the city's agricultural base, chopping down fruit trees and vines, and razing the "elegant and beautiful" settlements outside the walls. The next day, he led his army down to the coast, capturing a "very large number of captives and many flocks," and then turned back for some forty kilometers to the Kydnos or Hierax River, now Tarsus Cay. He did not then attack the city of Tarsos at the mouth of the river, instead returning to imperial territory by way of the Cilician Gates (now the pass of Gülek) that connect the coastal plains with the high plateau of central Anatolia.

When the jihadists still then besieging Mistheia heard of the devastating incursion into their own territory, they turned back and "gained nothing in either place" because they could not catch Nikephoros Phokas as he led his returning army through the mountains.[33]

This was a model operation, everything turned out just right for the Byzantines, and as raids went it was on a very large scale, but the essential concept was valid on any scale and it was often applied: it is harder to defend than to attack, because the defenders must be strong enough to confront the enemy offensive; if they go on the offensive themselves, they can pick their targets and attack where the enemy is weak, as Nikephoros Phokas did. Yet by so doing, the enemy may be forced to call off the large offensive that the defense was too weak to resist.

It is notable that this first victorious Phokas did not even try to invest Adana and Tarsos, both with potential booty and potential slaves but which were also fortified walled cities, as all cities in Cilicia necessarily were. It is unlikely that his men had the necessary skills, or his baggage train the necessary tools, to manufacture siege engines (no fast-moving raiding force could carry them), but it is also irrelevant, because the Byzantine raiders could not linger to besiege towns, given the imminent return of the failed jihadists.

When the author turns to siege operations, it is enemy sieges of Byzantine towns that he has in mind.[34] He baldly assets that many fortified towns are impregnable and have no reason to fear a siege—he must mean poor places both hardly worth taking and hard to attack because of their inaccessible locations, a description that fit many towns in the frontier zone.

As for cities both richer and more accessible and hence worth besieging, the author urges that food be stocked for at least four months, and that proper care be taken of water cisterns. For the rest, however, the author refers the reader to prior books on siegecraft, declaring that he will discuss only the skirmishing adjuncts to a siege: sorties to attack enemy encampments around the besieged city at night, diversionary actions to resupply cities under siege, and the thorough destruction of whatever can help the besiegers, including houses beyond the wall circuits.

Frontier warfare could be epic, but it was not chivalrous—our author at one point tells us that when a force must move fast, prisoners should be killed unless they can be sent ahead. There is no summing up at the end of this valuable work, only a Trinitarian invocation and the note: "With God's help, the end of the *taktikon*."

De Re Militari (Campaign Organization)

The diametric opposite of *De Velitatione,* dedicated to defensive operations against the Muslim Arabs on the Anatolian "eastern" front, was an equally practical manual dedicated to offensive operations against the Muslim Arabs as well, but more for the "northern front" against Bulgarians, Pechenegs, and the proto-Russians of Kievan Rus'. Previously published as *De Re Militari,* and with a manuscript title of "Anonymous Book on Tactics," the treatise has been newly edited as *Campaign Organization.*[35]

There are references to the "ancients," but *De Re Militari* also is a largely original work. The emperor is depicted as present and in command of an expeditionary army, and our author is giving him advice in distinctly unservile tones. The authoritative editor agrees with a controverted prior suggestion that the emperor is actually Basil II (976–1025), future conqueror of the Bulgarians but then still in his younger and less victorious days around 991–995. Everything in the text is consistent with a very late tenth-century or early eleventh-century date.

The manual starts with the setting up of a temporary camp for a large expeditionary force of sixteen taxiarchies—some sixteen thousand men at full strength. Both part-time thematic units and full-time *tagmata* are present. Infantry taxiarchies contain five hundred sword-and-shield heavy-infantry men *(hoplitas),*[36] two hundred "javelin-throwers," and three hundred archers.

The editor notes that Byzantine writers were apt to prefer the Platonic ideal to the prosaic actual, and that the camp described could be of that kind—perfect and perfectly imaginary.[37] Perhaps so, but it had been a standard Roman operating procedure to construct marching camps even more elaborately protected, even for a single night.

Detailed directions prescribe the layout of the camp (a square is favored) to ensure protection from without and avoid congestion within in the event of an emergency call to arms; a very good surveyor is essential—the author uses a Latin word of sorts, *mensurator*—it was a *mensor* in the Roman Legion.

At the center, there is the emperor's inner camp with his palace guards and elite forces:[38] the immortals *(athanatoi),* first raised by John Tzimiskes (969–976), named after their Achaemenid predecessors who fought at Thermopylae almost fifteen hundred years before, but now of cavalry; the premier guard regiment called *megale* (great) *hetaireia*—literally "companion entity"—as opposed to the *mese* (middle) *hetaireia* and *mikre* (little) *hetaireia* (little company) and the tagma of the *scholai*, the oldest of the palace guards.

Inner circles have their own inner circle, in this case the emperor's own personal bodyguard: the *manglavitai*, named after the mace used, gently one trusts, in their normal duty of clearing the way for the emperor as he walked through palace halls and corridors crowded with eager courtiers and petitioners.[39] It is further specified that a large empty space be left around the imperial tent to allow the men remaining on duty at night to move about—an appropriate security measure and not only against enemy infiltrators.

The emperor's camp was also his imperial court when on campaign. Two of its highest officials are named: there is a tent for the *protovestiarios,* the eunuch of the robes, and the *epi tes trapezes,* the eunuch in charge of the banquet table, both made powerful by proximity. One can safely assume that other, unnamed, high officials were there also, eager to remain close to the imperial power, or else compelled to do so. The imperial office being neither reliably dynastic nor elective but "occupative," emperors knew that friends had to be kept close but enemies closer so they might not plot back in the capital. The elegant Ottoman tents now in museums in Berlin, Cracow, and Vienna, booty of the failed 1683 siege of Vienna, had belonged to just such a courtly camp.

Guides *(doukatores)* were needed close by, to allow the emperor to get his information firsthand; but they were not his own palace guards, and they had to range far and wide to perform their duties, perhaps coming into contact with the enemy or the emperor's enemies. Hence the author offers a distinctly non-Platonic suggestion that bespeaks a shrewd sense of prudence: "The *doukatores* should be located with the *proximos*—a palace staff officer—or with someone else in whom the holy emperor has full confidence."[40]

Next the manual turns to military camps in general, prescribing "as was the custom among the ancients" encampment by units, not mixed. For expedient barriers, each infantryman should have about eight caltrops, and each unit of ten *(dekarchy)* was to have an iron stake to which the caltrop rope was to be tied; small pits with sharp stakes in them, "the sort called foot-breakers," and a perimeter of strings with bells were also recommended.[41] The third section, on posting guards at night, is of interest only because it directs that the *hoplitarches*—chief infantry officer—should supervise guarding arrangements against enemy infiltrations or raids; he was set over the taxiarchs as a branch inspector, in modern parlance, not an as operational combat commander, because that had to be a joint cavalry-infantry post.[42] The camp theme is pursued in the fourth section, about outposts (the editor calls them watch posts), infantry outposts with as few as four men and cavalry outposts with six horsemen farther out. In daylight only cavalry outposts are needed, much farther out. More follows on sizing the encampment, with additional calculations for a smaller proportion of cavalry (section 5), for an expeditionary force with only twelve rather than sixteen taxiarchies (section 6), and for cases where the terrain imposes two camps to avoid both congestion and a low-lying position below heights "from which missiles can easily rain down upon the tents" (section 7).

In this part of the work, calculations are replicated rather ritualistically with the variant numbers, but this does not betoken armchair strategy—on the contrary, it is the very thing that military professionals often do when idly contemplating alternative tables of organization and strength. Of that, section 8 is a perfect example: "if the mounted fighting force numbers 8,200 men, it should be divided into twenty-four units of up to three hundred men each . . . [which] should make up four groupings . . . each with six combat units."[43] There is a tactical purpose, to cover the rear and the sides with three left over for the front. Smaller cavalry forces should result in fewer rather than smaller units—sound advice—but there is a limit to reduction: "the emperor must not set out on campaign with such a small force."

Recent combat experiences against both eastern and northern enemies emerges in the next two sections, which owe nothing to ancient sources: the movement of the entire army from the camp, starting with the careful step-by-step dismantling of the tents, beginning with the emperor's, and the expeditionary march itself.

Instead of recycled Attic texts, we read of the *saka*, a rear guard (from the Arabic *saqat*), under an officer senior enough to receive his orders directly from the emperor—because to let troops pass and then attack them from the rear can be a highly effective technique. Later in the text, and later in the expedition, a special arrangement is suggested to relieve the units assigned to *saka* duty, because "they have to bear more than their share of trouble," while leaving their combat-experienced commander in place to guide the newly assigned units.

"To ward off the very bold onslaughts of Arabs [= Bedouins], and *Tourkous* [not Turks but Magyars, the future Hungarians, and like the Bedouin specialized in light-cavalry raids]. It is a good idea to assign about a hundred and fifty foot archers . . . to each of the twelve battle units on the outside."[44] The emperor has to be careful too, because raiders can penetrate even powerful forces once they are strung out on the march: "Let the emperor have with him as many archers as he wishes. Let him also have some Rhos [Varangian guards], and *malartioi*," presumably carriers of a specific weapon by that name, at least originally (as per the Grenadier Guards), but neither it nor they are otherwise known, or identifiable.[45]

This section closes with excellent advice for the emperor on campaign: at the end of the day's march, "unless some other task demands his attention" the emperor with his entourage should not go to his place within the camp but rather watch carefully all the units as they march inside, until the *saka* arrives.

It is not all action, there must be reaction too, because there is enemy action. Night attacks against the camp are best ambushed on their way, but prudently near the camp, and when repelled in whichever way, pursuit is to be avoided as useless and risky. Enemy forces can also attack the army on the march, and if very large, they should not be repelled off the march; instead the march must be stopped to put down the baggage and draw up in proper battle order.

Great care is needed when marching through areas in which there is no water: "It is a terrible thing to have to engage in two battles. I mean the one against the enemy and the one against the heat when water is lacking."[46]

Better choose a much longer route if it has water. The author no doubt remembered the cataclysmic defeat of the Byzantine army in the scorching July heat of 636 at the Yarmuk, a river withal. The eastern enemies of Byzantium, Sasanian Persians, and later the Muslim Arabs, Seljuk Turks, and finally the Ottomans were more familiar with desert warfare than men from Constantinople could ever be, for the city has an exceptionally well-watered hinterland, and faces the greenest part of Anatolia.

The author would have approved of the British army's adage that time spent on reconnaissance is seldom wasted. He calls for experienced and intelligent guides *(doukatoras),* who must be treated very well—not as lowly rankers—and who must know not only the terrain but also how to calculate the army's movements and needs within that terrain. Guides can only observe, they cannot probe enemy forces, and their scouting cannot safely reach very deep into enemy territory. To do that, and to ascertain the enemy's strength by launching small probing attacks (reconnaissance, in modern terms), is the role of small units of fast-riding light cavalry, called *trapezitai* or *tasinari* (from Armenian) but more commonly *chosaroi*. That new Greek word was taken from the Magyar *huszar*, which was taken in turn from the Old Serbian *husar*, itself derived from the Greek *prokoursator* or its Latin precursor, *procursator*, "he who runs forward"—a good description of light cavalry (the word thus went full circle back into Greek, while traveling west as *Hussar*—still in use today for armored reconnaissance troops).

To go even deeper than even the most agile light cavalry can go, spies are needed. No suggestions are added to manage them, as in the *De Re Strategica* (42.20), but it is specified that they are needed "not only among the Bulgarians [nearby] but also among all the other neighboring peoples, for example, in Patzinakia [the shifting domain of the Pechenegs], in *Tourkia* [the domain of the Magyars], in *Rosia* [Kievan Rus'], so that none of their plans will not be known to us."[47] If prisoners are captured with their families, they can be sent back to spy on their comrades to redeem the hostages.

Care is advised in crossing mountain passes even if they are not occupied by the enemy—that must have been recognized as the greatest danger in fighting the Bulgarians ever since the catastrophic defeat of emperor Nikephoros I and his vast army in 811. The Bulgarians had a particular ability to move cross-country to occupy and block mountain passes behind advancing enemy forces, to deny their retreat.[48]

Sound tactics come next, starting with several days of preliminary reconnaissance by "guides, spies, and hussars," followed by an advance guard with more archers and javelin throwers than less-agile heavy infantry. As the array begins to move into the pass, the commander is to seize the highest points ahead, to overlook possible lateral approaches to his line of advance. The main force marches into the pass only after it has been scouted, secured, and overwatched. Two infantry taxiarchies should march ahead of the cavalry, with tools to improve the road. Every time they reach a particularly difficult passage that an infiltrating enemy could seize, some foot soldiers are to be left behind to hold the place until the entire army has passed through. Much more is needed by way of preparations if the pass is occupied by the enemy. If his forces are strong, the best course is simply to go elsewhere, to advance through another pass, even if distant. Otherwise there is no remedy but to attack in great strength, hoping to precipitate flight but ready for battle, after softening up with missile attacks by archers, javelin throwers, and slingers.

On siege warfare the author is again well informed and far from optimistic.[49] He predicts failure in taking strongly fortified cities unless their agricultural supply base has first been destroyed by a prolonged campaign of raids to chop down fruit trees and vines, burn crops, and seize livestock, so that they cannot outlast the besiegers, who cannot carry with them more than a twenty-four-day supply of barley for the horses. Only if the passage of mule trains and "if feasible, wagons" can be assured by strong escorts against the enemy's own deep raiders, can the besiegers achieve a superiority in supply and prevail by hunger. One context explicitly identified is the struggle to retake the cities of Syria from the Muslim Arabs, a country naturally fertile but devastated by war; the other is the "country of the Bulgarians," in which there is "a total lack of necessities"—obviously because of the secular disruption of its agriculture, not natural infertility.

The greatest Arab-held city successfully taken was Antioch, on October 28, 969, under Nikephoros II Phokas (963–969), who in 962, when still an officer, had already looted and briefly held Say al-Dawlah's capital of Aleppe, the ancient Berroia, later Aleppo and now Halab. Along with Antioch, some sixty lesser cities were also taken by the Byzantines in the war zone that extended across northern Syria from east Anatolia to Mesopotamia, evoking a jihadist reaction:

> The capture of Antioch and the other cities . . . was an affront to the Saracens [= Muslim Arabs—the term *Arab* meant Bedouin] all over the world and to the other nations who shared their religion: Egyptians, Persians, Arabs, Elamites [= Kurds], together with the inhabitants of Arabia Felix and Saba [Yemen]. They came to an agreement and made an alliance, whereupon they assembled a great army from all parts and put the Carthaginians [Tunisia] in charge of it. Their commander was Zochar, a man of vigor and military skill with an accurate understanding of land and sea operations. Once all the forces had been brought together, they marched out against the Romans, numbering one hundred thousand fighting-men. They approached Antioch from Daphne [a wooded and watered park] and laid vigorous siege to it but those within resisted courageously and with excellent morale, so the siege dragged on for a long time. When this concentration of peoples was made known to the Emperor he quickly dispatched letters [ordering reinforcements to engage] the myriad of barbarians [who were] put to flight and dispersed in a single battle.[50]

Attempts were to be made to lure the defenders outside to defeat them in open-field battle. Failing that, and given the reliable arrival of supply trains, the besiegers needed their own protective trenches and ramparts against enemy sallies. There were foraging parties—literally, for forage, grass for the horses—and there was the need to safeguard them and the pasturing horses against enemy sallies, which could also be lured out by soldiers disguised as unarmed grooms.

Raids and counter-raids attend the siege; in assembling a large force to raid an enemy relief force a day's journey away, there were horsemen, javelin-throwers, archers, mounted heavy infantry—on "better horses," significantly—and mounted Rhos, who were singled out again to lead the column. Evidently, the Rhos, that is, the Norsemen of the Varangian guard, were then considered elite soldiers, indeed the best of several different elite guard forces.

The author then finally turns to actual siege operations against fortifications. Till now only the need for special precautions to protect siege machines from enemy raids, and the need to camp beyond the range of enemy stone-throwers *(petrobolos)*, have been mentioned.

After declaring that siege operations require "great inventiveness," the author lists mining, battering rams, tortoises, stone-throwing machines, again unspecified *petroboloi* (simple but powerful traction trebuchets are the most likely candidates, rather than tension catapults or yet more complicated torsion machines), ropes, wooden towers, ladders, and earth ramps. Engines "are built" but we are not told which ones or how they were built, because "the ancient authorities have written excellent and very practical things in their books."[51]

When it comes to training, the author appeals to the usual authority: "The ancients have passed on to us the necessity of training and organizing the army. . . . They would train not only the army as a unit, but they would also teach each individual soldier and have him practice how to use his weapon skillfully. In actual combat, then, bravery, assisted by experience and skill in handling weapons, should make him invincible. There is, assuredly, a need for exercises and careful attention to weapons. For many of the Romans and Greeks of old with small armies of trained and experienced men put to flight armies of tens of thousands of troops."[52]

That sounds wistful, implying that in his day Byzantine troops were quite untrained, but it really depended on the geography, its strategic depth, and day-to-day security, and the resulting sense of urgency about military training, or the lack of it. After complaining that the part-time thematic soldiers were no longer training and instead were "selling their combat gear and their best horses and buying cows," so that if the enemy attacks, "nobody will be found who can do the work of a soldier," the author immediately goes on to acknowledge that those who live in border areas "and have our enemies as neighbors" are "vigorous and brave. In view of their training and going on campaign . . . it is fitting that they should be honored as defenders of the Christians." A very natural difference, then, between life in secure rear areas that no enemy has penetrated in living memory, and where training is disregarded as a useless chore, and the frontier zone where soldier-farmers train in earnest. It is the same in contemporary Israel, where local defense units of overage soldiers released from reserve duty are largely inactive in cities but very alert along the frontiers.

The possible author Nikephoros Ouranos was himself a very successful general in the service of the equally successful fighting emperor Basil II—he utterly defeated the Bulgarians at the river Spercheios in 997 and participated in the reconquest of northern Syria from the Muslim Arabs. Whether the author was the protagonist or not, these successes were won not by poor peasants with clubs and slings but by soldiers thoroughly trained, both the part-time thematic troops on the frontiers who had to be ready to resist constant raids and frequent offensives and the elite, full-time, salaried troops of the tagmata. At the time, they were a substantial force by themselves, the mounted *scholai* of the *Exkoubitoi* and *Vigla*, all converted from earlier foot guards and newer *hikanatoi* ("worthies"), as well as the infantry of the foot guards, Wall guards, and *noumera,* who doubled as gendarmes and prison guards.[53] If the author was not Nikephoros Ouranos, he would still have been his contemporary, and as such cognizant of the army's accomplished training, without which its decisive contemporary victories over both Muslim Arabs and the Bulgarians would not have been possible.

The author next offers us a rare glimpse of Byzantine military administration—Personnel Department—at work. It is paperwork, of course, encouraged by the availability of actual paper as opposed to far more expensive vellum. The entire army must be registered in comprehensive *(katholika)* lists, to determine how many men are mustered, how many have been left at home, how many have run away, how many were exempted because of weakness, how many were found to be deceased.

The listings then turn to qualitative factors, implying systematic evaluations: how many keep their horses and fighting gear in good condition; who is working hard and who is lazy; how many are valiant— because "men who have risked death and capture should not be lined up [with] the lazy and slothful."[54] Then each man should receive his reward.

There is no closing flourish. Instead there is a very interesting aside. "We had also wanted to explain something about raids and the manner of conducting them in the land of the Agarenes [Arabs] and to set forth suitable and efficient ways of devastating their country. But . . . we consider it superfluous to write about something which everyone already knows."[55]

The *Praecepta Militaria* of Nikephoros II Phokas

The field manual conventionally known as *Praecepta Militaria,* whose Greek title means "Presentation and Composition on Warfare of the Emperor Nikephoros," is indeed attributed to the fighting emperor Nikephoros II Phokas (963–969) by its authoritative new editor.[56] The context is offensive warfare against the Muslims, specifically the decreasingly victorious Ali ibn Hamdan, Sayf ad-Dawlah, independent ruler of what is now Syria and beyond under the nominal authority of the Abbasid caliph. Previously a very successful jihadi whose forces raided deep into Anatolia, Sayf ad-Dawlah was repeatedly defeated by Nikephoros and his field commander and successor John Tzimiskes, losing the fertile lands of Cilicia and the important city of Antioch.

This work is therefore the exact counterpart to *De Velitatione,* which had the same geographic setting in eastern Anatolia and mostly fertile Cilicia and the same antagonists, but whose orientation was entirely defensive strategically, though it proposed vigorously offensive tactics.

Praecepta Militaria begins by prescribing what is needed for the infantry—that is, "heavy" infantry for close combat, *hoplitas* in the text: Roman or Armenian recruits under forty years of age and of large stature who are to train properly with shields and spears, and who are to serve under officers of ten (dekarchs), fifty (pentekontarchs), and a hundred (hekatonarchs), as with modern company commanders, except that it seems from the overall context that these were rather like sergeants, senior withal.

There is a proper concern with unit cohesion: the men should stay together with their friends and relatives in *kontoubernia,* the old Roman army's tent-group of eight, which varied in Byzantine use from five to sixteen, but the point is that the men should live, march, and fight together.

The author has a very specific field army in mind with exactly 11,200 heavy-infantry men, not counting the light infantry. Their equipment is economical, indeed poor, with padded tunics *(kabadia)* instead of metal or at least boiled-leather breastplates, high boots "if possible" but otherwise "sandals, that is, *mouzakia* or *tzerboulia,*" known as the lighter footwear of the poor, women, and monks.[57] There is no metal helmet, only a thick felt cap—this particular "heavy" infantry is so defined by its tactics and is not armored at all, as opposed to Roman legionary troops, for example.

But there is no skimping on weapons. The Hamdanid forces that were the intended enemy contained much cavalry, both light for skirmishing and raiding (*Arabitai* from their Bedouin origin), and armored cavalry *(kataphraktoi)* for the charge. Accordingly, *Praecepta Militaria* prescribes that the heavy infantry be armed with thick, sturdy, and long lances [*kontaria*] of some 25 to 30 spithamai (= 5.85 to 7.02 meters).[58] The editor calls these lengths "improbable."[59] That is certainly true for the upper end of the range, but as with the *sarissa* of Philip and Alexander, length could best keep charging cavalry at bay, while awkwardness on the march could be mitigated (as with the sarissa) by assembling the weapon from two half-lengths joined by fasteners in a collar—to which the mentioned *kouspia* might also refer.[60] The list of prescribed weapons continues with "swords girded at the waist, axes or iron maces, so that one man fights with one weapon, the next with another according to the skill of each one."[61] It is also specified that they should have slings in their belts, so that they can launch harassing shots from a distance before coming into close-combat range with their spears or swords; slings generally complemented the bows of the missile infantry, and were especially usefully in wet weather. The author prescribes large shields "six spithamai" (= 1.4 meters) "and if possible . . . even larger." That reflects the lack of body armor and the need to protect from the great number of arrows that Sayf ad-Dawlah's men could launch.

With all this equipment, the heavy infantry would indeed be heavy, too heavy. Accordingly the text prescribes that "each group of four [heavy-infantry men] must have one man *(antropon)* whose responsibility it is in time of battle to watch over their animals, baggage, and provisions."[62]

Alongside the 11,200 heavy infantry, the army is to have 4,800 "proficient archers." In *Praecepta Militaria* the light infantry is equated with archers, indeed proficient archers, and it is specified that they should have "two quivers each, one with forty arrows, the other with sixty, as well as two bows each, four bowstrings and small handheld shields, swords girded at the waist, and axes, and they must likewise carry slings in their belts."[63]

In spite of its apparent meaning, this passage must describe issued equipment, not what the men were actually supposed to carry with them in combat—with both sword and axe, two bows, and one hundred three-foot-long arrows on him, the infantryman would hardly be light or agile. It is much more likely that some of the equipment was carried by the pack animals and attendants of the baggage train.

At this point the author cites a mixed cavalry-infantry tactical array in use since antiquity, in which twelve separate infantry formations leave corridors between them from which small cavalry units of ten to fifteen can sally out and return. In addition, if there are javelin-men, foreign recruits as noted above, with connotations of expendability—they can stand behind an infantry square in readiness to block a corridor against enemy cavalry. The archers and slingers who are not optional—for without them there would be no missile strength—stand behind the heavy infantry of each formation. Dynamically, when enemy cavalry pursues the Byzantine cavalry in between the formations, the javeliners are to step out and block the enemy with missile support from the archers and slingers. That allows the Byzantine cavalry to attack without regard to its own defense, for the infantry provides protection when needed.

The files that make up each formation are seven deep, with three archers sandwiched between two heavy-infantry men at one end and two at the other, so that the resulting formation can face both ways. The taxiarchs (commanders of a thousand) have a hundred such files in their formation, and the remaining three hundred men consist of two hundred javelin-men and slingers—a cheaper, less skilled light infantry armed with cheap weapons—and one hundred men who are the very opposite kind of infantry, carefully selected soldiers armed with a weapon that had its own peculiar importance in the Roman past: the heavy thrusting spear, or pike, the Latin *hasta,* Greek *menavlion* (pl. *menavlia*).[64]

Its particular function was to protect infantry formations from cavalry charges—a role that pikemen would retain in European infantry regiments until the introduction of the bayonet.

Its more general function was to serve as the sturdy weapon of particularly sturdy men, formed into units of *menavlatoi* to hold the line under severe attack, or to the contrary, to thrust forward against stiff enemy resistance. As such, as the pike would also be, the menavlion was the weapon of elite soldiers—men who would stand bravely to confront charging heavy cavalry—who might also have a higher social standing, as was often true of pikemen. In Shakespeare's *Henry the Fifth,* the hard-drinking Pistol asks the disguised Henry V who he is. "I am a gentleman of a company," he replies, meaning a gentleman-volunteer. "Trailst thou the puissant [powerful] pike?" asks Pistol. The king replies, "Even so." It was the weapon of steadier men, long more prestigious than firearms.

As the *hasta* it had armed the troops of the most mature third echelon *(triarii)* in the legions of the Republic long before, and under the properly classical if misleading name *sarissa* it was mentioned as an issued weapon of the infantry in the sixth century,[65] but it received special emphasis in the *Praecepta Militaria,* which specified that the *menavlion*

> must not be made from wood cut into sections, but from saplings of oak, cornel or the so-called *atzekidia.* If saplings in one piece cannot be found, then let them be made from wood cut into sections but they must be made of hard wood and just so thick that hands can wield them. The *menavlatoi* themselves must be brave and stalwart.[66]

The length of the weapons is given in chapter 56 of the encyclopedic *Taktika* of Nikephoros Ouranos: one and a half or two ourguiai for the shaft, one and a half or two spithamai for the spearhead or point, that is, 2.7 to 3.6 meters and 35 to 47 centimeters.[67] Again, the specific purpose was to resist charges by the heavy cavalry, in this case the heavy cavalry of the Hamdanid army; the passages incidentally illustrate the difference between the sturdier weapon and ordinary lances [*kontaria*]:

> The *menavlatoi* must take their place in the front line of the infantry . . . if it should happen, and we hope it does not, that the . . . [lances] of the infantry are smashed by the enemy *kataphraktoi,* then the *menavlatoi,* firmly set, stand their ground bravely to receive the charge of the *kataphraktoi* and turn them away.[68]

The more general purpose was to add strength in frontal attacks—the "push of pike" that could still be decisive in the English Civil War, or to stabilize entire infantry formations in adverse circumstances:

> When the fighting begins, . . . [units] can form up without hindrance or disturbance [behind the protection of the *menavlatoi*]. . . . On the other hand, men worn out with fatigue and the wounded return [to] find relief under [their] protection.[69]

That again makes it clear that the menavlion was for threatening and thrusting, not for throwing, hence was entirely different from the classical Roman *pilum,* a heavy throwing spear, let alone javelins under whatever name. There are ineluctable distinctions between thrusting long weapons too heavy to be thrown (pikes, *menavlion, hasta*), missiles too flimsy for thrusting or to deter charging cavalry (javelins, *akontia*, *monocopia*), thrusting lances for the cavalry (*contus*, etc.), and the short-range heavy throwing spear of the legions *(pilum)*, of marginal use for thrusting, and then only because the *gladium* sword was so short.[70]

In the prescribed battle array, only three out of ten are bowmen, and in the envisaged army there 4,800 archers as against 11,200 heavy-infantry men—the same proportion as in the envisaged encampment of *Campaign Organization and Tactics.*

Evidently these are forces structured for the offensive, chiefly propelled by the shock action of the cavalry, with much less use for archery than forces on the defensive. In the Roman imperial army, archery was marginal for the same reason, and of course it did not have a really powerful bow. But while archery counted for much less in the tenth-century Byzantine army than in the sixth-, it was still important enough to warrant specific arrangements for the resupply of arrows. Given the discharge rates of well-trained bowmen, the one hundred arrows carried by each archer would not last long. Accordingly, another fifteen thousand, or fifty per bowman, are to be carried by animals that follow the forces into battle (not the main baggage train), and it is revealing that a chiliarch (= commander of a thousand), an officer of in the rank of a lieutenant colonel in modern terms, is enjoined

> to count them out beforehand and bind together each bundle of fifty, then put them away in their containers. . . . Eight or ten men in each unit [of a thousand] should be detailed to supply arrows to the archers so as not to take them away from their [battle positions].[71]

Fifty extra arrows per bowman does not seem like much, given their hundred already issued, but in battle not all bowmen could usefully discharge arrows all the time—they had to be positioned where the enemy came within range, which need not happen at all for some or possibly many of them. Thus the fifteen thousand extra arrows could count for much once allocated to the active bowmen, rather than just handed out fifty per man.

Next we hear of the special weapons that the commander of the army must also have: "small *cheiromangana*, three *elakatia*, a swivel tube with liquid fire and a hand pump." These are not equivalent as supporting weapons to modern machine-guns and mortars, which are just as versatile as rifles; they are comparable rather to such weapons as anti-tank rockets and grenade launchers, each narrowly specialized, of great use in particular circumstances but mostly idle in battle, waiting for their moment. Greek fire, which water could not extinguish, could burn and terrorize enemies within the very short range of siphons or hand-pumped projectors, say ten meters at most; hence it could be employed only when the attacking enemy was about to reach the battle line—even then affecting only those who came within its short range.

As for the *cheiromangana*, the modern editor tentatively defines it as a portable arrow launcher, similar to a gastraphetes or heavy cross-bow.[72] But the nomenclature of Roman and Byzantine artillery was notoriously unstable—during the fourth-century, *catapult* went from "stone-thrower" to "arrow-launcher" while *ballista* switched meanings on the opposite trajectory, if the pun be allowed—and it is more likely that it was a small, mobile, traction trebuchet.[73]

Because it has the virtue of specificity, that later French name has become the conventional term for a device that Byzantine texts describe with a variety of names, some carried over from the mechanically very different torsion and tension artillery of the ancient engineers that the trebuchet made largely obsolete: *helepolis, petrobolos, lithobolos, alakation, lambdarea, manganon, manganikon petrarea, tetrarea,* as well as the *cheiromangana* itself. Trebuchets could be big enough to demolish the best-built stone walls from tactically useful distances of two hundred meters and more, well beyond bowshot, or small enough to be easily mobile and to be operated by one man, as was most probably the case with the *cheiromangana* itself. The authority on the subject has suggested that the Byzantines realized the usefulness of small, mobile cheiromangana after the battle of Anzen in July 838, in which Abbasid forces employed traction trebuchets to hurl stones upon the Byzantine troops, scattering them in panic, after a rainstorm had incapacitated the bows of their Turkic mounted archers.[74]

Either way, this weapon consisted of a beam that pivoted around an axle supported by a relatively high frame, with uneven long and short arms. The missile was placed in a receptacle, or in a flexible sling attached at the end of the long arm, while pulling ropes were attached to the short arm. To launch a projectile, the short arm was abruptly pulled downward by human traction, by gravity through the release of a counterweight, or by a combination of both. It is generally believed that tenth-century Byzantine trebuchets were traction activated or hybrids, while more powerful gravity trebuchets were first habitually built and employed by John II Komnenos (1118–1143).[75]

It was a very long evolution, or a rather a very slow diffusion if it is true that the Chinese used trebuchets long before—in fact the Avars responsible for its first recorded use may have learned to build them from the Chinese before they came west, although Theophylact Simocatta reports the story of a captured Byzantine soldier Busas who taught the Avars how to build a *helepolis,* which the authority on the subject translates as trebuchet.[76] But it could mean any engine of war at all, including the original *helepolis,* a mobile siege tower. Besides, Simocatta portrays technically incompetent Avars, while the *Strategikon,* as we saw, repeatedly advises the use of Avar technology. In any case, the Avars used fifty trebuchets with devastating effects in their siege of Thessalonike in 597, when their existence is first attested in the celebrated memoir of its Archbishop John I:

> These *petroboloi* [= rock throwers = trebuchets] had quadrilateral [frames] that were wider at the base and became progressively narrower toward the top. Attached to these machines were thick axles plated with iron at the ends, and there were nailed to them pieces of timber like beams of a large house. Hanging from the back side of these pieces of timber were slings and from the front strong ropes, by which, pulling down and releasing the sling, they propel the stones up high and with a loud noise. And on being discharged they sent up many great stones so that neither earth nor human constructions could withstand the impacts.
>
> They also covered those quadrilateral-shaped *petroboloi* with planks on three sides, so that those inside launching them might not be wounded by arrows [shot] by those on the city walls. And since one of these, with its planks, had been consumed by fire from an incendiary arrow, they returned, carrying off the machines. On the following day they again brought up these trebuchets *(petroboloi)* covered with freshly skinned hides and planks, and placing them closer to the city walls, shooting, they hurled mountains and hills against us. For what else might one term these immensely large stones?[77]

Next the author reverts to the hoplites, the heavy infantry, to note that each pair of men should have a mule to carry their shields, spears, and provisions, and each group of four must have a man (*anthropou,* not a servant or a soldier) to watch over these possessions when the soldiers are in combat. The next comment is a reminder that battles must be fought where there is a water source.[78] These scattered observations are typical of a text that amounts to a set of practical notes left by a practitioner for his successors.

The tenth-century cavalry of the text is not as dominant as in the sixth-century *Strategikon.* The reason plainly is that an army structured to win and hold territory, rather than to outmaneuver and contain the enemy, must have heavy infantry that can stand its ground. The tenth-century cavalry moreover was greatly diversified, as compared to the archer-lancer cavalry of the *Strategikon*, which was certainly versatile but also homogeneous. The reason is again plain: in the east the Byzantines faced an enemy whose own cavalry was sharply differentiated. There were Bedouin light horsemen with swords and spears who were agile raiders and could be employed less reliably for scouting and reconnaissance; Turkic mounted archers who were increasingly displacing the Arabs and the Bedouin with them as the protagonists of jihad; and armored cavalry taken over from the Sasanian army, which the Romans had previously imitated with their *clibanarii*.

The first type of cavalry mentioned in the *Praecepta Militaria* are the *prokoursatores*, the current word for light cavalry meant for scouting, raiding, reconnaissance—and countering enemy efforts to do the same. It is specified that they should wear *klibania*, a word whose meaning had evidently changed over the centuries, because instead of plate or lamellar or any other heavy armor, the klibania must have been made of leather or tightly woven cloth or some other light protection, because the prokoursatores are defined as not "heavily armored and weighted down, but light and agile."[79]

Scouting, by definition, is limited to observation only, with no intentional combat, and it was an implicit role for prokoursatores, however important. But their more demanding role was reconnaissance in modern terms, that is, the deliberate engagement of enemy forces, even if cautiously, to induce them to reveal themselves, to probe their strength, to capture prisoners for interrogation, and to weaken them with surprise attacks or ambushes. Their other role was to fight off enemy prokoursatores engaged in scouting or reconnaissance themselves. In each case, if faced by superior strength or any form of determined set-piece action, their task was not to fight and die but to extricate themselves, because they could do more good synergistically by keeping the army well informed and the enemy less informed.

We can confidently determine the above because of the organization and equipment that the author prescribes for the prokoursatores. He envisages a total force of 500, of which 110 to 120 are to be proficient archers with body armor and helmets (*klibania* or *lorikia*) as well as swords and maces, while all the rest are to be lancers—the ideal weapon for the raiding light horsemen. Each is to have an extra horse actually with him when out raiding (not in set-piece battles), that being a practice learned long before from the nomads of the steppe and especially useful for getting away after a fight. The author is defining an actual tactical formation that on occasion might fight complete under a single commander, and not just an administrative entity—and in fact he notes that if the army is smaller, the prokoursatores should consist of three hundred men, including sixty archers.[80]

The second type of cavalry are even more specialized kataphraktoi, armored horsemen on protected horses, structured in a compact mass to inflict shock. The text recommends, if the army is large enough, a triangular wedge formation of 504 men with files twelve deep, with 20 horsemen in the first rank, 24 in the second, 28 in the third, 32 in the fourth, 36 in the fifth, and so on until the twelfth rank with 64 horsemen, for the total of 504. If there are fewer kataphraktoi, it is specified how they should be formed into a smaller wedge of 384 cavalrymen.

Those are not small numbers, because these are not ordinary troops. Expensively armored cavalry men on expensive big horses are the equivalent of armored vehicles in modern terms—as of this writing the entire British army has only 382 tanks. In favorable terrain, a determined charge by 504, or for that matter 384, armored horsemen could be terrifying, apt to scatter any but the most determined enemy by purely psychological shock effects, even without an actual clash of arms.

But the kataphraktoi were fully armed for the closest form of combat as well, for the very first weapon listed is not the lance but the classic weapon of the melee:

> iron maces [*sidhrorabdia* = iron staffs] with all-iron heads—the heads must have sharp corners . . . —or else other [straight] maces or sabers *(parameria)*. All of them must have swords *(spathia)*. They should hold their iron maces or sabers in their hands and have other iron maces either on their belts or saddles. . . . The first line, that is, the front of the formation, the second, third, and fourth lines must have the same complement, but from the fifth line on back the *kataphraktoi* on the flanks should set up like this—one man armed with a lance and one armed with a mace or else one of the men carrying a saber.[81]

All of the above makes perfect tactical sense, in fact it outlines a synergistic combination of weapons. The heavy iron maces, actually flanged at acute angles, are for hand-to-hand combat with enemies who could be heavily armored as well, and thus protected against lighter blows. The "other maces" are lighter variants but with blades embedded that could also be thrown *(vardoukion, matzoukion);* they could be formidable weapons in very skilled hands, and may habitually have been used for mounted hunting as well, presumably in close country (a scene in the Madrid Scylitzes, the famous twelfth-century illustrated manuscript of the historian, depicts Basil I killing a wolf on a hunt with a *vardoukion* that split its head). That is why it is prescribed that the men should have more than one mace on their belts or saddles—it would make no sense to carry more than one mace otherwise.

Sabers, *parameria,* single-edge for slashing and probably curved to avoid entanglements, were for those uncomfortable with the heavy mace and lacking the special skill needed for the throwing mace.

All had to have swords, *spathia,* a word that always meant a long weapon of one yard at least, and therefore useful in the charge as well.

The lances *(kontaria)* are not otherwise mentioned, but they must have been issued to all, because they were the weapon par excellence of the charge, and not too burdensome—in the *Strategikon,* they are light enough to be strapped on the back.

The kataphraktoi themselves have no missiles other than a few throwing maces, but to leave their formation with no missiles would limit it excessively. The author accordingly prescribes the inclusion of mounted archers—the third kind of cavalry—in the number of 150 for the formation of 504 kataphraktoi, or 80 for 384. They are to be sheltered by being positioned behind the fourth rank of armored horsemen.[82] That way, the formation could participate in the battle before the hand-to-hand stage—for example, by bringing forward the archers it sheltered to within effective range, while the horsemen in the forward ranks were protected from enemy missiles by their armor.

That is the virtue of armored forces in any age of war: superior *battlefield* mobility, that is, the ability to move in spite of enemy fire, in this case arrows, which allowed physically slower armored horsemen to advance faster than unencumbered light horsemen, for those had to hold back to stay out of the effective range of enemy arrows; it is the same with today's slow tanks that can advance faster than the fastest light vehicles when bullets are flying.

The armor in question is defined very precisely in the text. Each man must wear a *klibanion* with sleeves down to the elbows, skirts and armguards made of "coarse-silk or cotton as thick as can be stitched together," protected with *zabai,* scale armor.[83] It is evident that these *klibania* are made of metal armor, lamellar or otherwise, because armless *epilorika* of coarse silk or cotton are also recommended; that is not just because the word means "on top of armor"—not a reliable proof with all the shifts in meaning going on—but because it is metal armor that needs rust protection in wet weather.

The helmets are of iron and heavily reinforced so as to cover their faces with *zabai* two or three layers thick "so that only their eyes appear," and leg guards are also prescribed along with shields. Armor protection did not have to be perfect or complete to be valuable in combat, because even weak protection could prevent injuries from arrows discharged at long range with little momentum, screening out progressively more powerful arrows as protection increased. The armored horseman could fight dismounted to defend himself, but his offensive purpose required live horses and they too had to be protected from arrows, in fact their uppers had to be "covered in armor" of felt and boiled leather down to their knees, leaving only "eyes and nostrils" exposed, with optional chest protection made of bison hides—that would be the European bison, or wisent, *Bison bonasus,* then still common in the Caucasus as well as in forests across Europe.[84]

Naturally, less armor is prescribed for the archers—they positively need to keep out of close combat to use their weapons usefully—but they also are to have *klibania* and helmets, and their horses should be protected with padded cloths *(kabadia).*

The author envisages different combinations of the three kinds of cavalry, which are indeed further differentiated because among the kataphraktoi only some can assume the role of lancers. For all of them the basic combat unit—and building block of larger tactical formations—is the *bandon* of fifty men, brought together according to kinship and friendship, "who must share the same quarters and daily routine in every way possible."[85]

Like every serious military leader, the author knows that fifty men plus unit cohesion generate many times more fighting power than fifty individualistic warriors, and he also knows how it can be cultivated, by sharing everyday life for good or ill. Fifty, incidentally, is just about the biggest number that can achieve familial sentiments and maximum cohesion—in all modern armies the basic combat unit is the platoon of thirty men or so. It is obviously important to keep the unit together, even if the number makes an awkward fit when it is a bit more or a bit fewer men that are really needed.

Cohesion come first. As the author recommends different tactical arrays for different circumstances, the building blocks are always 50-men *banda,* so that the commander's own battle force, here as in *De Velitatione* identified with the Germanic term *foulkon,* is to have 150 men, three *banda* out of the total force of 500; if the total force is only 300 men, the *foulkon* is to have 100—two *banda*.[86] In both cases, all the other men, or rather *banda,* are to be assigned to reconnaissance—because numbers on all sides were normally too small to man continuous fronts, Byzantine armies and their enemies spent most of their campaigning days looking for each other.

Likewise in defining the main battle array, the outflankers on the right are to number one hundred, both lancers and archers, two *banda* evidently; the left wing is also to have one hundred men "to fend off enemy outflankers"; the main blocs are to have five hundred men each, three hundred lancers and two hundred archers, six *banda* and four therefore.[87] Only the formation of kataphraktoi of 504 does not quite fit the rule of fifty.

Homogeneity in the combat units, each of which consists of but one specialty for the sake of unit cohesion, coexists in the army of the *Praecepta Militaria* with the heterogeneity of its formations of heavy and light infantry, light horse, mounted archers, and armored cavalry, whose very differentiation creates opportunities for powerful synergies.

For example, the formidable triangular wedge of 504 kataphraktoi can charge the enemy's battle array, perhaps successfully breaking its ranks, inducing a panic flight of his cavalry, which only the fast prokoursatores can really exploit, by chasing down and stabbing the men with their spears and slashing at them with their sabers. If the enemy infantry runs as well, then the kataphraktoi themselves can do great execution with their swords and maces, while the mounted archers can use their own long swords.

Those would indeed be splendid successes for the kataphraktoi, and they were achieved on important occasions; but those were unusual occasions, of course—smashing victories always are. Much more commonly, however, the 504 kataphraktoi, or 384 for that matter, could achieve a more prosaic but still very useful result: they could compel the enemy to remain in very tight formations with pikes and spears poised to repel the charge, or rather to dissuade it, for cavalrymen will not normally charge a solid array of sturdy-looking infantry with sharp long weapons pointed straight at them. The tighter the enemy formations, the better the target for the bowmen (unless they faced a Roman *testudo* of upraised shields in perfect form), who no longer needed to pause to aim at individual targets and could instead volley rapidly and with effect at ranges of up to 200 yards for the best composite reflex bows and bowmen, killing few but wounding many, and incapacitating many horses too; at ranges of up to 100 yards the best bows and arrows could pierce most forms of armor, greatly increasing lethality.

Nikephoros Phokas II, or whoever wrote the text, had a sound understanding of combat psychology. It is often a good idea to frighten the enemy with "shouts and battle cries" as recommended in a particular case by *De Velitatione*. Drums, trumpets, firecrackers (by the Chinese), and piercing screams were much employed to frighten the enemy in ancient battles—and even in the Second World War the profusion of deafening explosions and rapid-fire detonations was not quite enough for some—the German *Luftwaffe* rigged some of its Junkers 87 *Sturzkampfflugzeug* ("Stuka") dive-bombers with wind-sirens that made an unearthly wailing sound, while the Red Army's Katyusha rockets came complete with a piercing scream that German troops learned to hate.

Noise frightens, and may contribute to breaking the enemy's morale. But so can silence, in the right circumstances, when it becomes a deadly silence. That is what the text prescribes: "As the enemy draws near, the entire [army] must say the invincible prayer proper to Christians, 'Lord Jesus Christ, our God, have mercy on us, Amen' and in this way let them begin their advance against the enemy, calmly proceeding in formation at the prescribed pace without making the slightest commotion or sound."[88] One can well imagine the effect: a force of armored horsemen advancing in perfect order and total silence will seem all the more inexorable.

The *Praecepta Militaria* contains the most concentrated expression of the Byzantine style of war. It is not Homeric combat for personal glory, nor the grand heroic warfare of Alexander, nor the relentless destruction of the enemy of classic Roman warfare. The Byzantine field commander depicted in the text is neither a devotee of holy war equally content with glorious victory or glorious martyrdom, nor an adventurer hoping for success. His task is to campaign successfully, occasionally by fighting battles but mostly not; he is to fight only victorious battles, an aim that can be achieved by carefully avoiding anything resembling a fair fight: "Avoid not only an enemy force of superior strength but also one of equal strength."[89]

Scouting, spies, and light-cavalry reconnaissance are to be employed amply and repeatedly to estimate the enemy's material and moral strength, the latter—as always—being more important, three times more so according to Napoleon, with whom the author would have agreed. Stratagems and ambushes are the alternatives to battles that must not be fought, because they can best demoralize the enemy over time, eventually allowing battle to be joined for assured victory.[90]

The *Taktika* of Nikephoros Ouranos

Nikephoros Ouranos, undisputed author of the last of the works of the tenth-century military renaissance, was not an emperor like the authors of the field manuals reviewed so far, but he was prominent enough. He is first mentioned as a court military advisor *(Vestes)* sent to Baghdad to negotiate the extradition of the pretender Bardas Skleros, who had fled to the Arabs after he was defeated in 979.[91]

Nikephoros Ouranos was entrapped by an interminable negotiation that ended with his own imprisonment, from which he had to be ransomed in 986; but in spite of this failure he was promoted, and it was as a *magistros* (senior general) in the very high post of *domestikos ton scholon* for the west, literally commander of the guards but in effect field commander, that he won a spectacular victory at the river Spercheios in 997. For two years Tsar Samuel had led his Bulgarians in successful plundering expeditions through Greece down to Attica after crushing the Byzantine garrison of Thessalonike, and it was as a returning conqueror laden with plunder that he camped his army by the river, still called Spercheios in Thessaly, central Greece. The Byzantine troops under Nikephoros reached the far bank of the river by forced marches:

> There was torrential rain falling from the sky; the river was in flood and overflowing its banks so there was no question of an engagement taking place. But the Magister [Nikephoros], casting up and down the river, found a place where he thought it might be possible to cross [i.e., a ford]. He raised his army by night, crossed the river and made a totally unexpected assault upon the sleeping troops of Samuel. The better part of them were slain, nobody daring to give a thought to resistance. Samuel and Romanus his son both received severe wounds and they only got away by hiding among the dead.[92]

As Dain noted, it was the sort of thing that the field manuals were constantly recommending, skillful maneuver to obtain total surprise that temporarily paralyzes the enemy, and can thus nullify even large differences in numbers and combat power; but the manuals only promised tactical advantages, while at the river Spercheios Nikephoros Ouranos won a strategic victory.[93] Bulgarian losses were so heavy that Greece was never again threatened by Samuel, who was decisively weakened, even though his military strength and kingdom were not finally destroyed until seventeen years later in the battle of the Kleidion Pass of July 1014.

The work of Nikephoros Ouranos was on an equally grand scale: no fewer than 178 chapters, which would come to five hundred large pages in Greek and more in English; but the work as a whole has never been published, although Dain reconstructed the text from eighteen different manuscripts of various dates from 1350 onward. One of them, *Constantinopolitanus gr. 36,* which contains thirty-three chapters, was revealed to the world only in 1887 by the German scholar Frederick Blass, who found it misplaced in the Topkapi Sarayi, the *Seraglio* of fevered harem imaginings, but in fact merely the residence and headquarters of the Ottoman sultans until 1853 (it was their departure to the modern Dolmabahçe Palace that allowed a foreign scholar to search through the library).[94] It is from that particular manuscript that the title *Taktika* is derived, that being the opening word of a paragraph-long title that lists the contents of the work.

The first part, comprising chapters 1 to 55, is a paraphrase of the *Tacticae Constitutiones* of Leo VI "The Wise" (886–912) mentioned above, itself based in large part on earlier works, so that the opening of the *Taktika* of Nikephoros on the qualities of the good general is based on Leo's second constitution, which is itself taken from the *Strategikos* (the general) of Onasander mentioned above, as Dain neatly illustrates with four side-by-side columns.[95]

Chapters 56 to 74 are paraphrases of the *Praecepta Militaria,* of which only six chapters survive; the lost text can be reconstructed in part from Nikephoros Ouranos, for which there is a modern edition of chapters 56 to 65 and another of chapters 63 to 74.[96] They describe the infantry, the cavalry, and especially the kataphraktoi, the armored cavalry, before proceeding to prescribe military operations of familiar style: skirmishing, raids, sieges—and, because head-on battles in open terrain cannot be altogether avoided, instructions on how to win battles.

The third and largest part of the *Taktika* of Nikephoros, comprising chapters 75 to 175, includes chapters 112–118 on how to correspond secretly, 119–123 on naval warfare, published by Dain himself in his 1943 *Naumachica,* and chapters 123–171, which contain an interminable sequence of antique examples of ruses of war, ultimately derived from the *Strategika* of Polyaenus.

In the year 999, Nikephoros Ouranos was placed in command of Cilicia and nearby Syria with his seat in Antioch, the modern Antakya, which had once been the third city of the Roman empire and had been conquered from the Arabs in 969. The greatly strengthened Byzantines did not have to fear the annual jihadi invasions of the past, but there was no peace on the frontier, with raids and counter-raids and larger incursions. They are the subject of chapters 63–65 of the *Taktika,* which reflect the practical military knowledge of an experienced field commander.

Raids differ from offensives because they are not intended to gain territory, but just like full-scale offensives they do require full-scale intelligence preparation, because raiding forces are small and therefore inherently vulnerable, and depend on their survival on the ability to surprise the enemy without ever being surprised themselves. That in turn requires superior knowledge of the enemy and his dispositions:

> The commander of the army must first make investigations through spies, prisoners, and [defectors] and find out the situation of each enemy area, of their villages and fortresses, as well as the size and nature of their [warlike] forces.[97]

Having collected his intelligence and formed his plan, with due consideration for the subtropical climate of the region, whose hottest low-lying regions are to be avoided in summer and whose springtime torrents block passage, secrecy is the very next priority.

Secrecy is always said to be important but is habitually compromised to gather superior strength—it is often better to have additional forces on the line of battle even if their move up to the front cannot be entirely disguised from the enemy. The forewarned enemy may inflict more casualties, but the additional forces can tip the balance and thus secure victory. Raids and incursions are different. There are no such trade-offs: if an enemy has enough advance knowledge to prepare an ambush of adequate strength in the path of advance, total destruction is very probable. Absolute secrecy is therefore not merely an aim, as with larger operations, but an absolute requirement.

To express nothing is essential but not enough, because preparations can still be spotted and intentions can always be guessed; therefore active deception is needed to divert attention from true information or to obscure its meaning by suggesting false interpretations *(suppresio veri, suggestio falsi);* and deception in turn requires countersurveillance:

> [The commander] must make absolutely sure to divulge neither his intentions nor which region he is about to invade to anyone at all, not even to one of those [normally] privy to his secrets. He should instead spread word

that he is planning to depart for some other place, and he should undertake the march as though heading for the region announced by him while keeping his intentions hidden. When he sees that no one is paying any heed, he must make all the proper preparations and then suddenly set off with haste for the region where he intends to go.[98]

In structuring a larger incursion, the infantry and baggage train are to form the rear, protected only by the relatively few kataphraktoi, armored cavalrymen. The lighter cavalry units with their mounted archers are the main strength of the incursion, and it must not be diminished by holding back cavalry in the rear to protect the infantry, and indeed their number should be increased:

> [The commander] should give orders to the officers that each of them must detach one hundred or 150 of his light infantrymen and convert them into cavalrymen *(kaballarios)* to accompany the cavalry force. . . . Similarly, forty or fifty of the *kataphraktoi* should be set apart, who will leave their heavy armor and that of their horses with the baggage train and head out with the other light horsemen.[99]

The text states that the reason is to share out the booty more fairly, but the implied tactical scheme was to have as many troops as possible in the light cavalry to seize plunder and prisoners—the immediate aim of the incursion—while the infantry and kataphraktoi are the cost of doing business, so to speak: they provide a main battle force to support the light cavalry when needed.

If the cavalry raiders are confronted by determined opposition that blocks their progress, the main battle force can move ahead to break through it with close combat; if the enemy counterattacks, the light cavalry can seek protection with the main battle force, which also protects the mules and wagons of the baggage train with its reserve of arrows, spare equipment, and food.

Intelligence was essential to plan the incursion, but fresh intelligence is needed as it begins:

> When the army approaches the enemy region . . . the *doukator* [commander] must above all be quick and alert to send men out . . . [to seize tongues] to acquire accurate knowledge. Invasions made unexpectedly against enemy territory frequently bring many hazards upon the army. For it often happens that one or two days before the raid is launched, a body of reinforcements from somewhere else comes to the enemy.[100]

That invalidates the intelligence previously gathered, even if it was only a week or four days old, and is anything but accidental: when the army moves, the enemy hears about it and sends reinforcements to strengthen defenses in its presumed path.

If the incursion with its raiding horsemen is successful, and the enemy nevertheless fails to mobilize and send his forces into the region under attack, then a second "return" raid is in order. First, animals and men have to rest for three days or more. Then the army of Byzantium must continue traveling on its path back home to allay suspicion, before the moment comes when it quickly turns around to attack again.

The incursion is an offensive operation, but it too needs a defensive dimension for its security: there are to be light-cavalry front guards and flank guards for the pillagers, with the main battle force following up behind along the vector of advance, ready to send reinforcements to light-cavalry forces under pressure. The main battle force must remain in close order for quick communications, so that detachments can quickly be sent ahead into action as needed.

The seizure of plunder and prisoners was an important incentive to fight for the soldiers, and Nikephoros Ouranos duly prescribes that the loot should be shared out fairly by the pillagers, to reward the troops that stayed behind to secure their rear, the camp and the baggage train.

But the strategic purpose of the incursion was to demoralize the Arabs and to weaken them materially, in part by vandalizing their agricultural economic base: "On your way through hostile territory you should . . . burn the dwellings, the crops, and the pastures."[101] This preempted enemy use, with pastures especially important, given the contemporary primacy of cavalry in the armies of both sides. No pastures, or burned-out pastures, meant no horses, which meant no army at the time, because the infantry had a decidedly secondary role, while the cavalry could not carry more than two days' worth of fodder, if that, by stuffing saddlebags and loading spare horses and mules to capacity.

So far strategic surprise has been assumed, but of course it can happen that the incursion is anticipated by the enemy:

> If the enemy is close to the encampment, if [his] army is [large] and they . . . seek battle, it is not good for our army to break camp and begin to march. It should instead remain in camp and the infantry units . . . will prepare for battle. . . . The cavalry units should move out from the encampment and deploy for battle . . . The javeliners, bowmen and slingers who are on foot should stand behind the cavalry units, but not at a great distance from the [heavy] infantry.[102]

If the enemy remains stationary and is weak, there is no need to keep back the infantry to secure the camp, and the entire army can move out to engage and disperse the enemy. If the enemy army is weak and remains as a distance, then its presence should not divert the army from its intended line of advance but the march must still be conducted securely with advance guards, flank guards, and a rear guard (the *saka* again) to form a moving defensive perimeter against hit-and-run attacks around the main battle force, and the unmentioned baggage train too, one may presume. The outer layer of this advancing perimeter is to be formed by light-cavalry units, of course, with infantry forming the inner layer. In so doing, the infantry is not to be burdened with personal armor and the heavier weapons, which are to be loaded on accompanying donkeys, horses, or mules. On the move, the men must march with their own unit, under its commander of ten, fifty, and a hundred *(dekarch, pentekontarch, hekatonarch)*:

> so that if there is a sudden attack . . . each man will be at his assigned station . . . , they will swiftly take up their equipment and all will stand in their formation, each man in his place.[103]

In this case the enemy is weak, too weak to induce the army to change its line of advance, but not so weak that it cannot launch sudden raids that could inflict painful losses, if the proper precautions are neglected. Light-cavalry patrols and pickets are needed all around the marching column to detect enemy attacks and intercept them if possible, while warning the main body to take up battle positions off the march.

Special precautions are prescribed for transiting narrow passes, even if there is no sign of the enemy. In essence the infantry must secure both entrance and exit before the cavalry ventures in, for it is inherently more vulnerable to ambushes. Things are much harder of course if the enemy actually defends a mountain pass that cannot be bypassed. It can be taken for granted that if there is an enemy army present, its main battle force of infantry and cavalry will fight in front of the pass—it cannot fight inside it, where there is no room to deploy, and the cavalry is of little use anyway. If that main battle force is defeated and scattered into flight, the lesser force actually standing within the pass or overlooking the roads through it from higher ground may turn to flee themselves. If they do not, a difficult fight will be inevitable:

> If [the enemy troops] are high up on steep ridges and guarding the roads down below, send javeliners, archers and slingers [= light infantry], and if

possible, some of the *menavlatoi* to encircle these steep places and approach them directly from the level, flat areas.[104]

The aim is to induce the enemy to retreat to avoid encirclement, abandoning tactically superior but isolated positions that will become so many traps if the Byzantines can get through the pass anyway, or around it.

As for the menavlatoi, they seem out of place on steep ridges, for their main role is to wield the pike against charging cavalry, but these sturdy men with sturdy weapons have their uses in mountain warfare too: a few of them can stand against the many if there is a sudden assault against the light infantry in terrain where the cavalry cannot ride to their rescue; and the menavlatoi can add their impressive appearance to the encircling forces that are meant to induce the flight of the defenders.

But if the enemy troops firmly stand their ground, or rather if they remain holding their steep and naturally strong positions, there are to be no frontal attacks: "Do not press on into battle and heedlessly engage them, since the terrain is of aid to the enemy, but go at them from various points and disrupt them with the aforementioned javeliners, archers and slingers."[105] Once again maneuver is the answer rather than the attrition of frontal engagements, but if that too fails, there is no alternative but to fight it out with infantry units, assisted by God of course.

Siege operations are the subject of chapter 65, which is much longer than the previous chapters. Even though the term used is the generic *kastron,* it is implicit that the target is no mere stronghold but rather a major fortress or more likely a fortified city. If the fortress is strongly built and has a very large garrison, there is to be no immediate attack. Instead a campaign of raids is recommended to burn harvests and destroy crops within one or two days' march of the fortress. This must continue until the fortress is weakened by the shortage of provisions and by the resulting decline in its garrison. Only then should the army approach the fortress to receive its surrender, or to take it by assault.

All this is taking place in northern Syria where the Byzantines are on the offensive against the Muslim Arabs, and Nikephoros Ouranos instructs the frontier strategoi—officers in charge of the different sectors—to guard against any inflow of supplies to the enemy. Even though much of the enthusiasm of earlier times is gone, it is still jihad and all the more intense on the defensive:

> For the enemy, oppressed by the lack of provisions, send to the inner regions of Syria and to the towns and communities, and proclaim to the

faithful [*matabadas,* from the Arabic *muta'abida* or pl. *muta'abiddun*] in the mosques the calamities which have befallen them and the pain of starvation oppressing them. . . . They tell them such things as, "should our fortress fall into the hands of the Romans, it will be the ruin of all the lands of the Saracens" whereupon the Saracens rise to the defense of their brethren and their faith . . . and they gather the "donation" as they call it, money, large quantities of grain, and other provisions, . . .—in particular, they send them a great deal of money.[106]

The money is the main threat: if paid enough, "one nomisma (4.5 grams of gold) for two or three modia of grain or even just one modion (12.8 kilos)," even the good Christians living within imperial boundaries, even of high station, will find ways of smuggling grain into the fortified city, and cheese too and sheep, and in great quantities. Intimidation and severe penalties are suggested to deter this treasonous commerce, while caravans bringing foodstuffs from inner Syria are to be diligently intercepted by the frontier forces.

Bribery is also a problem—from the text we can infer that even Byzantine officers were susceptible; there are no invocations to loyalty of faith, instead the suggested remedy is to outbid the enemy:

It is necessary to bolster the morale of the officers guarding the roads and their subordinates, and offer them promises, rewards and gifts so that all will work unstintingly . . . lest the men guarding the roads . . . allow foodstuffs to get through. . . . Those who do the opposite of these tasks, out of sympathy for the enemy or out of negligence, will be liable to severe penalties and punishments.[107]

From the above we can infer that there are also divided loyalties. The fortress is a town or city with a substantial, or even majority, Christian population under Muslim rule, and it too is being starved by the blockade; the Byzantine thematic troops guarding the roads are also local, and may have relatives on the other side, or at any rate empathize with the population under siege.

The fortress is important and a prolonged siege is expected; therefore there must be precautions against attacks from without by enemy forces that come to relieve the siege.

If a major enemy attack is expected, the troops cannot be scattered all around the fortress under siege. There must be an organized camp with a water supply and an entrenched perimeter, if possible further secured against cavalry charges by caltrops and tripods *(triskelia)* with spearlike barbs *(tzipata),* analogous to the chevaux-de-frise used as late as the American Civil War, in the simple form of logs perforated by two sets of spikes at 90-degree angles, so that two points of the four would always be projecting out at a 45-degree angle.

If an enemy relief army is approaching, a last attempt is to be made to obtain the surrender of the fortress, with a show of strength and a declaration. First, each formation, unit, and subunit *(thema, tagma, tourmai, banda)* is to be turned out in its assigned position around the fortress. Then comes the offer: "If you are willing to surrender the fortress to us by your own choice, you will keep your possessions. The first among you will receive gifts from us. If you do not do this [now] and afterward consent to do so, your petition will not be accepted, but the Roman army will carry off both your possessions and your persons as slaves."[108] To add pressure, a further announcement is to be made: "All the *Magaritai,* Armenians and Syrians in this fortress who do not cross over to us before the fortress is taken will be beheaded." Logically, these categories who are to be beheaded rather than enslaved had to be traitorous Christians rather than jihadis—who could redeem themselves by coming over to join the besiegers; in extant documents, starting with a celebrated Greek-Arabic papyrus of the year 642 now in Vienna,[109] the word *Magaritai* means Muslim warriors, not necessarily ex-Christian converts; no doubt the word changed its meaning over time. Nikephoros Ouranos was much too experienced to expect much from mere words, but even if there is no surrender there is another benefit: "disagreement and dissension among them."

Siege operations require specialized equipment as well as artillery, and it is evident that they were not brought in the baggage train, no doubt because of their volume and weight. So the army must manufacture what it needs on the spot, starting with *laisai* made of woven vines or branches for protection against arrows, as already mentioned above, but here wanted in the elaborate form of houses with peaked roofs, a screened porch and two doors "for fifteen or twenty men," rather than just flat screens. Yet more elaborate laisai with four doorways are mentioned, but they must still be easily portable, as indeed they would be if made out of vines and thin branches. They could not actually stop arrows unless their energy was already spent, but they could conceal individual soldiers so that enemy bowmen could not aim at them, and of course they would deflect most arrows.

These portable houses are to be brought right up to the walls—five or ten *orguiai,* that is, 9 or 18 meters away, which seems very close, indeed too close—so that the troops inside them can attack the defenders with their bows and slings. Other troops are to use stone-throwers (trebuchets) against the walls or attack them directly with sledgehammers and battering rams.

Mining is also to be started to dig under the fortress walls to collapse them. The tunnel must be deep to guard against countertunneling. If dug through loose earth, the tunnel's ceiling must be secured with mats supported by posts. The standard method is to be used to collapse the fortress wall at the right moment: first thick wooden posts are to replace dislodged foundation stones, then the cavity is to be filled with dry wood that can be set on fire when all is ready to burn the posts and bring down the wall. The assault must continue around the clock day and night, by dividing the army into three teams, two of which can rest at any one time while the third keeps up the fight.

Nikephoros Ouranos was obviously acquainted with the classical texts of siegecraft, which describe the elaborate equipment of the Hellenistic age, mobile towers, swinging siege ladders, tortoises, and more, "which our generation has never even seen." But after adding quite inconsistently that all those devices had recently been tried out, he asserts that mining is more effective than all of them if done properly.

If all goes well with the siege operations, the besieged enemy will seek to surrender on terms, giving up the fortress but leaving unmolested. The offer is only to be accepted if a relief army is approaching and the garrison is large and strong. Otherwise the fortress is to be taken by assault, in order to demoralize the defenders of other forts and fortresses: "The tidings will circulate everywhere, and other fortresses in Syria which you intend to attack will . . . [surrender] . . . without a struggle."[110]

Next, in chapter 66 Nikephoros Ouranos examines the tactics of the light infantry: archers, javelin throwers, and slingers who are usually positioned behind the heavy infantry, to both have their protection and to support them with their missiles. But the light infantry can also be positioned on the flank to counter encirclements, hold broken terrain, or seize it as the case might be. They will also have to be placed on one flank or both if the heavy infantry is in deep files, so that missiles launched from behind them could inflict fratricidal casualties. Finally, the light infantry can briefly step ahead of the heavy infantry to fight off the enemy cavalry with their missiles.[111]

A key tactical choice in positioning cavalry for battle is the rank/file ratio. A deep–narrow deployment can conceal the true size of the force, and power a breakthrough attack against the enemy line; it can also cross restricted passages less noticeably. A formation with no depth, composed of a single line, is useful to take prisoners and loot undefended places, but is quite useless in battle.[112]

We encounter *viglatores* in the text, meaning watchmen in this case rather than members of the imperial guard regiment named *Vigla* (from the distinctly non-elite *Vigiles*, the municipal guards and firemen of Rome). These watchmen are to light fires at a good distance from the camp, so that they stay closer to it in the dark and spot approaching enemies; they should have castrated horses, which are quieter. Because watchmen are outside the camp by themselves, they may be targeted for capture by the enemy and must be prepared for that.[113]

Very few battles end because the loser is physically destroyed or entirely encircled with no choice but surrender or death. In most cases it is the loser who decides the outcome of battles by choosing to withdraw to avoid further losses when any sort of success no longer seems possible. That is most likely to happen when the battle has lasted for some time, the forces on both sides are depleted and exhausted, and one side is unexpectedly reinforced by fresh forces even if few. (A modern example was the epic holding battle of the Israeli Seventh Brigade against the advance of four successive Syrian divisions in the 1973 October war: after some seventy hours of nonstop fighting, the Syrians abruptly started to withdraw when a mere seven tanks arrived to reinforce the Israelis, who by then could not even stay awake.)

What tips the balance is sometimes material and sometimes psychological, but more often both, and that is why all wise generals always keep a force in reserve, however small, even if that weakens the remaining battle force; the entry of fresh forces into the fight, especially if unexpected, can achieve much greater results than keeping those same forces in action from the start.

Nikephoros Ouranos recommends a stratagem that exploits the difference. If the strategos (here, army commander) is waiting for reinforcements that fail to arrive, he is to detach a contingent and send it off some distance surreptitiously. With battle under way, the detachment can be summoned back to join the fight "with ardor." The enemy will think that reinforcements have arrived and may withdraw from the fight.[114]

The advice for the general on the eve of battle echoes Onasander's literary concoction but passed through the filter of the author's very considerable combat experience. During the night, the strategos should send some cavalry units (*tagmata* in the text but obviously in the sense of detachments) to the rear of the enemy, so that in the morning the foe will be discomfited by seeing them.[115]

The *Strategikon* of Kekaumenos

According to Alphonse Dain and De Foucault, the final work here examined, the eleventh-century *Strategikon* of Kekaumenos, does not even belong to the canon of the Byzantine strategists, but instead is of the genre of all-purpose advice books because it is only in part concerned with military matters.[116] That is true, and the strategic title was added by the first modern editor and is not in the unique manuscript.[117]

The lack of other testimonies to the text is especially unfortunate because the manuscript was copied by a monkish scribe who plainly did not understand what he was writing, so that the successive editors have had to confront many gross errors and even sequences of meaningless characters.[118]

The composition of the text is also problematic in itself, because it meanders from theme to theme and back again with much repetition; on the other hand, it is written in the contemporary colloquial Greek unburdened by the forced classicism of so many Byzantine texts, with the resulting obscurities.

But the most interesting characteristic of this book is its point of view: like other manuals it is addressed to the strategos, whether as field commander or in charge of a theme, but uniquely its concern is not the power and glory of the empire but rather the career and personal honor of the strategos—it is a work of avuncular advice to a younger man, a relative.[119]

For example, its version of the familiar homily begins conventionally. Yes, the strategos must be cautious, but there is no excuse for pusillanimity in claiming concern for the safety of the troops. "If you wanted to safeguard your army, why did you go into enemy territory?"[120]

But the advice then offered is not to seek a middle way between caution and boldness—the safe advice of Onasander and such—but rather to focus on what really matters: one's personal reputation, one's own honor.[121] The strategos should avoid blame for being too timid or too audacious by mounting some seemingly risky but cunningly planned operations that will give him a reputation as one to be feared. It is implicit that these would be lesser operations, while the bulk of the troops would be cautiously handled.

And again, when discussing how the strategos, when in charge of a theme, should handle his nonmilitary role, Kekaumenos is categorical: "Never ever accept a position whose duties include tax collection—you cannot serve both God and Mammon."[122] But it was the empire that needed the taxes.

The same motivation is present when Kekaumenos offers his reading suggestions: "Read books, histories, ecclesiastical texts. And do not object 'what advantage can a soldier derive from the dogmas and books of the church?'—they will certainly be useful." Kekaumenos then points out that the Bible is filled with strategic advice and even in the New Testament there are precepts. But then he moves on to the real motive: "I want you to evoke everyone's admiration for your bravery, prudence, and for your culture."[123]

Kekaumenos was not an Enlightenment cynic but a Byzantine one, so that the text contains its full share of heartfelt and certainly sincere invocations to the deity, but there is no doubt that he calmly writes as if the empire's fate is less important than the personal success of his pupil—and is very much aware of that: "These counsels are not found in any other strategic treatise."[124]

Perhaps the explanation is personal, but it could be circumstantial, a reflection of the fallen condition of the empire. The most recent editor dates the work very closely, between 1075 and 1078, because it refers to Michael VII Doukas (1071–1078) as the ruling emperor, and to the patriarch of Constantinople, Xiphilinus (1065–1075), as already dead.[125]

Half a century earlier, when Basil II died in 1025 he left an amply victorious empire that had expanded in all directions, north and west into the Balkans and Italy, east into Mesopotamia and the Caucasus—an invasion of Sicily was in prospect. But fifty years is a very long time in international politics, especially for the Byzantines, who were forever exposed to the latest arrivals from Central Asia. For centuries Turkic nations had been moving westward in the steppe corridor north of the Black Sea, threatening and sometimes overrunning the Danube frontier, but more recently they had been moving south toward the riches of Iran and Mesopotamia, converting to Islam along the way. Some had joined the jihad against the Christian empire as mercenaries, *ghilman* slave soldiers, or enthusiasts under Arab leadership. But then came a changing of the guard as Turkic warrior-chiefs gradually seized power from Arab rulers across the central lands of Islam, from Afghanistan to Egypt. Alp Arslan of the Seljuk ruling family of Oğuz Turks already dominated Iran and Mesopotamia from the Oxus (Amu Darya) to the Euphrates when he defeated and captured Romanos IV Diogenes (1067–1071) at Mantzikert in August 1071, opening the way into Anatolia for his great many Turkic camp followers.

If the dating 1075–1078 is correct, Kekaumenos was therefore writing in catastrophic times, for Anatolia *was* the empire in large degree, and all but its western edges had fallen under Seljuk rule.[126]

In spite of its decidedly nonliterary style, perhaps the work of Kekaumenos is only literature derived from literature, and not from life. But from the opening passages of the strategic part (at section 24), the book appears to reflect real military experience: intelligence on enemy capabilities and intentions is absolutely essential—without it, it is impossible to achieve good results.

The strategos in charge of the army is therefore enjoined, first of all, to hire a great number of "reliable and dynamic" spies. The term used is *konsarios,* from *cursatores* or *protokoursatores,* which means a scouting or raiding light cavalry in earlier texts, but in this instance the men are also supposed to act as covert agents—indeed the text next specifies that each must work individually with no knowledge of the others, lest all be lost if one is captured (and successfully interrogated).[127] Spies are not enough, there must also be scouts *(sinodikoi)* in units of eight, nine, ten, or more. Only spies have a hope of penetrating enemy headquarters or the ruler's palace to steal or at least overhear war plans, but it takes scouts to detect, monitor, and report enemy actions on the ground that are already under way. The strategos is advised to be generous with gifts to his scouts when they perform well, and to talk with them often—to observe who among them is straightforward and who lies. But he must not share his plans with them.

The next bit of advice reflects bitter experience, including Mantzikert—though it was treason and defection that decided the battle, there was also a lack of information:

> Do everything possible to find out, on a daily basis, where enemy the enemy is and what he is doing.

Followed by another piece of rueful advice:

> Even if the enemy is not cunning, do not underestimate him—act as if he were ingenious.[128]

When it comes to the operational method of war, Kekaumenos is strictly orthodox, repeating the advice of all the previous field manuals since the *Strategikon* of Maurikios, the advice that contains the very essence of the distinctive Byzantine style of war: constantly gather intelligence all around, campaign vigorously, but fight only in small doses, avoiding all-out battles for decisive victory, because there is no such thing—only a brief respite until the arrival of the next enemy, when yesterday's losses *on both sides* would be regretted.

But once engaged in battle, there is no retreating, for that would demoralize the troops. Therefore before the battle the enemy must be probed and tested with lesser attacks, both to determine his strength and also to find out how he fights—for it could be an entirely new enemy that the empire has never fought before—another echo of Mantzikert. Kekaumenos was familiar with the military classics and expected his readers to know them as well—at any rate that is his stated reason for not writing on battle dispositions.[129]

Instead the strategos is urged to find out the ethnicity of the enemy before deploying his forces, because some peoples traditionally fight in a single phalanx, others in two, others still in open order. The author writes that the best battle formation is the Roman one, but gives no reasons for this judgment, most likely because Roman superiority in all military things was taken for granted.

The tone throughout is benign, but on one point the author is fierce: he favors the death penalty for a commander who is surprised by an enemy incursion against his encampment. Many sentries are to be posted all around, even where attack seems most unlikely, for the strategos should never have to say, "I did not expect an attack in that part [of the perimeter]" to which the reply is, "You had an enemy? If so, how could you not think of the worst contingency?"[130] This passage echoes previous manuals but may also reflect personal experience (as in my own case): it is easier to post sentries than to ensure that they remain awake night after night, even though nothing ever happens—except, of course, on the one night when the sentries are asleep.

A series of injunctions follows (sections 32–33): "Do not underestimate the enemy because they are barbarians [*ethnikon* in the text, not *barbaroi*], because they too have the power of reason, innate wisdom, and cunning"; and "If meeting the unexpected, behave bravely to give courage to your subordinates. If you are overcome by panic, who will be able to drive and encourage the army?" As for dealing with envoys visiting the encampment, Kekaumenos echoes the familiar Byzantine procedures:

> They should camp in a low-lying place, and a reliable man must stay with them so that they cannot spy on your army. They must not stroll about nor talk with anyone without permission. Moreover, if there is something really important do not let them in at all, instead read their letter, reply and send them off with magnificent gifts . . . they will praise you.[131]

These are all very familiar themes, but there is also a very interesting appeal for original thought, the greatest commonplace for modern man who is forever being told how clever it is to reinvent everything all the time, but most unusual in Byzantine culture:

> If you are sure that the chief *(archenos)* of the people against which you are fighting is cunning, be watchful for devilish expedients. You too should come up with countermeasures—not just those learned from the ancients—invent new ones. And do not object that it has not been passed down from the ancients.

Kekaumenos respected the ancients as much as the next man, but evidently felt it necessary to disenthrall his readers.[132]

Kekaumenos again takes on unspecified ancients—not Onasander this time—when he contradicts the advice that an army that has fled from battle should be kept out of war for three years:

> I say instead: "If right away, on the same day as the debacle you can collect even just one fourth of your army do not be fearful like the rabbit-hearted, get those whom you have been able to gather and throw yourself against the enemy. Not frontally of course but initially from the rear or the flank, both that day and that night."

That is excellent advice because, as Kekaumenos remarks, the enemy having won the battle will not be watchful, can be caught by surprise, and thus defeat might be turned into victory.[133] (That is how the German army, increasingly outgunned and outnumbered from 1943, prolonged its resistance to the advancing Red Army. Just as Kekaumenos prescribed, it habitually counterattacked immediately after suffering defeats. Nothing is more difficult materially or psychologically—the troops are demoralized, units disorganized, supplies short—and nothing is more effective, because it takes away the enemy's momentum. Red Army troops dashing forward after a breakthrough would run into counterattacks from troops last seen fleeing in panic. Officers with pistols drawn would form retreating troops into improvised *alarmheiten* to mount counterattacks that often inflicted disproportionate casualties. Thus the German army's excellent officers gained more time for their colleagues in the extermination branch, who killed most of their victims in the final year of a war prolonged by sheer tactical skill.)

This piece of advice alone shows that whoever he was, and whether he did or did not have military experience, Kekaumenos did understand the dynamics of combat.

The cold-blooded realism that pervades these pages is best illustrated by the advice on how to deal with enemy envoys asking for gold under the threat of attack: pay them, because the loss will be less than the damage that would be inflicted on imperial territory, and besides, battle is always a gamble.[134] As we saw, even when the Romans were very strong, they were always willing to buy off enemies if it was cheaper than fighting them—not for them the "Millions for defense, but not one cent for tribute" slogan of U.S. Representative Robert Goodloe Harper; since he said that on June 18, 1798, some millions have indeed been spent. Romans and Byzantines were less romantic, but Kekaumenos goes much further: "Refuse enemy demands to give up territory, unless . . . he accepts to be your liege and pay tribute"; a case of feudal subsidiarity, if such terms be allowed, or—it gets worse—"if there is pressing necessity." In other words, do what you can.[135]

Kekaumenos is much more optimistic on the offensive, when the enemy remains ensconced in a fortified city: "If you do not know the magnitude of his force, believe me, he has few men and insufficient forces." His advice is to send out konsarios—the latest version of the same old cursatores/prokoursatores, to find a way into the fortified town—"and do not credit anyone who says that there is no way in—how can such a large area be kept fully under watch?" Once the way in is found, do not go there but remain arrayed in front of the enemy while sending in some well-led men, who once inside are to signal by smoke or fire. Then attack.[136]

Among the Byzantine strategists, Kekaumenos is a lesser figure, but his book, for all its meanderings, shows that there was still a live military culture in which men of affairs where supposed to read military treatises, or even to write one. This gave the Byzantines a real advantage in facing the vicissitudes of their unending wars: a broader range of procedures, tactics, and operational methods than their enemies had, so that they were less often surprised and could themselves more often surprise their enemies, with one more tactic, operational method, or unknown stratagem.

CHAPTER 15

Strategic Maneuver:
Herakleios Defeats Persia

The deepest and boldest theater-level maneuver in the whole of Byzantine history was launched in desperate circumstances to rescue the empire from imminent destruction. It ended with the total defeat of Sasanian Persia.

In the year 603, Khusrau (Chrosoes) II (590–628) had launched the most ambitious and the most successful of all Sasanian offensives. All previous wars since the establishment of the Sasanian dynasty in 224, by Ardashir, grandson of the Zoroastrian priest Sasan, were fought over the control of the borderlands between the two empires: historic Armenia, now mostly in northeast Turkey, the Caucasian lands, and—of greater importance for both empires—upper Mesopotamia on both sides of the Tigris and the Euphrates, now in southeast Turkey. On the Mesopotamian front, the strongly fortified trading cities of Edessa (modern Şanlıurfa, Urfa), Nisibis (Nusaybin), Dara (Oğuz), and Amida (Diyarbakir) changed hands repeatedly from war to war. The balance of the evidence is that in spite of boasts and claims—according to Ammianus Marcellinus, Shapur II (309–379) wrote to Constantius II to claim Persian control as far as the river Strymon and the borders of Macedonia by right of ancient conquest—most Sasanian rulers were actually moderate in their war aims.[1] In spite of intense suspicion, they too recognized the Roman and Byzantine empires as their civilized neighbors that were not to be destroyed—so they were mostly content with limited gains in Mesopotamia when they went to war.

But Khusrau II was entirely more ambitious. His declared aim was to remove and replace emperor Phokas (602–610), whom he condemned

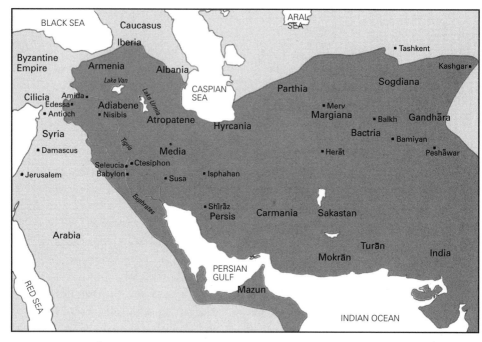

Map 12. The Sasanian empire, ca. 226–ca. 651

as an upstart and usurper—both of which Phokas certainly was, having seized power by mutiny when a *hekatontarchos,* commander of a hundred—in modern terms a captain or perhaps a company sergeant-major. Revenge was another declared motive, for the murder of Phokas's predecessor Maurikios (582–602), whom Khusrau claimed, again correctly, as his own patron and political father: as a young man, he had been sheltered at the imperial court in Constantinople from the deadly intrigues of Sasanian palace politics. Finally, it was also the proclaimed aim of Khusrau to propagate the ancient Zoroastrian religion of Persia and Iran, the dualist cult of Ahura Mazda, "God of Light and Goodness," which had once been the closest competitor to Christianity within the Roman empire as the old pagan cults were fading.

The scale of Khusrau's victories matched his ambitions.

In 610–611, Sasanian armies entered Syria and conquered Antioch, one of the largest cities of the empire.[2] By then they had taken the rich trading center of Edessa—its churches reportedly yielded a booty of 112,000 pounds of silver.[3] In 613 the Sasanians seized Emesa (Homs, Hims) and Damascus, then descending to capture Jerusalem in May 614, where they seized a celebrated fragment of the "true cross." Egypt, the largest single source of Byzantine tax revenues and grain supplies, was next: by 619 Alexandria had fallen, completing the conquest.

Sasanian armies threatened the survival of the empire even more directly by penetrating into its core territory of Anatolia. By 611 they won a major victory at Caesarea of Cappadocia (Kayseri), and in 626 a Sasanian army would reach all the way west to the Asiatic shore directly opposite Constantinople across the Bosporus, at that point less than a mile wide.

Nor could the Byzantines concentrate all their forces against the Persians, because another formidable enemy had advanced across the Balkans into Thrace and its peninsula, which contains Constantinople itself. It was in response to Hun incursions of two centuries before that the Theodosian Wall had been built to guard the city with its moat, walls, and fighting towers. An insurmountable barrier for many an invader until then, the Theodosian Wall offered no such guarantee against the enemy that camped before it in July 626.

The qaganate of the Avars had already defeated several Byzantine field forces and had already captured well-fortified cities before invading Thrace in 618–619; and they had already been bought off in 620 and 623, before attacking Constantinople once again in 626. Like the Huns before them, Avar mounted archers could penetrate targets at long ranges with their composite bows, but they were much more than just light cavalry—they could also fight as heavy cavalry by charging with the lance. They could therefore execute two-step attacks, first driving their enemies into packed close-order formations by threatening to charge or actually charging, and then using their bows to launch volleys of arrows into the dense mass. Moreover, in addition to their major improvements on Hun cavalry equipment and tactics, which the Byzantines studiously imitated, the Avars were also expert in siegecraft and in the construction and operation of artillery, or at any rate of highly effective trebuchets. In the contemporary *Chronicon Paschale* we read that in besieging Constantinople in 626 the Avars deployed

> a multitude of siege engines close to each other. . . . He [the "God-abhorred Chagan"] bound together his stone-throwers [for stability on releasing heavy shot] and covered them outside with hides [as protection against arrows] and . . . prepared to station 12 lofty [mobile] siege towers, which were advanced almost as far as the outworks, and he covered them with hides.

There was a countermeasure to such siege towers: navy sailors had joined the defenders on the wall, and one of them "constructed a mast and hung a skiff on it, intending by means of it to burn the enemies' siege towers."[4]

We later learn from the same source that the Avars also constructed a palisade as a form of circumvallation, to deny easy counterattack opportunities to the besieged, and erected mantelets, that is, wooden frames covered in hides, to shield the besiegers from missiles.[5] That is evidence enough that, unlike most nomads, the Avars had the technology needed to defeat fortifications.

Moreover, like Attila before him, and like other successful steppe potentates after him, the qagan of the Avars who reached the walls of Constantinople in 626 had gathered a much larger mass of other warriors around the elite core of his own Avar cavalry, in this case some Germanic Gepids and Slavs in great numbers. Finally the qagan evidently had his own talents of intelligence and diplomacy, because he arrived to attack Constantinople from the European side just when a Sasanian army that had advanced all the way to the western edge of Anatolia was camped on the Asiatic shore directly across the Bosporus in front of the city. The *Chronicon Paschale* records that when a delegation from the city went to negotiate with the Avars,

> the Chagan brought within their sight three Persians dressed in pure silk who had been sent to him from Salbaras [Shahrbaraz, chief of that Sasanian army] and he arranged that they should be seated in his presence, while our envoys should stand. And he said, "Look, the Persians have sent an embassy to me, and are ready to give me 3,000 men in alliance. Therefore if each of you in [Constantinople] is prepared to take no more than a cloak and a shirt, we will make a compact with Salbaras, for he is my friend: cross over to him and he will not harm you; leave me your city and property. For otherwise it is impossible for you to be saved [given that Avars and Persians controlled all the land on both sides of the Bosporus], unless you become fish and depart by sea, or birds and ascend the sky.[6]

Ever since he had seized power from Phokas in 610, Herakleios had been trying to fight against the offensives of Khusrau II, sometimes winning and sometimes losing battles, and twice being forced to retreat altogether to confront the Avars. The net result was that by 622 all that remained of the empire beyond its capital, Greece, and the uninvaded part of Anatolia were scattered islands, coastal tracts, and thinly garrisoned cities in North Africa, Sicily, Italy, and the Dalmatian coast—none of them really valuable as sources of revenue or recruits. The result was the exhaustion of the imperial treasury, further depleted by futile attempts to appease the Avar qagan with tribute. In the face of the deadly and immediate threat of the combined attack on Constantinople by the Avars with their mass of Slav warriors and the Sasanian Persians, the money needed to keep fighting was running out. Under the year 6113 since the creation (622 CE) the *Chronicle of Theophanes* records the extreme measures taken by Herakleios:

> He took on loan the moneys of religious establishments and he also took the candelabra and other vessels of the holy ministry from the Great Church, which he minted into a great quantity of gold and silver coin [*nomismata*, gold coins, 72 to the pound; and *milaresia*, silver coins, exchanged at 12 to the *nomisma*].[7]

Ever since the Sasanian invasion had started nineteen years before in 603, Byzantine forces had been battered again and again, in defeats, retreats, and the outright collapse of frontier and city defenses. But evidently surviving units, fragments of units, individual veterans, and new recruits rallied around Herakleios, whose ability to lead them to victory was entirely unproven, but who could certainly pay them:

> Having found, then, the army in a state of great sluggishness, cowardice, indiscipline, and disorder, and scattered over many parts of the earth, he speedily gathered everyone together.[8]

The opportunity to raise morale in the usual way, by stressing the iniquity of the enemy, was not missed, according to Theophanes, who cites Herakleios speaking to the troops in very familiar manner unusual for an emperor, if only to incite them by arousing their religious resentment:

> You see, O my brethren and children, how the enemies of God have trampled upon our land, have laid our cities waste, have burnt our sanctuaries . . . and . . . how they defile with their impassioned pleasures our churches.

But training was the main thing, with the learning of individual combat skills leading to full-scale battle exercises by complete formations, whose realism evidently made a great impression on the source that informed Theophanes—unless he is relaying what he witnessed of the army of his own time (he died in 818), which certainly exercised in the same realistic manner.

> [Herakleios] . . . formed two armed contingents, and the trumpeters, the ranks of the shield-bearers and men in armor stood by. When he had securely marshalled the two [sides] he bade them attack each other: there were violent collisions . . . and a semblance of war was to be seen. One could observe a frightening sight, yet one without the fear of danger, murderous clashes without blood.[9]

Later that year, the new army of Herakleios won some minor battles, or possibly only skirmishes, in fighting against the Sasanian forces in southeast Anatolia, but in 623 another Avar advance on Constantinople forced him to return, only to be almost captured himself when he tried to negotiate with the Avar qagan.

The suburbs of the city were devastated by Avar looting and there was certainly a land blockade, but no determined assaults, as far as we know. In any case, on March 25, 624, Herakleios set out to launch his first serious counteroffensive against Sasanian Persia.

By then everything else had been tried, including an attempt to come to terms with a dominant Persia by a negotiated surrender. According to the *Chronicon Paschale,* in the 615, after major defeat before the walls of Antioch, the loss of Syria, and the fall of Jerusalem, when the first Persian incursion through Anatolia had reached the Sea of Marmara shore opposite Constantinople, a letter was sent to Khusrau II, practically accepting his overlordship, whereby Byzantium would become a client-state under the traditional Persian system of indirect rule:

> We . . . having . . . confidence in . . . God and your majesty, have sent [to you] your slaves Olympius the most glorious former consul, patrician and praetorian prefect, and Leontius the most glorious former consul, patrician and city prefect, and Anastasius the most God-loved presbyter [of Hagia Sophia]; we beseech that they may be received in appropriate manner by your superabundant Might. We beg too of your clemency to consider Heraclius, our most pious emperor, as a true son, one who is eager to perform the service of your Serenity in all things.[10]

According to the Armenian chronicle attributed to Sebeos, Herakleios himself sent his own letter addressed to Shahin, the Sasanian commander on the scene, stating his own willingness to accept whomever Khusrau should appoint: "If he should say: I shall install a king for you, let him install whom he wishes, and we shall accept him."[11]

As it turned out, these peace initiatives failed, but there was no attack on Constantinople in 615 because Khusrau's armies were diverted to instead invade Egypt, of greater economic value than battered and besieged Constantinople, and much easier to conquer.

So it was not until 622 that a supposed reply by Khusrau is recorded. The text contained in the Armenian history of Sebeos seems designed to evoke religious indignation; perhaps it was deliberately distorted, if not outrightly forged, by Herakleios himself as a piece of propaganda, to stiffen Byzantine resistance.

That it was propaganda, or at least that it contained disinformation, is suggested by the quotations from Isaiah and the Psalms—language most unlikely to be used by Khusrau, who had endowed his war with a strongly religious dimension, as the struggle of the Zoroastrian worshippers of Ahura Mazda, "the God of Light," against the Christian empire:[12]

> [I, Khusrau], honored among the gods, lord and king of all the earth, and offspring of the great Aramazd [Ahura Mazda], to Heraclius our senseless and insignificant servant. . . .
>
> Having collected an army of brigands, you give me no rest. You claim to trust in your god. Why did he not save Caesarea [of Cappadocia] and Jerusalem and the great Alexandria [of Egypt] from my hands? Do you not know that I have subjected to myself the sea and the dry land? So it is only Constantinople that I shall not be able to erase?[13]

Further increasing the likelihood of forgery, the next passage in the letter seems designed to enhance the authority of Herakleios as a war leader, because its offer of generous terms for himself implies that his campaigning is selfless:

> However, I shall forgive you all your trespasses. "Arise, take your wife and children and come here. I shall give you estates, vineyards and olive trees whereby you may make a living" [Isaiah, 36.16–17]. . . . Let not your vain hope deceive you.[14]

Diplomacy having failed, and defensive operations having failed also, allowing Avar and Sasanian forces to converge on Constantinople, Herakleios set out on March 25, 624, with his newly trained army to launch a counteroffensive.

The safe course would have been to push back the Sasanian armies step by step across the length of Anatolia and back into Mesopotamia—except that all the candelabra and vessels of the churches of Constantinople could not have paid for an army large enough to advance by sheer strength in a frontal offensive.

Moreover, even if successful initially—unlikely given the ratio of forces—a frontal offensive could not have succeeded for long, because it would have allowed Khusrau full warning and ample time to recall the Sasanian garrisons scattered in Egypt and Syria to reinforce his armies facing Herakleios.

Accepting the huge risk of leaving Constantinople to defend itself, Herakleios took his forces in a deep-penetration offensive, or if one prefers, a strategic raid, all the way east through what was then Armenia and is now northeast Turkey, to reach the original Sasanian heartland in what is now northwest Iran. Great boldness was rewarded by complete surprise.

The army of Herakleios seems to have been little resisted as it went through Theodosiopolis (Erzerum) and what is now the province of Ayrarat, capturing and looting Dvin, before reaching and destroying the great Zoroastrian temple at Takht-I-Suleiman, extinguishing the eternal fire of *Vshnasp,* as Sebeos called it, or more correctly Adur Gushnasp, near Ganzak, the Greek Gazaca, capital of Media Atropatene, near Takab in modern western Azerbaijan, now still within Iran.[15]

This was no doubt revenge for the burning of the Jerusalem churches in 614, but it is impossible to believe that it was not also a calculated move designed to provoke Khusrau into a frantic and ill-prepared response, because Adur Guhnasp was very much his dynasty's sanctuary.[16] That is what actually happened: separate and uncoordinated Sasanian armies were sent to intercept Herakleios, and his men defeated at least one of them, led by the most distinguished of Sasanian field commanders, Shahrbaraz, before encamping for the winter.

In March 625 Herakleios retreated rapidly from Armenia to the warmer plains of southeast Anatolia, crossing through the mountain pass of the Cilician Gates (modern Turkey's Gülek Boğazi). At one point he was again pursued by Shahrbaraz, but all Persian attempts to combine forces against his own highly mobile army failed. It was an exceptionally sustained and especially successful implementation of the operational method recommended for outnumbered forces by the manual *Peri Strategikes,* discussed in Chapter 10.

This war of movement was conducted within Anatolia, imperial territory that is.[17] Much of it is mountainous, but even there interrupted by fertile, well-watered valleys, while in southern Anatolia there were the richer coastal plains of Cappadocia and Cilicia. That explains how the army of Herakleios could survive at all. It was clearly not the treasury in Constantinople, starved of imperial revenues, that was feeding his armies, but locally collected taxes, contributions from churches and monasteries in those parts, and, no doubt, forceful requisitions. Moreover, all the sources concur that the marching and fighting on both sides came to an end each year with the approach of winter. It can be very harsh in Anatolia, not only in rugged mountain country but in the plains as well, at least in the interior where even the plains are mostly fairly high plateaus.

Troops could be hardened for winter campaigning—it was done at times by Roman commanders, who acquired fame as disciplinarians for keeping men in tents in all weathers. That was not, however, a practice that Herakleios was likely to imitate if he could help it: his predecessor once removed, arguably his direct legitimate predecessor, Maurikios (582–602), had been overthrown and killed by mutineers when he ordered the army to set out for a winter campaign against the Avars and the Slavs, after long months of fighting. Their leader, Phokas (602–610), seized the imperial power as many had done before him, but lacked the political talent to fabricate an atmosphere of legitimacy, and the resulting turmoil provided the opportunity for Khusrau's invasion.

Outnumbered as he was, and chased by more than one Persian force, Herakleios could not have simply stopped when the cold weather arrived, if the Sasanian troops had not already stopped before him—as they must have done. It was not that the Sasanian troops were less hardy than the Byzantine, but their horses needed fodder to survive after October, when green pasture runs out in the mountainous parts of Anatolia, and so they had to retreat to winter quarters where they had stored fodder, staying there more or less immobile till the spring.

This logistic detail became a factor of great importance in the ensuing events. An expectation had been built up that whatever else Herakleios might do, he would retreat to imperial territory in Anatolia and take shelter in well-stocked quarters before winter set in—just as he had done in 624 and then again in 625.

That created the conditions for strategic surprise in 627, when he continued to advance right through the winter. Byzantine horses were no different from those of the Sasanians, but very different horses did exist in the world, and they would soon arrive on the scene in the most dramatic way possible.

Herakleios was still away from Constantinople on June 29, 626, when the city came under the convergent attack of the Avars with their siege engines, their Slav followers, and the Sasanian army of Shahrbaraz. According to the *Chronicon Paschale*:

> [Some Avars] approached the venerated church of the Holy Maccabees (in Galata, across the Golden Horn from Constantinople); they made themselves visible to the Persians, who had congregated in the region of Chrysopolis [Üsküdar, on the Asiatic ashore directly opposite Constantinople] and they made their presence known to each other by fire signals.[18]

Both Avars and Persians had been there before, but separately. That they concerted their moves in 626 to attack the city at the same time is probable—Theophanes stated it: "As for Sarbaros [= Shahrbaraz], he [Khusrau] dispatched him with his remaining army against Constantinople with a view to establishing an alliance between the Western Huns [Avars] and the . . . Slavs, and so advancing on the City and laying siege to it."[19]

Even if the two sides were perfectly coordinated at the political level, that did nothing for the operational coordination of the two armies—and it was chiefly armies that were attacking the maritime city of Constantinople, which projected into the sea with only a narrow landward side, and that too strongly fortified. Neither the Sasanians nor the Avars had ships with them, let alone warships. The qagan's solution was to send his Slav subjects in their small boats, *monoxyla* (= one tree = dugouts), to attack the seaward side of the city that faces the Golden Horn, where there was a protecting seawall but much weaker than the triple Theodosian Wall. The same small boats were to ferry Sasanian troops across. They "filled the gulf of the Horn with an immense multitude, beyond all number, whom they had brought from the Danube in [*monoxyla*]."[20] The Slav boats were no match for the galleys and skiffs of the Byzantine navy with their skilled crews and embarked bowmen. All the catastrophic defeats that had deprived the empire of its most valuable lands, and much of its army, could not have done equal damage to the navy of an empire that still had coastal possessions in North Africa, southern Spain, Sicily, Italy, Crete, Cyprus, and many Aegean islands.

The Byzantine navy waxed and waned, but it was powerful enough from July 29 to August 7, 626, the day when both Avars and Sasanians gave up their sieges. In the *Chronicon Paschale* we read: "70 of our skiffs sailed up towards Chalae, even though the wind was against them, so as to prevent the [*monoxyla*] from crossing over."[21] These were not warships, but they were boats of some size, not common "skiffs," flat-bottomed rowing boats. The revealing detail that they sailed against the wind implies skilled crews and well-rigged sails as well as a combat capability with bowmen.

The Armenian history of Sebeos reports "a battle at sea, from which the Persian army returned in shame. They had lost 4,000 men with their ships."[22] The *Chronicon Paschale* describes the fate of the Slavs:

They sank [the boats] and slew all the Slavs found in the canoes. And the Armenians [infantry] too came out from the [sea] wall . . . and threw fire into the portico which is near St. Nicholas. And the Slavs who had escaped by diving from the canoes thought, because of the fire, that those positioned by the sea were Avars, and when they came out [from the water] . . . they were slain by the Armenians.[23]

When the siege was abandoned, the Sasanian troops of Shahrbaraz retreated into eastern Anatolia to pursue Herakleios once more, and the qagan dismantled his siege engines under a truce while threatening to return, though many of his Slavs were departing in discord; as we saw, their dissent may have been encouraged and rewarded by the Byzantines.

Having failed to seize the city, having certainly eaten out the surroundings during the monthlong siege, Avars and Slavs had to raid elsewhere for their food. In normal circumstances, Byzantine armies on the move could summon convoys of carts loaded with food as the Romans had done before them, and could therefore mount long sieges lasting months even after the country roundabout had been stripped bare.

The Avars had no such logistics based on tax collection—they depended on tribute and simple extortion. That obviously worked well enough for them, and besides with their many horses for warriors, for family members and servants too, the Avars could forage in a wide radius to sustain long sieges. The Slavs had fewer horses—few indeed, it seems—and were far too numerous to be fed by propitiatory food gifts from the besieged. So they left, and that alone would have forced the qagan to give up the siege, because while the Avars were tactically so dominant that they could overawe many Slavs, they were too few to besiege the six kilometers of the Theodosian Wall.

The war of movement that Herakleios had been conducting ever since March 624 changed drastically in the autumn of 627.[24]

Once again he advanced eastward into the Caucasus, and was no doubt expected to retreat once again with the approach of winter. But this was no raid, however "strategic"; it was a full-scale, deep-penetration offensive. It was made possible by the powerful reinforcement of Herakleios's by now experienced, highly mobile, but necessarily small army. Riding the small, hardy horses (or ponies) that first appeared with the Huns and would return to Europe for the last time six hundred years later with the Mongols, mounted archers from the steppe lands had arrived by way of the Caspian plains, "40,000 brave men" according to Theophanes in the year 6117 since the creation.[25]

They were brought by a much grander Turkic qagan than his Avar counterpart: "Ziebel" in Theophanes but undoubtedly the Tong Yabghu who was chief ruler of the western qaganate of the vast Türk empire that stretched all the way from China to the Black Sea, and that was then disintegrating, or head of its emerging successor, the Khazar qaganate—or both of these things, given that the Khazars certainly came out of the larger qaganate.[26] Be that as it may, the people of Tong Yabghu were former allies and present enemies of Sasanian Persia, whose influence in Central Asia naturally collided with the Turkic interest, under whatever name—the ancestral (and continuing) competition between Iran and Turan. Moreover, they were the hereditary enemies of the Avars, who had originally ruled the Türks and then had fled westward from them. To the Türks, the Avars who had just failed to take Constantinople were "slaves . . . who had fled their masters . . . [fit only to be] trampled under the hooves of our horses, like ants."[27]

Even this boldest of Byzantine offensives could not be a military action alone. It was preceded, accompanied, made possible, and followed by energetic efforts to secure allies and divide enemies by all possible means.

The arrival of the mounted archers was not a fortuitous event. The Byzantines had been negotiating with the western Türk qaganate for decades by sending envoys in long and perilous journeys. And *De Administrando Imperio* claims that the envoys of Herakleios were instrumental in the political separation of the future Serbs and Croats from the undifferentiated mass of the Slav followers of the Avars, who were then persuaded to actively oppose the Avars before fleeing from them northward to where they still reside.[28]

Evidently Herakleios was a very great field commander, but without concurrent efforts to persuade, induce, and dissuade, he could hardly have won his war. He failed to appease Khusrau, or the Avar qagan, but he did much better by inducing Serb and Croat tribes to defect from the Avars, and above all by recruiting Tong Yabghu.

According to Theophanes, Herakleios did even more, successfully subverting Shahrbaraz, Khusrau's leading general, when he was within easy reach during the siege of Constantinople.[29] Subsequent events, however, leave us uncertain about Shahrbaraz's loyalties, so the complicated intrigue recorded by Theophanes may be romance.

With thousands of formidable mounted archers added to his force, Herakleios could obviously maneuver far more freely, because Sasanian pursuit forces were more likely to turn away than to fight against such poor odds. The Turkic allies also brought with them another advantage, or rather rode upon it: while the horses of the Byzantine and Sasanian cavalry had to be fed in winter at least, their Mongol horses (or ponies) could survive in almost any terrain that had almost any vegetation, even under the thin layer of icy snow typical of the windy steppe in winter—and of the hill country of northwest Iran where Herakleios was headed.

Setting out from Tbilisi—now the capital of Georgia—in September 627, Herakleios led his small army with his formidable allies in a vast turning movement hinged on Lake Urmia, now in northwest Iran, to then move south, crossing the Greater Zab River to reach Nineveh by the Tigris River, once the great Assyrian capital mentioned in Genesis, and now the very large and unprepossessing Iraqi city of Mosul. Khusrau had sent a large army under Roch Vehan to pursue Herakleios, though the Persians could not overtake the Byzantines to force them into battle. But on December 12, 627, it was Herakleios who chose to give battle, suddenly turning to confront the Sasanian army. The Armenian history of Sebeos offers a glimpse of the battle, enough to recognize the tactical style of Herakleios as field commander—first maneuver to confuse the enemy to achieve surprise, only then attack:

> Joining forces [the Sasanians] pursued Herakleios. But Herakleios drew them on as far as the plain of Nineveh; then he turned back to attack them with great force. There was mist on the plain and the Persian army did not realize that Herakleios had turned against them until they encountered each other . . . [the Byzantines] massacred them to a man.[30]

It was important to deplete Sasanian strength, but the decisive new fact was that the forces of Herakleios had penetrated into the deep rear of the hugely expanded territory conquered by Khusrau's armies, and could strike at the vital centers of Sasanian power in what is now central Iraq. The empire was definitely Persian, but the Sasanian capital was in Mesopotamia, at Ctesiphon on the Tigris River, less than twenty miles south of modern Baghdad. It was certainly one of the largest cities in the world, if not the largest—and now it was exposed to Byzantine attack, for Khusrau's past victories and past conquests had created an unsolvable strategic problem: Sasanian armies were scattered across a wide arc from remote Egypt to Syria and into distant Anatolia—all of them much too far away to come back in time and intervene quickly enough to stop Herakleios before he did more damage. Had the Sasanians not been so sure that Herakleios would again retreat in winter as he had done each year till then, they could certainly have withdrawn enough troops from Egypt and Syria to defend their core territory.

For Herakleios, the victory at Nineveh meant first of all the solution of all his logistic problems. Khusrau had many palaces instead of just one or two, a Sasanian habit that reemerged with Saddam Hussein and for the same reason—each was a simulacrum of power to overawe the surroundings. Khusrau's palaces, moreover, were built in the classic Persian style, with large, paradisaical gardens rather than especially large buildings, including zoological gardens filled with both exotic and domestic animals—all meat on the hoof for hungry troops:

> He found therein in one [palace] enclosure 300 corn-fed ostriches, and in another about 500 corn-fed gazelles, and in another 100 corn-fed wild assess [onagers], and all of these he gave to his soldiers. And they celebrated 1 January there [not our January 1]. They also found sheep, pigs and oxen without number, and the whole army rested contentedly.[31]

Next Herakleios continued south toward Ctesiphon, the capital, crossing the Lesser Zab by the end of January 628 before advancing some two hundred miles across the river Diyala to seize another, much larger palace at Dastagard. Theophanes relishes the result:

> In his palace of [Dastagard], the Roman army found 300 Roman standards which the Persians had captured at different times. They also found . . . a great quantity of aloes . . . much silk and pepper. More linen shirts than one could count, sugar, ginger, and many other goods. . . . They also found in this palace an infinite number of ostriches, gazelles, wild asses, peacocks and pheasant, and in the hunting park huge live lions and tigers.[32]

There is also evidence of how swiftly Herakleios moved, and of the extent of the surprise he pulled off, because Theophanes records that many palace officials were captured.

In itself, as purely military events, the defeat at Nineveh, and even the subsequent advance of the Byzantine army down the Tigris Valley toward Ctesiphon, need not have been catastrophic for Khusrau. He still had large intact forces at his command that were still occupying the vast territories newly won, which obviously required substantial garrisons. His field commander Shahrbaraz was in Syria, with a large army that could have returned to protect the capital Ctesiphon.

What happened next shows that the constant warfare that had lasted for a quarter of a century had finally exhausted the tolerance of the Sasanian ruling elite and of Khusrau's own family. It is possible that Herakleios stimulated the process by sending him a letter offering peace for propaganda purposes, which Khusrau supposedly rejected: "I am pursuing you as I hasten towards peace. For it is not of my free will that

I am burning Persia, but constrained by you. Let us therefore throw down our arms even now and embrace peace. Let us extinguish the fire before it consumes everything." Theophanes continues: "But Chosroes did not accept these proposals, and so the hatred of the Persian people grew against him."[33]

Instead Khusaru mobilized his last "retainers, noblemen and servants," sending these palace nonfighters to fight the highly experienced veterans of Herakleios. It was a last throw of the dice. On February 23, 628, according to Theophanes, when Herakleios seemed to be on the verge of entering Ctesiphon and finishing off the empire, Khusrau was overthrown and killed in a coup d'état by his own son Kawadh-Siroy, who opened peace negotiations and offered a prisoner exchange.

What ensued was not a capitulation but a negotiation—there were still large Sasanian armies in the field whose return could have tipped the balance. Instead of entering Ctsesiphon—at some thirty square kilometers, its sheer size was probably intimidating for his small army—Herakleios moved more than three hundred miles northeast, returning to the familiar terrain of Takht-i-Suleiman (now Ganzak) in the foothills of the Zagros Mountains in April 628.[34]

That did not stop the deadly sequence of palace politics in Ctesiphon—if one man could seize the throne, why not another? Kawadh-Siroy was himself overthrown by Shahrbaraz in a military coup d'état—he being the same field commander with whom Herakleios had talked more than once in the past. Shahrbaraz duly started negotiating for terms, eventually reaching an agreement. All the lost provinces, from Egypt to Syria and Anatolia's Cilicia, were returned to Byzantine rule, but the Sasanians seemingly kept their very first conquests on their side of the river—actually reconquests for them, because those had earlier been Sasanian territories.[35]

As the inheritor of an imperial culture already ancient with experience, Shahrbaraz knew how to negotiate: conversion to Christianity was part of his package of concessions—a conversion of course soon revoked.

Herakleios had risen to power when the empire was in immediate danger of being extinguished by Khusrau's invasions, which reached much deeper than any Sasanian invasion of the previous four centuries of intermittent warfare, and by the Avar offensives that directly attacked Constantinople in the summer of 626. He did not have the military strength that would have been needed to drive back the Sasanian Persians or the Avars with their camp followers, let alone both. He had just enough strength both on land and at sea simply to resist them at the very walls of the city, and on the waters just in front of it—not to restore an empire submerged by enemy forces. It is not a wild exaggeration to say that July 626 could have been May 1453, as the beleaguered defenders of Constantinople awaited the end.

The solution devised by Herakleios combined diplomacy and subversion (within both enemy camps) with a high-risk, *relational* maneuver on a theater-wide scale—an historical rarity in itself. What was "relational" about it, and very profitably too, was that successive seasonal raids by Herakleios had habituated Khusrau and his advisors to expect bold, deep, but ultimately inconclusive raids that would last a few months till winter, and would leave the strategic situation unchanged. Yes, the damage was sometimes painful, as with the destruction of the Zoroastrian temple at Takht-I-Suleiman, a real blow to the prestige of Khusrau and his dynasty. It claimed priestly authority—it was named after Sasan or Sassan, great priest of the Temple of Anahita and grandfather of Ardashir the founder—and its rulers were consecrate before that same "royal" fire of Adur Gushnasp.

But Khusrau evidently decided that even very damaging raids did not justify the cost of the only fully reliable remedy, which was to withdraw Sasanian forces from the newly conquered lands of Syria and Egypt, to instead guard the old borders of the empire and its core territory in Mesopotamia. It would have meant abandoning Khusrau's great achievement—his unprecedented conquests of Byzantine territories.

In essence, in the crucial year 627 the Sasanians did not withdraw from the west because they were sure that once his raiding was done, Herakleios would again withdraw from the east. He did not, and the result was the end of a dynasty and an empire that had endured over four centuries.

What happened next was the loss of the Levant, Egypt, and eventually North Africa to the Muslim conquest, but that scarcely nullified the epic victory of Herakleios, because it was the empire itself that Khusrau had wanted, and had claimed as the avenger of his benefactor Maurikios, and not just the lands lost and regained only to be lost again.

As the empire entered its most miserable years, in the gloom there still shined the bright memory of what had been accomplished by the expeditionary army of Herakleios. For that we have the testimony of our major source, Theophanes, who died in 818: it is evident from his prose that the memory of those glorious events had remained undimmed by the passage of two centuries.

Conclusion: Grand Strategy and the Byzantine "Operational Code"

All states have a grand strategy, whether they know it or not. That is inevitable because grand strategy is simply the *level* at which knowledge and persuasion, or in modern terms intelligence and diplomacy, interact with military strength to determine outcomes in a world of other states, with their own "grand strategies."

All states must have a grand strategy, but not all grand strategies are equal. There is coherence and effectiveness when persuasion and force are each well guided by accurate intelligence, and then combine synergistically to generate maximum power from the available resources. More often, perhaps, there is incoherence so that the fruits of persuasion are undone by misguided force, or the hard-won results of force are spoiled by clumsy diplomacy that antagonizes neutrals, emboldens enemies, and disheartens allies.

The Byzantines had no central planning staffs to produce documents in the modern manner, including the recent innovation of formal statements of "national strategy" that attempt to define "interests," the means to protect and enhance them, and the alignment of the two in rational or at least rationalized terms. The Byzantines never called it that—even "strategy" is only a Greek-sounding word not used by ancient or Byzantine Greeks. But they assuredly had a grand strategy, even if it was never stated explicitly—that is a *very* modern and indeed rather dubious habit—but certainly it was applied so repetitively that one may even extract a Byzantine "operational code."

First, however, two matters must be defined: the identity of the protagonists and the nature of strategy, or rather of the paradoxical logic of strategy.

Identity

The Byzantine ruling elite faced the outside world and its unending dangers with a strategic advantage that was neither diplomatic nor military but instead psychological: the powerful moral reassurance of a triple identity that was more intensely Christian than most modern minds can easily imagine, and specifically Chalcedonian in doctrine; Hellenic in its culture, joyously possessing pagan Homer, agnostic Thucydides, and irreverent poets—though *Hellene* was a word long avoided, for it meant pagan; and proudly Roman as the *Romaioi*, the living Romans, not without justification for Roman institutions long endured, at least symbolically.[1]

But until the Muslim conquest took away the Levant and Egypt from the empire, this triple identity was also a source of local disaffection from the ruling Constantinopolitan elite, for of the three only the Roman identity was universally accepted.

To begin with, the speakers of Western Aramaic and Coptic, who accounted for most of the population of Syria and Egypt, including the Jews in their land and beyond it, did not partake in the Hellenic culture—except for their own secular elites, which were organically part of the Byzantine regime and were indeed often attacked by nativists as "Hellenizers." For the rest, the masses either did not know that Homer ever lived, or were easily led by unlettered fanatical priests to vehemently hate what they were too ignorant to enjoy.

Moreover, the zone that rejected Hellenism, as it had rejected the Roman habit of bathing as too sensual, also rejected the excessively intellectual Chalcedonian definition of the dual nature of Christ, both human and divine, insisting on the more purely monotheistic conception of the single, divine nature of Christ.

That is the Monophysite creed still upheld by the Christians of the Coptic churches of Egypt and Syria, the Orthodox churches of Ethiopia and Eritrea, the Jacobite and Malankara Orthodox churches of India, and in much more nuanced fashion by the Armenian Orthodox Apostolic Church. In these ecumenical days, Orthodox Christians are no longer deeply committed to their sides of the Chalcedonian dispute, but the Byzantine empire of the sixth and seventh centuries was dangerously divided by the Chalcedonian persecution on one side and, on the other,

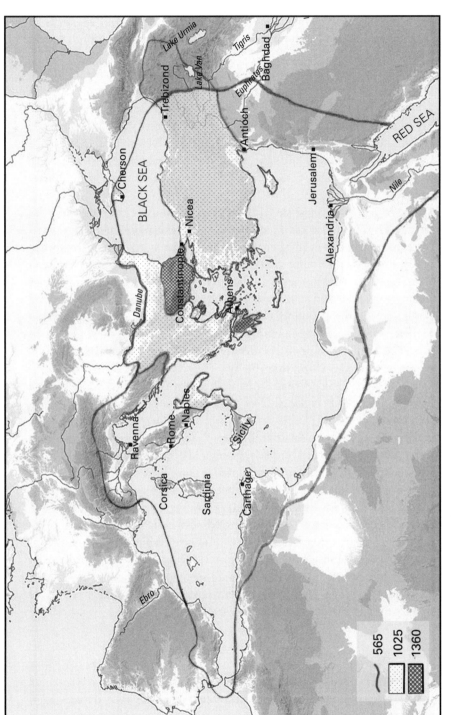

Map 13. The empire in 565, 1025, and 1360

by Monophysite vehemence, which rejected all imperial attempts at doctrinal compromise, notably the monoenergism and monotheletism of Herakleios.

None of this should have diminished the willingness of Monophysite Christians to fight for the empire against non-Christians, but it did, and for good reasons: most non-Christians who attacked the empire were not doctrinally anti-Christian, and several pagan enemies converted to Christianity, notably the Bulghars, Magyars, Kievan Rus', Serbs, and Croats; the doctrinally anti-Christian Muslims, on the other hand, were as purely monotheistic as the Jews, and more so than the Chalcedonians.

The Muslim conquest saved the empire from these deep divisions by cutting away its most vehement dissidents. It was by no means linguistically homogeneous even after that—there were many Armenian-speakers in the east, and many Slav-speakers in the west, while in between there long survived autochthonous languages, such as the Thracian, or Bessic, spoken within sight of the Theodosian Wall and recorded among monks within them. But none of that interfered with participation in the Hellenic culture for those who wanted it, and a different original language comported none of the divisiveness of the doctrinal fracture. It might be said, therefore, that the loss of Syria and Egypt, unlike Latin-speaking and Chalcedonian North Africa, was a mixed curse for the empire: it brought the blessing of religious harmony, and increased cultural unity.

The Muslim onslaught found a very large but disunited empire that might have disintegrated entirely, given a little more bad luck, and left a smaller, much poorer, but more united empire hardened to successfully withstand six centuries of wars. (There are sound reasons not to abuse the word *crusade* to describe Byzantine warfare, yet it had a holy war dimension, certainly on the frontier as described in *De Velitatione*.)

This too is an explanation for the immense resilience of the imperial ruling class in times of acute crisis, and during agonizingly protracted periods of extreme insecurity: when all seemed bleak and hopeless, the Christian faith, the culture of ancient Greece, and Roman pride combined to reject surrender and inspire tenacity.[2]

The Logic of Strategy

"Strategy" is one of those Greek words that the Greeks never knew, for the pan-Western word *strategy* (*strategie, strategia,* etc.) derives from *strategos,* often mistranslated as "general," but in historical fact a combined politico-military chief, and thus a better source-word for an activity that is equally broad. The logic of strategy is not quite so simple.

"Men do not understand . . . [the coincidence of opposites]: there is a 'back-stretched connection' like that of the bow."[3] Thus Herakleitos or Heraclitus of Ephesus, thought very obscure by the ancients, but for us entirely transparent after the experience of the paradoxes of nuclear deterrence, whereby the peaceful had to be constantly ready to attack in retaliation, aggressors had to be meekly prudent, and nuclear weapons could be useful only if they were not used. Deterrence unveiled for all to see the paradoxical logic of strategy with its apparent contradictions, turning the "back-stretched" connection that unites opposites into a commonplace, except for those incurable innocents who fail to see that safety could be the sturdy child of terror.

With that, Herakleitos, the first Western strategic thinker ("War is the father of all and king of all and so he renders some gods, others men, he makes some slaves, others free"[4]), was finally vindicated, though long before him many a cunning fighter had won by instinctively applying the paradoxical logic to surprise his enemy, a thing possible only when the better ways of fighting, hence the expected ways, are deliberately renounced. When a reacting enemy is present, the straightest and broadest and best-paved highway is the worst road upon which to attack an enemy, because it is the best road, while a bad road could be good. It is by that same logic in dynamic action and reaction that the victories of an advancing army can bring defeat once they exceed the culminating point of success, indeed victory *becomes* defeat by the prosaic workings of overextension. Likewise, warfare itself can yield peace by burning out the strength and the will to continue fighting.

Indeed, all that is formed and forged in the crucible of conflict eventually turns into its opposite, if only it persists long enough, a dynamic version of the *coincidentia oppositorum* of Nicola de Cusa or Nikolaus von Kues. It is not necessary to be philosophically inclined to apprehend the logic, nor does one need to know of its existence in order to apply it—but none that ever built an empire in war as the Romans did, or preserved one over the centuries as the East Romans did, could do so but by obeying the logic. It starts with the simple, static contradiction of *si vis pacem para bellum* (if you want peace, prepare war) and proceeds to dynamic contradictions: if you defend every foot of a perimeter, you are not defending the perimeter; if you win too completely, destroying the enemy, you make way for another; and so on.

The only additional complication is that conflict unfolds at separate levels—grand strategic, theater-strategic, operational, tactical—which interpenetrate downward much more easily than upward. To cite a modern example, Adolf Hitler's choice of the wrong allies and the wrong enemies at the level of grand strategy—he had Italy and Japan with him, America, Russia, and the British empire against him—could not be overcome by the many German tactical, operational, and even theater-level victories, notably over France in 1940; and the final outcome could not have been changed by even larger battlefield victories. Had the D-Day landings been repulsed, Germany would still have lost the war, it is only that the first target of the fission bomb would have been Berlin instead of Hiroshima. And even if there had been no fission bomb, Americans, Russians, and the British empire would still have won the war in a few more years. Tactical brilliance, operational ingenuity, even theater-level victories cannot outweigh a defective grand strategy, while by contrast a coherent grand strategy requires only mere adequacy in theater strategy, operational methods, and tactics.

Given overwhelming superiority, material or moral or any combination thereof, wars can be won and peace kept without need of strategy. Antagonists too weak to react significantly are, in effect, mere objects. War may still present huge difficulties, for reasons of distance, terrain, and so forth. But to overcome physical problems, it is not the paradoxical logic of strategy that is wanted but rather the "linear" logic of sound common sense and functional procedures.

Hence it is those fighting against the odds, the outnumbered, the beleaguered, and overambitious gamblers, who have tried to exploit the logic of strategy to the fullest, accepting the resulting risks—sometimes achieving victories disproportionate to their resources, sometimes collapsing ignominiously. Naturally, therefore, more often than not, the great names of strategy—Napoleon, most notably—ultimately failed. Herakleitos succeeded, as we saw, the ingenuous Belisarios ultimately did not; but most Byzantine commanders, while admiring both, preferred the prudent ways illustrated in the manuals, in which the paradoxical logic was exploited—but only up to the limit of risks prudently acceptable. We have seen how preoccupied the Byzantines were with the need to maximize every possible tactical and operational advantage, while studiously trying not to depend on military strength any more than they had to.

In all their infinite variety, grand strategies can be compared by the extent of their reliance on costly force, as opposed to the leveraging of *potential* force by diplomacy ("armed suasion"), inducements (subsidies, gifts, honors), and deception and propaganda. The lesser the actual force content, the greater the possibility of transcending the material balance of strength, to achieve more with less. The Byzantines had many precursors in this, but they became and perhaps remain the unsurpassed masters. Before his death while retreating from a foolhardy attempt to invade India, Alexander had already earned millennial glory by conquering Achaemenid Iran, the only superpower for the Greeks. His grand strategy certainly conformed to the paradoxical logic: while his tactics were "hard"—frontal attacks by the infantry phalanx, and all-out cavalry charges—Alexander's diplomacy was "soft" and inclusive, as symbolized by the encouragement of Macedonian-Iranian marriages, to win over Achaemenid satraps and vassal peoples. Only the attempt to extend his invention of consensual empire to India overshot his culminating point of success, for there was still an original Macedonian base in his military strength, and it was by then excessively diluted.

As we have seen, the East Roman empire by us called Byzantine was least Roman in its strategy, certainly after Justinian's attempt at total reconquest. Successively threatened from the east by Sasanian Persia, the Muslim Arabs, and finally the Seljuk and Ottoman Turks, and from the north by waves of steppe invaders, the Huns, Avars, Bulghars, Pechenegs, Magyars, and Cumans, and from the west too by the ninth century, the Byzantines could not hope to subdue or annihilate all comers in the classic Roman manner.

To wear out their own forces, chiefly of expensive cavalry, in order to utterly destroy the immediate enemy would only open the way for the next wave of invaders. The genius of Byzantine grand strategy was to turn the very multiplicity of enemies to advantage, by employing diplomacy, deception, payoffs, and religious conversion to induce them to fight one another instead of fighting the empire. Only their firm self-image as the only defenders of the only true faith preserved their moral equilibrium. In the Byzantine scheme of things, military strength was subordinated to diplomacy instead of the other way around, and used mostly to contain, punish, or intimidate rather than to attack or defend in full force.

The Byzantine "Operational Code"

It was an attractive Byzantine habit to convey what would now be called "methodology" in the direct form of brisk injunctions or even avuncular advice—do this, do not do that, require fewer and less cumbersome phrases than third-person hortatory language. That is the format I use below to define most succinctly the essential norms of Byzantine strategic culture. They are nowhere stated in any Byzantine source as comprehensively as in what follows, but can be imputed legitimately on the basis of observed behavior, as well as the diverse recommendations of the Byzantine guidebooks and field manuals here examined. A normative summary that minimizes duplication is one way of defining an operational code.[5]

I. *Avoid war by every possible means in all possible circumstances, but always act as if it might start at any time.* Train both individual recruits and complete formations intensively, exercise units against each other, prepare weapons and supplies to be ready for battle at all times—but do not be eager to fight.

The highest purpose of maximum combat readiness is to increase the probability of not having to fight at all.

II. *Gather intelligence on the enemy and his mentality, and monitor his movements continuously.* Patrols and reconnaissance probes by light cavalry units are always necessary, but not sufficient. Spies are needed inside enemy territory to provide early warning of war threats, or at least to report preparations for war, and help divine enemy intentions. In between reconnaissance by combat units and espionage in civilian garb, the middle layer of intelligence gathering is often the most productive: clandestine (= hidden in nature) scouting, that is, passive observation and reporting. Efforts to scout and prevent enemy scouting are seldom wasted.

III. *Campaign vigorously, both offensively and defensively, but attack mostly with small units; emphasize patrolling, raiding, and skirmishing rather than all-out attacks.* Avoid battle, and especially large-scale battle, except in very favorable circumstances—and even then avoid it if possible, unless the enemy has somehow fallen into a condition of complete inferiority, as in the case of a fleet badly damaged by storms.

IV. *Replace the battle of attrition with the "nonbattle" of maneuver.* On the defensive, do not confront greatly superior forces; instead keep close to invading armies, remaining just beyond their reach, to quickly pounce on outnumbered detachments, baggage trains, and looting parties. Prepare ambushes large and small in the path of enemy forces, and lure them into ambushes by feigned retreats. On the offensive, mount raids or, better, probes that withdraw promptly if they encounter stiff resistance. Rely on constant activity, even if each action is small in scale, to demoralize and materially weaken the enemy over time.

V. *Strive to end wars successfully by recruiting allies to change the overall balance of power.* Diplomacy is therefore even more important in war than at peace—not for the Byzantines the foolish aphorism that when the guns speak, diplomats fall silent. In recruiting allies to attack the enemy, *his* allies are the most useful recruits because they are nearest and know best how to fight the enemy's forces. Enemy commanders successfully subverted to serve the imperial interest are even better allies, and the best of all might be found at the enemy's court, or within his family. But even peripheral allies that can only help a little are to be recruited if at all possible.

VI. *Subversion is the best path to victory.* It is so cheap as compared to the costs and risks of battle that it must always be attempted, even with the most unpromising targets infused with hostility or religious ardor. When facing an imminent jihadi offensive, the strategos is advised to befriend the emirs of the frontier castles, sending them "gift baskets."[6] No exception was to be made for known fanatics: by the tenth century, the Byzantines had certainly discovered that religious fanatics can also be bribed, and indeed often more easily—they are creative in inventing religious justifications for taking bribes ("the ultimate victory of Islam is inevitable anyway . . .").

VII. *When diplomacy and subversion are not enough and there must be fighting, it should done with "relational" operational methods and tactics that circumvent the most pronounced enemy strengths and exploit enemy weaknesses.* To avoid consuming the major combat forces, it may be necessary to patiently whittle down the enemy's moral and material strength. That may require much time. But there is no urgency because as soon as one enemy is no more, another will surely take his place for all is constantly changing as rulers and nations rise and fall. Only the empire is eternal.

Note: The operational code outlined here allows for no historical evolution. Having claimed at the start that the construct here called "Byzantine strategy" was invented during the fifth century in response to the specific circumstances of the time, I recognize that the very different circumstances of subsequent centuries left their marks on Byzantine strategy. After all, the Byzantines learned from experience as well as any of us: they did make the same mistake twice or three times or four times, but after that they were unlikely to repeat the same mistake again, or not in the same way at any rate. It is a lesser claim that is here advanced—that there was enough continuity to define an "operational code," and in thus conflating eight centuries I am reassured by eminent predecessors.[7]

Appendix

Emperors from Constantine I to Constantine XI

Glossary

Notes

Works Cited

Index of Names

General Index

Appendix: Was Strategy Feasible in Byzantine Times?

An earlier book of mine, which has continued to attract inordinate attention, was both praised and criticized for attributing a grand strategy to the Romans of the first three centuries CE.[1] Noting the absence of military or civil planning staffs, and of such elementary requirements as accurate maps, some challenged the entire notion that the Romans could think strategically at all, or even define their frontiers coherently.[2] At least the lack of accurate maps was no great obstacle: large-scale survey techniques were in habitual use, while itineraries could be quite accurate—there were even distance-measuring machines.[3]

As for the much larger question of the feasibility of strategy in Roman and Byzantine times, the argument rests on how strategy is defined. The skeptics evidently see it as an essentially modern and bureaucratic activity, the result of explicit calculations and systematic decisions, destined for equally systematic application. Their insistence on the importance of geographic knowledge even suggests a confusion between strategy and the discredited pseudo-science of Hausoferian "geopolitics."[4] I hold that strategy is not about moving armies, as in board games, but rather comprehends the entire struggle of adversarial forces, which need not have a spatial dimension at all, as with the eternal competition between weapons and countermeasures. Indeed, the spatial dimension of strategy is rather marginal these days, and in some ways it always was.

It is the struggle of adversarial forces that generates the paradoxical logic of strategy, which is diametrically opposed to the commonsense linear logic of everyday life. In strategy, contradictions are pervasive: bad roads are good because their use is unexpected, victories are transformed into defeats by overextension, and much more of the same. Hence strategy is not transparent and never was, but it always determines outcomes, whether men know of its existence or not.[5] Nonpractitioners, by contrast, seem to accept the comforting official version that presents strategy as a form of systematic group thinking guided by rational choices, which reflect a set of "national interests," whose results are then itemized in official documents. It is true that decisions driven by the paradoxical logic, shaped by the culture, and motivated by the urges of power are nowadays *rationalized* in that way, but that is all.

Nor is strategic practice the mere application of techniques that could be applied anywhere and by anyone—it is always the expression of an entire culture. Accordingly, in this book I have tried to evoke the strategic culture of Byzantium—which I believe is in part applicable even today, or perhaps especially today.

Emperors from Constantine I to Constantine XI

The Western Empire

474–475   Julius Nepos
475–476   Romulus Augustulus

The Eastern Empire

395–408   Arkadios
408–450   Theodosios II
450–457   Marcian
457–474   Leo I
473–474   Leo II
474–491   Zeno
491–518   Anastasios I
518–527   Justin
527–565   Justinian I
565–578   Justin II
578–582   Tiberios II
582–602   Maurikios
602–610   Phokas
610–641   Herakleios
641–668   Constans II
668–685   Constantine IV
685–695   Justinian II (banished)
695–698   Leontios
698–705   Tiberios III
705–711   Justinian II (restored)
711–713   Bardanes
713–716   Anastasios II
716–717   Theodosios III
717–741   Leo III
741–775   Constantine V
775–780   Leo IV
780–797   Constantine VI
797–802   Irene
802–811   Nikephoros I
811   Straurakios
811–813   Michael I
813–820   Leo V
820–829   Michael II
829–842   Theophilos
842–867   Michael III
867–886   Basil I
886–912   Leo VI
913   Leo VI and Alexander
919–944   Romanos I Lekapenos
945–959   Constantine VII Porphyrogennetos

959–963	Romanos II
963–1025	Basil II and Constantine VIII
963	Regency of Theophano
963–969	Nikephoros II Phokas
969–976	John Tzimiskes
1025–1028	Constantine VIII (alone)
1028–1034	Romanos II Argyros
1034–1041	Michael IV
1041–1042	Michael V
1042	Zoe and Theodora
1042–1055	Constantine IX Monomakhos
1055–1056	Theodora (alone)
1056–1057	Michael VI Bringas
1057–1059	Isaac I Komnenos
1059–1067	Constantine X Doukas
1067–1071	Romanos IV Diogenes
1071–1078	Michael VII Doukas
1078–1081	Nikephoros III Botaniates
1081–1118	Alexios I Komnenos
1118–1143	John II Komnenos
1143–1180	Manuel I
1180–1183	Alexios II
1183–1185	Andronikos I
1185–1195	Isaac II
1195–1203	Alexios II
1203–1204	Isaac II with Alexios IV
1204	Alexios V Doukas Mourtzouphlos

April 13, 1204 *Fall of Constantinople*

Lascarid Dynasty in Nicea

1204–1222	Theodore I Lascaris
1222–1254	John III Doukas Vatatzes
1254–1258	Theodore II Laskaris
1258–1261	John IV Laskaris

1261 *Recovery of Constantinople*

1259–1282	Michael VIII Palaiologos
1282–1328	Andronikos II
1328–1341	Andronikos III
1341–1376	John V Palaiologos
1347–1354	John V Palaiologos with John VI Kantakouzenos
1341–1354	John VI

1376–1379	Andronikos IV
1379–1391	John V (restored)
1390	John VII
1391–1425	Manuel II
1425–1448	John VIII
1449–1453	Constantine XI Dragases

May 29, 1453 *Fall of Constantinople*

Glossary

AG = Ancient Greek term in continued literary use; L = Latin

Agarenoi   Hagarenoi (from Abraham's concubine Agar) Muslim Arabs

agens in rebus   Junior official, sometimes a messenger, not an "agent"

akontion   Javelin, throwing spear (AG)

akontistès   Javeliner (AG)

akritas   Frontier warrior, raider, scout, robber

akropolis   The high point and citadel of a city (AG)

ala, alae   Auxiliary cavalry unit(s) (L)

Alania   Caucasian region partly in modern Georgia (Ossetia)

Alans   Mounted warriors of Iranic origins

Albania   Caucasian region, mostly in modern Azerbaijan

Anatolia   Asia Minor; within modern Asiatic Turkey

Anatolikoi   Byzantine theme

Antae, Antes   Probably Slavs, moved west from Pontic steppe with Avars

anthypatos   Title in ninth to twelfth centuries, not a functional position

Arab   Until the twentieth century: a Bedouin, a nomad

Arabitai   Bedouin light horsemen, raiders, skirmishers, robbers

archon   Generic "ruler," from classical term for high city official (AG)

Arian   Believer in the doctrine of the earthly substance of Christ

arithmos   Translation of *numerus*, a unit; also used for the *tagma* of the *Vigla*

Armenia   Caucasian region with expansive boundaries

Armeniakon   Byzantine theme; attested 667; located in eastern Anatolia *a sekretis*   *A secretis,* an imperial secretary

Asia   Roman province in western Anatolia

Athinganoi   Judaizing sect, especially in Phrygia

augustus   Greek: "Sebastos"; ruling emperor with or without co-emperors

autocephalous   "Self-headed"; the patriarchal sees of the Orthodox Church

autokrator   The emperor was *basileus kai autokrator* (= *Latin imperator*)

ballista   Torsion-powered artillery

bandon   From the German, a unit—in the *Strategikon,* of three hundred

brachialion   Defensive outerwork beyond the main wall

boukellarios   A soldier, originally in the personal service of a commander

caesar/kaisar   Title reserved for an emperor's son

caliph   *Khalifa,* placeholder, deputy for the dead prophet Muhammad

caltrop   Multispike object much used to impede cavalry

candidatus   Member of the imperial bodyguard

Cappadocia   Region of central Anatolia, mostly rather arid

carat   One twenty-fourth of a solidus or nomisma

Chalcedonian   Believer in the twin-nature doctrine of the Council of 451

cheiromanganon   Arrow-launching artillery

chelandion   Originally a horse-transport ship; later a transport

chiliarchy   Unit of a thousand

chrysobull   Document sealed with the emperor's gold seal, a decree

Cibyrrhaeotai   Kibyrrhaeotai, coastal theme in southwest Anatolia

codex   A bound book of pages, as opposed to a scroll

cohors   Auxiliary infantry unit; pl., *cohortes*

come/komes   Literally: "companion" (of the emperor) field-grade officer

consulate   One-year magistracy; after 541 reserved for emperors

defensores   Heavy infantry employed in close ranks to hold ground

dekarch   Junior officer in charge of ten soldiers

domestikos   Very high rank, civil or more often military

domestikos ton scholon   Commander of imperial guard units (L.: *scholae*)

doration   Throwing spear, javelin (AG)

doukatores   Scouts, from Latin, those who lead the way

doux   Latin *dux;* a commander, originally provincial, later lower

dromon   Warship with a fighting deck over the hull; pl. *dromones*

dromos   The imperial horse-relay post

droungarios   Army or naval commander

droungarios tou ploimou   Commander of the Constantinople fleet

dualist   Believer in the conflict between good and evil deities

ektaxis   Battle array (AG)

eparchos   City prefect, responsible for public order

epilorikion   Light overcoat of cotton or silk worn over armor

exarchos   Plenipotentiary (viceroy); in Carthage and Ravenna

exkoubita   "Outside the bedchamber" palace guard unit

federati   Originally barbarian troops; a unit in the Anatolikoi theme

gastald   Lombard city governor

gastraphetès   "Belly bow," a large bow with reloader, heavy crossbow (AG)

Genikon   Fiscal department under the *logothetes tou genikou*

gladius   The short legionary sword, for close-order fighting

Golden Gate   Of the Theodosian Wall, where the Via Egnatia ended

Hagarene   Muslim, from Agar, Abraham's concubine

hagia   Holy

Hebdomon   Suburb of Constantinople on the Sea of Marmara

hekatontarch   Officer in charge of a hundred soldiers

hetaireia   Escort company, tagmatic palace unit

hikanatoi   Tagmatic palace unit

hippagogos   Horse-transport ship

hippotoxotès   Mounted archer (AG)

hoplitès   Infantryman with body armor and/or a heavy shield (AG)

hypatos   Greek: consul

hypostrategos   Literally "undergeneral," senior military title

Iberia   Caucasian region within modern Georgia

icon   Literally, a "likeness," a religious image

iconoclasm   Destruction of idolatrous images, eighth/ninth century

iconodule   Defender of the liturgical use of icons

indiction   Fiscal year, from September to August, in a fifteen-year cycle

Isauria   Region in southwest Anatolia

Ishmaelite   Muslim Arab, from descendant of Ishmael, son of Hagar

Jacobite   Monophysite (Church, doctrine); after Jacob Baradaeus

kabadia   Thick felt tunic or leg drape worn for protection

kandidatos   Originally an imperial bodyguard, later a nonfunctional title

karabisianoi   Sailors or marines

katalogos   Muster list (AG)

kataphraktos Armored cavalryman

katepano From ninth century, civil or military official, later governor

khagan Qagan, Turkic chief of chiefs (qans, or khans)

Khazars Ruling people centered in north Caucasus, seventh to tenth centuries

kleisoura Defile or mountain pass; a command thereof

klibanion Sleeveless cuirass of scale or lamellar armor

kontos Thrusting spear, pike, or cavalry lance (AG)

kontoubernium From Lat. *contubernium;* tent-group; unit of ten, synonym of dekarchia

koubikoularios Of the (imperial) bedchamber; senior eunuch

kouropalates High title for imperial relatives and foreign princes: majordomo

laisa Anti-arrow barrier or shelter made of intertwined vines or thin branches

lithobolos Stone-thrower; artillery (AG)

liturgy Designated duty, also in church services

logothetes Official, originally fiscal; later, highest administrative rank

logothetes tou dromou Originally of the post, later with many other duties

logothetes ton angelon In charge of stables, herds, and remounts

lorikion Mail vest or cuirass, other chest armor

Macedonia Region between Epirus on the Adriatic and Thrace

Magaritai Muslim warriors, Christian apostates

magister militum Master of soldiers, highest military commander

magister militum per Armeniam Commander of the forces in northeast Anatolia

magister militum per Orientem Commander of the forces facing and in Mesopotamia

magister militum per Thracias Commander of the forces in Thrace and beyond

magister militum praesentalis Commander of the forces in or near Constantinople, the imperial field army

magister officiorum Master of offices, head of the civil bureaucracy

magister utriusque militiae Highest commander of both infantry and cavalry

menavlatos Heavy elite infantryman, armed with a menavlion

menavlion Pike

Manichean Follower of the dualist Mani

Maronite Levantine church in communion with Rome, Syriac liturgy

Melkite Chalcedonian churches, both Orthodox and Catholic

meros Formation of three *moiras,* six to seven thousand men in the *Strategikon*

milion Constantinople milestone from which distances were measured *milliaresion*   Silver coin worth one-twelfth of a solidus or nomisma

minsouratores   From Lat. *mensuratores;* military surveyors

moira   Formation of three bandons, one to two thousand men in the *Strategikon*

Monophysites   Believers in the single, divine nature of Christ; also "Jacobite" and Copt

monotheletism   Christological compromise: one will in unspecified natures

naumachia   Sea battle, naval tactics (AG)

nomisma   Byzantine gold coin, successor to the solidus

Opsikion   Theme in central Anatolia, centered on Ancyra (Ankara)

Optimatoi   Theme in northwest Anatolia, centered on Nikomedeia (Kocaeli)

ourguia   Length measure, 1.8 meters

palatine   Of the palace; as with units of palace guards

parakoimomenos   Senior eunuch, head of the imperial bedchamber staff

patriarch   Of Rome, Constantinople, Alexandria, Antioch, Jerusalem

patrikios, patricius   Very high rank, later diminished below *anthypatos*

Paulicians   A supposedly dualist, certainly anticlerical, heresy

pentekontarch   Infantry commander of fifty

phoulkon   Or *fulkon*, Germanic, battle force, commander's escort

phylarch   Tribal chief, mostly Arab, commander of auxiliaries

Pontic steppe   Inland from the northern shore of the Black Sea

Porphyrogennetos   "Born in the purple [bedchamber]," offspring of an emperor

praepositus sacri cubiculi   Overseer of the bedchamber; highest palace official

praipositos   Same in Greek, eunuch high official

proskynesis   Ritual full-length prostration before the emperor

prokoursatores   Light-horse scouts or skirmishers (Lat: forward runners)

protektor   Bodyguard; later under the domestikos ton scholon

protospatharios   Military rank; from the eighth century, high title with Senate membership

Rhos, Rhosoi   Originally Scandinavian, later Slavic inhabitants of *Rhosia*, Kievan Rus'

Saracen   Originally a Sinai tribe; from the seventh century, Arab Muslim; later, any Muslim

sarissa   Long spear, basic weapon of the Macedonian phalanx (AG)

scholai   Originally Lat.: *scholae palatinae;* imperial guard units based in Constantinople; tagmatic, full-time

Scythia   The Pontic steppe and beyond; late Roman province

spatharios   Lit. swordsman; an imperial bodyguard, later a title

sphendonè   A sling (AG)

spithame   Length measure, 23.4 centimeters

strategos   Commander; commander in chief of a theme

stratelates   High rank, Greek, for magister militum; later a title

stylite   Christian ascetic who lived on top of a pillar

tagma/tagmata   Cavalry formations, full-time, not thematic; elite units

taktika   Official lists of offices and titles, ninth and tenth centuries

tattoo   To array; to deploy (AG)

Tauroscythians   Originally inhabitants of Crimea; later of Kievan Rus' (literary)

taxiarchos   Commander of a thousand

tetrarch   Commander of four soldiers; equivalent to a lieutenant corporal

thema/themata   Military/administrative districts garrisoned by part-time forces

toparches   Rulers of petty states

toporetes   Literally, lieutenant, deputy commander of a tagma

tourma   Large military unit; themes often had three

tribune   Commander of a *bandon*, equivalent to a *comes, komes*

tzerboulia   Heavy boots, for infantrymen

vexillation   Originally a legionary detachment; later an army formation

vigla   From Latin *vigiles,* the watch; tagmatic guard unit in Constantinople

Vizir   Of Iranian origin; the chief administrator of Muslim rulers

voivode   Southern Slav chief or "prince"

zabai   Segments of mail worn as supplementary armor

zupan/zoupan   Southern Slav ruler

Notes

Abbreviations

Ammianus Marcellinus—Rolfe
Ammianus Marcellinus, 3 vols., trans. John C. Rolfe (Cambridge, Mass.: Harvard University Press, 1935)

Anecdota (of Prokopios)
H. B. Dewing, ed. and trans., *Procopius* (Cambridge: Harvard University Press, 1969), vol. 6

Chronicon Paschale
Chronicon Paschale, 284–628 AD, trans. Michael Whitby and Mary Whitby (Liverpool: Liverpool University Press, 1989)

De Cerimoniis (Book of Ceremonies)
Constantine Porphyrogennetos, *De Cerimoniis Aulae Byzantinae*, in J. Reiske, ed., *Corpus Scriptorum Historiae Byzantinae* (Bonn: Weber, 1829)

DAI
De Administrando Imperio, ed. Gy. Moravcsik, trans. R. J. H. Jenkins (Washington, D.C.: Dumbarton Oaks Center for Byzantine Studies / Harvard University, 1967)

DAI—Commentary
R. J. H. Jenkins, F. Dvornik, B. Lewis, Gy. Moravcsik, D. Obolensky, and S. Runciman, eds., *De Administrando Imperio*, vol. 2: *Commentary* (London: Athlone Press, 1962)

Getica
Jordanes, *Getica: De origine actibusque Getarum;* translated by Charles C. Mierow as *The Gothic History of Jordanes* (Cambridge: Speculum Historiale, 1915); Latin text: http://www.thelatinlibrary.com/iordanes1.html.

Marcellinus Comes

> *Marcellinus Comes*, trans. Brian Coke, *Byzantina Australiensia* 7 (Sydney: Australian Association for Byzantine Studies, 1995)

ODB

> *The Oxford Dictionary of Byzantium*, ed. Alexander P. Kazhdan and Alice-Mary Talbot with Anthony Cutler, Timothy E. Gregory, Nancy P. Ševčenko (New York: Oxford University Press, 1991)

Sebeos

> *The Armenian History Attributed to Sebeos*, trans. R. W. Thompson, vols. 1–2 (Liverpool: Liverpool University Press, 1999)

Strategikon—Dennis

> George T. Dennis, *Maurice's Strategikon: Handbook of Byzantine Military Strategy* (Philadelphia: University of Pennsylvania Press, 1984)

Theodosian Code

> http:ancientrome.ru/ius/library/codex/theod/liber16.htm8.

Theophanes

> Theophanes, *The Chronicle of Theophanes Confessor, AD 284–813*, trans. Cyril A. Mango and Roger Scott with Geoffrey Greatrex (Oxford: Clarendon Press, 1997)

I. The Invention of Byzantine Strategy

1. As vigorously argued by Bryan Ward-Perkins, *The Fall of Rome and the End of Civilization* (2005). For the broader debate, see Guy Halsall, *Barbarian Migrations and the Roman West, 376–568* (2007), pp. 17–18, 422–447.
2. There were twenty-five to thirty legions of some 5,500 to 6,000 men each with roughly as many auxiliary light infantry and cavalry, for a total of some 275,000 to 360,000 men; there were also fleets; H. M. D. Parker, *The Roman Legions* (1928/1985). G. L. Chessman, *The Auxilia of the Roman Imperial Army* (1914/1975).
3. Hugh Elton, *Warfare in Roman Europe, A.D. 350–425* (1997), from p. 199; Edward N. Luttwak, *The Grand Strategy of the Roman Empire: From the First Century A.D. to the Third* (2007), see the appendix on the grand strategy and its critics.
4. See the surveys in Charalambos Papasotiriou, "Byzantine Grand Strategy" (1991), from p. 93; Mark Whittow, *The Making of Byzantium, 600–1025* (1996), pp. 15–37; and John H. Pryor, *Geography, Technology and War: Studies in the Maritime History of the Mediterranean, 646–1571* (1988), pp. 1–24.
5. Ze'ev Rubin, "The Sasanid Monarchy" (2001), from p. 638; but in the same he lists all the wars started by the Sasanians. See A. D. Lee, *Information and Frontiers: Roman Foreign Relations in Late Antiquity* (1993), from p. 21.

6. From an inscription by Shapur I (240–270) cited in Touraj Daryaee, "Ethnic and Territorial Boundaries in Late Antique and Early Medieval Persia (Third to Tenth Century)" (2005), p. 131.

7. H. Grégoire, "Imperatoris Michaelis Palaeologi *De vita sua*" (1959–1960), p. 462. Michele Amari in *La Guerra del Vespro Siciliano* (1851) discounted the "supposed plots" of Giovanni da Procida, crediting local initiative instead (p. 90). But I follow Steven Runciman, *The Sicilian Vespers* (1960), pp. 226–227 (on p. 313 he gently rebukes the great Amari).

8. On the Romans' lack of cartographic conceptions, see Pietro Janni, *La Mappa e Il periplo: Cartografia antica e spazio odologico* (1984).

9. See the discussion in Lee, *Information and Frontiers*, from p. 81. The Roman army used itineraries: "itineraria prouinciarum," N. P. Milner, trans., *Vegetius: Epitome of Military Science* (1996), III.6.

10. *Excerpta de Legationibus Romanorum ad gentes,* 14, in *The History of Menander the Guardsman*, trans. R. C. Blockley (1985), p. 175.

11. Michael F. Hendy, *Studies in the Byzantium Monetary Economy, c. 300–1450* (1985), from p. 157.

12. J. F. Haldon, *Byzantium in the Seventh Century: The Transformation of a Culture* (1990), from p. 173; compare Salvatore Cosentino, "Dalla tassazione tardoromana a quella bizantina: Un avvio al medioevo" (2007), pp. 119–133.

13. Nicolas Oikonomides, "The Role of the Byzantine State in the Economy" (2002), pp. 973–1058.

14. *The Chronicle of Theophanes Confessor, A.D. 284–813,* trans. Cyril Mango et al. (1997), no. 303, AM 6113, pp. 435–436.

15. Jordanes, *Getica (De origine actibusque Getarum),* line 261; translated by Charles C. Mierow as *The Gothic History of Jordanes* (1915), p. 126. The Latin text can be found at www.thelatinlibrary.com/iordanes1.html.

16. The most elegant rendition remains Denis Van Berchem, *L'armée de Dioclétien et la réforme Constantinienne* (1952).

17. For the broader context, see Fergus Millar, *A Greek Roman Empire: Power and Belief under Theodosius II (408–450)* (2007).

18. *Hòuhànshū* [Book of the Later Han], partial translation by John E. Hill at http://depts.washington.edu/silkroad/texts/hhshu/hou_han_shu.html.

19. See the indispensable Peter B. Golden, *Introduction to the History of the Turkic Peoples: Ethnogenesis and State Formation in Medieval and Early Modern Eurasia and the Middle East* (1992), from p. 87. Professor Golden (March 23, 2008) corrected my dismissal of the connection, referencing Miklós Érdy, "Hun and Xiongnu Type Cauldron Finds throughout Eurasia" (1995), pp. 3–26; and Étienne de la Vaissière, "Huns et Xiongnu" (2005), pp. 3–26.

20. Henry Yule and A. C. Burnell, *Hobson-Jobson: A Glossary of Colloquial Anglo-Indian Words and Phrases,* ed. Rev. E. Crooke (1985), p. 947.

21. For a brief overview of the vast literature on ethnogenesis, see Halsall, *Barbarian Migrations,* pp. 14–16, 457–470; Halsall concurs with Walter

Pohl, as in (briefly) Pohl's "Conceptions of Ethnicity in Early Medieval Studies" (1998), pp. 15ff. On the original concept of ethnogenesis, see Peter B. Golden, "Ethnicity and State Formation in Pre-Cinggisid Turkic Eurasia" (2001).

22. Otto J. Maenchen-Helfen, *The World of the Huns* (1973), from p. 386. He also squashed Altheim's speculations (p. 385 n. 82).

23. *In Eutropium* I, 250: "Taurorum claustra, paludes flos Syriae seruit"; see Averil Cameron Claudian, *Poetry and Propaganda at the Court of Honorius* (1970), from p. 124.

24. Theodoret, bishop of Cyrrhus, Syria, 423–457; not seen by me; cited in Maenchen-Helfen, *The World of the Huns*, pp. 58–59.

25. Golden, *History of the Turkic Peoples*, p. 108.

1. Attila and the Crisis of Empire

1. For a brief overview, see C. R. Whittaker, *Frontiers of the Roman Empire* (1994), from p. 132.

2. *Liber Pontificalis*, 47.7, in Raymond Davis, trans., *The Book of Pontiffs* (1989), p. 39; money changed hands.

3. Ambrosii, *Expositio Evangelii secundum Lucam*, X, 10; Migne, *Patrologia Latina*, vol. 15, cols. 1806–1809.

4. *Wulfhere sohte ic ond Wyrmhere . . .*: From the Exeter Book, ca. 975. See Kemp Malone, ed., *Widsith*(1936/1962), pp. 118–121.

5. Theodore M. Andersson, *A Preface to the Nibelungenlied* (1987).

6. E. A. Thompson, *The Huns* (1996), e.g.: "In material civilization [the Huns] belonged to the Lower Stage of Pastoralism" (p. 47)—the capitalization reveals the doctrine; Otto J. Maenchen-Helfen, *The World of the Huns* (1973), p. 226. Peter Heather disagrees in his *Fall of the Roman Empire* (2006), from p. 300.

7. XXX I.2.1., in *Ammianus Marcellinus,* trans. John C. Rolfe (1935) (hereafter cited as *Ammianus Marcellinus*–Rolfe), pp. 3:381.

8. *Peri Ktismaton, De aedificis*, IV.v.1–7; translated by H. B. Dewing (with G. Downey) as *Buildings* (1971), pp. 7:266–267.

9. Peter B. Golden, *Introduction to the History of the Turkic Peoples: Ethnogenesis and State Formation in Medieval and Early Modern Eurasia and the Middle East* (1992), p. 92.

10. The horses: "duris quidem sed deformibus" of *Ammianus Marcellinus,* XXXI.2.6.

11. Ibid., XXXI.2.9, *Ammianus Marcellinus*–Rolfe, p. 385.

12. Maenchen-Helfen, *The World of the Huns*, p. 240, cites many cases, including Josephus, *The Jewish War,* VII.4, 249–250, on the near capture of Tiridates I of Armenia by Alan warriors ca. 72 CE.

13. *Panegyric on Anthemius*, 266–269. In *Sidonius,* trans. W. B Anderson (1963), p. 2:31.

14. *Panegyric on Avitus,* 236; Anderson, *Sidonius,* p. 139; Maenchen-Helfen asserts that Sidonius was echoing Claudian, whose *iacula* were arrows.

15. Gad Rausing, *The Bow* (1967), from p. 140. I owe the hand-copied text to the kindness of Hans and Marit Rausing.
16. Heather, *Fall of the Roman Empire*, p. 156, has 130 centimeters.
17. "This magnificent instrument" etc.: E. W. Marsden, *Greek and Roman Artillery* (1969), pp. 8–10.
18. A German master craftsman, Markus Klek, has revealed the intricacies of the manufacturing process by actually making a tendon and bone composite bow; see http://www.primitiveways.com/pt-composite_bow.html.
19. Maenchen-Helfen, *The World of the Huns*, p. 226.
20. Under "The Fall of Hervor and the Gathering of Angantyr's Army," pt. 14 of *The Saga of Hervor and King Heidrek the Wise,* trans. Peter Tunstall (2005). http://www.northvegr.org/lore/oldheathen/018.php.
21. Homer, *The Odyssey,* bk. 24, 167–178, trans. A. T. Murray (1946), p. 2:415.
22. As noted by Hans Rausing, personal communication, January 9, 2008.
23. Fred Isles, "Turkish Flight Arrows," (1961). Online at http://www.atarn.org/mongolian/mongol_1.htm.
24. http://www.hermitagemuseum.org/html_En/12/b2003/hm12_3_1_5.html. For a contrary view, see Gongor Lhagvasuren, http://www.atarn.org/mongolian/mongol_1.htm.
25. McLeod, "The Range of the Ancient Bow," (1965), p. 8; McLeod, "The Range of the Ancient Bow: Addenda," (1972), p. 80. On armor types, see A. D. H. Bivar, "Cavalry Tactics and Equipment on the Euphrates" (1972), p. 283.
26. Edward N. Luttwak, "The Operational Level of War," (Winter 1980–1981), pp. 61–79; Luttwak, *Strategy* (2001), from p. 112.
27. *Ammianus Marcellinus*–Rolfe, p. 383; XXXI.2.6: "verum equis prope affixi."
28. *Panegyric on Anthemius,* 262–266, in Anderson, *Sidonius,* p. 31.
29. XXXI, 2, 8, in *Ammianus Marcellinus*–Rolfe, p. 385.
30. H. D. F. Kitto's cheerfully nonliteral translation of 6 D 5W ("Thracian" for *Saian*), in *The Greeks* (1951), p. 88; preceded by the invaluable "I hate a woman thick about the ankles."
31. John Haldon, *Warfare, State and Society in the Byzantine World, 565–1204* (1999), pp. 163–165. On carts, see *The Oxford Dictionary of Byzantium* (1991), pp. 1:383–384, s.v. "Carts."
32. Prokopios, *Anecdota,* bk. XXX, 4–7, in H. B. Dewing, ed. and trans., *Procopius* (1969), pp. 6:347–349.
33. Adapted from Adrian Keith Goldsworthy, *The Roman Army at War, 100 BC–AD 200* (1996), p. 293, including table 5. More broadly, see John F. Haldon, "The Organization and Support of an Expeditionary Force: Manpower and Logistics in the Middle Byzantine Period" (2007), from p. 422; and Haldon, "Introduction: Why Model Logistical Systems?" (2005), from p. 6.
34. Donald R. Morris, *The Washing of the Spears* (1966), pp. 312–313, with South African data on ox wagons (Goldsworthy's data is from India). Per-

sonal experience in tropical Bolivia (www.amazonranch.org) confirms the Morris data.

35. Haldon, *Warfare, State and Society*, p. 164.

36. War Department, *Handbook on German Military Forces* (1945/1990), p. 297.

37. *Strategikon* of Maurikios (see Part III below), XII, 2.9.

38. *De Velitatione;* see Part III below.

39. See Jordanes, *Getica*, translated by Charles C. Mierow as *The Gothic History of Jordanes* (1915) (hereafter cited as *Getica*), XLII.221, p. 113, regarding the siege of Acquileia in 452. On specific engines at Naissus (Nish, Serbia) in 447, see Priskos of Panium, 6.2, in R. C. Blockley, *The Fragmentary Classicizing Historians of the Later Roman Empire* (1983), pp. 2:230–233.

40. On the controverted chronology, see Constantine Zuckerman, "L'Empire d'Orient et les Huns" (1994), pp. 165–168. But Maenchen-Helfen, *The World of the Huns,* remains convincing (from p. 108) in aligning the evidence on a sequence of conquests from 441 to 447; almost fully endorsed by Blockley, *Fragmentary Classicizing Historians* (p. 1:168 n. 48).

41. Letter LXXVII, 8, "ad Oceanum de morte Fabiolae," in *Selected Letters of St. Jerome,* trans. F. A. Wright (1954), pp. 328–331.

42. Maenchen-Helfen, *The World of the Huns*, pp. 57–58.

43. From the *Liber Chalifarum* (not seen by me) cited in Maenchen-Helfen, *The World of the Huns*, p. 58.

44. *Getica* XXXV.180. The Visigoths ex Thervingi, or Vesi, were renamed by Cassiodorus to obtain a matched pair with Ostrogoths. Translation from pp. 101–102, adapted by author ("domain" for "tribes" etc).

45. *Excerpta de Legationibus Romanorum ad gentes,* 3, lines 40–65, in Blockley, *Fragmentary Classicizing Historians*, pp. 2:284–285.

46. Ibid., lines 373–378, in Blockley, *Fragmentary Classicizing Historians*, pp. 2:264–265.

47. Ibid., lines 378–385, pp. 2:265–267. On the name, see Maenchen-Helfen, *The World of the Huns*, pp. 388–389.

48. Thompson, *The Huns*, p. 226.

49. *Excerpta de legationibus Romanorum ad gentes,* 5, lines 585–618, in Blockley, *Fragmentary Classicizing Historians*, pp. 2:276–279.

50. That is, the prevailing interpretation of Priskos, who records traveling on a "level road" over a plain and crossing the "navigable rivers" Drecon, Tigas, and Tiphesas. In Jordanes (XXXIV, 178) they are "ingentia si quidem flumina, id est Tisia Tibisiaque et Dricca"; see Thompson, *The Huns*, appendix F, "The Site of Attila's Headquarters," pp. 276–277; compare Robert Browning, "Where Was Attila's Camp?" (1953), pp. 143–145, who places the camp across the lower Danube in Wallachia; but Blockley, *Fragmentary Classicizing Historians* (p. 2:384 n. 43), endorses Thompson.

51. [Flavius] Areobindos, future *magister militum, Argagisklos* = Arnegisklos.

52. *The Chronicle of Theophanes Confessor, A.D. 284–813,* trans. Cyril Mango et al. (1997), no. 103, AM 5942, p. 159.

53. Constantine Zuckerman, "L'Empire d'Orient et le Huns" (1994), p. 169. Zuckerman otherwise disputes Maenchen-Helfen's chronology.

54. Fifteenth indiction, consulship of Ardabur and Calepius [Mommsen 447], in Brian Coke, trans., *The Chronicle of Marcellinus Comes* (1995), p. 19 and p. 89 for 447.5.

55. See the discussion in Heather, *Fall of the Roman Empire*, from p. 334.

56. Eminent: J. B. Bury, no less, "Justa Grata Honoria," 1919, pp. 1–13; curtly dismissive: Maenchen-Helfen, *The World of the Huns*, p. 130; credulous: Thompson, who builds much on it, *The Huns*, from p. 145; it is just juicy court gossip for Heather, *Fall of the Roman Empire*, p. 335.

57. *Getica* XXXV.182, p. 102.

58. Sidonius, *Carmina VII, Panegyric on Avitus*, 319–330, in Anderson, *Sidonius* (1963), pp. 1:145–147.

59. On the role of Aetius, see, most recently, Peter Heather, "The Western Empire, 425–476" (2001), from p. 5. In this, the story of Honoria's betrothal is credited. See also A. H. M. Jones, *The Later Roman Empire, 284–602* (1973), pp. 1:176–177.

60. *Getica* XXXVI.191, p. 105.

61. *Getica* XL.212, p. 110.

62. *Getica* XLI.216, p. 111.

63. Thompson, *The Huns*, pp. 155, 156–157; Maenchen-Helfen, *The World of the Huns*, p. 132, robustly disagreed.

64. Thompson's chapter is entitled "The Defeats of Attila," *The Huns*, p. 137.

65. *Getica* XLI.217, p. 112; Maenchen-Helfen, *The World of the Huns*, p. 132, talks of very heavy losses among the Huns specifically—with no evidence, of course.

66. Evidence from the acts of the Council (not seen by me) cited in Maenchen-Helfen, *The World of the Huns*, p. 131 nn. 613–615.

67. *Ammianus Marcellinus*–Rolfe, XXI.11,2; and 12,1, p. 2:141.

68. Herodian, bk. 8, 4, 6–7; text from Edward C. Echols, trans., *Herodian of Antioch's History of the Roman Empire* (1961), pp. 203–204. Also at www.Tertullian.Org/Fathers/Herodian_08_Book8.Htm.

69. *Getica* XLII.221, p. 113.

70. Maenchen-Helfen, *The World of the Huns*, pp. 137–139; Thompson, *The Huns*, p. 161—the class hatred he directs at Avienus, dead some nine hundred years by 1948, is absurd and hilarious.

71. Maenchen-Helfen, *The World of the Huns*, p. 141; Thompson, *The Huns*, pp. 161–163, is of the same view.

72. *Getica* XLIII.225, p. 114; Priskos, *Excerpta de Legationibus Romanorum ad gentes*, 6, in Blockley, *Fragmentary Classicizing Historians*, pp. 2:315–317.

2. The Emergence of the New Strategy

1. For an overview, see A. D. Lee, "The Eastern Empire: Theodosius to Anastasius" (2001), pp. 34ff.

2. R. C. Blockley, *East Roman Foreign Policy* (1992), from p. 56.

3. C. Toumanoff, "Armenia and Georgia" (1966), pp. 593ff.

4. Theophanes, *The Chronicle of Theophanes Confessor, AD 284–813,* trans. Cyril A. Mango et al. (1997) (hereafter cited as *Theophanes*), no. 82, AM 5906, p. 128.

5. See Averil Cameron, "Vandal and Byzantine Africa" (2001), pp. 553ff.

6. *Marcellinus Comes,* trans. Brian Coke (1995) (hereafter cited as *Marcellinus Comes*), p. 17.

7. *Theophanes,* no. 101, AM 5941, p. 157.

8. Ibid., no. 103, AM 5942, p. 159.

9. *Liber Pontificalis,* 47.6, in Raymond Davis, trans., *The Book of Pontiffs* (1989), p. 39.

10. Otto J. Maenchen-Helfen, *The World of the Huns* (1973), p. 125.

11. In an immense literature, see Anatoly M. Khazanov, *Nomads and the Outside World* (1994), from p. 69, on the "non-autarky of the pastoral economy."

12. *Excerpta de Legationibus gentium ad Romanos,* 3, in R. C. Blockley, *The Fragmentary Classicizing Historians of the Later Roman Empire* (1983), pp. 2:237–239.

13. E. A. Thompson, *The Huns* (1996), p. 214.

14. *Excerpta de Legationibus Romanorum ad gentes,* 6, in Blockley, *Fragmentary Classicizing Historians,* p. 2:423.

15. Paul Stephenson corrected my own misinterpretation; private communication, February 16, 2008.

16. See John [F.] Haldon, "Blood and Ink: Some Observations on Byzantine Attitudes towards Warfare and Diplomacy" (1992), from p. 281.

17. In a vast literature, the foundation remains A. H. M. Jones, *The Later Roman Empire* (1973), from p. 1:608.

18. Prokopios, *History of the Wars* (hereafter cited as *Wars*): *The Persian War,* bk. 1, i, 8–15, in H. B. Dewing, ed. and trans., *Procopius* (1962–1978), pp. 1:6–7.

19. The "pseudo-Avar" question (in Theophylact Siimocatta from VII.7.10, which reads as a romanced version of Menander's fragment 19.1) is of scant interest; see Peter B. Golden, *Introduction to the History of the Turkic Peoples* (1992), pp. 109ff. Professor Golden was more definitive verbally, October 23, 2003, on the Avar–Jou jan connection. See Walter Pohl, *Die Awaren* (2002), p. 158.

20. *Excerpta de Legationibus gentium ad Romanos,* 1, in *The History of Menander the Guardsman,* trans. R. C. Blockley (1985), p. 49.

21. = Greek *Outrigouroi* from *Utur* or *Otur Oghur* = the thirty Oghurs—clan or tribes; and *Koutrigouroi* from *Quturghur,* from *Toqur Oghur* = the nine Oghurs; personal communication from Professor Peter B. Golden, April 15, 2008.

22. Michael Whitby and Mary Whitby, *The History of Theophylact Simocatta* (1986), bk. 8, 3.2–6, p. 212.

23. See Walter Emil Kaegi Jr., "The Contribution of Archery to the Turkish Conquest of Anatolia" (2007), pp. 237–267.

24. See Nike Koutrakou, "Diplomacy and Espionage: Their Role in Byzantine Foreign Relations, 8–10th Centuries" (2007), p. 137, reprinted from *Graeco-Arabica* pp. 6:125–144.

25. *Anecdota,* bk. XXX, 12–16, in Dewing, ed. and trans., *Procopius* (1969), pp. 6:351–353.

26. "When Chrysaphius had demanded gold [for his appointment], Flavian sent him sacred vessels to humiliate him." In Michael Whitby, trans., *The Ecclesiastical History of Evagrius Scholasticus* (2000), bk. II, 2, p. 61.

27. *Excerpta de Legationibus gentium ad Romanos,* 5, in Blockley, *Fragmentary Classicizing Historians,* p. 2:245.

28. The hostile Theophanes Confessor describes Chrysaphius's techniques in AM 5940, paras. 98–100. *Theophanes,* pp. 153–155.

29. *Excerpta de Legationibus Romanorum ad gentes,* 3, in Blockley, *Fragmentary Classicizing Historians,* pp. 2:255. Iran's contemporary post-shah barbarians have been less civilized.

30. Ibid., p. 2:295.

31. Cyril Mango, "The Water Supply of Constantinople" (1995), p. 13.

32. Ibid., p. 16.

33. J. Durliat, "L'approvisionnement de Constantinople" (1995), p. 20.

34. G[ilbert] Dagron, "Poissons, pêcheurs et poissonniers de Constantinople" (1995), p. 59; Diocletian's limit (no doubt exceeded) was 24 denarii per pound of first-quality fresh fish.

35. Most recently collated in John H. Pryor and Elizabeth M. Jeffreys, *The Age of the Dromon: The Byzantine Navy ca. 500–1204* (2006); compare the earlier Hélène Ahrweiler's *Byzance et la Mer: La marine de guerre, la politique et les institutions maritimes de Byzance aux VIIe–XVe siecles*(1966).

36. *Marcellinus Comes,* p. 19.

37. For an overview that incidentally corrects common misconceptions, see Paul E. Chevedden, "Artillery in Late Antiquity" (1999), pp. 131–173.

38. J. F. Haldon, "Strategies of Defence, Problems of Security: The Garrisons of Constantinople in the Middle Byzantine Period" (1995) p. 146.

39. S.v. "Long Wall," *The Oxford Dictionary of Byzantium,* ed. Alexander P. Kazhdan et al. (1991), p. 2:1250. Also J. G. Crow, "The Long Walls of Thrace" (1995), pp. 109ff. Considerable parts survive. See James Crow, Alessandra Ricci, and Richard Bayliss, University of Newcastle Anastasian Wall Project, at http://longwalls.ncl.ac.uk/AnastasianWall.htm.

40. Bk. III, 38, in Whitby, *Evagrius Scholasticus,* p. 183.

41. Novel 26, *De Praetore Thraciae: Praefatio: "In Longo enim muro duos quosdam sedere vicarios."* Vicars were, in effect, deputy praetorian prefects; conveniently from http://web.upmf-grenoble.fr/Haiti/Cours/Ak/Corpus/Novellae.htm.

42. Golden, *History of the Turkic Peoples,* pp. 98–100; and private communication from Professor Golden, March 23, 2008.

43. Prokopios, *Wars,* bk. V, xxvii, 27–29, in Dewing, *Procopius,* p. 3:261.

44. Compare the appraisal in Averil Cameron, "Justin I and Justinian" (2001), pp. 67ff. (I wrote the above before reading Cameron.)

45. Prokopios, *Wars,* bk. 3, XI, 2, in Dewing, *Procopius,* from p. 2:101.

46. Ibid., bk. 5, xxvii, 27, p. 3:261.

47. Ibid., bk. 5, 9, 12, from p. 3:87.

48. Ibid., bk. 5, 14, 1, 2, pp. 3:141, 142.

49. John Haldon, *The Byzantine Wars: Battles and Campaigns of the Byzantine Era* (2001), pp. 37–44.

50. Novel 30, *De Proconsule Cappadociae,* chap. XI.2.

51. See below for Long Wall and Dara. For the 320,000 pounds of gold, see *Anecdota,* bk. 19, 7, in Dewing, *Procopius,* p. 6:229.

52. *Buildings,* from bk. 1, 1, 24, in Dewing, *Procopius,* from p. 7:11.

53. John Moorhead, "The Byzantines in the West in the Sixth Century"(2005), p. 127.

54. Not even Halsall's very recent *Barbarian Migrations and the Roman West, 376–568* (2007), which says that the plague merely "erodes morale" (p. 504). But now see Lester K. Little, ed., *Plague and the End of Antiquity: The Pandemic of 541–750* (2006).

55. Dewing, *Procopius,* pp. 1:451–465.

56. Ibid., p. 1:465.

57. Chaps. 48 and 49, in Charles Forster Smith, ed., *Thucydides: History of the Peloponnesian War* (1951), pp. 1:343, 345.

58. Notably by J. Durliat, "La peste du VIme siècle, pour un nouvel examen de sources byzantins" (1989), pp. 107–119. For the historiographical context, see Lester K. Little, "Life and Afterlife of the First Plague Pandemic" (2006), pp. 3–32; on Durliat, see p. 17.

59. See, for example, P. Allen, "The 'Justinianic' Plague" (1979), pp. 5–20.

60. Whitby, *Evagrius Scholasticus,* pp. 229–231.

61. *Pseudo-Dionysius of Tel-Mahre: Chronicle, Known Also as the Chronicle of Zuqnin,* pt. 3, trans. Witold Witakowski (1996), pp. 74–75.

62. Ibid., pp. 80–81.

63. Michael Whitby, "Recruitment in Roman Armies from Justinian to Heraclius (c. 565–615)" (1995), p. 93.

64. Cameron, "Justin I and Justinian," pp. 76, 77, order reversed.

65. A biovar similar to *Orientalis:* see I. Wiechmann and G. Grupe, "Detection of *Yersinia pestis* in Two Early Medieval Skeletal Finds from Aschheim (Upper Bavaria, 6th century A.D.)" (2005), pp. 48–55. It was earlier presumed that the biovar was *antiqua* (so named because of the 541 pandemic), which persists and is less lethal.

66. According to William F. Ruddiman, "The anthropogenic greenhouse era began thousands of years ago": http://courses.eas.ualberta.ca/eas457/Ruddiman2003.pdf. See also the debate in *Real Climate,* December 5,

2005, "Debate over the Early Anthropogenic Hypothesis," at http://www
.realclimate.org/index.php/archives/2005/12/early-anthropocene-
hyppothesis/. Regarding predation, the last lion in Anatolia was killed in
1870; the last Caspian tiger was killed in 1959; a few Cheetahs survive.

67. Hugh N. Kennedy, "Justinianic Plague in Syria and the Archaeological Evidence" (2006), p. 95.

68. Richard Alston, "Managing the Frontiers: Supplying the Frontier Troops in the Sixth and Seventh Centuries" (2002), p. 417, writes that in *Anecdota* xxiv, 12, Prokopios is malevolent (true) and exaggerating (not).

69. Agathias Scholasticus, *Histories* V.2. J. B. Bury (born 1861), on whom we happily depend, must be allowed his titillation—it is the volume's longest citation: *History of the Later Roman Empire: From the Death of Thodosius I to the Death of Justinian* (1958), p. 2:305.

70. *Excerpta de Legationibus Romanorum ad gentes,* 1, lines 13–30, in Blockley, *History of Menander,* pp. 43–45. On the Kutrigurs and Utrigurs, see Golden, *History of the Turkic Peoples,* pp. 98–100.

71. Bk. IV, 13.7–13, in Whitby and Whitby, *History of Theophylact Simocatta*(1986), pp. 121–122. See Walter E. Kaegi, *Byzantium and the Early Islamic Conquests* (1992), p. 32.

II. Byzantine Diplomacy

1. For overviews, see Alexander Kazhdan, "The Notion of Byzantine Diplomacy," p. 17, and Jonathan Shepard, "Byzantine Diplomacy, AD 800–1204: Means and Ends," pp. 42–45, both in Shepard and Franklin, *Byzantine Diplomacy* (1992).

2. Coincidentally a byzantinist, Mabillon (1632–1707) was a pupil of Charles du Fresne, sieur Du Cange (1610–1688), the greatest early scholar of Byzantium.

3. Garrett Mattingly, *Renaissance Diplomacy* (1988), p. 61.

3. Envoys

1. Not "Gok," which is modern Turkish, nor from blue = sky = Tengri Ulgen, shamanistic Sky God, beloved of secular nationalists keen on pre-Islamic Turkdom; Peter B. Golden, *Introduction to the History of the Turkic Peoples* (1992), p. 117, and personal communication, December 6, 2007: "Kök Türk is never used regarding the Western Türk Qaghanate."

2. R. C. Blockley, trans., *The History of Menander the Guardsman* (1985), pp. 116–117.

3. On Byzantine-Sasanian relations, see R. C. Blockley, *East Roman Foreign Policy: Formation and Conduct from Diocletian to Anastasius* (1992), pp. 121–127; and James Howard-Johnston, "The Two Great Powers in Late Antiquity: A Comparison" (1995), pp. 3:157–226.

4. Golden, *History of the Turkic Peoples*, p. 129; Blockley, *History of*

Menander, p. 264 n. 129. Blockley favors the Tekes valley, perhaps the locus of the Ch'ien Ch'uan ("Thousand Springs") of Chinese sources.

5. *Excerpta de Legationibus Romanorum ad gentes,* 8, in Blockley, *History of Menander,* p. 119.

6. Ibid., pp. 121–123.

7. *Excerpta de Legationibus gentiun ad Romanos* 9, in Blockley, *History of Menander,* p. 127.

8. Blockley, *History of Menander,* p. 153.

9. *The Anonymous Byzantine Treatise on Strategy,* sec. 43; previously known as *Peri Strategikes* or, more commonly, *De re strategica.* In George T. Dennis, ed. and trans., *Three Byzantine Military Treatises* (1985), pp. 125–127.

10. Prosper: Prosperi Tironis, *Epitoma chronicon* (1892/1980), p. 1:465.

11. *Vita Germani,* 28. From Andrew Gillett, *Envoys and Political Communication in the Late Antique West, 411–533* (2003), p. 121.

12. *Carmina,* VII, lines 308–311. "Avite, novas; saevum tua pagina regem lecta domat; iussisse sat est te, quod rogat orbis; credent hoc umquam gentes populique futuri? Littera Romani cassat quod, barbare, vincis." Text in W. B. Anderson, *Sidonius: Poems and Letters* (1963), p. 1:144.

13. *Excerpta de Legationibus gentium ad Romanos,* 3, in Blockley, *History of Menander,* p. 51.

14. Ibid., 1, in Blockley, *History of Menander,* p. 49.

15. *Chronicon Paschale,* indiction 11, year 13 [year 623], in Michael Whitby and Mary Whitby, trans., *Chronicon Paschale, 284–628 A.D.* (1989), p. 165.

16. *Theophanes,* no. 302, AM 6110, p. 434.

17. *Excerpta de Legationibus gentium ad Romanos,* 7, in Blockley, *History of Menander,* p. 115.

18. *Patzinakitai* in contemporary Greek sources; Golden, *History of the Turkic Peoples,* from p. 264.

19. *De Administrando Imperio,* ed. Gy. Moravcsik, trans. R. J. H. Jenkins (1967) (hereafter cited as *DAI*). Mischievously reintroduced by Ihor Sevcenko, "Re-reading Constantine Porphyrogenitus" (1992), pp. 167ff.

20. *DAI,* vol. 2: *Commentary,* ed. R. J. H. Jenkins et al. (1962) (hereafter cited as *DAI—Commentary*), sec. 8, pp. 55–57. On "Zakana" (lines 8–17), compare modern Russian *zakon* = law, derived from archaic *kon* = border, limit. *DAI—Commentary,* sec. 38, introductory note, pp. 145–146.

21. *Excerpta de Legationibus Romanorum ad gentes,* 3, lines 305–390, in Blockley, *History of Menander,* pp. 71–75.

22. Ibid., lines 408–423, in Blockley, *History of Menander,* p. 77. I follow Blockley (p. 255 n. 47): the procedure is not a textual "doublet" but authentic—and certainly prudent.

23. Theodosian Code VI, *De Agentibus in rebus,* 27.23; Justinian's Code XII.20.3.

24. Otto Seeck, *Notitia Dignitatum accedunt Notitia Urbis Constantinopolitanae Laterculi Prouinciarum* (1876), pp. 31–33. I adhere to Peter Brennan's skeptical interpretation—see "Mining a Mirage," a subheading

in his introduction to his *Notitia* (forthcoming); he kindly sent me the text on March 31, 2008.

25. Five in the Diocese of Oriens; four in Pontica; one in Asiana; two in the Diocese of the two Thraces; and four in the Diocese of Illyricum.

26. J. B. Bury, "The Imperial Administrative System in the Ninth Century with a Revised Text of the Kletorologion of Philotheos" (1911), 32.

27. Prokopios, *Anecdota,* bk. XXX, 4–7, in H. B. Dewing, ed. and trans., *Procopius* (1962–1978), pp. 6:347–349.

28. Louis Bréhier, *Les Institutions de l'empire Byzantin*(1949/1970), pp. 263–268; there was also an optical telegraph, pp. 268–270.

29. Hunt papyrus data in the John Ryland Library, but I follow Peter Heather, *The Fall of the Roman Empire: A New History of Rome and the Barbarians* (2006), p. 105.

30. John Lydus [John the Lydian], *On the Magistracies of the Roman Constitution* trans. T. F. Carney (1965), bk. II.10.5.

31. N. J. E. Austin and N. B. Rankov, *Exploratio: Military and Political Intelligence in the Roman World from the Second Punic War to the Battle of Adrianople* (1995), p. 152.

32. D. A. Miller, "The Logothete of the Drome in the Middle Byzantine Period" (1966), pp. 438–470.

33. C. D. Gordon, "The Subsidization of Border Peoples as a Roman Policy of Imperial Defense" (1948).

34. Michael F. Hendy, *Studies in the Byzantium Monetary Economy, c. 300–1450* (1985), p. 261.

35. Blockley, *East Roman Foreign Policy,* pp. 149–150; Blockley, "Subsidies and Diplomacy: Rome and Persia in Late Antiquity" (Spring 1985), pp. 62–74.

36. Blockley, *East Roman Foreign Policy,* pp. 151ff.

4. Religion and Statecraft

1. John Meyendorff, *Byzantium and the Rise of Russia*(1989), from p. 173.

2. Robert de Clari, *La Prise de Constantinople* (1873/1966), chap. 73, pp. 57–58.

3. *Buildings,* I.i.61–64, in H. B. Dewing, ed. and trans., *Procopius* (1962–1978), p. 7:33.

4. Saxo Grammaticus, *Gesta Danorum,* bk. 12.7.4, http://www2.kb.dk/elib/lit//dan/saxo/lat/or.dsr/index.htm.

5. The title of Ioli Kalavrezou's essay in Henry Maguire, ed., *Byzantine Court Culture from 829 to 1204* (1997), pp. 53–79.

6. On the number of relics in Constantinople, see Kalavrezou, "Helping Hands," p. 53, citing O. Meinardus, "A Study of the Relics of the Greek Church" *Oriens Christianus,* no. 54 (1970) pp. 130–133.

7. Samuel H. Cross and Olgerd P. Sherbowitz-Wetzor, *The Russian Primary Chronicle: Laurentian Text* (1953), p. 111.

8. Ibid., year 6495 (987), p. 112.

9. "Tsar," a contraction of "tsesar" from the original Roman title *Caesar,* not from its devalued Byzantine derivative *Kaisar,* subordinate to the Basileus or emperor proper, as the late-Roman Caesar had been subordinate to Augustus in Diocletian's Tetrarchy, 293–306, of two senior emperors (Augusti) and two junior emperors (Caesares); from the thirteenth century, the title *Kaisar* was outranked by *Despotes* and *Sebastokrator.*

10. Cross and Sherbowitz-Wetzor, *Russian Primary Chronicle,* p. 112.

11. John Wortley, *Ioannis Scylitzes: A Synopsis of Histories (811–1057 AD), a Provisional Translation,* unpublished manuscript, Centre for Hellenic Civilization, University of Manitoba, Winnepeg; text kindly lent by Prof. John Wortley. "Basil II Bulgharoctonos," chap. 17, p. 181.

12. Sigfus Blondal, *The Varangians of Byzantium*(1978), from p. 43. In Kievan Rus' also, "Varangian" meant foreign recruit. Simon Franklin and Jonathan Shepard, *The Emergence of Rus, 750–1200* (1996), p. 197.

13. Dimitri Obolenskly, *The Byzatine Commonwealth: Eastern Europe, 500–1453* (1971), from p. 272,"factors in cultural diffusion."

5. The Uses of Imperial Prestige

1. Jordanes, *Getica: De origine actibusque Getarum,* XXVIII.142, 143, in Charles C. Mierow, trans., *The Gothic History of Jordanes* (1915), p. 91.

2. Famously ridiculed by Liutprand of Cremona. See Gerard Brett, "The Automata in the Byzantine 'Throne of Solomon'" (July 1954).

3. It was updated in part under Nikephoros II Phokas (963–969). Constantine VII Porphyrogennetos, *De Cerimoniis Aulae Byzantinae,* in *Corpus Scriptorum Historiae Byzantinae,* ed. J. Reiske (1829) (hereafter cited as *De Cerimoniis*).

4. Admirably studied in Eric McGeer, *Sowing the Dragon's Teeth: Byzantine Warfare in the Tenth Century* (1995), pp. 233–236.

5. *De Cerimoniis,* from p. 571; but here from Arnold Toynbee, *Constantine Porphyrogenitus and His World* (1973), from p. 500.

6. On how the robes reflected rank, see Elisabeth Piltz, "Middle Byzantine Court Costume" (1997), pp. 39–51.

7. Toynbee, *Constantine Porphyrogenitus,* p. 502 (I could not reconcile his Reiske page references).

8. David Ayalon, *Eunuchs, Caliphs and Sultans: A Study in Power Relationships* (1999), appendix F, p. 347.

9. Ibid., appendix J., pp. 345–346.

10. Liliana Simeonova, "In the Depth of Tenth-Century Byzantine Ceremonial: The Treatment of Arab Prisoners of War at Imperial Banquets" (2007), p. 553.

11. Toynbee, *Constantine Porphyrogenitus,* p. 503.

12. Nadia Maria El-Cheikh, "Byzantium Viewed by the Arabs" (1992), from p. 173.

13. Paul Magdalino, "In Search of the Byzantine Courtier" (1997), pp. 141–165.

14. Alexander P. Kazhdan and Michael Mcormick, "The Social World of the Byzantine Court" (1997), pp. 167–197.

15. Prokopios, *Anecdota,* bk. VI, 2, in H. B. Dewing, ed. and trans., *Procopius* (1962–1978), p. 6:69.

16. Snorri Sturluson, *Heimskringla: The Chronicle of the Kings of Norway* (1907).

17. Nicolas Oikonomides, "Title and Income at the Byzantine Court" (1997), pp. 199–215.

18. Even Constantine has now admitted that is was silly "mumbo jumbo" via his medium: Ihor Sevcenko, "Re-reading Constantine Porphyrogenitus" (1992), p. 182.

19. *De Administrando Imperio,* ed. Gy. Moravcsik, trans. R. J. H. Jenkins (1967), sec. 13, lines 25–70.

20. *De Cerimoniis,* bk. II, cols. 46–47, pp. 1:679–686. But here from Paul Stephenson: http://homepage.mac.com/paulstephenson/trans/decer2.html.

21. *De Cerimoniis,* bk. I, cols. 47–48, from p. 680. But here from Paul Stephenson's Web site.

22. See Walter Emil Kaegi Jr., "The Frontier: Barrier or Bridge?" (2007), pp. 269–293, with ample bibliographic notations and rich in Arabic sources.

6. Dynastic Marriages

1. Ruth Macrides, "Dynastic Marriages and Political Kinship" (1992), from p. 263.

2. Theophanes, *The Chronicle of Theophanes Confessor, AD 284–813,* trans. Cyril A. Mango et al. (1997) (hereafter cited as *Theophanes*), no. 410, AM 6224, p. 567.

3. *De Cerimoniis Aulae Byzantinae,* I, 17. Cited and translated in Elisabeth Piltz, "Middle Byzantine Court Costume" (1997), p. 42.

4. *De Administrando Imperio,* ed. Gy. Moravcsik, trans. R. J. H. Jenkins (1967), sec. 13, lines 105–170, pp. 70–71. See also R. J. H. Jenkins, ed., *De Administrando Imperio,* vol. 2: *Commentary* (1962), p. 67, see 13/107 and 13/121–122.

5. *Vita Karoli Magni* 28, translation from http://www.fordham.edu/halsall/basis/einhard.htmlNo.Charlemagne. For the context, see Robert Folz, *The Coronation of Charlemagne, 25 December 800* (1974), from p. 132.

6. *Theophanes,* no. 473, AM 6288, p. 649.

7. *Theophanes,* no. 475, AM 6293, p. 653.

8. Folz, *Coronation of Charlemagne,* p. 174.

9. *Relatio de legatione Constantinapolitana* (Narrative of the embassy to Constantinople), in F. A. Wright, trans., *The Works of Liutprand of Cremona* (1930).

10. "John Tzimiskes," in *Ioannis Scylitzes: A Synopsis of Histories,* trans. John Wortley, chap. 21, p. 168.

11. Overviews: David Morgan, *The Mongols*(1986); John J. Saunders, *The History of the Mongol Conquests* (1971).

12. Private communication from Peter B. Golden, March 23, 2008.

13. "The Secret History of the Mongols," redacted in 1240 before the great conquests, has nothing on the Cinggisid states but much on the Mongol mentality in prose and lyric. See Marie-Dominique Even and Rodica Pop, *Histoire secrète des Mongols = Mongghol-un ni'uca tobciyan: Chronique mongole du XIIIe siècle* (1964); or in English, see Igor de Rachewiltz, *The Secret History of the Mongols: A Mongolian Epic Chronicle of the Thirteenth Century*, 2 vols. (2004); and the anachronistically old-fashioned but wonderful Francis Woodman Cleaves, *The Secret History of the Mongols*, vol. 1 (1982).

7. The Geography of Power

1. The dilettante orientalist and political propagandist Edward Said started an evil fashion: e.g., Francois Hartog, *The Mirror of Herodotus: The Representation of the Other in the Writing of History* (1988); and Edith Hall, *Inventing the Barbarian* (1989): the "Other" is always "invented," prejudicially of course. For how it actually was, see Gilbert Dagron, "Ceux d'en face: Les peuples étrangers dans les traités militaires byzantins" (1987), pp. 207–232.

2. Constantine Porphyrogennetos, *De Cerimoniis Aulae Byzantinae*, II, 48, in *Corpus Scriptorum Historiae Byzantinae*, ed. J. Reiske (1829), pp. 1:686–692. But here from Paul Stephenson's translation: http://homepage.mac .com/paulstephenson/trans/decer2.html.

3. Gilbert Dagron, "Byzance et ses Voisins: Étude sur certain passages du livre de cérémonies" (2000), II, 15 and 46–48, and map on p. 357.

4. For the tenth-century context and its historical background, see Bernardette Martin-Hisard, "Constantinople et les Archontes du Monde Caucasien dans le Livre de Cérémonies" (2000), II, 48.

5. C. Toumanoff, "Armenia and Georgia" (1967), pp. 593–637.

6. Dagron, "Byzance et ses Voisins."

7. For a summary, see Eckhard Muller-Mertens, "The Ottonians as Kings and Emperors" (1999), pp. 233–254.

8. Hebrew: *Kuzari*, pl. *Kuzarim;* Persian, Arabic: *Xazar, Qazar;* Slav: *Khozar;* Chinese: *Ho-sa, K'o-sa* but invariably qualified by *Tujue* (Turk) = *Tujue Kesa.* Peter B. Golden, *Introduction to the History of the Turkic Peoples: Ethnogenesis and State Formation in Medieval and Early Modern Eurasia and the Middle East* (1992), pp. 233–244; and Golden, *Khazar Studies: An Historio-Philological Inquiry into the Origins of the Khazars*(1980).

9. Compare Thomas Noonan, "Byzantium and the Khazars: A Special Relationship?" (1992), p. 109, but also p. 129: "The looming clash between Khazars and Byzantines in the southern Caucasus never took place."

10. Theophanes, *The Chronicle of Theophanes Confessor, AD 284–813*, trans. Cyril A. Mango et al. (1997), no. 315, AM 6117, p. 446.

11. Jie Fei, Jie Zhou, and Yongjian Hou, "Circa A.D. 626 Volcanic Eruption, Climatic Cooling, and the Collapse of the Eastern Turkic Empire" (April 2007). The authors rely on the Jiu Tang Shu (Old Book of Tang) by Liu Xu, ca. 945, i.e., the Tang dynastic history.

12. R. J. H. Jenkins et al., eds., *De Administrando Imperio,* vol. 2: *Commentary* (1962) (hereafter cited as *DAI-Commentary*), s.v. "9/I," pp. 20–23. The same applies to the state that the Rus established.

13. *Annales Bertiniani,* ca. 839, ed. G. Waitz (1883), pp. 19–20. See Simon Franklin and Jonathan Shepard, *The Emergence of Rus, 750–1200* (1996), pp. 29–32.

14. In his *Bibliotheca* or *Myriobiblon* ("A Thousand Books"), Photios reviewed and summarized 279 books (an online edition is under way at www.ccel .org/p/pearse/morefathers/photius_03bibliotheca.htmNo.34). Extract from Cyril Mango, trans., *The Homilies of Photius,* in Deno John Geanakoplos, *Byzantium: Church, Society, and Civilization Seen through Contemporary Eyes* (1984), p. 351.

15. See Andras Rona-Tas, *Hungarians and Europe in the Early Middle Ages: An Introduction to Early Hungarian History* (1999), from p. 319; Golden, *History of the Turkic Peoples,* p. 258; and the summary in Peter B. Golden, "The Peoples of the Russian Forest Belt" (1990), pp. 229–248, which presents the famous work of Gyula Németh, *A honfoglaló magyarság kialakulása* (in my parental library).

16. Rona-Tas, *Hungarians and Europe,* p. 311.

17. As other tribes were absorbed into the Magyar identity; only the Székely, or Szeklers, of Romania speak Magyar but are not Magyars, retaining distinctive customs.

18. Golden, *History of the Turkic Peoples,* from p. 264.

19. *De Administrando Imperio,* ed. Gy. Moravcsik, trans. R. J. H. Jenkins (1967) (hereafter cited as *DAI*), from line 15, sec. 1, pp. 48–49.

20. Ibid., sec. 2. pp. 49–50.

21. Ibid., sec. 2, p. 50.

22. Ibid., sec. 3, p. 51.

23. Ibid., sec. 8, p. 57

24. Ibid., sec. 5. p. 53.

25. Ibid., sec. 7, p. 55

26. Ibid., sec. 6, p. 53

27. Ibid., sec. 7, p. 55.

28. Ibid., sec. 37, p. 167; for the context, see *DAI-Commentary,* s.v. "37," p. 143.

29. *DAI,* sec. 4, p. 51.

30. They were engulfed by Mongol rule; Golden, *History of the Turkic Peoples,* from p. 270.

31. John F. Haldon, *The Byzantine Wars: Battles and Campaigns of the Byzantine Era* (2001), pp. 112–127.

32. The epic Oğuz songs are still known among the Turks of modern Turkey and Iran (officially described as "Azeri").

33. Bk. 8, V [204–205], translated by Elizabeth A. Dawes (London, 1928), http://www.fordham.edu/halsall/basis/annacomnena-alexiad08.html.

34. *DAI*, sec. 31, p. 149. There is skepticism about this formulation: see *DAI-Commentary*, sec. 31/8–9, p. 124.

35. *DAI*, sec. 32, p. 153. But see the interpretation in *DAI-Commentary*, p. 131.

36. *DAI*, sec. 30, p. 145; *DAI-Commentary*, sec. 30/90–93, pp. 120–122.

37. See the reconstruction in Charles R. Bowlus, *Franks, Moravians and Magyars: The Struggle for the Middle Danube, 788–907* (1995).

38. *The Oxford Dictionary of Byzantium*, ed. Alexander P. Kazhdan et al. (1991), pp. 1:310–311.

39. J. Innes Miller, *The Spice Trade of the Roman Empire, 29 BC to AD 641* (1969). The cutoff date is the death of Herakleios and the fall of Alexandria.

40. *The Gothic War*, bk. 8, XVII, 1–8, in H. B. Dewing, ed. and trans., *Procopius* (1962–1978), pp. 5:227–231.

41. Lin Yin, "Western Turks and Byzantine Gold Coins Found in China" (July 2003), online at http://www.transoxiana.org/0106/lin-ying_turks_solidus .html.

42. Notably the *Jiu Tang shu*, "Old Book of Tang," attributed to Liu Xu, ca. 945.

43. From Friedrich Hirth, *China and the Roman Orient: Researches into Their Ancient and Mediaeval Relations as Represented in Old Chinese Records* (1885), pp. 65–67. Edited by J. S. Arkenberg in Paul Halsall's Internet East Asian History Source Book, http://www.fordham.edu/halsall/eastasia/ 1372mingmanf.html.

8. Bulghars and Bulgarians

1. See the excellent surveys in Paul Stephenson, *Byzantium's Balkan Frontier: A Political Study of the Northern Balkans, 900–1204* (2000); and Florin Curta, *Southeastern Europe in the Middle Ages, 500–1250* (2006), pp. 147–179.

2. Peter B. Golden, *Introduction to the History of the Turkic Peoples: Ethnogenesis and State Formation in Medieval and Early Modern Eurasia and the Middle East* (1992), from p. 244.

3. Chap. CXX, 47, 48. From R. H. Charles, *The Chronicle of John, Bishop of Nikiu* (1916). Online: http://www.tertullian.org/fathers/index.htmNo .John_of_Nikiu.

4. Golden, *History of the Turkic Peoples*, p. 246.

5. Theophanes, *The Chronicle of Theophanes Confessor, AD 284–813*, trans. Cyril A. Mango et al. (1997) (hereafter cited as *Theophanes*), no. 374, AM 6196, p. 374.

6. Ibid., no. 375, AM 6198, p. 523.

7. Nikephoros, Patriarch of Constantinople, *Short History [Breviarium Historicum]*, ed. Cyril A. Mango (1990), no. 2, pp. 103–105; *Theophanes*, no. 376, AM 6200, p. 525.

8. *Theophanes,* no. 376, AM 6200, p. 525.

9. "He asked once again help from Terbelis, chief of the Bulgarians, who sent him three thousand men." Nikephoros, *Short History,* no. 45, p. 111.

10. *Theophanes,* no. 382, AM 6204, p. 532.

11. Ibid., no. 387, AM 6209, p. 546.

12. Andrew Palmer, *The Seventh Century in the West-Syrian Chronicles* (1993), text no. 13, no. 158, p. 215.

13. Ibid., nos. 159–160, p. 216.

14. Ibid., text no. 12, p. 80.

15. On Krum, see Florin Curta, *Southeastern Europe in the Middle Ages, 500–1250* (2006), pp. 149–153.

16. *Theophanes,* nos. 490–492, AM 6303, pp. 672–674.

17. *Scriptor Incertus,* fragment 1, ed. I. Dujcev (1965), 205–254 at 210–216.

18. See *De thematibus,* ed. and trans. Agostino Pertusi (1952); Agostino Pertusi, "La formation de themes byzantins" (1958), I, 1–40. *The Oxford Dictionary of Byzantium,* ed. Alexander P. Kazhdan et al. (1991) (hereafter cited as *ODB*), pp. 3:2034–2035, introduces the literature.

19. Revised translation, November 2004, Paul Stephenson, http://homepage .mac.com/paulstephenson/trans/scriptor1.html.

20. John F. Haldon, *Byzantine Praetorians: An Administrative, Institutional and Social Survey of the Opsikion and Tagmata, c. 580–900* (1984), from p. 209.

21. That must refer to Krum's capital of Pliska, some two hundred miles from Constantinople, because by July 26 Nikephoros had conquered and had been conquered.

22. "Defensive Mountain Warfare," in Carl von Clausewitz, *On War,* ed. Michael Howard and Peter Paret (1976), from p. 417.

23. "Constantine VII," in Scylitzes, *Synopsis of Histories,* no. 3, p. 110; *ODB,* s.v. "Symeon of Bulgaria," p. 3:1984.

24. Nicholas I, Patriarch of Constantinople, *Letters,* ed. and trans. R. J. H Jenkins and L. G. Westerink (1973), letter 19, p. 216.

25. Paul Stephenson, *Byzantium's Balkan Frontier: A Political Study of the Northern Balkans, 900–1204* (2000), p. 18, dissents.

26. J. Darrouzes, *Épistoliers byzantins du Xe siecle* (1960), letters 5–7, pp. 94.

27. Theophanes, *Theophanes Continuatus,* ed. I. Bekker (1838), pp. 388–390, no. 10, translation by Paul Stephenson at http://homepage.mac.com/ paulstephenson/trans/theocont2.html.

28. Paul Stephenson, *The Legend of Basil the Bulgar-Slayer* (2003), from p. 12. In what follows, I rely on this masterly new analysis.

29. Bk. X.8, in Alice-Mary Talbot and Denis F. Sullivan, trans., *The History of Leo the Deacon: Byzantine Military Expansion in the Tenth Century* (2005), pp. 214–215.

30. Scylitzes, *Synopsis of Histories,* "Basil [II] and Constantine [VIII]," sec. 26, p. 184.

31. Stephenson, *Legend of Basil,* is skeptical about the annual raiding and doubts that Basil II intended the annihilation of Bulgaria.

32. Ibid., p. 13.

33. Scylitzes, "Basil [II] and Constantine [VIII]," sec. 35, p. 187.

34. Stephenson, *Legend of Basil,* pp. 4–6. He duly notes that the *Strategikon* of Kekaumenos corroborates the tale he persuasively deconstructs; but when Kekaumenos was writing, recollections of great victories were much needed. See Kekaumenos, *Raccomandazioni e consigli di un galantuomo: Stratēgikon,* ed. and trans. Maria Dora Spadaro (1998), pp. 19ff.

35. What follows is entirely derived from N. G. Wilson, *Scholars of Byzantium,* rev. ed. (1996), pp. 3–4. The text is based on the MSS Patmos 178.

36. Paul Stephenson, *Byzantium's Balkan Frontier: A Political Study of the Northern Balkans, 900–1204* (2000), p. 18.

9. The Muslim Arabs and Turks

1. There are five: the avowal (*shahada*), "There is no god but God, and Mohammad is the Messenger of God"; the five daily prayers *(salat); * charity *(zakat); * daylight fasting during Ramadan *(sawm); * and the pilgrimage to Mecca *(hajj),* which reconciled the Meccans to the new religion by perpetuating the pre-Islamic pilgrimage to the black stone, the Ka'ba.

2. In a vast literature, see, most recently, Michael Bonner, *Jihad in Islamic History: Doctrines and Practice* (2007). Compare Valeria F. Piacentini, *Il Pensiero Militare nel Mondo Mussulmano*(1996), pp. 191–221, 290–301.

3. The Alevis are more Bektashi than anything else; their Shi'ism is mostly nominal, while their distinctive practices are mostly shamanistic.

4. That is the classic opinion. It is now more often argued that internal division was fatally undermining Byzantine rule; see Walter E. Kaegi Jr., *Byzantium and the Early Islamic Conquests* (1992), from p. 47.

5. High Kennedy, *The Armies of the Caliphs* (2001), from p. 19.

6. Kaegi, *Byzantium,* p. 119.

7. On Byzantine taxes, see J. F. Haldon, *Byzantium in the Seventh Century: The Transformation of a Culture* (1990), from p. 173.

8. See the short overview in A. H. Jones, *The Later Roman Empire, 284–602: A Social and Economic Survey* (1973), from p. 1:411.

9. Michael Whitby, trans., *The Ecclesiastical History of Evagrius Scholasticus* (2000), bk. III, 38, p. 183.

10. Seleucid year 809 (497/498 CE). Frank R. Trombley and John W. Watt, trans., *The Chronicle of Pseudo-Joshua the Stylite* (2000), no. 31, pp. 30–31.

11. Prokopios, *Anecdota,* bk. XIX, 7, In H. B. Dewing, ed. and trans., *Procopius* (1962–1978), p. 6:229.

12. Theophanes, *The Chronicle of Theophanes Confessor, AD 284–813,* trans. Cyril A. Mango et al. (1997) (hereafter cited as *Theophanes*), no. 303, AM 6113, p. 435.

13. Folio 46b. Tractate Nedarim, 46/b. Most conveniently from www.come-and-hear.com/nedarim/nedarim_46.html. Reassured by Maurice Sartre,

D'Alexandre à Zénobie: Histoire du Levant antique, IVe siècle avant Jésus-Christ–IIIe siècle après Jésus-Christ (2001), I risk giving more weight to the Talmudic evidence than Ze'ev Rubin does in "The Reforms of Khusro Anushirvan" (1995), from p. 232.

14. But for the authoritative Ze'ev Rubin the matter is controversial. Rubin, "Reforms," p. 231.

15. Under "Mention of the Holders of Power in the Kingdom of Persia after Ardashir b. Babak," subheading "Resumption of the History of Kisra Anusharwan," sec. 960–962. In C. E. Bosworth, *The History of al-Tabari* (1999), pp. 5:256–257.

16. Abdullah Yusuf Ali, translation of *Al-Tawba* (repentance), Sura 9.29.

17. "The infidel remains standing . . . his head bowed and his back bent. The infidel must place money on the scales, while the collector holds him by his beard and strikes him on both cheeks" (Al-Nawawi); or, "Jews, Christians, and Majians must pay the jizya . . . on offering up the jizya, the dhimmi must hang his head while the official takes hold of his beard and hits [the dhimmi] on the protruberant bone beneath his ear [i.e., the mandible]" (Al-Ghazali). http://en.wikipedia.org/wiki/Dhimmi Sect. 4.7.1.

18. They were denigrated as "Melkites" by the Monophysites—that is, as followers of the *malko* Syriac for emperor, specifically Marcian (450–457), who presided over the Council. In the eighteenth century the insult became the name—Melkite Greek Catholic of Chalcedonian churches that retain a Byzantine liturgy but give their allegiance to the pope.

19. *Pseudo-Dionysius of Tel-Mahre: Chronicle, Known Also as the Chronicle of Zuqnin,* pt. 3, trans. Witold Witakowski (1996), pp. 19–21.

20. Andrew Palmer, *The Seventh Century in the West-Syrian Chronicles* (1993), text no. 13, no. 53, p. 148.

21. Walter E. Kaegi Jr., *Heraclius: Emperor of Byzantium* (2003), pp. 269–271.

22. The Maronite church of Lebanon long had the distinction of being the only monothelite church—until a Chalcedonian affiliation to the French-protected papacy became valuable with the rise of French influence during the nineteenth century.

23. R. H. Charles, *The Chronicle of John, Bishop of Nikiu* (1916), chap. CXX, sec. 72, and chap. CXXI, sec. 1. http://www.tertullian.org/fathers/index .htmNo.John_of_Nikiu.

24. Martin Luther reacted even more furiously, calling for their incineration when the Jews inexplicably refused to become Lutherans.

25. *Doctrina Jacobi nuper baptizati,* ed. and trans. V. V. Déroche (1991), pp. 70–218.

26. *Encyclopedia Judaica* (1973), s.v. "Pumbedita," p. 13:1383.

27. Frederick C. Conybeare, "Antiochus Strategos' Account of the Sack of Jerusalem (614)" (July 1910). For the broader context, see Kaegi, *Byzantium,* from p. 220.

28. Theodosian Code, XVI, 18.24. Latin text and translations, Amnon Linder, *The Jews in Roman Imperial Legislation* (1987), no. 45, pp. 281–282.

29. Ibid., no. 46, pp. 284–285. The text of XVI, 8, is now online: http://ancientrome.ru/ius/library/codex/theod/liber16.htm8.

30. Codex Justinianus, I, 5, 13, in Linder, *The Jews*, no.56, from p. 356.

31. See the authoritative summary in Haldon, *Byzantium in the Seventh Century*, pp. 63–64, and the earlier assessment in George Ostrogorsky, *History of the Byzantine State* (1969), pp. 123–125.

32. *Theophanes*, no. 354, pp. 493–494.

33. Ibid., no. 355, p. 496.

34. Palmer, *West-Syrian Chronicles*, text 13, nos. 161, 162, 163, pp. 217–218.

35. The continuing de-urbanization debate is slowly being illuminated by archaeology; see Haldon, *Byzantium in the Seventh Century*, from p. 93. On Constantinople, see *The Oxford Dictionary of Byzantium*, ed. Alexander P. Kazhdan et al. (1991) (hereafter cited as *ODB*), p. 1:511, col. 1.

36. Peter B. Golden, *Introduction to the History of the Turkic Peoples: Ethnogenesis and State Formation in Medieval and Early Modern Eurasia and the Middle East* (1992), from p. 216; the *Dede Qorqut* epic of the Oğuz is the earliest literature of contemporary Turks, in part their descendants.

37. See Walter Emil Kaegi Jr., "The Contribution of Archery to the Turkish Conquest of Anatolia" (2007).

38. Their ancestral capital and the final redoubt of Ghazna has retained one feature of an imperial capital: it is the most multi-ethnic town of Afghanistan but for Kabul.

39. In modern Turkey's Karz Province on the Armenian border. It was the depredations of Kurdish nomads centuries later that ruined the city.

40. Qur'an, Yusuf Ali translation; e.g., Sura 52 At-Tur: 17: "As to the Righteous, they will be in Gardens, and in Happiness." 18: "Enjoying the (Bliss) which their Lord hath bestowed on them." 19: (To them will be said:) "Eat and drink ye, with profit and health, because of your (good) deeds." 20: "They will recline (with ease) on Thrones (of dignity) arranged in ranks; and We shall join them to [female] Companions, with beautiful big and lustrous eyes." 22: "And We shall bestow on them, of fruit and meat, anything they shall desire." 24: "Round about them will serve, (devoted) to them. Youths [handsome] as Pearls well-guarded."

41. *ODB*, p. 3:1086, col. 1. See the brief but authoritative summary in John F. Haldon, *The Byzantine Wars: Battles and Campaigns of the Byzantine Era* (2001), p. 114.

42. This author was once present when a unit misdirected by a navigation error blundered into an enemy command post, then inflicting much damage.

43. *ODB*, p. 2:739, col. 2.

44. Ibid., p. 1:658, col. 1.

45. Haldon, *The Byzantine Wars*, p. 126.

46. *ODB*, p. 1:63, col. 1.

47. Golden, *History of the Turkic Peoples*, p. 223.

48. John Kinnamos, *Deeds of John and Manuel Comnenus*, trans. Charles M. Brand (1976), bk. V, 3, pp. 156–157.

49. Haldon, *The Byzantine Wars*, from p. 139.

III. The Byzantine Art of War

1. Bk. I; George T. Dennis, ed., *Das Strategikon des Maurikios,* trans. Ernst Gamillscheg (1981); earlier: Haralambie Mihăescu, ed. and trans., *Arta militară: Mauricius* (1970). Here cited from George T. Dennis, trans., *Maurice's Strategikon Handbook of Byzantine Military Strategy* (1984), p. 16.
2. G. L. Chessman,*The Auxilia of the Roman Imperial Army* (1914/1975), pp. 90–92.

10. The Classical Inheritance

1. The indispensable survey remains Alphonse Dain, "Les stratégistes byzantins," ed. J. A. de Foucault (1967), pp. 317–392.
2. Publius Flavius Vegetius Renatus, *Epitoma Rei Militari.* Latin intra-text: http://www.intratext.com/X/LAT0189.HTM by Eulogos SpA—may they prosper. N. P. Milner, trans., *Vegetius: Epitome of Military Science* (1996), pp. xxxvii–xli on dating.
3. Bk. I, 15, Milner, *Vegetius,* p. 15.
4. Sextus Julius Frontinus, *Strategemata,* trans. C. E. Bennet (1980), p. 7.
5. Ibid., IV, vii, p. 309
6. Ibid., p. 3.
7. Peter Krentz and Everett L. Wheeler, eds. and trans., *Stratagems of War: Polyaenus* (1994), bk. I, 6–8, p. 1:5 para. 1. The preface informs us that Wheeler dedicates the work to his cat J.B., "who loves to read Greek."
8. A commendable effort to put the 30,000-odd entries of the *Suda* online in searchable form is now under way at http://www.stoa.org/sol-bin/search.pl.
9. Greek text and translation in James G. DeVoto, *Flavius Arrianus* (1993).
10. E.g., *Cl. Aeliani et Leonis Imp. tactica sive de instruendis aciebus,* ed. Meursius (1613).
11. *Ala I Augusta Colonorum, Ala I Ulpia Dacorum, Ala II Gallorum, Ala II Ulpia Auriana.*
12. *Cohors IIII Raetorum equitata, Cohors III Augusta Cyrenaica sagittariorum equitata, Cohors I Raetorum equitata. Cohors Ituraeorum sagittariorum equitata, Cohors I Numdiarum equitata.*
13. E. A. Thompson, *A Roman Reformer and Inventor: Being a New Translation of the Treatise De Rebus Bellicis* (1952).
14. Frank Granger, *Vitruvius on Architecture* (1983), p. xv.
15. Later the meaning of the two terms was reversed: stone-throwing catapults and arrow-firing ballistae.
16. *Hygini Gromatici Liber de Munitionibus Castrorum* (1887/1972), with an analysis *(Die Lagerordnung),* from p. 39. There is a valuable intra-text at http://www.intratext.com/X/LAT0347.html, and a recent English translation by Cathrine M. Gilliver, "The *de munitionibus castrorum,*" *Journal of Roman Military Equipment Studies* 4 (1993), pp. 33–48 (not seen by me).

17. See also John Earl Wiita, *The Ethnika in Byzantine Military Treatises* (1977), p. 101.
18. *Aeneas Tacticus, Asclepiodotus, Onasander,* ed. and trans. W. A. Oldfather (1977).
19. Pyrrhus, Alexandros, Clearque(!), Pausanias, Evangelos, Eupolemos, Iphicrates, Posidonius, Bryon; Dain, "Les Stratégistes Byzantins," p. 321.
20. "Biton's Construction of War Engines and Artillery," text in E. W. Marsden, *Greek and Roman Artillery: Technical Treatises* (1971), from p. 67.
21. Polybius, *The Histories,* trans. W. R. Paton (1979), bk. VIII.4, 4–9, p. 3:455.
22. Marsden, *Artillery: Technical Treatises,* from p. 107.
23. Ibid., diagram 9, p. 179; on Chinese repeater crossbows, see p. 178 n. 106.
24. Dain, "Les stratégistes byzantins," ed. J. A. de Foucault (1967).
25. Ibid., p. 206.
26. See the survey in Stephen Edward Cotter, "The Strategy and Tactics of Siege Warfare in the Early Byzantine Period: From Constantine to Heraclius" (1995), from p. 99.
27. Asclepiodotus," in Oldfather, *Aeneas Tacticus, Asclepiodotus, Onasander,* from p. 229.
28. Dain, "Les Stratégistes Byzantins," pp. 328–329.
29. "Onasander," in Oldfather, *Aeneas Tacticus, Asclepiodotus, Onasander,* p. 389.
30. Jean-René Vieillefond, *Les "Cestes" de Julius Africanus: Étude sur l'ensemble des fragments avec édition, traduction et commentaries* (1970). English translation with commentary: Francis C. R. Thee, *Julius Africanus and the Early Christian View of Magic* (1984). See the different perspectives in Martin Wallraff et al., *Julius Africanus und die christliche Weltchronistik* (2006).
31. Bk. VII, I, 20. I follow Giovanni Amatuccio, *Peri Toxeias l'arco di guerra nel mondo bizantino e tardo-antico* (1996), pp. 34, 51–53.
32. The text is abruptly inserted under the title *Ourbikioi Epitedeyma* ("Adaosul lui Urbicius") in Haralambie Mihăescu, ed. and trans., *Arta militară: Mauricius* (1970), from p. 368; also Dain, "Les Stratégistes Byzantins," p. 341.
33. Dain, "Les Stratégistes Byzantins," p. 342. The authoritative edition is Dain's own *Naumachica* (1943), but I have relied on F. Corazzini, *Scritto sulla Tattica Navale, di anonimo greco* (1883); from Roma Aeterna / Domenico Carro, www.romaeterna.org. There is now a superior text in Pryor and Jeffreys, *Age of the Dromon,* pp. 457–481.
34. George T. Dennis, ed. and trans., *Three Byzantine Military Treatises* (1985), from p. 11; Dain, "Les Stratégistes Byzantins," p. 343.
35. 565–895 are the outer limits; Salvatore Cosentino, "The Syrianos's 'Strategikon': A 9th Century Source?" (2000), p. 266, comprehensively reviews the debate and favors a late Syrianos.
36. V. Kucma and Col. O. L. Spaulding Jr. cited by Dennis, *Three Byzantine Military Treatises,* p. 3.

37. Sec. 4, in Dennis, *Three Byzantine Military Treatises,* p. 21.
38. *The Oxford Dictionary of Byzantium,* ed. Alexander P. Kazhdan et al. (1991), p. 1:588, col. 1; Prokopios wrongly attributes its principal structures to Justinian, in *Buildings,* II, I.4ff., in Dewing, ed. and trans., *Procopius* (1962–1978), from p. 7:99.
39. From a fragment of the lost work *De temporum qualitatibus et positionibus locorum* appended to Mommsen's edition of the *Chronicle of Marcellinus* [*Comes*], in Brian Coke, trans., *Marcellinus Comes* (1995), p. 40.
40. Dennis, *Three Byzantine Military Treatises,* pp. 3, 47 n.1.
41. *The Armenian History Attributed to Sebeos,* trans. R. W. Thompson (1999), sec. 127, chap. 39, p. 1:84.
42. *Peri Strategikes,* 19, 38–51, in Dennis, *Three Byzantine Military Treatises,* pp. 63–65.
43. Ibid., 24, 10–11, p. 79.
44. Ibid., 33, 42–47, p. 105.
45. Ibid., 36, 4–8, p. 109.
46. Ibid., 39, p. 117.
47. Ibid., from 42.20, p. 123.
48. From Diodorus Siculus (bk. II, 45.3), who so relished the fantasy of dominant, if mutilated, females that he re-created it with imaginary Libyans as well (bk. III, 52.2,3) whose men "like our married women, spent their days about the house, carrying out the orders which were given them by their wives," and who seared the breasts of their infant girls, of course. C. H. Oldfather, trans., *Diodorus of Sicily* (1935), pp. 2:33, 249.
49. *Peri Strategikes,* sec. 46, "Training for Power of Fire," in Dennis, *Three Byzantine Military Treatises,* p. 133.
50. *Peri Toxeias L'arco di guerra,* pp. 78, 79.
51. *The Persian War,* bk. I, xviii.32–34, in Dewing, *Procopius,* pp. 1:169–171.
52. *Peri Strategikes,* sec. 47, in Dennis, *Three Byzantine Military Treatises,* p. 135.
53. Dain, "Les Stratégistes Byzantins," p. 344.
54. Eric McGeer, private communication, January 14, 2008; I had followed Dain, "Les Stratégistes Byzantins," uncritically; see Gilbert Dagron, "Byzance et le modèle islamique au Xe siècle," (April–June 1983), pp. 219–224.
55. Eric McGeer, "Two Military Orations of Constantine VII," in John W. Nesbitt, ed., *Byzantine Authors: Literary Activities and Preoccupations* (2003), p. 112.

11. The *Strategikon* of Maurikios

1. E.g., the 1535 Paris edition with 120 elegantly bizarre illustrations (with Frontinus, Aelianus, and Modestus). Renati Fl. Vegetii, *Viri illustris: De re militari libri quatuor* (1535).
2. *Arriani Tactica et Mauricii Artis militaris libri duodecim* (1664). Alphonse Dain, "Les stratégistes byzantins," ed. J. A. de Foucault (1967), p. 344.

3. George T. Dennis, *Maurice's Strategikon: Handbook of Byzantine Military Strategy* (1984)(hereafter cited as *Strategikon*–Dennis), p. 8.

4. Ibid., p. xvi.

5. Henrik Zilliacus, *Zum Kampf der Weltsprachen im oströmischen Reich* (1935), pp. 134–135; the regiment: Kronprinsens husarer, p. 137 n.

6. "B Infantry Formations," bk. XII, 16, 14, in *Strategikon*–Dennis, pp. 146, 145.

7. Ibid., bk. I, 1, p. 11.

8. Prokopios, *The Wars*, bk. I, i, 13–15, in H. B. Dewing, ed. and trans., *Procopius* (1962–1978), p. 1:7.

9. Caroline Sutherland, "Archery in the Homeric Epics" (2001), http://www .ucd.ie/cai/classics-ireland/2001/sutherland.html.

10. *Iliad,* trans. E. V. Rieu (1950), bk. XI, 385–392, p. 207.

11. Bk. I, 1, 5. Haralambie Mihăescu, ed. and trans., *Arta militară* (1970), p. 51, translates *kontarion* into Romanian *sulita* = javelin, an error; "spear" (*Strategikon*–Dennis, p. 11) would also imply a throwing weapon, but it is the familiar stabbing *contos* of the Roman heavy cavalry adopted from the Sarmatians.

12. Bk. I, 2, *Strategikon*–Dennis, p. 13.

13. *Corpus Inscriptionum Latinarum,* vol. 8, 18042 = Dessau *Inscriptiones Latinae Selectae,* 2487. In Naphtali Lewis and Meyer Reinhold, trans., *Roman Civilization Source Book II: The Empire* (1966), p. 509.

14. Bk. XII, B, "Infantry Formations," 2, *Strategikon*–Dennis, p. 138.

15. Ibid., 3, p. 138.

16. Taxiarchis G. Kolias, *Byzantinische Waffen: Ein Beitrag zur byzantinischen Waffenkunde von den Anfängen bis zur lateinischen Eroberung* (1988).

17. Justinian's Novel 85 of AD 539, Caput V: "contos et quolibet modo vel figura factas lanceas, et quae apud Isauros nominantur monocopia."

18. D. Nishimura, "Crossbow, Arrow-Guides, and the Solenarion" (1988), pp. 422–435. Compare G. T. Dennis, "Flies, Mice, and the Byzantine Crossbow" (1981), pp. 1–5.

19. Giovanni Amatuccio cites a possible depiction of a *solenarion* in Bari; see "Lo Strano arciere della porta dei Leoni" (2005), http://www.arcosophia .net/database/N1/amatuccio_leoni.htm_edn6.

20. John Haldon, *Warfare, State and Society in the Byzantine World, 565–1204* (1999), p. 216.

21. Bk. I, 2, *Strategikon*–Dennis, p. 12.

22. On the trebuchet, see Paul E. Cheveddent, "The Invention of the Counterweight Trebuchet: A Study in Cultural Diffusion" (2000), pp. 72–116; and, more recently, Stephen McCotter, "Byzantines, Avars and the Introduction of the Trebuchet" (2003), http://www.deremilitari.org/resources/articles/ mccotter1.htm. On the *skala,* see *Strategikon*–Dennis, bk. I, 2, p. 13.

23. Maurice Keen, *Chivalry* (1986), p. 23.

24. R. P Alvarez, "Saddle, Lance and Stirrup" (1998).

25. Peter Connolly and Carol van Driel-Murray, "The Roman Cavalry Saddle" (1991), pp. 33–50; I owe the reference to Eric McGeer.

26. See the valuable intra-text at http://www.intratext.com/X/LAT0212.htm.

27. J. W. Eadie, "The Development of Roman Mailed Cavalry" (1967), pp. 161–173.

28. In *Medieval Technology and Social Change* (1966).

29. Bk. I, 2, *Strategikon*–Dennis, p. 12.

30. A. Pertusi, ed. and trans., *De Thematibus* (1952), pp. 133–136. For the federati, see also A. H. Jones, *The Later Roman Empire, 284–602: A Social and Economic Survey* (1973), pp. 1:665–667.

31. Jones, *Later Roman Empire*, pp. 1:663–666.

32. In Eric McGeer, *Sowing the Dragon's Teeth: Byzantine Warfare in the Tenth Century* (1995), pp. 1:51–52; p. 15 para. 6.

33. Bk. VII, 15, *Strategikon*–Dennis, p. 69.

34. Ibid., bk. I, 2, p. 12.

35. Prokopios, *The Wars*, bk. II, xxvii, 33–35, in Dewing, *Procopius*, p. 1:511.

36. Justinian's Code, 11.47.0, whereby any nonimperial use of weapons is prohibited ("Ut armorum usus inscio principe interdictus sit"), which repeated a law already in the Theodosian code (XV, 151) that dated back to Valentin and Valens in 364.

37. Bks. X, XXXIII, and XXXIV. In Betty Radice, trans., *Pliny Letters and Panegyricus* (1975), pp. 2:206–209.

38. Bk. VII, "Before the Day of Battle," *Strategikon*–Dennis, p. 65.

39. *Strategikon*–Dennis, pp. 143–144.

40. See Michael Whitby, "Recruitment in Roman Armies from Justinian to Heraclius (c. 565–615)" (1995), pp. 3:61–124.

41. Prokopios, *The Wars*, bk. I, xviii, 39, 40, in Dewing, *Procopius*, p. 1:173.

42. Giorgio Ravegnani, *Soldati di Bizanzio in Eta Giustinianea* (1988), p. 53.

43. Bk. VIII, 1, 25, *Strategikon*–Dennis, p. 81.

44. Ibid., 2, 92, p. 91.

45. Ibid., 86, p. 90.

46. Ibid., 1, 8, p. 80.

47. Ibid., 2, 29, p. 85.

48. Ibid., 1, 17, p. 80.

49. Ibid., 20, p. 81.

50. Bk. VII, *Strategikon*–Dennis, p. 64.

51. Gilbert Dagron, "Ceux d'en face: Les peuples étrangers dans les traités militaires byzantins" (1987), p. 209.

52. In 1973, the Egyptian General Staff, otherwise not incompetent, set the date of its well-planned surprise attack on October 6, the eleventh day of the seventh month of Tishrei in the Jewish calendar, Yom Kippur, the Day of Atonement—and the very best day of the year for an emergency mobilization because all reservists are at home or at prayer, with no traffic allowed. The Egyptians did everything possible to delay the deployment of mobilized Israeli reserve forces at the front—but evidently nobody knew of Yom Kippur.

53. A. D. Lee, *Information and Frontiers: Roman Foreign Relations in Late Antiquity* (1993), pp. 101–102, draws the contrast.

54. Dagron, "Ceux d'en face," p. 209.

55. Bk. XI, 1, *Strategikon*–Dennis, pp. 113–115.
56. Bk. XIX, 7, 2, *Ammianus Marcellinus*–Rolfe, p. 1:503.
57. Bk. XI, 2, *Strategikon*–Dennis, from p. 116.
58. Robert de Clari, *La Conquête de Constantinople,* as cited in John Earl Wiita, *The Ethnika in Byzantine Military Treatises* (1977), p. 144.
59. *Strategikon*–Dennis, p. 65.
60. By E. A. Thompson, *The Huns,* revised by Peter Heather (1996), p. 65, citing Dain's edition, *Leonis VI Sapientis Problemata* (1935) as well as (p. 287 n. 57) Urbicius vii, I, from *Arriani Tactica et Mauricii Artis militaris libri duodecim* (1664).
61. Bk. XI, 3, *Strategikon*–Dennis, pp. 119–120.
62. *Wars,* bk. VI, xxv, 1–4, in Dewing, *Procopius,* p. 4:85.
63. Ibid., VII, Xiv, 25–26, p. 4:271.
64. Bk. II, 1, *Strategikon*–Dennis, pp. 23–24.
65. Bk. XII from A, 1, *Strategikon*–Dennis, from p. 127.
66. Ibid., 7, p. 134.
67. Ibid., p. 135.
68. Salvatore Cosentino, "Per una nuova edizione dei *Naumachica* ambrosiani: Il *De fluminibus traiciendis* (Strat. XII B, 21)" (2001), pp. 63–105, includes the text and an Italian translation.
69. Bk. I, 2, *Strategikon*–Dennis, pp. 13–14.
70. Ibid., bk. V, 1, p. 58.

12. After the *Strategikon*

1. Alphonse Dain, "Les stratégistes byzantins," ed. J. A. de Foucault (1967), mentions an *antipoliocerticum* on resisting sieges (p. 349), a *corpus nauticum,* a *Tactica perdita* (p. 350), and cites Dain, *Le Corpus perditum* (1939).
2. Dain, "Les stratégistes byzantins," p. 354, for the *Problemata,* and from p. 354 for the *Taktika.*
3. Henrik Zilliacus, *Zum Kampf der Weltsprachen im oströmischen Reich* (1935), pp. 134–135 ("Kommandoworte bei der Reiterei and fussvolk").
4. George T. Dennis, typescript very kindly shown to the author; his edition is eagerly awaited.
5. Eric McGeer, *Sowing the Dragon's Teeth: Byzantine Warfare in the Tenth Century* (1995), pp. 233–236.
6. Though Gilbert Dagron ("Byzance et le modèle islamique au Xe siècle: À propos des Constitutions tactiques de l'empereur Léon VI," April–June 1983, p. 220) views the text as ill-informed: "[Il] trahit une assez mauvaise information"; see also Dagron, "Ceux d'en face: Les peuples étrangers dans les traités militaires byzantins" (1987), p. 223.
7. Compare *De Velitatione,* sec. 7.
8. Eric McGeer, "Two Military Orations of Constantine VII" (2003), pp. 113, 116–117, where the campaign of 964 is plausibly mentioned.

9. Ibid., p. 117.
10. Ibid., p. 119.
11. Ibid.
12. Ibid., p. 120.
13. Ibid.
14. Alphonse Dain, ed., *Sylloge Tacticorum* (1938), from p. 116.
15. Eric McGeer, "Infantry versus Cavalry: The Byzantine Response," in John F. Haldon, ed., *Byzantine Warfare* (2007), pp. 336–337. Although I subsequently consulted Dain's text, I was directed to the above by the kindness of Eric McGeer (private communication, January 14, 2008).
16. See John F. Haldon, "Some Aspects of Early Byzantine Arms and Armour" in Haldon, *Byzantine Warfare,* p. 375; a reprint from David Nicolle, ed., *A Companion to Medieval Arms and Armor* (2002), pp. 65–87.
17. Alphonse Dain, *La Tradition du Texte d'Héron de Byzance* (1933).
18. Denis F. Sullivan, *Siegecraft: Two Tenth-Century Instructional Manuals by "Heron of Byzantium"* (2000), p.vii.
19. See the difference between drawings a, b, and c in Sullivan, *Siegecraft,* from p. 281; as directed by Denis F. Sullivan, personal communication, March 29, 2008.
20. Athenaeus Mechanicus, *On Machines,* ed. and trans. David Whitehead and P. H. Blyth(2004).
21. Chap. 2; Sullivan, *Siegecraft,* pp. 29, 33 n. 1.
22. *The Chronicle of Pseudo-Joshua the Stylite,* trans. Frank R. Trombley and John W. Watt (2000), paras. 296–297, p. 89.
23. Chaps. 12 and 16, in Sullivan, *Siegecraft,* pp. 45, 51.
24. Ibid., chaps. 39, 30, p. 85, commentary on ingenuity at p. 208.
25. Ibid., chaps. 50, 51, pp. 100–101.
26. Ibid., chap. 52, pp. 103–104.
27. Ibid., chaps. 55–57, pp. 109–113, commentary on pp. 241–243.
28. Denis F. Sullivan, "A Byzantine Instructional Manual on Siege Defense: The *De obsidione toleranda*" (2003), pp. 140–266. The text was previously edited by Hilda Van Den Berg, *Anonymous de obsidione toleranda* (1947).
29. Sullivan, "Byzantine Instructional Manual," p. 151.
30. Ibid., pp. 153–154.
31. Ibid., p. 177.
32. Ibid., p. 179.
33. Ibid., p. 157.
34. Ibid., p. 161.
35. Ibid., p. 163.
36. Ibid., p. 165.
37. Ibid., p. 179.
38. Ibid., p. 181.
39. Ibid., pp. 197–199.
40. Ibid., p. 201.
41. Ibid., p. 225.

42. Ibid., p. 229.
43. Alphonse Dain, "Memorandum inédit sur la défense des places" (1940), pp. 123–136; but here cited from Sullivan, "Byzantine Instructional Manual."
44. I was corrected on this by Denis F. Sullivan, private communicaton, March 29, 2008.
45. Sullivan, "Byzantine Instructional Manual," p. 147.
46. Ibid., Injunction 18, p. 147.
47. Ibid., Injunction 30, p. 148.
48. Ibid., p. 149.
49. For the three texts, see Alphonse Dain, *Les stratégistes byzantines* (2000), pp. 359–361. *Praecepta imperatori* is an appendix to the first volume (from p. 444) of *De Cerimoniis*, pp. 1:444–454.

13. Leo VI and Naval Warfare

1. For a summary, see John H. Pryor and Elizabeth M. Jeffreys, *The Age of the Dromon: The Byzantine Navy, ca. 500–1204* (2006), p. 448; for the inception, from p. 123; for tenth-century data, from p. 175.
2. *De Administrando Imperio,* ed. Gy. Moravcsik, trans. R. J. H. Jenkins (1967) (hereafter cited as *DAI*), sec. 13, from line 73, pp. 69–70. On line 73, *siphonon* is unaccountably translated as "tubes."
3. Theophanes, *The Chronicle of Theophanes Confessor, AD 284–813,* trans. Cyril A. Mango and Roger Scott with Geoffrey Greatrex (1997) (hereafter cited as *Theophanes*), no. 353, AM 6164, p. 493. John Haldon, Andrew Lacey, Colin Hewes "Greek Fire Revisited: Recent and Current Research" (2006), pp. 291–325. For its use by Arabs and Latins, seePryor and Jeffreys, *Age of the Dromon,* appendix 6, pp. 607–631.
4. *Theophanes*, no. 354, AM 6165, p. 494.
5. Andrew Palmer, *The Seventh Century in the West-Syrian Chronicles* (1993), pt. 2, p. 194 n. 476.
6. Haldon et al., "Greek Fire Revisited," pp. 297–316.
7. *DAI*, para 53, pp. 493–510—as cited in ibid., p. 292.
8. J. Haldon and M. Byrne, "A Possible Solution to the Problem of Greek Fire" (1977), pp. 91–100; now overtaken by Haldon et al., "Greek Fire Revisited," p. 310.
9. See the discussion in Pryor and Jeffreys, *Age of the Dromon,* pp. 609–612.
10. Ibid., p. 612, citing John Kaminiates, *De expugnatione Thessalonicae*, 34.7.
11. See the calculations in J. F. Guilmartin, *Gunpowder and Galleys: Changing Technology and Mediterranean Warfare at Sea in the Sixteenth Century*(1975), pp. 62–63.
12. Now there is a superior text and translation in Pryor and Jeffreys, *Age of the Dromon* (*Naumakia Leontos Basileus*, pp. 485–519), unused by me.
13. Lionel Casson, *Ships and Seamanship in the Ancient World* (1995), p. 93.
14. Analogous to the "golden bridges" condemned by Clausewitz for making war ultimately more destructive.

15. Vassilios Christides, "Naval Warfare in the Eastern Mediterranean (6th–14th Centuries): An Arabic Translation of Leo VI's *Naumachica*" (1984), pp. 137–148.

16. Derived from John F. Haldon, "Theory and Practice in Tenth-Century Military Administration: Chapters II, 44 and 45 of the Book of Ceremonies" (2000) and from the version thereof in Pryor and Jeffreys, *Age of the Dromon,* appendix 4, pp. 547–570.

17. The terminology for the different kinds of ships was unstable; see Pryor and Jeffreys, *Age of the Dromon,* pp. 188–192.

18. Kolias, *Byzantinische Waffen* (1988), has "lydischen Waffenproduktion," p. 95; characteristically, nothing else is learned of the *schild* in question.

19. Ibid., from p. 554.

20. = single trunk. But they were not dugout canoes, which would have been too heavy to be carried—in this instance down to the Golden Horn; they had a dugout keel and lower hull base with built-up planking on the sides.

21. *Theophanes,* no. 316, AM 6117, p. 447.

22. *The Armenian History Attributed to Sebeos,* trans. R. W. Thompson (1999), p. 1:79.

23. Indiction 14, year 16; year 626; *Chronicon Paschale, 284–628 AD,* trans. Michael Whitby and Mary Whitby(1989), p. 178.

24. *Theophanes,* no. 353, AM 6164, p. 493.

14. The Tenth-Century Military Renaissance

1. More than that, according to Eric McGeer: "Nikephoros Phokas composed the *Praecepta Militaria,* his notes formed the basis for the *De Velitatione,* and his call for a treatise on campaigning in the west inspired the *De re militari*"; Eric McGeer, *Sowing the Dragon's Teeth: Byzantine Warfare in the Tenth Century* (1995), p. 178.

2. See "Tradition et modernité dans le corpus de tacticiens" in Gilbert Dagron and Haralambie Mihăescu, *Le traité sur la guérilla (De velitatione) de l'empereur Nicéphore Phocas (963–969)* (1986), from p. 139.

3. John F. Haldon, ed. and trans., *Constantine Porphyrogenitus: Three Treatises on Imperial Military Expeditions* (1990), pp. 95–97. Haldon's reconstruction of the editorial process (p. 53) is wholly persuasive.

4. *De Velitatione,* in George T. Dennis, ed., *Three Byzantine Military Treatises* (1985)(hereafter cited as *De Velitatione*). In noting the work, Alphonse Dain, "Les stratégistes byzantins" (1967), p. 369, concludes: "Il mériterait une traduction et un commentaire." Dennis duly supplied a translation; Dagron and Mihăescu added a detailed commentary to their newly edited text.

5. For an overview, see Ralph-Johannes Lilie, "The Byzantine-Arab Borderland from the Seventh to the Ninth Century," in Florin Curta, ed., *Borders, Barriers, and Ethnogenesis: Frontiers in Late Antiquity and the Middle Ages* (2005).

6. Clinton Bailey, *Bedouin Poetry from Sinai and the Negev* (1991), from p. 253 includes modern Bedouin war poems.

7. Dennis, *Three Byzantine Military Treatises*, p. 147.

8. *De Velitatione*, 1, p. 151 n. 1: 1 Byzantine mile = 1,574 meters = 0.978 standard miles.

9. Ibid., 2, p. 153; comment, p. 153 n. 1. Justinian's Digest, 47.18, is in *De effractoribus et expilatoribus*.

10. *De Velitatione*, 2, p. 153. On Armenian themes, see Dagron and Mihǎescu, *Le traité*, from p. 247.

11. *De Velitatione* 2, p. 153.

12. Ibid., 4, pp. 157–158.

13. Jak Yakar, *Ethnoarchaeology of Anatolia: Rural Socio-Economy in the Bronze and Iron Ages* (2000).

14. *De Velitatione*, 4, pp. 157–159.

15. Bk. ii.2, in Alice-Mary Talbot and Denis F. Sullivan, trans., *The History of Leo the Deacon: Byzantine Military Expansion in the Tenth Century* (2005), p. 72.

16. Ibid., p. 74 n. 26.

17. Bk. ii.4, ibid., pp. 74–75.

18. Dagron and Mihǎescu, *Le traité*, from p. 198, even illustrates the recommended operational methods and tactics *("dispositifs de combat")*.

19. *De Velitatione*, 7, p. 163.

20. For an overview, see Daniel Pipes, *Slave Soldiers and Islam: Genesis of the Military System* (1981).

21. *De Velitatione*, 7, p. 162.

22. The dark surcoats: *Epanoklibana* in the text, *epilorika* elsewhere = what comes on top of the *klibanion;* originally a breastplate, later any body armor, or *lorika*, originally sewn-on scale armor *(lorica squamata),* or interlinked lamellar armor, or chain mail *(lorica hamata)*. Ibid., p. 167 n. 2.

23. Ibid., 9, p. 169.

24. Ibid., p. 173. See Philip Rance, "The Fulcum, the Late Roman and Byzantine *Testudo:* The Germanization of Roman Infantry Tactics?" (2004), 265–326.

25. *De Velitatione*, 10, p. 175.

26. Ibid., 11, p. 183.

27. Ibid., 12, p. 187.

28. Ibid., 16, p. 201.

29. Ibid., 17, pp. 206–207.

30. Ibid., 19, p. 215.

31. Ibid., p. 217.

32. H. J. Scheltema, "Byzantine Law," in J. M. Hussey, ed., *The Cambridge Medieval History*, vol. 4: *The Byzantine Empire*, pt. 2 (1967), from p. 73.

33. *De Velitatione*, 20, p. 221.

34. Ibid., 21, p. 223.

35. *Campaign Organization* (formerly *De re militari*), in George T. Dennis, ed., *Three Byzantine Military Treatises* (1985) (hereafter cited as *Campaign Organization*). See also Rudolf Vary, *Incerti Scriptoris Byzantini Liber de rei militari* (1901).

36. In line 12 of the Greek text, p. 246; translated as "regular soldiers" in Dennis, *Three Byzantine Military Treatises,* p. 247.
37. Dennis, *Three Byzantine Military Treatises,* p. 242.
38. From line 99 of the Greek text, p. 250; *Campaign Organization,* no. 1, p. 251.
39. Dennis, *Three Byzantine Military Treatises,* p. 256 n. 12. Harald III Sigurdsson "Hardrada" received the title *manglavites* from Michael IV the Paphlagonian (1034–1041) for his exploits in Sicily.
40. Alpha 121. *Campaign Organization,* no. 1, p. 253.
41. Beta from 18. Ibid., p. 263.
42. But the eminent editor wrote (p. 265 n. 1): "The hoplitarch seems to have commanded the infantry on campaign."
43. *Campaign Organization,* no. 8, p. 275.
44. Iota (Zeta is repeated) 31. Ibid., no. 10, p. 281; and Lamda Alpha, 5–9, ibid., no. 31, p. 325.
45. Not quite an *hapax legomenon:* the editor can cite (p. 283 n. 2) a single 1079 document in which *malartioi* are unhelpfully equated with *kontaratoi* (lance carriers) but in the meaning of "soldiers" in general.
46. *Campaign Organization,* no. 13, p. 285.
47. Iota eta, from 22. Ibid., no. 18, p. 293.
48. Ibid., no. 19, pp. 293–295.
49. Ibid., no. 21, from p. 303.
50. Ioannis Scylitzes, *A Synopsis of Histories, 811–1057 AD,* trans. John Wortley; "John Tzimiskes," sec. 4, p. 155.
51. *Campaign Organization,* no. 27, pp. 317–319.
52. Ibid., no. 28, pp. 319–321.
53. John F. Haldon, *Byzantine Praetorians: An Administrative, Institutional and Social Survey of the Opsikion and Tagmata, c. 580–900* (1984), from p. 256.
54. *Campaign Organization,* no. 29, p. 321.
55. Lamda beta from 16. *Campaign Organization,* no. 32, p. 327. The reference may be to *De Velitatione.*
56. McGeer, *Sowing the Dragon's Teeth,* composition p. 178, text from p. 13, with a valuable commentary and analysis. It was previously available only in a 1908 edition of the sole extant manuscript (Mosquensis Gr. 436) by the Byzantinist J. A Kulakovsky, now not unjustly claimed by the Ukrainians.
57. McGeer, *Sowing the Dragon's Teeth,* p. 62 nn. 20–23; *tzerboulianoi* became a pejorative.
58. One spithame = 23.4 centimeters.
59. McGeer, *Sowing the Dragon's Teeth,* p. 63 nn. 29–31.
60. Ibid., p. 63 n. 32.
61. *Praecepta* I.25–26, in McGeer, *Sowing the Dragon's Teeth,* p. 15 para. 3.
62. Ibid., II.2–4, p. 23 para. 1.
63. Ibid., I.33–38, p. 15 para. 4.
64. Ibid., I.62ff, p. 17. McGeer's notes, from note 83 on p. 64, are indispensable to clarify the text corrupt in places; for instance, the length of the *menavlia*

is stated as two or two and a half *spithamai* (0.46 to 0.58 meters), not nearly long enough.

65. Giorgio Ravegnani, *Soldati di Bizanzio in Eta Giustinianea* (1988), p. 46.

66. *Praecepta* I.119–125, in McGeer, *Sowing the Dragon's Teeth*, p. 19 para. 11.

67. *Taktika, 56,* 82–85, in McGeer, *Sowing the Dragon's Teeth,* pp. 93 and pp. 210, where clarity is imposed.

68. *Praecepta* I.116–119, in McGeer, *Sowing the Dragon's Teeth,* p. 19 para. 10.

69. Ibid., I.75ff., p. 17 para. 8.

70. Contra: M. P. Anastasiadis, "On Handling the Menavlion" (1994), pp. 1–10, cited at p. 349 n. 86. But see J. F. Haldon, *Byzantium in the Seventh Century: The Transformation of a Culture* (1990), p. 218.

71. Praecepta I.137, in McGeer, *Sowing the Dragon's Teeth,* p. 21 para. 14.

72. McGeer, *Sowing the Dragon's Teeth,* pp. 65, 150–155 nn.

73. Paul E. Chevedden, "Artillery in Late Antiquity: Prelude to the Middle Ages" (2007), p. 137.

74. Paul E. Cheveddent, "The Invention of the Counterweight Trebuchet: A Study in Cultural Diffusion" (2000), pp. 72–116; on the identification of the cheiromangana of *Praecepta Militaria,* p. 79; also p. 10 n. 144.

75. Ibid., p. 114.

76. Michael Whitby and Mary Whitby, *The History of Theophylact Simocatta* (1986), bk. 2, 16.10–11; Cheveddent, "Invention of the Counterweight Trebuchet," p. 75 and n. 9.

77. In his *Miracula St. Demetrii.* Here minimally adapted from Cheveddent, "Artillery in Late Antiquity," p. 74, itself from P. Lemerle, ed., *Les plus anciens recueils des miracles de saint Démétrius et la pénétration des Slaves dans les Balkans* (1979), p. 1:154.

78. *Praecepta* II.2–13, in McGeer, *Sowing the Dragon's Teeth,* p. 23 para. 1–2.

79. Ibid., II.20–23, p. 23 para. 3.

80. Ibid., IV.7–11, p. 39 para. 1–2.

81. Ibid., III.53ff., p. 37 para. 7.

82. Ibid., III.46–53, p. 37 para. 6.

83. Tentatively translated as mail ("chain-mail") in McGeer, *Sowing the Dragon's Teeth,* p. 70 n. 29, and in the glossary, p. 370, but the silk or cotton cloth is evidently backing for scale armor, because chain needs no backing (III.27–30, p. 35 para. 4). For another view, see Timothy Dawson, "Kremasmata, Kabadion, Klibanion: Some Aspects of Middle Byzantine Military Equipment Reconsidered" (1998), pp. 38–50.

84. *Praecepta* III.36–46, in McGeer, *Sowing the Dragon's Teeth,* p. 37 paras. 4–6.

85. Ibid., IV. 1, p. 39 para. 1.

86. Ibid., IV.17–22, p. 41 para. 2.

87. Ibid., IV.29ff, p. 41 para. 3

88. Ibid., IV.106ff., p. 45 para. 11.

89. Greek text lines 204–205 in McGeer, *Sowing the Dragon's Teeth,* p. 50; English text: p. 51 para. 19.

90. The authoritative new editor of *Praecepta Militaria* has also contributed a

valuable tactical analysis, in which the specific ways in which synergies were to be exploited according to the text are carefully reconstructed. See McGeer, *Sowing the Dragon's Teeth*, esp. from p. 294.

91. Alphonse Dain, *La "Tactique" de Nicephore Ouranos* (1937), p. 134.

92. Skylitzes, "Basil II Bulgharoctonos," sec. 23, p. 183.

93. "A la suite d'une habile manoeuvre dont il pouvait voir le thème dans le manuels de tactique qui lui étaient familiers . . . ," Dain, *La "Tactique,"* p. 135.

94. Blass listed all the classical texts he found: "Die griech. und lat. Handschriften in alten Serail zu Konstantinopel" (1887), pp. 219–233, including Constantinopolitanus Graecus 36 (p. 225).

95. He adds the *Sylloge Tacticorum*; Dain, *La "Tactique,"* pp. 42–43.

96. For chaps. 56–65, see McGeer, *Sowing the Dragon's Teeth*, pp. 79–163; chaps. 63–74 edited and translated by J.-A. De Foucault, "Douze chapitres inédits de la Tactique de Nicéphore Ouranos" (1973), pp. 286–311.

97. *Taktika*, from 63.1, in McGeer, *Sowing the Dragon's Teeth*, p. 143, adapted.

98. Ibid., 63.12, p. 143.

99. Ibid., from 63.24, p. 143.

100. Ibid., 63.5, p. 145, where the expression is described as sinister (p. 165 n. 36); to "seize tongues" was the identical Russian Army expression for the capture of German troops for the interrogators.

101. Ibid., 63.84, p. 147.

102. Ibid., 64.3, p. 147.

103. Ibid., 64.42, p. 149.

104. Ibid., 64.94, p. 153.

105. Ibid.

106. The term *Saracen*, originally denoting a tribe of northern Sinai in Ptolemy, here replaces *Agarene* (offspring of Hagar, Abraham's repudiated concubine) to describe the Arabs—for the term *Arab* usually meant Bedouin at the time, and long after among the Arabs themselves. Ibid., 65.20–33, p. 155.

107. Ibid., 65.55–64, p. 157.

108. Ibid., 65.73–79, pp. 158–159.

109. Papyrus Erzherzou Rainer 558.

110. Ibid., 65.162–165, p. 163.

111. *Taktika* 66, in De Foucault, "Douze chapitres," pp. 302–304.

112. Ibid. 68, pp. 306–307.

113. Ibid. 69, pp. 306–307.

114. Ibid., 72, p. 308. According to Dain's reconstruction (Dain, *La "Tactique,"* p. 21), this passage echoes *Praecepta Militaria*, itself from Onasander 22.

115. *Taktika* 73, De Foucault, "Douze chapitres," pp. 308–310.

116. Alphonse Dain, "Les stratégistes byzantins," ed. J. A. de Foucault (1967), p. 373.

117. Mosquensis Gr. 436, purchased in 1654 from the Iviron monastery on Mount Athos and kept in Moscow ever since.

118. I rely on the latest edition: Maria Dora Spadaro, ed. and trans., *Kekaumenos: Raccomandazioni e consigli di un galantuomo: Stratēgikon* (1998). The first editors (B. Wassiliewsky and V. Jernstedt, *Cecaumeni "strategikon" et incerti scriptoris "de officiis regiis libellus,"* St.Petersburg, 1896) were in the first wave of Russian Byzantine scholarship that ended with the Bolshevik revolution; the second editor (G. G. Litavrin, *Sovety I rasskazy Kekavmena Socinenie vizantijskogo polkovodka XI veka*, Moscow, 1972) was part of the post-1945 revival.

119. Alexios G. C. Savvides, "The Byzantine Family of Kekaumenos" (1986–1987), pp. 12–27.

120. Secs. 26, 36–37; Spadaro, *Kekaumenos*, pp. 66–68.

121. Paul Magdalino, "Honour among the Romaioi: The Framework of Social Values in the World of Digenes Akrites and Kekaumenos" (1989), pp. 183–218.

122. Sec. 52, 4–5, Spadaro, *Kekaumenos*, p. 88.

123. Sec. 54, 19–24, p. 88.

124. Sec. 54, 26–27, p. 88.

125. Ibid., p. 88.

126. For a much broader perspective *(Cecaumeno e la societa' bizantina)*, see Spadaro, *Kekaumenos*, from p. 19.

127. Sec. 24, Spadaro, *Kekaumenos*, p. 65.

128. Sec. 25, p. 67.

129. Sec. 28, p. 69.

130. Sec. 30, p. 70.

131. Sec. 36, p. 75.

132. Sec. 34, pp. 73–74.

133. Secs. 41.14–42.29, pp. 78–79.

134. Sec. 46, p. 83.

135. Sec. 47, p. 83.

136. Sec. 49, p. 85.

15. Strategic Maneuver: Herakleios Defeats Persia

1. *Ammianus Marcellinus,* trans. John C. Rolfe (1935), bk. VII, 5; on Sasanian moderation, see the discussion in A. D. Lee, *Information and Frontiers: Roman Foreign Relations in Late Antiquity* (1993), from p. 21.

2. I follow the chronology in Walter E. Kaegi Jr., *Heraclius: Emperor of Byzantium* (2003), summarized at pp. 324–326.

3. Andrew Palmer, *The Seventh Century in the West-Syrian Chronicles* (1993), text no. 13; pp. 133–134 for Dionysius reconstituted; p. 134 n. 303, Michael the Syrian has 120,000 pounds.

4. Indiction 14, year 16. *Chronicon Paschale, 284–628 AD,* trans. Michael Whitby and Mary Whitby (1989), p. 174.

5. *Chronicon Paschale,* p. 179.

6. Ibid., p. 175.

7. Theophanes, *The Chronicle of Theophanes Confessor, AD 284–813,* trans. Cyril A. Mango et al. (1997) (hereafter cited as *Theophanes*), no. 303, p. 435; no. 304, p. 436.

8. Ibid., no. 303, p. 436.

9. Ibid., no. 304, p. 436.

10. *Chronicon Paschale,* p. 161.

11. *The Armenian History Attributed to Sebeos,* trans. R. W. Thompson, vols. 1–2 (1999) (hereafter cited as *Sebeos*), chap. 38.122, pp. 1:78–79.

12. Ibid., p. 2:214.

13. Ibid., chap. 38.123, pp. 1:79–80.

14. Ibid., chap. 38.123, p. 1:80.

15. Ibid., chap. 38.124, p. 1:81; ibid., p. 2:215; Kaegi, *Heraclius,* from p. 12.

16. Mary Boyce, "Adur Gušnasp" (1985); Kaegi, *Heraclius,* p. 122 passim.

17. Kaegi, *Heraclius,* p. 122 passim.

18. *Chronicon Paschale,* p. 171.

19. *Theophanes,* no. 315, AM 6117, p. 446.

20. Ibid., no. 316, AM 6117, p. 447.

21. *Chronicon Paschale,* p. 178.

22. *Sebeos,* pt. I, p. 79.

23. *Chronicon Paschale,* p. 178.

24. Kaegi, *Heraclius,* from p. 156.

25. *Theophanes,* no. 316, AM 6117, p. 447.

26. Ibid., p. 158, has Yabghu Xak'an and "Kok" Turk; but see Peter B. Golden, *An Introduction to the History of the Turkic Peoples* (1992), pp. 135, 236, and a complementary personal communication.

27. Menander, *Excerpta de Legationibus Romanorum ad gentes* 14, frag. 19, in R. C. Blockley, trans., *The History of Menander the Guardsman* (1985), p. 175.

28. *De Administrando Imperio,* sec. 31, p. 149.

29. *Theophanes,* no. 324, AM 6118, pp. 452–453.

30. *Sebeos,* pt. I, p. 84.

31. *Theophanes,* no. 321, AM 6118, p. 451.

32. Ibid., no. 322, AM 6118, p. 451.

33. *Theophanes,* no.325, AM 6118, p. 453. See the persuasive and fascinating reconstruction of the campaign in Kaegi, *Heraclius,* from p. 156; and post-Nineveh, p. 172. Kaegi notes that Theophanes may have constructed the tale—the rules prescribe a rejected chance of salvation before the coming doom.

34. Ibid., pp. 177–178.

35. *Sebeos,* p. 2:224.

Conclusion

1. Among innumerable writings, I have found the following most useful: Agostino Pertusi, *Il pensiero politico bizantino,* ed. Antonio Carile (1990),

with a three-part periodization (Justinian; after Justinian; from the 1261 reconquest); Alexander Kazhdan and Giles Constable, *People and Power in Byzantium: An Introduction to Modern Byzantine Studies* (1982), esp. on religion, from p. 76; and Cyril Mango, *Byzantium: The Empire of New Rome* (1980), esp. on Hellenism and its boundaries, from p. 13.

2. Tia M. Kolbaba, "Fighting for Christianity: Holy War in the Byzantine Empire" (2007), pp. 43–70, mentions but does not confront the ideological dimension of frontier warfare—the jihad was real enough for all the robbery of both sides; compare G. T. Dennis, "Defenders of the Christian People: Holy War in Byzantium" (2007), pp. 71–79.

3. Number 51 in H. A. Diels and Rev. W. Kranz, *Die Fragmente der Vorsokratiker* (2004); or number 27 in the stupendous M. Marcovich, *Heraclitus: Greek Text with Commentary* (1967), rightly dubbed *Editio Maior*.

4. Number 53 in Diels and Kranz, *Die Fragmente*; number 29 in Marcovich, *Heraclitus*.

5. "Operational code" here signifies a set of observed and imputable behavioral norms; from Natan Constantin Leites, *The Operational Code of the Politburo* (1951). Compare the "Byzantine agreement" problem in computer science: L. Lamport, R. Shostak, and M. Pease, "The Byzantine Generals Problem" (1982).

6. *De Velitatione*, in George T. Dennis, ed., *Three Byzantine Military Treatises* (1985), 7, p. 162.

7. George T. Dennis, "Some Reflections on Byzantine Military Theory" (2007);Dennis, "Byzantium at War (9th–12th C.)" (1997); and Walter Emil Kaegi Jr., *Some Thoughts on Byzantine Military Strategy* (1983). *Note:* Charalambos Papasotiriou, in "Byzantine Grand Strategy" (1991), offers a periodization: Justinian's overambitious conquest conditioned by internal equilibria with the monophysites; from Herakleios to the second Arab siege—a strategy of containment executed with the new thematic organization; territorial recovery followed by consolidation, and then moderate expansion powered by demographic and economic improvement; "the triumph of diplomacy, 843–959," including the successful alliance with the Khazars—at a time when western Europe was under the concurrent attacks of the Scandinavian sea raiders ("Vikings"), Magyar cavalry incursions, and Arab corsairs and invaders. Then there comes the "triumph of force, 959–1025," made possible by the decay of Arab power and the resulting interval of relative tranquility until the arrival of the Seljuks, and by professional field forces. Nothing wrong with that, but I hold that the foundation was in place before Justinian.

Appendix

1. *The Grand Strategy of the Roman Empire: From the First Century AD to the Third* (1976/2007).

2. Pro: P. A. Brunt; Ernst Badian; Stephen L. Dyson, *The Creation of the Roman Frontier* (1985). Detailed critiques: John C. Mann, "Power, Force and the Frontiers of the Empire" (1979); at greater length, Benjamin Isaac, *The Limits of Empire: The Roman Army in the East* (1990, from p. 372. Also, among many others, Luigi Loreto, "La Storia della grand strategy un dibattito Luttwak?" (2006) and "Il paradosso Luttwakiano power projection, low intensity e funzione del limes" (2006); Mickaël Guichaoua, "Lecture critique de Luttwak: La Grande Stratégie de l'Empire romain" (2004); and Karl-Wilhelm Welwei, "Probleme römischer Grenzsicherung am Bespiel der Germanienpolitik des Augustus" (2004) from p. 675. Contra, among others: C. R. Whittaker, *Frontiers of the Roman Empire: A Social and Economic Study* (1994); and Susan P. Mattern, *Rome and the Enemy: Imperial Strategy in the Principate* (1999). Neither accepts the autonomy of military history, or the consequentiality of military action—a common stance but not among practitioners and scholars of wider reach.

3. Moritz Cantor, *Die römischen Agrimensoren und ihre Stellung in der Geschichte der Feldmesskunst: Eine historisch-mathematische Untersuchung* (1875).

4. A. D. Lee, *Information and Frontiers: Roman Foreign Relations in Late Antiquity* (1993), p. 87, and n. 29.

5. Edward N. Luttwak, *Strategy: The Logic of War and Peace* (2001).

Works Cited

Cl. Aeliani et Leonis Imp. tactica sive de instruendis aciebus. Ed. Meursius. Lugduni Batavorum [Leiden]: Ludovicum Elzevirium, 1613.

Aeneas Tacticus, Asclepiodotus, Onasander. Ed. and trans. W. A. Oldfather. Cambridge, Mass: Harvard University Press, 1977.

Ahrweiler, Hélène. *Byzance et la mer: La marine de guerre, la politique et les institutions maritimes de Byzance aux VIIe–XVe siècles.* Paris: Presses universitaires de France, 1966.

Allen, Pauline. "The 'Justinianic' Plague." *Byzantion* no. 49 (1979): 5–20.

Alston, Richard. "Managing the Frontiers: Supplying the Frontier Troops in the Sixth and Seventh Centuries." In Paul Erdkamp, ed., *The Roman Army and the Economy.* Amsterdam: J. C. Gieben, 2002.

Alvarez, R. P. "Saddle, Lance and Stirrup." *International Newsletter for the Fencing Collector* 4, nos. 3–4 (July 15, 1998).

Amari, Michele. *La guerra del vespro siciliano.* Turin: Pomba, 1851.

Amatuccio, Giovanni. *Peri Toxeias l'arco di guerra nel mondo bizantino e tardo-antico.* Bologna: Planetario, 1996.

Amatuccio, Giovanni. "Lo Strano arciere della porta dei Leoni." *Arcosophia* 1, supplement to *Arco* 1, no. 1 (2005). http://www.arcosophia.net.

Ambrosii [Saint Ambrose]. *Expositio Evangelii secundum Lucam,* X, 10. In Migne, *Patrologia Latina,* vol. 15, cols. 1806–1809.

Ammianus Marcellinus. 3 vols. Trans. John C. Rolfe. Cambridge, Mass.: Harvard University Press, 1935.

Andersson, Theodore M. *A Preface to the Nibelungenlied.* Stanford: Stanford University Press, 1987.

Annales Bertiniani (ca. 839). In G. Waitz, *Monumenta Germaniae Historica: Scriptores rerum Germanicarum in usum scholarum.* Hannover: Hahn, 1883.

The Anonymous Byzantine Treatise on Strategy (formerly *Peri Strategikes, De re strategica*). In George T. Dennis, ed. and trans., *Three Byzantine Military Treatises.* Washington, D.C.: Dumbarton Oaks, 1985.

Arriani Tactica et Mauricii Artis militaris libri duodecim. Uppsala: Joannes Schefferus, 1664.

Arrianus, Fl. *The Expedition against the Alans [Acies contra Alanos].* In James G. DeVoto, ed., *Flavius Arrianus.* Chicago: Ares, 1993.

Athenaeus Mechanicus. *On Machines.* Ed. and trans. D. Whitehead and P. H. Blyth. *Historia Einzelschriften,* 182. Stuttgart: Franz Steiner, 2004.

Austin, N. J. E., and N. B. Rankov. *Exploratio: Military and Political Intelligence in the Roman World from the Second Punic War to the Battle of Adrianople.* London: Routledge, 1995.

Ayalon, David. *Eunuchs, Caliphs and Sultans: A Study in Power Relationships.* Jerusalem: Magnes Press, 1999.

Bivar, A. D. H. "Cavalry Equipment and Tactics on the Euphrates Frontier." *Dumbarton Oaks Papers* no. 26 (1972): 271–291.

Blockley, R. C. *East Roman Foreign Policy: Formation and Conduct from Diocletian to Anastasius.* Leeds: F. Cairns, 1992.

——. *The Fragmentary Classicizing Historians of the Later Roman Empire: Eunapius, Olympiodorus, Priscus and Malchus.* 2 vols. Liverpool: Liverpool University Press, 1983.

——. *The History of Menander the Guardsman.* Liverpool: F. Cairns, 1985.

——. "Subsidies and Diplomacy: Rome and Persia in Late Antiquity." *Phoenix* 39, no. 1 (Spring 1985): 62–74.

Bonner, Michael. *Jihad in Islamic History: Doctrines and Practice.* Princeton: Princeton University Press, 2007.

Bowlus, Charles R. *Franks, Moravians and Magyars: The Struggle for the Middle Danube, 788–907.* Philadelphia: University of Pennsylvania Press, 1995.

Boyce, M[ary]. "Adur Gušnasp." In *Encyclopaedia Iranica,* vol. 1, ed. Y. Yarshater. Costa Mesa, Calif.: Eisenbrauns, 1985.

Brennan, Peter. "Mining a Mirage." In *Notitia.* Liverpool: Liverpool University Press, forthcoming.

Bréhier, Louis. *Les institutions de l'empire Byzantin* (1949). Paris: Albin Michel, 1970.

Brett, Gerard. "The Automata in the Byzantine 'Throne of Solomon.'" *Speculum* 29, no. 3 (July 1954): 477–487.

Browning, Robert. "Where Was Attila's Camp?" *Journal of Hellenic Studies* 73 (1953): 143–145.

Bury, J. B. *History of the Later Roman Empire: From the Death of Thodosius I to the Death of Justinian.* New York: Dover, 1958.

——. *The Imperial Administrative System in the Ninth Century with a Revised Text of the Kletorologion of Philotheos.* Oxford: British Academy, 1911.

———. "Justa Grata Honoria." *Journal of Roman Studies* 9 (1919): 1–13.

Cameron, Averil, ed. *The Byzantine and Early Islamic Near East.* 3 vols. Princeton: Darwin Press, 1995.

———. *Claudian: Poetry and Propaganda at the Court of Honorius.* Oxford: Clarendon Press, 1970.

———. "Justin I and Justinian." In A. Cameron, B. Ward-Perkins, and M. Whitby, eds., *Late Antiquity: Empire and Successors, A.D. 425–600.* Vol. 14 of *The Cambridge Ancient History.* Cambridge: Cambridge University Press, 2001.

———. "Vandal and Byzantine Africa." In A. Cameron, B. Ward-Perkins, and M. Whitby, eds., *Late Antiquity: Empire and Successors, A.D. 425–600.* Vol. 14 of *The Cambridge Ancient History.* Cambridge: Cambridge University Press, 2001.

Cameron, Averil, Bryan Ward-Perkins, and Michael Whitby, eds. *Late Antiquity: Empire and Successors, A.D. 425–600.* Vol. 14 of *The Cambridge Ancient History.* Cambridge: Cambridge University Press, 2001.

Campaign Organization (formerly *De re militari*). In George T. Dennis, trans., *Three Byzantine Military Treatises.* Washington, D.C.: Dumbarton Oaks, 1985.

Cantor, Moritz. *Die römischen Agrimensoren und ihre Stellung in der Geschichte der Feldmesskunst: Eine historisch-mathematische Untersuchung.* Leipzig: Teubner, 1875. PDF available from Harvard online library.

Casson, Lionel. *Ships and Seamanship in the Ancient World.* Baltimore: Johns Hopkins University Press, 1995.

Charles, R. H. *The Chronicle of John, Bishop of Nikiu: Translated from Zotenberg's Ethiopic Text* (1916). http://www.tertullian.org/fathers/index .htmNo.John_of_Nikiu.

Chessman, G. L. *The Auxilia of the Roman Imperial Army.* Oxford: Clarendon Press, 1914. Reprint, Chicago: Ares, 1975.

El-Cheikh, Nadia Maria. "Byzantium Viewed by the Arabs." Dissertation, Harvard University, 1992; Cambridge, Mass.: Distributed by Harvard University Press for the Center for Middle Eastern Studies of Harvard University, 2004.

Chevedden, Paul E. "Artillery in Late Antiquity." In A. Corfis and M. Wolfe, eds., *The Medieval City under Siege.* Woodbridge, UK: Boydell and Brewer, 1999.

———. "The Invention of the Counterweight Trebuchet: A Study in Cultural Diffusion." Dumbarton Oaks Papers no. 54, pp. 72–116. Washington, D.C., 2000.

Christides, V[assilios]. "Naval Warfare in the Eastern Mediterranean (6th–14th Centuries): An Arabic Translation of Leo VI's *Naumachica.*" *Graeco-Arabica* 3 (1984): 137–148.

The Chronicle of Pseudo-Joshua the Stylite. Trans. Frank R. Trombley and John W. Watt. Liverpool: Liverpool University Press, 2000.

Chronicon Paschale, 284–628 A.D. Trans. Michael Whitby and Mary Whitby. Liverpool: Liverpool University Press, 1989.

Cl. Aeliani et Leonis Imp. tactica sive de instruendis aciebus. Ed. Meursius. Lugduni Batavorum [Leiden]: Ludovicum Elzevirium, 1613.

Clausewitz, Carl von. *On War.* Ed. and trans. Michael Howard and Peter Paret. Princeton: Princeton University Press, 1976.

Cleaves, Francis Woodman. *The Secret History of the Mongols.* Vol. 1. Cambridge, Mass.: Harvard University Press, 1982.

Connolly, Peter, and Carol van Driel-Murray. "The Roman Cavalry Saddle." *Britannia* 22 (1991): 33–50.

Constantine Porphyrogennetos. *De Administrando Imperio.* Ed. Gy. Moravcsik. Trans. R. J. H. Jenkins. Washington, D.C.: Dumbarton Oaks Center for Byzantine Studies / Harvard University Press, 1967.

———. *De Cerimoniis Aulae Byzantinae.* In J. Reiske, ed., *Corpus Scriptorum Historiae Byzantinae.* Bonn: Weber, 1829.

———. *De Thematibus: Introduzione, testo critico, commento.* Ed. and trans. Agostino Pertusi. Vatican City: Biblioteca apostolica vaticana, 1952.

Conybeare, Frederick C. "Antiochus Strategos' Account of the Sack of Jerusalem (614)." *English Historical Review* 25, no. 99 (July 1910): 502–517. Online, http://www.fordham.edu/halsall/source/strategos1.html.

Corazzini, F. *Scritto sulla Tattica Navale, di anonimo greco.* Leghorn: Pia Casa del Refugio, 1883. Online, Domenico Carro, www.romaeterna.org.

Cosentino, Salvatore. "Dalla tassazione tardoromana a quella bizantina: Un avvio al medioevo." In M. Kajava, ed., *Gunnar Mickwitz nella storiografia europea tra le due guerre: Acta Institutum Romanum Finlandiae* (Rome) 34 (2007): 119–133.

———. "Per una nuova edizione dei *Naumachica* ambrosiani: Il *De fluminibus traiciendis* (Strat. XII B, 21)." *Bizantinistica* ser. 2, year 2 (2001): 63–105.

———. "The Syrianos's 'Strategikon': A 9th Century Source?" *Bizantinistica* ser. 2, year 2 (2000): 243–280.

Cotter, Stephen Edward. "The Strategy and Tactics of Siege Warfare in the Early Byzantine Period: From Constantine to Heraclius." Ph.D thesis, Queen's University, Belfast, 1995.

Crow, J[ames]. G. "The Long Walls of Thrace." In Cyril A. Mango and Gilbert Dagron, eds., *Constantinople and Its Hinterland.* Aldershot, UK: Variorum, 1995.

Crow, J[ames] G., Alessandra Ricci, and Richard Bayliss. University of Newcastle Anastasian Wall Project. Online, http://longwalls.ncl.ac.uk/AnastasianWall.htm.

Curta, Florin, ed. *Borders, Barriers, and Ethnogenesis: Frontiers in Late Antiquity and the Middle Ages.* Turnhout, Belgium: Brepols, 2005.

———. *Southeastern Europe in the Middle Ages, 500–1250.* Cambridge: Cambridge University Press, 2006.

Dagron, Gilbert. "Byzance et le modèle islamique au Xe siècle: À propos des Constitutions tactiques de l'empereur Léon VI." *Comptes rendus de l'Académie des Inscriptions et Belles-Lettres* (April–June 1983): 219–243.

———. "Byzance et ses Voisins: Étude sur certain passages du livre de cérémonies." In *Travaux et mémoires du Centre de recherche d'histoire et civilisation byzantine,* no. 13, pp. 353–357. Paris: De Boccard, 2000.

———. "Ceux d'en face: Les peuples étrangers dans les traités militaires byzantins." In *Travaux et mémoires du Centre de recherche d'histoire et civilisation byzantine,* no. 10, pp. 207–232. Paris: De Boccard, 1987.

———. "Poissons, Pêcheurs et Poissonniers de Constantinople." In Cyril A. Mango and Gilbert Dagron, eds., *Constantinople and Its Hinterland.* Aldershot, UK: Variorum, 1995.

Dagron, Gilbert, with Haralambie Mihăescu. *Le traité sur la guérilla (De velitatione) de l'empereur Nicéphore Phocas (963–969).* Paris: Éditions du Centre national de la recherche scientifique, 1986.

Dain, Alphonse. *Leonis VI Sapientis Problemata.* Paris: Les Belles Lettres, 1935.

———. "Memorandum inédit sur la défense des places." *Revue des Études Grecques* 53 (1940): 123–136.

———. *Naumachica.* Paris: Les Belles Lettres, 1943.

———. "Les stratégistes byzantins." Ed. J. A. de Foucault. In *Travaux et mémoires du Centre de recherche d'histoire et civilisation byzantine,* no. 2, pp. 317–392. Paris: De Boccard, 2000.

———, ed. *Sylloge Tacticorum* (formerly *Inedita Leonis Tactica*). Annotations in Latin. Paris: Les Belles Lettres, 1938.

———. *La tradition du texte d'Héron de Byzance.* Paris: Les Belle Lettres, 1933.

Darrouzes, J. *Épistoliers byzantins du Xe siecle.* Archives de l'Orient Chrétien 6. Paris: Institut français d'études byzantines, 1960.

Daryaee, Touraj. "Ethnic and Territorial Boundaries in Late Antique and Early Medieval Persia (Third to Tenth Century)." In Florin Curta, ed., *Borders, Barriers, and Ethnogenesis: Frontiers in Late Antiquity and the Middle Ages.* Turnhout, Belgium: Brepols, 2005.

De Administrando Imperio. Ed. Gy. Moravcsik. Trans. R. J. H. Jenkins. Washington, D.C.: Dumbarton Oaks Center for Byzantine Studies, 1967.

de Clari, Robert. *La conquête de Constantinople.* Paris: Flammarion, 1956.

———. *La Prise de Constantinople.* Chapter 73 in Charles Hopf, ed., *Chroniques gréco-romanes inédites ou peu connues* (1873). Reprint, with translation by D. C. Munro, Brussels: Culture et Civilisation, 1966.

Cross, Samuel H., and Olgerd P. Sherbowitz-Wetzor. *The Russian Primary Chronicle: Laurentian Text.* Harvard Studies and Notes in Philology and Literature, no. 108. Cambridge, Mass: Medieval Academy of America, 1953.

De Foucault, J.-A. "Douze chapitres inédits de la Tactique de Nicéphore Ouranos." In *Travaux et mémoires du Centre de recherche d'histoire et civilisation byzantine,* no. 5. Paris: De Boccard, 1973.

de la Vaissière, Étienne. "Huns et Xiongnu." *Central Asiatic Journal* 49, no. 1 (2005): 3–26.

Dennis, G. T. "Byzantium at War (9th–12th c.)." In International Symposium 4, National Hellenic Research Foundation, Institute for Byzantine Research. Athens: Goulandri-Horn Foundation, 1997.

———. "Defenders of the Christian People: Holy War in Byzantium." In John F. Haldon, ed., *Byzantine Warfare*. Aldershot, UK: Ashgate, 2007.

———. "Flies, Mice, and the Byzantine Crossbow." *Byzantine and Modern Greek Studies* 7 (1981): 1–5.

———, trans. *Maurice's Strategikon Handbook of Byzantine Military Strategy*. Philadelphia: University of Pennsylvania Press, 1984.

———. "Some Reflections on Byzantine Military Theory." In R. S. Calinger and Thomas R. West, eds., *John K. Zender: A Festschrift*. Indianapolis: Perspectives Press, 2007.

———, ed. *Das Strategikon des Maurikios*. Trans. Ernst Gamillscheg. Corpus fontium historiae byzantinae, 17. Vienna: Verlag der österreichischen Akademie der Wissenschaft, 1981.

———. *Three Byzantine Military Treatises*. Washington, D.C.: Dumbarton Oaks, 1985.

De re strategica. In George T. Dennis, ed., *Three Byzantine Military Treatises* (pp. 1–136, "The Anonymous Byzantine Treatise on Strategy"). Washington, D.C.: Dumbarton Oaks, 1985.

De temporum qualitatibus et positionibus locorum. Fragment. In Brian Coke, trans., *The Chronicle of Marcellinus Comes. Byzantina australiensia* 7. Sydney: Australian Association for Byzantine Studies, 1995.

De Velitatione. In George T. Dennis, ed., *Three Byzantine Military Treatises* (pp. 137–239, "Skirmishing"). Washington, D.C.: Dumbarton Oaks, 1985.

Dewing, H. B., ed. and trans. *Procopius*. 7 vols. Loeb Classical Library. Cambridge: Harvard University Press, 1962–1978.

Diels, H. A., and Rev. W. Kranz. *Die Fragmente der Vorsokratiker*. Hildesheim: Weidmann, 2004.

Diodorus Siculus. *Diodorus of Sicily*. 12 vols. Trans. C. H. Oldfather. Cambridge, Mass.: Harvard University Press, 1935.

Doctrina Jacobi nuper baptizati. Ed. and trans. V. Dé Roc. In *Travaux et mémoires du Centre de recherche d'histoire et civilisation byzantine*, no. 11, pp. 70–218. Paris: De Boccard, 1991.

Durliat, J. "L'approvisionnement de Constantinople." In Cyril A. Mango and Gilbert Dagron, eds., *Constantinople and Its Hinterland*. Aldershot, UK: Variorum, 1995.

———. "La peste du VIme siècle, pour un nouvel examen de sources byzantins." In J. M. Lefort et al., eds., *Hommes et richesses dans l'Empire byzantin*, vol. 1, *IVe–VIIe siècles*. Paris: P. Lethielleux, 1989.

Dyson, Stephen L. *The Creation of the Roman Frontier*. Princeton: Princeton University Press, 1985.

Eadie, J. W. "The Development of Roman Mailed Cavalry." *Journal of Roman Studies* 57 (1967): 161–173.

Echols, Edward C. *Herodian of Antioch's History of the Roman Empire from the Death of Marcus Aurelius to the Accession of Gordian*. Berkeley: University of California Press, 1961.

Einhard. *Vita Karoli Magni*. Translation in Paul Halsall's Internet Medieval

Sourcebook, http://www.fordham.edu/halsall/basis/einhard.htmlNo.Charle-
magne.

Elton, Hugh. *Warfare in Roman Europe, A.D. 350–425.* Oxford: Clarendon
Press, 1997.

Encyclopedia Judaica. Ed. Cecil Roth. 16 vols. Jerusalem: Ketcr, 1973.

Érdy, Miklós. "Hun and Xiongnu Type Cauldron Finds throughout Eurasia."
Eurasian Studies Yearbook 67 (1995): 3–26.

Evagrius Scholasticus. *The Ecclesiastical History of Evagrius Scholasticus.* Trans.
Michael Whitby. Liverpool: Liverpool University Press, 2000.

Even, Marie-Dominique, and Rodica Pop, eds. *Histoire secrete des Mongols =
Mongghol-un ni'uca tobciyan: Chronique mongole du XIIIe siecle.* Paris:
Gallimard, 1994.

Fàn Yè. *Hou Han Shu* (Pinyin: *Hòuhànshū*; Book of the Later Han). Partial translation by John E. Hill, http://depts.washington.edu/silkroad/texts/hhshu/
hou_han_shu.html.

Fei, Jie, Zhou Jie, and Hou Yongjian. "Circa A.D. 626 Volcanic Eruption, Climatic Cooling, and the Collapse of the Eastern Turkic Empire." *Climatic
Change* 81, nos. 3–4 (April 2007): 469–475.

Folz, Robert. *The Coronation of Charlemagne, 25 December 800.* Trans. J. E.
Anderson. London: Routledge and Kegan Paul, 1974.

Franklin, Simon, and Jonathan Shepard. *The Emergence of Rus, 750–1200.* New
York: Longman, 1996.

Frontinus, Sextus Julius. *Strategemata.* Trans. C. E. Bennet. Cambridge, Mass:
Harvard University Press, 1980.

Geanakoplos, Deno John, ed. *Byzantium: Church, Society, and Civilization Seen
through Contemporary Eyes.* Chicago: Chicago University Press, 1984.

Jenkins, R. J. H., F. Dvornik, B. Lewis, Gy. Moravcsik, D. Obolensky, and S.
Runciman, eds. *De Adminstrando Imperio.* Vol. 2: *Commentary.* London:
Athlone Press, 1962.

Jordanes. *Getica: De origine actibusque Getarum.* Translated by Charles C.
Mierow as *The Gothic History of Jordanes.* Cambridge: Speculum
Historiale, 1915. Latin text: http://www.thelatinlibrary.com/iordanes1.html.

Gillett, Andrew. *Envoys and Political Communication in the Late Antique West,
411–533.* Cambridge: Cambridge University Press, 2003.

Gilliver, Cathrine M. "The de munitionibus castrorum." *Journal of Roman Military Equipment Studies* 4 (1993): 33–48.

Golden, Peter B. "Ethnicity and State Formation in Pre-Cinggisid Turkic Eurasia." *Central Eurasian Studies Lectures,* vol. 1. Bloomington: Indiana University Press, 2001.

———. *An Introduction to the History of the Turkic Peoples: Ethnogenesis and
State-Formation in Medieval and Early Modern Eurasia and the Middle
East.* Wiesbaden: Harrasowitz, 1992.

———. *Khazar Studies: An Historio-Philological Inquiry into the Origins of the
Khazars.* Budapest: Akademia Kiado, 1980.

———. "The Peoples of the Russian Forest Belt." In D. Sinor, ed., *The Cam-*

bridge History of Early Inner Asia. Cambridge: Cambridge University Press, 1990.

Goldsworthy, Adrian Keith. *The Roman Army at War, 100 B.C.–A.D. 200.* Oxford: Clarendon Press, 1996.

Gordon, C. D. "The Subsidization of Border Peoples as a Roman Policy of Imperial Defense." Ph.D. dissertation, University of Michigan, 1948.

Granger, Frank. *Vitruvius on Architecture.* 2 vols. Cambridge, Mass.: Harvard University Press, 1983.

Guichaoua, Mickaël. "Lecture critique de Luttwak: La Grande Stratégie de l'Empire romain." *Enquetes et Documents, Revue du Centre de recherches en histoire internationale et Atlantique* (CRHIA) 30 (2004).

Guilmartin, J. F. *Gunpowder and Galleys: Changing Technology and Mediterranean Warfare at Sea in the Sixteenth Century.* Cambridge: Cambridge University Press, 1975.

Haldon, John F. "Blood and Ink: Some Observations on Byzantine Attitudes towards Warfare and Diplomacy." In Jonathan Shepard and Simon Franklin, eds., *Byzantine Diplomacy: Papers from the Twenty-fourth Spring Symposium of Byzantine Studies.* Aldershot, UK: Variorum, 1992.

———. *Byzantine Praetorians: An Administrative, Institutional and Social Survey of the Opsikion and Tagmata, c. 580–900.* Bonn: Dr. Rudolf Habelt, 1984.

———, ed. *Byzantine Warfare.* Aldershot, UK: Ashgate, 2007.

———. *The Byzantine Wars: Battles and Campaigns of the Byzantine Era.* Charleston, S.C.: Tempus, 2001.

———. *Byzantium in the Seventh Century: The Transformation of a Culture.* Cambridge: Cambridge University Press, 1990.

———, ed. and trans. *Constantine Porphyrogenitus: Three Treatises on Imperial Military Expeditions.* Vienna: Verlag der österreichischen Akademie der Wissenschaft, 1990.

———. "Introduction: Why Model Logistical Systems?" In John F. Haldon, ed., *General Issues in the Study of Medieval Logistics: Sources, Problems, Methodologies.* Leiden: Brill, 2005.

———. "The Organization and Support of an Expeditionary Force: Manpower and Logistics in the Middle Byzantine Period." In J. F. Haldon, ed., *Byzantine Warfare.* Aldershot, UK: Ashgate, 2007. http://www.deremilitari.org/RESOURCES/ARTICLES/haldon1.htm#_ftn10.

———. "Some Aspects of Early Byzantine Arms and Armour." In David Nicolle, ed., *A Companion to Medieval Arms and Armor.* Woodridge, UK: Boydell and Brewer, 2002.

———. "Strategies of Defence, Problems of Security: The Garrisons of Constantinople in the Middle Byzantine Period." In Cyril A. Mango and Gilbert Dagron, eds., *Constantinople and Its Hinterland.* Aldershot, UK: Variorum, 1995. http://www.deremilitari.org/resources/articles/haldon2.htm.

———. "Theory and Practice in Tenth-Century Military Administration: Chapters 2, 44 and 45 of the Book of Ceremonies." In *Travaux et mémoires du*

Centre de recherche d'histoire et civilisation byzantine, no. 13, pp. 201–352. Paris: De Boccard, 2000.

———. *Warfare, State and Society in the Byzantine World, 565–1204.* London: UCL Press, 1999.

Haldon, John F., and M. Byrne. "A Possible Solution to the Problem of Greek Fire." *Byzantinische Zeitsschrift* 70 (1977): 91–100.

Haldon, John F., Andrew Lacey, and Colin Hewes. "Greek Fire Revisited: Recent and Current Research." In Elizabeth Jeffreys, ed., *Byzantine Style, Religion and Civilization: In Honour of Sir Steven Runciman.* Cambridge: Cambridge University Press, 2006.

Halsall, Guy. *Barbarian Migrations and the Roman West, 376–568.* Cambridge: Cambridge University Press, 2007.

Hartog, Francois. *The Mirror of Herodotus: The Representation of the Other in the Writing of History.* Trans. Janet Lloyd. Berkeley: University of California Press, 1988.

Heather, Peter. *The Fall of the Roman Empire: A New History of Rome and the Barbarians.* Oxford: Oxford University Press, 2006.

———. "The Western Empire, 425–476." In A. Cameron, B. Ward-Perkins, and M. Whitby, eds., *Late Antiquity: Empire and Successors, A.D. 425–600.* Vol. 14 of *The Cambridge Ancient History.* Cambridge: Cambridge University Press, 2001.

Hendy, Michael F. *Studies in the Byzantium Monetary Economy, c. 300–1450.* Cambridge: Cambridge University Press, 1985.

Hirth, Friedrich. *China and the Roman Orient: Researches into Their Ancient and Mediaeval Relations as Represented in Old Chinese Records.* Shanghai and Hong Kong, 1885. Edited by J. S. Arkenberg at http://www.fordham .edu/halsall/eastasia/1372mingmanf.html.

Homer. *The Iliad.* Trans. E. V. Rieu. Harmondsworth, UK: Penguin Books, 1950.

Homer. *The Odyssey.* Trans. A. T. Murray. Cambridge, Mass.: Harvard University Press, 1946.

Howard-Johnston, James. "The Two Great Powers in Late Antiquity: A Comparison." In Averil Cameron, ed., *The Byzantine and Early Islamic Near East.* Princeton: Darwin Press, 1995.

Hygini Gromatici Liber de Munitionibus Castrorum. Leipzig: von s. Hirzel, 1887. Reprint, Hildesheim: H. A. Gestenberg, 1972. Latin text: http://www .intratext.com/X/LAT0347.html.

Isaac, Benjamin. *The Limits of Empire: The Roman Army in the East.* Oxford: Clarendon Press, 1990.

Isles, Fred. "Turkish Flight Arrows." *Journal of the Society of Archer-Antiquaries* 4 (1961). http://margo.student.utwente.nl/sagi/artikel/turkish/.

Janni, Pietro. *La Mappa e Il periplo: Cartografia antica e spazio odologico.* Rome: Giorgio Bretschneider, 1984.

Jenkins, R. J. H., ed., with F. Dvornik, B. Lewis, Gy. Moravcsik, D. Obolensky, and S. Runciman. *De Administrando Imperio.* Vol. 2: *Commentary.* London: Athlone Press, 1962.

Jerome. *Selected Letters of St. Jerome*. Trans. F. A. Wright. Cambridge, Mass.: Harvard University Press, 1954.

Jiu Tang shu ("Old Book of Tang" of Liu Xu/Hsü). Taipei: Tai wan shang wu yin shu guan, 1983.

Jones, A. H. M. *The Later Roman Empire, 284–602: A Social and Economic Survey*. Oxford: Basil Blackwell, 1973.

Jordanes. *Getica: De origine actibusque Getarum*. Translated by Charles C. Mierow as *The Gothic History of Jordanes*. Cambridge: Speculum Historiale, 1915. Latin text: www.thelatinlibrary.com/iordanes1.html.

Josephus. *The Jewish War*. www.guternberg.org/files/2850/2850.txt.

Justinian. Novels. Online Latin texts: http://web.upmf-grenoble.fr/Haiti/Cours/Ak/Corpus/Novellae.htm.

Kaegi, Walter E., Jr. *Byzantium and the Early Islamic Conquests*. Cambridge: Cambridge University Press, 1992.

———. "The Contribution of Archery to the Turkish Conquest of Anatolia." In John F. Haldon, ed., *Byzantine Warfare*. Aldershot, UK: Ashgate, 2007.

———. "The Frontier: Barrier or Bridge?" In John F. Haldon, ed., *Byzantine Warfare*. Aldershot, UK: Ashgate, 2007.

———. *Heraclius: Emperor of Byzantium*. Cambridge: Cambridge University Press, 2003.

———. *Some Thoughts on Byzantine Military Strategy*. The Hellenic Studies Lecture. Brookline, Mass.: Hellenic College Press, 1983.

Kazhdan, Alexander. "The Notion of Byzantine Diplomacy." In Jonathan Shepard and Simon Franklin, eds., *Byzantine Diplomacy*. Aldershot, UK: Variorum, 1992.

Kazhdan, Alexander, and Giles Constable. *People and Power in Byzantium: An Introduction to Modern Byzantine Studies*. Washington, D.C.: Dumbarton Oaks, 1982.

Kazhdan, Alexander, and Michael Mcormick. "The Social World of the Byzantine Court." In Henry Maguire, ed., *Byzantine Court Culture from 829 to 1204*. Washington, D.C.: Dumbarton Oaks Research Library / Harvard University Press, 1997.

Keen, Maurice. *Chivalry*. New Haven: Yale University Press, 1986.

Kekaumenos. *Raccomandazioni e consigli di un galantuomo: Stratēgikon*. Ed. and trans. Maria Dora Spadaro. Alessandria: Edizioni dell'Orso, 1998.

Kennedy, Hugh. *The Armies of the Caliphs: Military and Society in the Early Islamic State*. New York: Routledge, 2001.

———. "Justinianic Plague in Syria and the Archaeological Evidence." In Lester K. Little, ed., *Plague and the End of Antiquity: The Pandemic of 541–750*. Cambridge: Cambridge University Press, 2006.

Khazanov, Anatoly M. *Nomads and the Outside World*. 2nd ed. Madison: University of Wisconsin Press, 1994.

Kinnamos, John. *Deeds of John and Manuel Comnenus*. Trans. Charles M. Brand. New York: Columbia University Press, 1976.

Kitto, H. D. F. *The Greeks*. Harmondsworth, UK: Penguin Books, 1951.

Klek, Markus. "Making an Asiatic Composite Bow." http://www.primitiveways
.com/pt-composite_bow.html.

Kolbaba, Tia M. "Fighting for Christianity: Holy War in the Byzantine Empire."
In John F. Haldon, ed., *Byzantine Warfare*. Aldershot, UK: Ashgate, 2007.

Kolias, Taxiarchis G. *Byzantinische Waffen: Ein Beitrag zur byzantinischen
Waffenkunde von den Anfängen bis zur lateinischen Eroberung*. Vienna:
Verlag der östereichischen Akademie der Wissenschaften, 1988.

Koutrakou, Nike. "Diplomacy and Espionage: Their Role in Byzantine Foreign
Relations, 8–10th Centuries." *Graeco-Arabica* 6 (1995): 125–144.

Lamport, L., R. Shostak, and M. Pease. "The Byzantine Generals Problem."
ACM Transactions on Programming Languages and Systems 4 (1982): 382–
401.

Lee, A. D. "The Eastern Empire: Theodosius to Anastasius." In A. Cameron, B.
Ward-Perkins, and M. Whitby, eds., *Late Antiquity: Empire and Successors,
A.D. 425–600*. Vol. 14 of *The Cambridge Ancient History*. Cambridge:
Cambridge University Press, 2001.

———. *Information and Frontiers: Roman Foreign Relations in Late Antiquity*.
Cambridge: Cambridge University Press, 1993.

Leites, Natan Constantin. *The Operational Code of the Politburo*. New York:
McGraw-Hill, 1951.

Lewis, Naphtali, and Reinhold Meyer. *Roman Civilization: The Empire*. New
York: Harper Torchbooks, 1966.

Liber Pontificalis. Translated by Raymond Davis as *The Book of Pontiffs: The
Ancient Biographies of the First Ninety Roman Bishops to A.D. 715*. Liverpool: Liverpool University Press, 1989.

Linder, Amnon. *The Jews in Roman Imperial Legislation*. Detroit: Wayne State
University Press, 1987.

Little, Lester K., ed. *Plague and the End of Antiquity: The Pandemic of 541–750*.
Cambridge: Cambridge University Press, 2006.

Liutprand of Cremona. *Relatio de legatione Constantinapolitana* (Narrative of
the Embassy to Constantinople). Translated by F. A. Wright as *The Works of
Liutprand of Cremona*. New York: Dutton, 1930.

Loreto, Luigi. "Il paradosso Luttwakiano power projection, low intensity e
funzione del limes." In *Per la storia militare del mondo antico*, ed. Luigi
Loreto, pp. 84–92. Naples: Jovene, 2006.

———. "La Storia della grand strategy un dibattito Luttwak?" In *Per la storia
militare del mondo antico*, ed. Luigi Loreto, pp. 67–81. Naples: Jovene,
2006.

Jones, A. H. *The Later Roman Empire, 284–602: A Social and Economic Survey*.
Oxford: Basil Blackwell, 1973.

Luttwak, Edward N. *The Grand Strategy of the Roman Empire: From the First
Century A.D. to the Third*. Baltimore: Johns Hopkins University Press, 1976,
2007.

———. "The Operational Level of War." *International Security* 5, no. 3 (Winter
1980–1981): 69–79.

———. *Strategy: The Logic of War and Peace.* Cambridge: Belknap Press of Harvard University Press, 2001.

Lydus, John [John the Lydian]. *On the Magistracies of the Roman Constitution.* Trans. T. F. Carney. Sydney: Wentworth Press, 1965.

Macrides, Ruth. "Dynastic Marriages and Political Kinship." In Jonathan Shepard and Simon Franklin, eds., *Byzantine Diplomacy.* Aldershot, UK: Variorum, 1992.

Maenchen-Helfen, Otto. J. *The World of the Huns: Studies in Their History and Culture.* Ed. Max Knight. Berkeley: University of California Press, 1973.

Magdalino, Paul. "In Search of the Byzantine Courtier." In Henry Maguire, ed., *Byzantine Court Culture from 829 to 1204.* Washington, D.C.: Dumbarton Oaks Research Library / Harvard University Press, 1997.

Maguire, Henry, ed. *Byzantine Court Culture from 829 to 1204.* Washington, D.C.: Dumbarton Oaks Research Library / Harvard University Press, 1997.

Mango, Cyril A. *Byzantium: The Empire of New Rome.* New York: Scribner's, 1980.

———. "The Water Supply of Constantinople." In Cyril A. Mango and Gilbert Dagron, eds., *Constantinople and Its Hinterland.* Aldershot, UK: Variorum, 1995.

Mango, Cyril A., and Gilbert Dagron, eds. *Constantinople and Its Hinterland.* Aldershot, UK: Variorum, 1995.

Mann, John C. "Power, Force and the Frontiers of the Empire." *Journal of Roman Studies* 69 (1979): 175–183.

Marcellinus Comes. Trans. Brian Coke. *Byzantina Australiensia* 7. Sydney: Australian Association for Byzantine Studies, 1995.

Marcovich, M., ed. *Heraclitus: Greek Text with a Short Commentary.* Merida, Venezuela: Los Andes University Press, 1967.

Marsden, E. W. *Greek and Roman Artillery: Historical Development.* Oxford: Clarendon Press, 1969.

———. *Greek and Roman Artillery: Technical Treatises.* Oxford: Clarendon Press, 1971.

Martin-Hisard, Bernardette. "Constantinople et les Archontes du Monde Caucasien dans le Livre de Cérémonies." In *Travaux et mémoires du Centre de recherche d'histoire et civilisation byzantine,* no. 13, pp. 361–521. Paris: De Boccard, 2000.

Mattingly, Garrett. *Renaissance Diplomacy.* New York: Dover, 1988.

McCotter, Stephen. "Byzantines, Avars and the Introduction of the Trebuchet." Dissertation, Queen's University of Belfast, 2003. Available online at http://www.deremilitari.org/resources/articles/mccotter1.htm.

McGeer, Eric. "Infantry versus Cavalry: The Byzantine Response." *Revue des Études Byzantines* 46 (1988): 135–145.

———. *Sowing the Dragon's Teeth: Byzantine Warfare in the Tenth Century.* Washington, D.C.: Dumbarton Oaks, 1995.

———. "Two Military Orations of Constantine VII." In John W. Nesbitt, ed., *Byzantine Authors: Literary Activities and Preoccupations; Texts and Trans-*

lations Dedicated to the Memory of Nicolas Oikonomides. Leiden: Brill, 2003.

McLeod, Wallace. "The Range of the Ancient Bow." *Phoenix* 19 (1965): 1–14.

———. "The Range of the Ancient Bow: Addenda." *Phoenix* 26, no.1 (1972): 78–82.

Menander. *The History of Menander the Guardsman.* Trans. R. C. Blockley. Liverpool: F. Cairns, 1985.

Mihăescu, Haralambie, ed. and trans. *Arta militară: Mauricius.* Bucharest: Academiei Republicii Socialiste, 1970.

Mihăescu, Haralambie, with Gilbert Dagron. *Le traité sur la guérilla (De velitatione) de l'empereur Nicéphore Phocas, 963–969.* Paris: Éditions du Centre national de la recherche scientifique, 1986.

Millar, Fergus. *A Greek Roman Empire: Power and Belief under Theodosius II (408–450).* Berkeley: University of California Press, 2006.

Miller, D. A. "The Logothete of the Drome in the Middle Byzantine Period." *Byzantion* no. 36 (1966): fasc. 2, pp. 438–470.

Miller, J. Innes. *The Spice Trade of the Roman Empire, 29 B.C. to A.D. 641.* Oxford: Clarendon Press, 1969.

Milner, N. P., trans. *Vegetius: Epitome of Military Science.* 2nd ed. Liverpool: Liverpool University Press, 1996.

Moravcsik, Gy., ed. *De Administrando Imperio.* Trans. R. J. H. Jenkins. Washington, D.C.: Dumbarton Oaks Center for Byzantine Studies / Harvard University Press, 1967.

Morgan, David. *The Mongols.* Oxford: Blackwell, 1986.

Moorhead, John. "The Byzantines in the West in the Sixth Century." In Paul Fouracre, ed., *The New Cambridge Medieval History: Vol. 1, c. 500–c. 700.* Cambridge: Cambridge University Press, 2005.

Morris, Donald R. *The Washing of the Spears.* London: Jonathan Cape, 1966.

Muller-Mertens, Eckhard. "The Ottonians as Kings and Emperors." In Timothy Reuter, ed., *The New Cambridge Medieval History: Vol. 3, c. 900–c. 1024.* Cambridge: Cambridge University Press, 1999.

Nicholas I, Patriarch of Constantinople. *Letters.* Ed. and trans. R. J. H Jenkins and L. G. Westerink. Washington, D.C.: Dumbarton Oaks, 1973.

Nikephoros, Patriarch of Constantinople. *Short History* [*Breviarium Historicum*]. Ed. Cyril A. Mango. Washington, D.C.: Dumbarton Oaks, 1990.

Nishimura, D. "Crossbow, Arrow-Guides, and the Solenarion." *Byzantion* no. 58 (1988): 422–435.

Noonan, Thomas. "Byzantium and the Khazars: A Special Relationship?" In Jonathan Shepard and Simon Franklin, eds., *Byzantine Diplomacy.* Aldershot, UK: Variorum, 1992.

Oikonomides, Nicolas. "The Role of the Byzantine State in the Economy." In Angeliki E. Laiou, ed., *The Economic History of Byzantium: From the Seventh through the Fifteenth Century.* Washington, D.C: Dumbarton Oaks Research Library and Collection, 2002.

———. "Title and Income at the Byzantine Court." In Henry Maguire, ed.,

Byzantine Court Culture from 829 to 1204. Washington, D.C.: Dumbarton Oaks Research Library / Harvard University Press, 1997.

Oldfather, C. H., trans. *Diodorus of Sicily.* 12 vols. Cambridge, Mass.: Harvard University Press, 1935.

Ostrogorsky, George. *History of the Byzantine State.* Revised ed. Trans. Joan Hussey. New Brunswick: Rutgers University Press, 1969.

The Oxford Dictionary of Byzantium. Ed. Alexander P. Kazhdan and Alice-Mary Talbot with Anthony Cutler, Timothy E. Gregory, and Nancy P. Šev\\5,99\\ enko. New York: Oxford University Press, 1991.

Palmer, Andrew, trans. *The Seventh Century in the West-Syrian Chronicles.* Liverpool: Liverpool University Press, 1993.

Papasotiriou, Charalambos. "Byzantine Grand Strategy." Ph.D. dissertation, Stanford University, 1991.

Parker, H. M. D. *The Roman Legions.* Oxford: Clarendon Press, 1928. Reprint, Chicago: Ares, 1985.

Pertusi, Agostino. "La formation de themes byzantins." *Berichte zum XI Internationalen Byzantinisten-Kongress* (Munich) 1 (1958): 1–40.

——. *Il pensiero politico bizantino.* Ed. Antonio Carile. Bologna: Pàtron Editore, 1990.

——, ed. and trans. *De thematibus: Introduzione, testo critico, commento.* Vatican City: Biblioteca apostolica vaticana, 1952.

Photios. *Bibliotheca, Myriobiblon.* Online, www.ccel.org/p/pearse/morefathers/ photius_03bibliotheca.htmNo.34.

——. *The Homilies of Photius I, Saint, Patriarch of Constantinople.* Trans. Cyril A. Mango. Cambridge, Mass.: Harvard University Press, 1958.

Piacentini, Valeria F. *Il Pensiero Militare nel Mondo Mussulmano.* Milan: FrancoAngeli, 1996.

Piltz, Elisabeth. "Middle Byzantine Court Costume." In Henry Maguire, ed., *Byzantine Court Culture from 829 to 1204.* Washington, D.C.: Dumbarton Oaks Research Library / Harvard University Press, 1997.

Pliny the Younger. *Pliny Letters and Panegyricus.* 2 vols. Trans. Betty Radice. Cambridge, Mass: Harvard University Press, 1975.

Pohl, Walter. "Conceptions of Ethnicity in Early Medieval Studies." In Lester K. Little and Barbara H. Rosenwein, eds., *Debating the Middle Ages: Issues and Readings.* Oxford: Blackwell, 1998.

——. *Die Awaren: Ein Steppenvolk in Mitteleuropa, 567–822 n. Chr.* Munich: C. H. Beck, 2002.

Polyaenus. *Stratagems of War: Polyaenus.* 2 vols. Ed. and trans. Peter Krentz and Everett L. Wheeler. Chicago: Ares, 1994.

Polybius. *The Histories.* 6 vols. Trans. W. R. Paton. Cambridge, Mass: Harvard University Press, 1979.

Praecepta imperatori. Appendix to J. Reiske, ed., *Corpus Scriptorum Historiae Byzantinae,* bk. 1, pp. 397–430. Bonn: Weber, 1829.

Prokopios. *De aedificis.* Vol. 7. Translated by H. B. Dewing with G. Downey as *Buildings.* Cambridge, Mass.: Harvard University Press, 1971.

———. *Anecdota*. Vol. 6. Translated by H. B. Dewing as *Procopius*. Cambridge, Mass.: Harvard University Press, 1969.

———. *The Wars*. Vols. 1–4. Trans. H. B. Dewing. Cambridge, Mass.: Harvard University Press.

Prosper of Acquitaine. *Epitoma chronicon* (Prosperi Tironis). In *Monumenta Germaniae Historica: Chronica Minora Saec. IV, V, VI, VII*. Ed. T. Mommsen. Vol. 1, pp. 341–501. Berlin: Weidmann, 1892. Reprint, Hannover: Hahnsche Buchhandlung, 1980.

Pryor, John H. *Geography, Technology and War: Studies in the Maritime History of the Mediterranean, 646–1571*. Cambridge: Cambridge University Press, 1988.

Pryor, John H., and Elizabeth M. Jeffreys. *The Age of the Dromon: The Byzantine Navy ca. 500–1204*. Leiden: Brill Academic, 2006.

Pseudo-Dionysius of Tel-Mahre. *Chronicle: Known Also as the Chronicle of Zuqnin*. Part 3. Trans. Witold Witakowski. Liverpool: Liverpool University Press, 1996.

Rachewiltz, Igor de. *The Secret History of the Mongols: A Mongolian Epic Chronicle of the Thirteenth Century*. 2 vols. Leiden: Brill, 2004.

Rance, Philip. "The Fulcum, the Late Roman and Byzantine *Testudo*: The Germanization of Roman Infantry Tactics?" *Greek, Roman, and Byzantine Studies* no. 44 (2004): 265–326.

Rausing, Gad. *The Bow: Some Notes on Its Origins and Development*. Acta Archaeologica Lundensia Series 8, no. 6. Bonn: Rudolf Habelt Verlag; Lund: CWK Gleerups Forlag, 1967.

Ravegnani, Giorgio. *Soldati di Bizanzio in Eta Giustinianea*. Rome: Jouvence, 1988.

Róna-Tas, András. *Hungarians and Europe in the Early Middle Ages: An Introduction to Early Hungarian History*. Budapest: Central European University Press, 1999.

Rubin, Ze'ev. "The Reforms of Khusro Anushirvan." In Averil Cameron, ed., *The Byzantine and Early Islamic Near East,* vol. 3. Princeton: Darwin Press, 1995.

———. "The Sasanid Monarchy." In A. Cameron, B. Ward-Perkins, and M. Whitby, eds., *Late Antiquity: Empire and Successors, A.D. 425–600*. Vol. 14 of *The Cambridge Ancient History*. Cambridge: Cambridge University Press, 2001.

Runciman, Steven. *The Sicilian Vespers*. Harmondsworth, UK: Penguin Books, 1960.

The Saga of Hervor and King Heidrek the Wise. Trans. Peter Tunstall. 2005. Online, http://www.northvegr.org/lore/oldheathen/018.php.

Sartre, Maurice. *D'Alexandre à Zénobie: Histoire du Levant antique, IVe siècle avant Jésus-Christ–IIIe siècle après Jésus-Christ*. Paris, Fayard, 2001.

Saunders, John J. *The History of the Mongol Conquests*. Philadelphia: University of Pennsylvania Press, 1971.

Scriptor Incertus. Frag. 1. Ed. I. Dujcev. In *Travaux et mémoires du Centre de re-*

cherche d'histoire et civilisation byzantine, no. 1, pp. 205–254, trans. Paul Stephenson. Paris: De Boccard, 1965. Revised November 2004, http://homepage.mac.com/paulstephenson/trans/scriptor1.html.

The Armenian History Attributed to Sebeos. Trans. R. W. Thompson. Vols. 1–2. Liverpool: Liverpool University Press, 1999.

Seeck, Otto. *Notitia Dignitatum accedunt Notitia Urbis Constantinopolitanae Laterculi Prouinciarum.* Berlin: Apud Weidmannos, 1876.

Sevcenko, Ihor. "Re-reading Constantine Porphyrogenitus." In Jonathan Shepard and Simon Franklin, eds., *Byzantine Diplomacy.* Aldershot, UK: Variorum, 1992.

Shepard, Jonathan, and Simon Franklin, eds. *Byzantine Diplomacy.* Aldershot, UK: Variorum, 1992.

Sidonius. *Poems and Letters.* 2 vols. Trans. W. B. Anderson. Cambridge: Harvard University Press, 1936.

Simeonova, Liliana. "In the Depth of Tenth-Century Byzantine Ceremonial: The Treatment of Arab Prisoners of War at Imperial Banquets." In John F. Haldon, ed., *Byzantine Warfare.* Aldershot, UK: Ashgate, 2007. Originally in *Byzantine and Modern Greek Studies* 22 (1998): 75–104.

Sinor, Denis, ed. *The Cambridge History of Early Inner Asia.* Cambridge: Cambridge University Press, 1990.

Scylitzes, Ioannis. *A Synopsis of Histories, 811–1057 A.D.* Trans. John Wortley. Unpublished typescript. Winnipeg: University of Manitoba, Centre for Hellenic Civilization.

Stephenson, Paul. "Byzantine Diplomacy, A.D. 800–1204: Means and Ends." In Jonathan Shepard and Simon Franklin, eds., *Byzantine Diplomacy.* Aldershot, UK: Variorum, 1992.

———. *Byzantium's Balkan Frontier: A Political Study of the Northern Balkans, 900–1204.* Cambridge: Cambridge University Press, 2000.

———. *The Legend of Basil the Bulgar-Slayer.* Cambridge: Cambridge University Press, 2003.

Sturluson, Snorri. *Heimskringla: The Chronicle of the Kings of Norway Saga of Harald Hardrade.* London: Norroena Society, 1907. Electronic edition, Douglas B. Killings DeTroyes, 1996, Online Medieval and Classical Library, release no. 15, http://omacl.org/Heimskringla/hardrade1.html.

Suda On Line: Byzantine Lexicography: http://www.stoa.org/sol-bin/search.pl.

Sullivan, Denis F. "A Byzantine Instructional Manual on Siege Defense: The *De obsidione toleranda;* Intro., English translation and annotations." In John W. Nesbitt, ed., *Byzantine Authors: Literary Activities and Preoccupations. Texts and Translations Dedicated to the Memory of Nicolas Oikonomides.* Leiden: Brill, 2003.

———. *Siegecraft: Two Tenth-Century Instructional Manuals by "Heron of Byzantium."* Washington, D.C.: Dumbarton Oaks, 2000.

Sutherland, Caroline. "Archery in the Homeric Epics." *Classics Ireland* 8 (2001); http://www.ucd.ie/cai/classics-ireland/2001/sutherland.html.

al-Tabari. *The History of al-Tabari.* Vol. 5. Trans. Ann. C. E. Bosworth. Albany: State University Press of New York Press, 1999.

Talbot, Alice-Mary, and Denis F. Sullivan, trans. *The History of Leo the Deacon: Byzantine Military Expansion in the Tenth Century.* Washington, D.C.: Dumbarton Oaks, 2005.

Thee, Francis C. R., trans. *Julius Africanus and the Early Christian View of Magic.* In *Hermeneutische Untersuchungen zur Theologie,* vol. 19. Tubingen: J. C. B. Mohr, 1984.

Theophanes. *The Chronicle of Theophanes Confessor, A.D. 284–813.* Trans. Cyril A. Mango and Roger Scott with Geoffrey Greatrex. Oxford: Clarendon Press, 1997.

———. *Theophanes Continuatus.* Ed. Immanuelis Bekker. Bonn: E. Weber, 1838. Online translation by Paul Stephenson, http://homepage.mac.com/ paulstephenson/trans/theocont2.html.

Thompson, E. A. *The Huns.* Revised by Peter Heather. Oxford: Blackwell, 1996.

———. *A Roman Reformer and Inventor: Being a New Translation of the Treatise De Rebus Bellicis.* Oxford: Oxford University Press, 1952.

Thucydides. *History of the Peloponnesian War.* 4 vols. Trans. Charles Forster Smith. Cambridge: Harvard University Press, 1951.

Travaux et mémoires du Centre de recherche d'histoire et civilisation byzantine. Paris: De Boccard.

Toumanoff, C. "Armenia and Georgia." In J. M. Hussey, ed., *The Cambridge Medieval History.* Vol. 4: *The Byzantine Empire,* part 1. Cambridge: Cambridge University Press, 1967.

Toynbee, Arnold. *Constantine Porphyrogenitus and His World.* London: Oxford University Press, 1973.

Urbicius. *Ourbikioi Epitedeyma (Adaosul lui Urbicius).* In Haralambie Mihăescu, ed. and trans., *Mauricius Arta Militară.* Bucharest: Academiei Republici Romania, 1970.

U.S. War Department. *Handbook on German Military Forces.* 15 March 1945. Reprinted, Baton Rouge: Louisiana State University Press, 1990).

Van Berchem, Denis. *L'armée de Dioclétien et la réforme Constantinienne.* Institut français d'archéologie de Beyrouth. Bibliothèque archéologique et historique 56. Paris: Librarie Orientaliste Paul Geuthner, 1952.

Van Den Berg, Hilda, trans. *Anonymous de obsidione toleranda.* Leiden: Brill, 1947.

Vegetius Renatus, Publius Flavius. *Epitoma Rei Militari.* Latin intratext: http:// www.intratext.com/IXT/LAT0189.HTM by Èulogos SpA—may they prosper.

———. *Viri illustris: De re militari libri quatuor.* Basel: Christian Wechel, 1535.

Vieillefond, Jean-René. *Les "Cestes" de Julius Africanus: Étude sur 1'ensemble des fragments avec édition, traduction et commentaries.* Publications de l'Institute Français de Florence. Ser. 1, vol. 20. Paris: Sansoni, 1970.

Wallraff, Martin, et al. *Julius Africanus und die christliche Weltchronistik.* Texte und Untersuchungen zur Geschichte der altchristlichen Literatur, vol. 15. Berlin: de Gruyter, 2006.

Ward-Perkins, Bryan. *The Fall of Rome and the End of Civilization.* Oxford: Oxford University Press, 2005.

Welwei, Karl-Wilhelm. "Probleme römischer Grenzsicherung am Beispiel der Germanienpolitik des Augustus." In Mischa Meier and Meret Strothmann, eds., *Res publica und Imperium: Kleine Schriften zur römischen Geschichte.* Historia Einzelschriften, 177, pp. 25–263. Stuttgart: Franz Steiner Verlag, 2004.

Whitby, Michael. "Recruitment in Roman Armies from Justinian to Heraclius (c. 565–615)." In Averil Cameron, ed., *The Byzantine and Early Islamic Near East*, vol. 3. Princeton: Darwin Press, 1995.

———, trans. *The Ecclesiastical History of Evagrius Scholasticus.* Liverpool: University Press, 2000.

Whitby, Michael, and Mary Whitby. *The History of Theophylact Simocatta.* Oxford: Clarendon Press, 1986.

———, trans. *Chronicon Paschale, 284–628 A.D.* Liverpool: Liverpool University Press, 1989.

White, Lynn, Jr. *Medieval Technology and Social Change.* Oxford: Clarendon Press, 1966.

Whitehead, David, and P. H. Blyth, eds. and trans. *On Machines*, by Athenaeus Mechanicus. Historia Einzelschriften, 182. Stuttgart: Franz Steiner, 2004.

Whittaker, C. R. *Frontiers of the Roman Empire: A Social and Economic Study.* Baltimore: Johns Hopkins University Press, 1994.

Whittow, Mark. *The Making of Byzantium, 600–1025.* Berkeley: University of California Press, 1996.

Widsith. Ed. Kemp Malone. London: Methuen, 1936. Rev. ed., Copenhagen: Rosenkilde and Bayyer, 1962.

Wiechmann, I., and G. Grupe. "Detection of *Yersinia pestis* in Two Early Medieval Skeletal Finds from Aschheim (Upper Bavaria, 6th Century A.D.)." *American Journal of Physical Anthropology* no. 126 (2005): 48–55.

Wiita, John Earl. "The Ethnika in Byzantine Military Treatises." Ph.D dissertation, University of Minnnesota, 1977.

Wilson, N. G. *Scholars of Byzantium.* Rev. ed. London: Duckworth, 1996.

Yin, Lin. "Western Turks and Byzantine Gold Coins Found in China." *Transoxiana* 6 (July 2003). Online, http://www.transoxiana.org/0106/lin-ying_turks_solidus.html.

Xu (Hsü), Liu. *Jiu Tang shu* (Old Book of Tang). Taipei: Tai wan shang wu yin shu guan, 1983.

Yule, Henry, and A. C. Burnell. *Hobson-Jobson: A Glossary of Colloquial Anglo-Indian Words and Phrases.* 2nd, new ed. Ed. Rev. E. Crooke. London: Routledge and Kegan Paul, 1985.

Yakar, Jak. *Ethnoarchaeology of Anatolia: Rural Socio-Economy in the Bronze and Iron Ages.* Tel Aviv: Tel Aviv University, 2000.

Zilliacus, Henrik. *Zum Kampf der Weltsprachen im oströmischen Reich.* Helsinki: Mercators Tryckeri Aktiebolag, 1935.

Zuckerman, Constantine. "L'Empire d'Orient et les Huns." In *Travaux et mémoires du Centre de recherche d'histoire et civilisation byzantine*, no. 12, pp. 165–168. Paris: De Boccard, 1994.

Index of Names

General Index

Abbasid caliphate, 126, 143, 188, 218, 224
Adrianople, battle of (378), 19
Adur Gushnasp, Vshnasp, eternal fire of, 400, 408
Agentes in rebus, 108, 111
Ahl-al-dhimma, 205
Ahura Mazda, 107, 394–399
Alamanni, 17
Alans, 5, 11, 18, 19, 20, 50, 59, 78, 160, 243–245
Aleppo, 135, 146, 191
Amida, Diyarbakir, Amed, siege of (359), 223, 259, 291, 393
Anatolikon, theme, 178, 182, 353
Ankara, battle of (1402), 234
Antae, 18, 60, 103, 296, 299
Antioch, Antakya, 360
Aquileia, 40, 46, 189
Arabitai, Beduin cavalry, 364
Arab sieges of Constantinople, 68, 75, 175, 212
Arkadiopolis Lüleburgaz, 40
Arkadioupolis, Constanța, Romania, 41
Armenia, 4, 35, 40, 243
Arsacid Parthia, 3, 241, 276
Athanatoi, 355
Avars, 4, 10, 52, 58–61, 69, 103–105, 156, 164, 172, 215, 244, 291, 395, 396

Banat, 40, 60
Bandon, 267, 300, 373
Bederiana, 86, 130

Bessic, 412
Bethlehem, 34, 115
Bison, 373
Blachernae, 72, 73, 75, 174, 185, 337
Black Death, 90
Boukellarion, theme, 178
Bulghars, 36, 69, 73, 157, 176, 191, 215
Burgundians, 17, 18, 43

Caltrops, 263, 315, 335, 356, 383
Campus Mauriacus, Catalaunian Fields, 43, 55
Cappadocia, 40, 395
Carthage, capital of *Africa,* 50, 71
Chalcedonian christology, 45, 205–206, 410
Cheiromaggana, 318, 367, 368
Chosaroi, huszar, hussar, 237, 359, 389
Chrovatia, Croatia, 163, 164
Chrysopolis, Usküdar, 69, 122
Cilician Gates, pass of Gülek, 353, 363, 400
Cinggisid states, 143
Clibanarii, 108, 276, 277, 278
Coincidentia oppositorum, 413
Corpus Juris Civilis, 84
Crete, Arab conquest of, 75, 171, 326
Ctesiphon, 35, 50, 199, 406
Cumans, Qipchaqs, Polovtsy, 4, 52, 143, 161, 162

Danastris, Dniester, 155
Dara, Oğuz, 83, 106, 202, 223, 259, 393
Dar al-harb, the land of war, 135, 220, 340